T0189701

# Communications
# in Computer and Information Science  **1385**

More information about this series at http://www.springer.com/series/7899

Wanling Gao · Kai Hwang ·
Changyun Wang · Weiping Li ·
Zhigang Qiu · Lei Wang ·
Aoying Zhou · Weining Qian ·
Cheqing Jin · Zhifei Zhang (Eds.)

# Intelligent Computing and Block Chain

First BenchCouncil International Federated Conferences, FICC 2020
Qingdao, China, October 30 – November 3, 2020
Revised Selected Papers

Springer

*Editors*
Wanling Gao 🅘
Institute of Computing Technology
Chinese Academy of Sciences
Beijing, China

Changyun Wang
Renmin University of China
Beijing, China

Zhigang Qiu
Renmin University of China
Beijing, China

Aoying Zhou
East China Normal University
Shanghai, China

Cheqing Jin
East China Normal University
Shanghai, China

Kai Hwang
The Chinese University of Hong Kong
Shenzhen, China

Weiping Li
Oklahoma State University
Stillwater, OK, USA

Lei Wang
Institute of Computing Technology
Chinese Academy of Sciences
Beijing, China

Weining Qian
East China Normal University
Shanghai, China

Zhifei Zhang
Capital Medical University
Beijing, China

ISSN 1865-0929          ISSN 1865-0937   (electronic)
Communications in Computer and Information Science
ISBN 978-981-16-1159-9          ISBN 978-981-16-1160-5   (eBook)
https://doi.org/10.1007/978-981-16-1160-5

This Springer imprint is published by the registered company Springer Nature Singapore Pte Ltd.
The registered company address is: 152 Beach Road, #21-01/04 Gateway East, Singapore 189721, Singapore

# Preface

This volume contains a selection of revised papers presented at FICC 2020: BenchCouncil Federated Intelligent Computing and Block Chain Conferences, held in Qingdao, Shandong, China from October 30th to November 3rd, 2020. The main theme and topic of FICC 2020 was AI and Block Chain Technologies for computer sciences, finance, medicine, and education, to foster communication, collaboration, and interplay among these communities, and propose benchmark-based quantitative approaches to tackle multi-disciplinary challenges.

FICC 2020 assembled a spectrum of affiliated research conferences into a week-long coordinated meeting held at a common time in a common place, consisting of seven individual conferences—Symposium on Intelligent Medical Technology (MedTech 20), Symposium on Intelligent Computers (IC 20), Symposium on Chips (Chips 20), Symposium on Evaluation, Simulation, and Test (ESTest 20), Symposium on Education Technology (EduTech 20), Symposium on Block Chain (BChain 20), and Symposium on Financial Technology (FinTech 20).

FICC 2020 solicited papers that address hot topic issues in federated intelligent computing and block chain. The call for papers for the FICC 2020 conference attracted a number of high-quality submissions. During a rigorous review process, each paper was reviewed by at least three experts. This book includes 38 accepted papers from the FICC 2020 conference, which were selected from 103 submissions. The acceptance rate is 36.9%. FICC 2020 had more than 200 keynote lectures and invited talks by distinguished experts from China, the Americas, and Europe, including Academicians of the Chinese Academy of Sciences/Engineering, Fellows of the Royal Academy of Engineering, Fellows of the European Academy of Sciences, Fellows of the National Academy of Sciences/Engineering, etc.

During the conference, BenchCouncil sponsored two kinds of awards to recognize important contributions in the area of federated intelligent computing and block chain. BenchCouncil Best Paper Award is to recognize a paper presented at the FICC conferences which demonstrates potential impact on research and practice in this field. This year, we had four best paper award recipients. Bo Zhou, Ping Zhang, and Runlin Zhou from Operations and Maintenance Department, National Computer Network Emergency Response Technical Team received the best paper award of IC 20, for their paper "Root Cause Localization from Performance Monitoring Metrics Data with Multidimensional Attributes." Li Lin, Jiewei Wu, Pujin Cheng, Kai Wang, and Xiaoying Tang from Sun Yat-sen University and Southern University of Science and Technology received the best paper award of MedTech 20, for their paper "BLU-GAN: Bi-Directional ConvLSTM U-Net with Generative Adversarial Training for Retinal Vessel Segmentation." Zhiyun Chen, Yinuo Quan, and Dongming Qian from East China Normal University received the best paper award of EduTech 20, for their paper "Automatic essay scoring model based on multi-channel CNN and LSTM." Wei-Tek Tsai, Weijing Xiang, Wang Rong, and Enyan Deng from Beihang University and

Beijing Tiande Technology received the best paper award of BChain 20, for their paper "LSO: A Dynamic and Scalable Blockchain Structuring Framework."

BenchCouncil Award for Excellence for Reproducible Research is to encourage reliable and reproducible research using benchmarks from all organizations. This year, we have three award recipients for excellence for reproducible research, including Jun Lu from Institute of Physics, Chinese Academy of Sciences for his paper "A reconfigurable electrical circuit auto-processing method for direct electromagnetic inversion," Jie Li and Ping Li from Beihang University for their paper "Dynamic Copula Analysis of the Effect of COVID-19 Pandemic on Global Banking Systemic Risk," and Wenrui Kang, Xu Wang, Jixia Zhang, Xiaoming Hu and Qin Li from Beijing Institute of Technology for their paper "Two-way Perceived Color Difference Saliency Algorithm for Image Segmentation of Port Wine Stains."

We are very grateful for the efforts of all authors related to writing, revising, and presenting their papers at the FICC 2020 conferences. We appreciate the indispensable support of the FICC 2020 Program Committee and thank them for their efforts and contributions in maintaining the high standards of the FICC 2020 Symposia.

January 2020

Wanling Gao
Kai Hwang
Changyun Wang
Weiping Li
Zhigang Qiu
Lei Wang
Aoying Zhou
Weining Qian
Cheqing Jin
Zhifei Zhang

# Organization

## General Chairs

| | |
|---|---|
| Demin Han | Capital Medical University |
| Yike Guo | Imperial College London |
| Tianzi Jiang | Institute of Automation, Chinese Academy of Sciences |
| Guangnan Ni | Chinese Academy of Engineering |
| Kai Hwang | The Chinese University of Hong Kong, Shenzhen |
| Tony Hey | Rutherford Appleton Laboratory |
| Aoying Zhou | East China Normal University |
| Changyun Wang | Renmin University of China |
| Weiping Li | Oklahoma State University |

## Program Chairs

| | |
|---|---|
| Yanchun Zhang | Victoria University |
| Qiyong Gong | West China Hospital |
| Chunming Rong | University of Stavanger (UiS) |
| Lei Wang | Institute of Computing Technology, Chinese Academy of Sciences |
| Tong Wu | National Institute of Metrology |
| Yunquan Zhang | National Supercomputing Center in Jinan |
| Weining Qian | East China Normal University |
| Zhigang Qiu | Hanqing Advanced Institute of Economics and Finance, Renmin University of China |
| Xueming Si | Fudan University |
| Jianliang Xu | Hong Kong Baptist University |

## Program Committee

| | |
|---|---|
| Ömer Faruk Alçin | Bingöl University |
| Hong An | University of Science and Technology of China |
| Wei Ba | The General Hospital of the People's Liberation Army |
| Varun Bajaj | IIITDM Jabalpur |
| Lingfeng Bao | Zhejiang University |
| Weide Cai | Beihang University |
| Jiannong Cao | The Hong Kong Polytechnic University |
| Yang Cao | Kyoto University |
| Enhong Chen | University of Science and Technology of China |
| Qingcai Chen | Harbin Institute of Technology |
| Tianshi Chen | Institute of Computing Technology, Chinese Academy of Sciences and Cambricon |

| | |
|---|---|
| Wenguang Chen | Tsinghua University |
| Yiran Chen | Duke University |
| Sherman S. M. Chow | Chinese University of Hong Kong, Hong Kong |
| Yeh-Ching Chung | Chinese University of Hong Kong, Shenzhen |
| Bin Cui | Peking University |
| Chen Ding | University of Rochester |
| Shuai Ding | Hefei University of Technology |
| Xuan Ding | Tsinghua University |
| Junping Du | Beijing University of Posts and Telecommunications |
| Shaojing Fu | National University of Defense Technology |
| Lin Gan | Tianjin University |
| Haoyu Gao | Renmin University of China |
| Zhipeng Gao | Beijing University of Posts and Telecommunications |
| Dan Grigoras | University College Cork |
| Thomas Hacker | Purdue University |
| Dake He | Xin Hua Hospital, School of Medicine, Shanghai Jiao Tong University |
| Debiao He | Wuhan University |
| Wanqing He | Alibaba |
| Haibo Hu | Hong Kong Polytechnic University |
| Ronghuai Huang | Beijing Normal University |
| Shiyang Huang | The University of Hong Kong |
| Yi Huang | The Graduate Institute, Geneva |
| Yunyou Huang | Guangxi Normal University |
| Md. Rafiqul Islam | University of Technology Sydney |
| Martin G. Jaatun | SINTEF |
| Dun Jia | Renmin University of China |
| Haipeng Jia | Institute of Computing Technology, Chinese Academy of Sciences |
| Zhen Jia | Amazon |
| Bo Jiang | Beihang University |
| Fuwei Jiang | Central University of Finance and Economics |
| Chenqing Jin | East China Normal University |
| Nan Jing | Shanghai University |
| Enamul Kabir | University of Southern Queensland |
| Dexing Kong | Zhejiang University |
| Boying Lei | Shenzhen University |
| Kenli Li | Hunan University |
| Ping Li | Beihang University |
| Ye Li | Shenzhen Institutes of Advanced Technology, Chinese Academy of Sciences |
| Zhangsheng Li | Guangzhou Medical University |
| Zhengdao Li | The Chinese University of Hong Kong, Shenzhen |
| Xiubo Liang | Zhejiang University |
| Yi Liang | Beijing University of Technology |
| Yuan Liang | Guangxi Normal University |

| | |
|---|---|
| Zhouchen Lin | Peking University |
| Shiyuan Liu | Shanghai Changzheng Hospital |
| Yan Lou | China Medical University |
| Gang Lu | Huawei |
| Jieping Lu | First Affiliated Hospital of China University of Science and Technology (Anhui Provincial Hospital) |
| Chunjie Luo | Institute of Computing Technology, Chinese Academy of Sciences |
| Yu Luo | Renmin University of China |
| Hairong Lv | Tsinghua University |
| Qinying Ma | The First Affiliated Hospital of Hebei Medical University |
| Renwen Ma | The Chinese University of Hong Kong, Shenzhen |
| Qiguang Miao | Xidian University |
| Faisal Nawab | UC Santa Cruz |
| Gaoxiang Ouyang | Beijing Normal University |
| Yi Pan | Georgia State University |
| Chaoyi Pang | Zhejiang University |
| George Angelos Papadopoulos | University of Cyprus |
| Qingqi Pei | Xidian University |
| Shaoliang Peng | Hunan University |
| Yun Peng | Hong Kong Baptist University |
| Zhe Peng | Hong Kong Baptist University |
| Xiaosong Qian | Soochow University Financial Engineering Research Center |
| Md. Mijanur Rahman | University of Newcastle |
| Rui Ren | Cyberspace Security Research Institute Co., Ltd. |
| Kuheli Sai | University of Pittsburgh |
| Abdulkadir Sengur | Firat University |
| Xuequn Shang | Northwestern Polytechnical University |
| Meng Shen | Beijing Institute of Technology |
| Yingjie Shi | Beijing Institute of Clothing Technology |
| Kefan Shuai | The Chinese University of Hong Kong, Shenzhen |
| Siuly Siuly | Victoria University |
| Ke Song | Renmin University of China |
| Jiaming Sun | Winning Health Technology Group Co., Ltd. |
| Jian Sun | Xi'an Jiaotong University |
| Le Sun | Nanjing University of Information Science and Technology |
| Yi Sun | ICT, CAS |
| Wenjun Qin | Northeastern University |
| Xiaoying Tang | Southern University of Science and Technology |
| Guohao Tang | Hunan University |
| Yuzhe Tang | Syracuse University |
| Md. Nurul Ahad Tawhid | University of Dhaka |

| | |
|---|---|
| Jianhui Tian | Longhua Hospital, Shanghai University of Traditional Chinese Medicine |
| Chaodong Wang | Xuanwu Hospital, Capital Medical University |
| Cho-Li Wang | The University of Hong Kong |
| Hefei Wang | Renmin University of China |
| Huadong Wang | Jinan University |
| Lei Wang | Institute of Computing Technology, Chinese Academy of Sciences |
| Maoning Wang | Central University of Finance and Economics |
| Ruixuan Wang | Sun Yat-sen University |
| Wei Wang | East China Normal University |
| Xinyan Wang | Air Force General Hospital |
| Dongqing Wei | Shanghai Jiaotong University |
| Yanjie Wei | Shenzhen Institutes of Advanced Technology, Chinese Academy of Sciences |
| Jian Wu | Zhejiang University |
| Ke Wu | Renmin University of China |
| Yonghe Wu | East China Normal University |
| Yong Xia | Northwestern Polytechnical University |
| Bin Xiao | Hong Kong Polytechnic University |
| Jiang Xiao | Huazhong University of Science and Technology |
| Biwei Xie | ICT, CAS |
| Juanying Xie | Shaanxi Normal University |
| Chunxiao Xing | Tsinghua Univerity |
| Cheng Xu | Simon Fraser University |
| Cheng-Zhong Xu | University of Macau |
| Guangxia Xu | Chongqing University of Posts and Telecommunications |
| Jun Xu | Nanjing University of Information Science and Technology |
| Qingxian Xu | National Chung Cheng University |
| Yinxing Xue | University of Science and Technology of China |
| Dejin Yang | Beijing Jishuitan Hospital |
| Xiaokang Yang | Shanghai Jiao Tong University |
| Yun Yang | Yunnan University |
| Zijiang James Yang | Western Michigan University |
| Zhaoxiang Ye | Tianjin Cancer Hospital |
| Ge Yu | Northeastern University |
| Guanzhen Yu | Shanghai University of Traditional Chinese Medicine |
| Lun Yu | Fuzhou University |
| Xinguo Yu | Central China Normal University |
| John Yuen | Hong Kong Polytechnic University |
| Demetrios Zeinalipour-Yazti | University of Cyprus |
| Fan Zhang | ICT, CAS |
| Ming Zhang | Peking University |

| | |
|---|---|
| Xu Zhang | Capital Medical University |
| Bin Zhao | Nanjing Normal University |
| Di Zhao | Institute of Computing Technology, Chinese Academy of Sciences |
| Gang Zhao | Air Force Medical University |
| Yongjun Zhao | Nanyang Technological University |
| Zhiming Zhao | University of Amsterdam |
| Chen Zheng | Institute of Software, Chinese Academy of Sciences |
| Jianqiao Zhou | Rui Jin Hospital, School of Medicine, Shanghai Jiao Tong University |
| Yajin Zhou | Zhejiang University |
| Jia Zhu | South China Normal University |
| Liehuang Zhu | Beijing Institute of Technology |
| Liying Zhu | Harbin Medical University |
| Yifeng Zhu | Central University of Finance and Economics |
| Yongxin Zhu | Shanghai Jiao Tong University |
| Albert Zomaya | University of Sydney |
| Beiji Zou | Central South University |
| Zhiqiang Zuo | Nanjing University |

# Contents

## AI and Block Chain

## AI and Education Technology

## AI and Financial Technology

# AI and Medical Technology

# BLU-GAN: Bi-directional ConvLSTM U-Net with Generative Adversarial Training for Retinal Vessel Segmentation

Li Lin[1,2], Jiewei Wu[1,2], Pujin Cheng[2], Kai Wang[1(✉)], and Xiaoying Tang[2(✉)]

[1] School of Electronics and Information Technology,
Sun Yat-Sen University, Guangzhou, China
`wangkai23@mail.sysu.edu.cn`
[2] Department of Electrical and Electronic Engineering,
Southern University of Science and Technology, Shenzhen, China
`tangxy@sustech.edu.cn`

**Abstract.** Retinal vascular morphometry is an important biomarker of eye-related cardiovascular diseases such as diabetes and hypertension. And retinal vessel segmentation is a fundamental step in fundus image analyses and diagnoses. In recent years, deep learning based networks have achieved superior performance in medical image segmentation. However, for fine vessels or terminal branches, most existing methods tend to miss or under-segment those structures, inducing isolated breakpoints. In this paper, we proposed Bi-Directional ConvLSTM U-Net with Generative Adversarial Training (BLU-GAN), a novel deep learning model based on U-Net that generates precise predictions of retinal vessels combined with generative adversarial training. Bi-directional ConvLSTM, which can better integrate features from different scales through a coarse-to-fine memory mechanism, is employed to non-linearly combine feature maps extracted from encoding path layers and the previous decoding up-convolutional layers and to replace the simple skip-connection used in the original U-Net. Moreover, we use densely connected convolutions in certain layers to strengthen feature propagation, encourage feature reuse, and substantially reduce the number of parameters. Through extensive experiments, BLU-GAN has shown leading performance among the state-of-the-art methods on the DRIVE, STARE, CHASE_DB1 datasets for retinal vessel segmentation.

**Keywords:** Retinal vessel segmentation · Densely-connected ConvLSTM U-Net · Generative adversarial training

## 1 Introduction

Retinal vascular examination is the only non-invasive way for ophthalmologists or other experts to observe and photograph vasculature in vivo [1]. Segmentation of retinal vessels serves as an important component in the diagnosis of retinal diseases as well as systemic diseases, such as hypertension, arteriolosclerosis, age-related macular degeneration (AMD), and diabetic retinopathy (DR). To address

© Springer Nature Singapore Pte Ltd. 2021
W. Gao et al. (Eds.): FICC 2020, CCIS 1385, pp. 3–13, 2021.
https://doi.org/10.1007/978-981-16-1160-5_1

the limitations of manual segmentation, such as high complexity, labor intensiveness and low reproducibility, various unsupervised and supervised methods have been proposed for automatically segmenting retinal vessels. One major challenge in the vessel segmentation task is that vessels are multi-scale with relatively low contrast from the background, especially for fine vessels or terminal branches in noisy images. Moreover, it will be more complicated if there exist lesions on fundus images such as hemorrhages and microaneurysms. Such lesions may cause mis-segmentation of vessels and make this task even more challenging.

Existing automatic vessel segmentation algorithms can probably be summarized into two categories. The first category is unsupervised methods, usually including matched filter methods and morphological processing methods [2]. For instance, Nguyen et al. [3] proposed a method based on multi-scale line operators for vessel detection. Singh et al. [4] developed a Gumbel probability distribution function based matched filter approach for this task.

The other category is supervised methods, in which data with ground truth are used to train a classifier based on predefined features [5,6]. The past recent years have witnessed the rapid development and outstanding performance of deep learning in various computer vision tasks. For example, Ronneberger et al. proposed the so-called U-Net [7] in an encoder-decoder structure with skip connections, allowing feature maps to maintain low-level features and efficient information propagation. Recently, many deep learning architectures inspired by U-Net, such as R2U-Net [8], LadderNet [9], and SSCF-Net [10], achieved superior performance on the vessel segmentation task. Despite their progress, these methods still have limitations, such as confusing blood vessels with interfering structures and failing to segment tiny vessels or vessels at terminal branches. The main disadvantage of the aforementioned U-Net based models is that the processing step is performed individually for the two sets of feature maps, and these features are then directly concatenated. On the other hand, models used in existing methods mainly rely on pixel-wise objective functions that compare ground truth images with model-predicted results. The results in [11] show that without changing the network structure, supplementing a pixel-wise loss with a topology loss can generate results with appropriate topological characteristics, resulting in a substantial performance increase. However, it is kind of unrealistic to artificially design different topological loss functions for different topological structures.

Generative Adversarial Network (GAN) is a framework consisting of two networks, discriminator and generator [12]. The discriminator is dedicated to distinguishing the predicted results generated by the generator from the ground truth images, whereas the generator tries to generate realistic outputs that are very similar to the ground truth images to confuse the discriminator. Therefore, we can apply the GAN framework, adding a discriminator after the generator network to serve as a supplemental automatically learned loss to the pixel-wise loss.

In such context, we propose a novel method for vessel segmentation, namely BLU-GAN (Fig. 1), which consists of a Bi-Directional ConvLSTM U-Net with densely connected convolutional layers as the generator and an image-level discriminator. Through extensive quantitative and qualitative evaluation experiments, our method shows leading performance among state-of-the-art models on DRIVE [13], STARE [14], and CHASE_DB1 [15] datasets. The results reveal that Bi-Directional ConvLSTM connections and densely connected convolutional layers can improve the network's capability and enable propagation of subtle features, thereby improving the overall segmentation results. Moreover, applying adversarial training and adding a discriminator can supplement the pixel-wise loss function, substantially improving the quality of the segmentation by training the generator to extract vessel prediction images that are indistinguishable from those annotated by ophthalmologists.

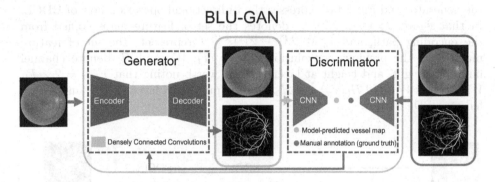

**Fig. 1.** The proposed BLU-GAN framework for vessel segmentation.

## 2 Method

### 2.1 Image Preprocessing

In each of our experimental datasets, the amount of training data is no larger than 20 and thus data argumentation and preprocessing are required. Existing studies [16, 17] have shown that employing image patches for training and testing can generate superior prediction results than using whole images. However, this process reduces the training and testing efficiency and it is not employed in this work. In our method, each whole image is augmented by left-right flip and rotation sampled every 4° from 4° to 360°. Then we adjust the brightness, contrast, and color of all images with random coefficients from 0.8 to 1.2. Finally, all images are normalized to z-score.

## 2.2   The Proposed Architecture

As demonstrated in Fig. 1, the proposed BLU-GAN consists of a generator network and a discriminator network. The generator takes a fundus image as input and generates a probability mask of retinal vessels with the same size as the input. For the discriminator, it needs to judge whether the input vessel map belongs to manual ground truth or the output of the generator according to the input fundus image and the corresponding vessel map. The two networks conduct adversarial training so that the vessel maps generated by the generator are as similar as possible to the manual annotations.

For the generator, we take full advantages of U-Net, bi-directional ConvL-STM (BDCL) and the mechanism of dense convolutions [18,19]. We use BDCLs to replace the simple skip-connections in U-Net, and non-linearly combine the feature maps extracted from the corresponding encoding path with the previous decoding up-convolutional layer. Figure 2 demonstrates the architecture of the generator and Fig. 3 (a) expresses the bidirectional operation flow of BDCL. In that figure, $x_e \in \mathbb{R}^{F_l \times W_l \times H_l}$ denotes the set of feature maps copied from the encoding path, and $x_d \in \mathbb{R}^{F_{l+1} \times W_{l+1} \times H_{l+1}}$ represents the set of feature maps from the previous convolutional layer ($F_l$, $W_l$, and $H_l$ denote channel number, weight, and height at layer $l$). It is worth noting that $F_{l+1} = 2 \times F_l$, $W_{l+1} = \frac{1}{2} \times W_l$, and $H_{l+1} = \frac{1}{2} \times H_l$. The output of BN layer ($\hat{x}_d^{tran}$) and $x_e$ are fed to a BDCL layer.

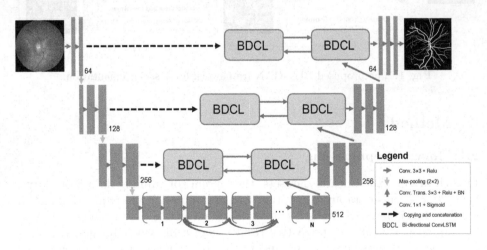

**Fig. 2.** The architecture of the generator in BLU-GAN.

ConvLSTM [20], consisting of an input gate $i_t$, an output gate $o_t$, a forget gate $f_t$, and a memory cell $c_t$, was proposed to employ convolution operations for performing input-to-state and state-to-state conversions. The calculation process in ConvLSTM is as follows (for convenience, bias terms are omitted):

$$i_t = \sigma \left( W_i^x * x_t + W_i^{\mathcal{H}} * \mathcal{H}_{t-1} \right), \tag{1}$$

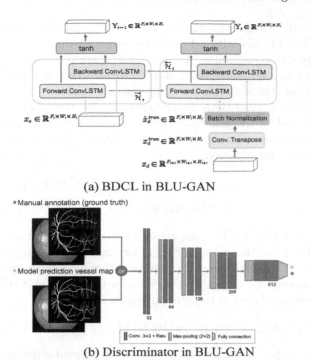

(a) BDCL in BLU-GAN

(b) Discriminator in BLU-GAN

**Fig. 3.** Illustration of some details in BLU-GAN. Panels (a) and (b) respectively denote the BDCL block in the generator and the structure of the discriminator.

$$f_t = \sigma \left( W_f^x * x_t + W_f^{\mathcal{H}} * \mathcal{H}_{t-1} \right), \tag{2}$$

$$o_t = \sigma \left( W_o^x * x_t + W_o^{\mathcal{H}} * \mathcal{H}_{t-1} \right), \tag{3}$$

$$c_t = f_t \circ c_{t-1} + i_t \circ \tanh \left( W_c^x * x_t + W_c^{\mathcal{H}} * \mathcal{H}_{t-1} \right), \tag{4}$$

$$\mathcal{H}_t = o_t \circ \tanh \left( c_t \right), \tag{5}$$

where $*$ and $\circ$ respectively denote the convolution operator and the Hadamard product. All the gates, memory cell, hidden state $\mathcal{H}$, and learnable weights $W$ are 3D tensors. $W_*^x$ and $W_*^{\mathcal{H}}$ are 2D convolution kernels corresponding to the input and hidden state. Information from both the forward and backward frames are important and complementary for segmenting vessels, and thus BDCL uses two ConvLSTMs to process the input data into two directions, and then determines the current input by analyzing the data dependencies in both directions. Therefore, we obtain two sets of parameters for both directions, and the output of the BDCL is calculated as

$$\mathbf{Y}_t = \tanh \left( \mathbf{W}_y^{\overrightarrow{\mathcal{H}}} * \overrightarrow{\mathcal{H}}_t + \mathbf{W}_y^{\overleftarrow{\mathcal{H}}} * \overleftarrow{\mathcal{H}}_t \right), \tag{6}$$

where $\overrightarrow{\mathcal{H}}$ and $\overleftarrow{\mathcal{H}}$ indicates the hidden states from forward and backward ConvLSTM units, and $\mathbf{Y}_t$ indicates the final output considering bi-directional spatio-temporal information. More details can be found in [18, 20].

As illustrated in Fig. 3 (b), the discriminator in BLU-GAN consists of ten $3 \times 3$ convolutional layers followed by batch normalization, rectified linear units (Relus) and spatial max-pooling. The last two layers of the network are a fully connected layer and a final sigmoid layer. Through the sigmoid layer, the discriminator outputs the classification of its judgments [28].

The objective of BLU-GAN is to minimize the probability of the paired samples (generated by the generator $G$) to be recognized while maximizing the probability of the discriminator $D$ making a mistake, which is formulated as the following minimax optimization:

$$loss_{GAN} = E_{x,y \sim p_{\text{data}}(x,y)}[\log D(x,y)] + E_{x \sim p_{\text{data}}(x)}[\log(1 - D(x, G(x)))], \quad (7)$$

$$G^* = \arg \min_{G} \max_{D} loss_{GAN}, \quad (8)$$

where $x$, $y$ represents a fundus image and a corresponding vessel ground truth. The discriminator D takes a pair of $(x, y)$ or $(x, G(x))$ to binary classification $\{0, 1\}$ where 0 and 1 denote the pair sample is either model-generated or manual-generated. Moreover, in the generator, we also utilize ground truth by taking binary cross-entropy between the ground truths and outputs,

$$loss_{SEG} = E_{x,y \sim p_{\text{data}}(x,y)} - y \cdot \log G(x) - (1 - y) \cdot \log(1 - G(x)). \quad (9)$$

By summing up Eq. (8) and Eq. (9), we obtain the global objective function as

$$G = G^* + \lambda loss_{SEG}, \quad (10)$$

where $\lambda$ is set to balance two loss functions.

## 3   Experiments

In this section, we introduce some specific parameter settings and implementation details of BLU-GAN and our experiments. Comparison and visualization results are demonstrated in Fig. 4, Fig. 5, and Table 1.

### 3.1   Datasets

We evaluate our method on three publicly available datasets, DRIVE, STARE, CHASE_DB1. Manual annotations by two ophthalmologists are included for each dataset. We employ the first ophthalmologist's annotations as the ground truth for training and testing, and use the ones from the second ophthalmologist as human performance. DRIVE consists of 40 fundus images with binary filed of view (FOV) masks. We use standard train/test split: 20 training images and 20 testing images. The STARE dataset consists of 20 images, and we divide the data into 10 images as the training set and 10 images as the testing set according to [21]. For CHASE_DB1, we follow the setting of [9,22], namely the first 20 are used as the training set and the remaining 8 as the testing set. The FOV masks of STARE and CHASE_DB1 are obtained through morphological transformation and thresholding operation.

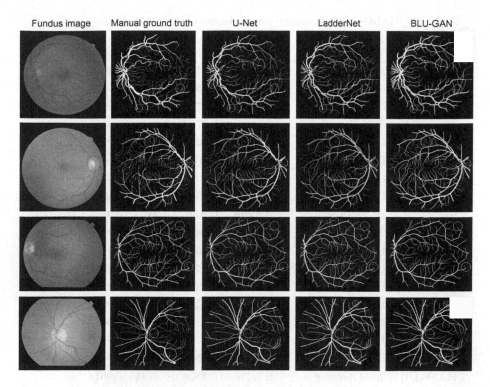

**Fig. 4.** Visualization of the segmentation results from different methods on DRIVE and STARE datasets.

## 3.2 Implementation Details

Our methods are implemented based on Keras library with tensorflow backend. In the densely connected convolutional layers of the generator, we set $N = 3$ to achieve a balance in network depth, feature extraction capabilities, and parameter quantities. Augmented images are divided into the training/validation sets with a ratio of 9:1. We conduct multiple rounds of training until convergence.

During training, the discriminator and generator are trained alternately for one epoch. We adopt Adam optimizer with the initial learning rate being $e^{-4}$ and reduce the learning rate on Plateau. Besides, the trade-off coefficient $\lambda$ in Eq. (10) is set to be 10. As for evaluation metrics, we employ true positive (TP), false positive (FP), false negative (FN), and true negative (TN) by comparing automated segmentations with the corresponding ground truth annotations. Then, accuracy (Acc), sensitivity (Se), specificity (Sp), F1-score, and area under ROC curve (AUC) are used to evaluate the performance of BLU-GAN.

10    L. Lin et al.

**Table 1.** Performance comparisons on three datasets.

| Datasets | Methods | F1-score | Se | Sp | Acc | AUC |
|---|---|---|---|---|---|---|
| DRIVE | 2nd Observer | 0.7881 | 0.7760 | 0.9725 | 0.9473 | - |
| | U-Net [8] | 0.8142 | 0.7537 | 0.9820 | 0.9531 | 0.9755 |
| | RU-Net [8] | 0.8155 | 0.7751 | **0.9816** | 0.9556 | 0.9782 |
| | R2U-Net [8] | 0.8171 | 0.7792 | 0.9813 | 0.9556 | 0.9784 |
| | LadderNet [9] | 0.8205 | 0.8081 | 0.9770 | 0.9561 | 0.9793 |
| | BCDU-Net [17] | 0.8224 | 0.8007 | 0.9784 | 0.9560 | 0.9789 |
| | Liu et al. [24] | 0.8225 | 0.8072 | 0.9780 | 0.9559 | 0.9779 |
| | DEU-Net [25] | 0.8270 | 0.7940 | **0.9816** | **0.9567** | 0.9772 |
| | BLU-GAN | **0.8296** | **0.8367** | 0.9810 | 0.9563 | **0.9813** |
| STARE | 2nd Observer | 0.7401 | **0.8951** | 0.9386 | 0.9350 | - |
| | U-Net [7] | 0.7595 | 0.6681 | 0.9915 | 0.9639 | 0.9710 |
| | Liskowski et al. [26] | - | 0.8554 | 0.9862 | **0.9729** | **0.9928** |
| | DRIU [27] | 0.7139 | 0.5949 | **0.9938** | 0.9534 | 0.9240 |
| | LadderNet [9] | 0.7319 | 0.6321 | 0.9898 | 0.9535 | 0.9609 |
| | DU-Net [23] | 0.7629 | 0.6810 | 0.9903 | 0.9639 | 0.9758 |
| | Liu et al. [24] | 0.8036 | 0.7771 | 0.9843 | 0.9623 | 0.9793 |
| | BLU-GAN | **0.8447** | 0.8527 | 0.9805 | 0.9670 | 0.9883 |
| CHASE_DB1 | 2nd Observer | 0.7978 | 0.8328 | 0.9744 | 0.9614 | - |
| | U-Net [8] | 0.7783 | 0.8288 | 0.9701 | 0.9575 | 0.9772 |
| | ResU-Net [8] | 0.7800 | 0.7726 | 0.9820 | 0.9553 | 0.9779 |
| | RU-Net [8] | 0.7810 | 0.7459 | 0.9836 | 0.9622 | 0.9803 |
| | R2U-Net [8] | 0.7928 | 0.7756 | 0.9820 | 0.9634 | 0.9815 |
| | LadderNet [9] | 0.7895 | 0.7856 | 0.9799 | 0.9620 | 0.9772 |
| | DEU-Net [25] | 0.8037 | 0.8074 | **0.9821** | **0.9661** | 0.9712 |
| | BLU-GAN | **0.8263** | **0.8422** | 0.9805 | 0.9643 | **0.9856** |

### 3.3  Results

We compare our models with some state-of-the-art (SOTA) ones, including U-Net, Recurrent U-Net (RU-Net), R2U-Net, DEU-Net, etc., on the three aforementioned datasets. All the quantitative comparison results are demonstrated in Table 1. Evaluation results indicate that BLU-GAN achieves leading performance on three datasets among SOTA methods with the highest F1-score and almost highest scores on other metrics.

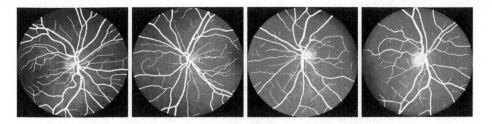

**Fig. 5.** Representative overlapped segmentation results from BLU-GAN (white) and corresponding ground truths (green) on CHASE_DB1. (Color figure online)

## 4  Conclusion

In this paper, we proposed a novel GAN framework by incorporating BDCL and densely connected convolutional layers into U-Net and an image-level discriminator for segmenting retinal vessels. The proposed structure can incorporate both spatial and temporal information (in our case bi-directional inter-samples correlative information) of vessels and improve the network's capability and propagation of subtle features, thereby improving the overall segmentation results. Additionally, the discriminator was designed to serve as a supplemental automatically learned loss to the pixel-wise loss, which further improved the segmentation performance. We demonstrated that the proposed method has leading performance among state-of-the-art algorithms on DRIVE, STARE, and CHASE_DB1 datasets. Future work will focus on how to improve the training efficiency of the proposed BLU-GAN framework and integrate patch sampling methods to further improve performance.

**Acknowledgement.** This study was supported by the National Natural Science Foundation of China (62071210), the Shenzhen Basic Research Program (JCYJ20190809120205578), the National Key R&D Program of China (2017YFC0112404), and the High-level University Fund (G02236002). The authors declare that they have no competing financial interests.

## References

1. Chatziralli, I.P., Kanonidou, E.D., Keryttopoulos, P., Dimitriadis, P., Papazisis, L.E.: The value of fundoscopy in general practice. Open Ophthalmol. J. **6**, 4 (2012). https://doi.org/10.2174/1874364101206010004
2. Singh, S., Tiwari, R.K.: A review on retinal vessel segmentation and classification methods. In: International Conference on Trends in Electronics and Informatics, pp. 895–900 (2019). https://doi.org/10.1109/ICOEI.2019.8862555
3. Nguyen, U.T., Bhuiyan, A., Park, L.A., Ramamohanarao, K.: An effective retinal blood vessel segmentation method using multi-scale line detection. Pattern Recogn. **46**(3), 703–715 (2013). https://doi.org/10.1016/j.patcog.2012.08.009
4. Singh, N.P., Srivastava, R.: Retinal blood vessels segmentation by using Gumbel probability distribution function based matched filter. Comput. Methods Programs Biomed. **129**, 40–50 (2016). https://doi.org/10.1016/j.cmpb.2016.03.001

5. Zhang, J., Chen, Y., Bekkers, E., Wang, M., Dashtbozorg, B., ter Haar Romeny, B.M.: Retinal vessel delineation using a brain-inspired wavelet transform and random forest. Pattern Recogn. **69**, 107–123 (2017). https://doi.org/10.1016/j.patcog.2017.04.008

6. Holbura, C., Gordan, M., Vlaicu, A., Stoian, I., Capatana, D.: Retinal vessels segmentation using supervised classifiers decisions fusion. In: IEEE International Conference on Automation, Quality and Testing, Robotics, pp. 185–190 (2012). https://doi.org/10.1109/AQTR.2012.6237700

7. Ronneberger, O., Fischer, P., Brox, T.: U-Net: convolutional networks for biomedical image segmentation. In: Navab, N., Hornegger, J., Wells, W.M., Frangi, A.F. (eds.) MICCAI 2015, Part III. LNCS, vol. 9351, pp. 234–241. Springer, Cham (2015). https://doi.org/10.1007/978-3-319-24574-4_28

8. Alom, M.Z., Hasan, M., Yakopcic, C., Taha, T.M., Asari, V.K.: Recurrent residual convolutional neural network based on U-Net (R2U-Net) for medical image segmentation. arXiv preprint arXiv:1802.06955 (2018)

9. Zhuang, J.: LadderNet: multi-path networks based on U-Net for medical image segmentation. arXiv preprint arXiv:1810.07810 (2018)

10. Lyu, J., Cheng, P., Tang, X.: Fundus image based retinal vessel segmentation utilizing a fast and accurate fully convolutional network. In: Fu, H., Garvin, M.K., MacGillivray, T., Xu, Y., Zheng, Y. (eds.) OMIA 2019. LNCS, vol. 11855, pp. 112–120. Springer, Cham (2019). https://doi.org/10.1007/978-3-030-32956-3_14

11. Mosinska, A., Marquez-Neila, P., Koziński, M., Fua, P.: Beyond the pixel-wise loss for topology-aware delineation. In: IEEE Conference on Computer Vision and Pattern Recognition, pp. 3136–3145 (2018). https://doi.org/10.1109/CVPR.2018.00331

12. Goodfellow, I., et al.: Generative adversarial nets. In: Advances in Neural Information Processing Systems, pp. 2672–2680 (2014). https://doi.org/10.5555/2969033.2969125

13. Staal, J., Abrmoff, M.D., Niemeijer, M., Viergever, M.A., Van-Ginneken, B.: Ridge-based vessel segmentation in color images of the retina. IEEE Trans. Med. Imaging **23**(4), 501–509 (2004). https://doi.org/10.1109/TMI.2004.825627

14. Hoover, A., Kouznetsova, V., Goldbaum, M.: Locating blood vessels in retinal images by piece-wise threshold probing of a matched filter response. IEEE Trans. Med. Imaging **19**(3), 203–210 (2000). https://doi.org/10.1109/42.845178

15. Owen, C.G., Rudnicka, A.R., Mullen, R., Barman, S.A., et al.: Measuring retinal vessel tortuosity in 10-year-old children: validation of the computer-assisted image analysis of the retina (CAIAR) program. Invest. Ophthalmol. Vis. Sci. **50**(5), 2004–2010 (2009). https://doi.org/10.1167/iovs.08-3018

16. Wang, C., Zhao, Z., Ren, Q., Xu, Y., Yu, Y.: Dense U-Net based on patch-based learning for retinal vessel segmentation. Entropy **21**(2), 168 (2019). https://doi.org/10.3390/e21020168

17. Azad, R., Asadi-Aghbolaghi, M., Fathy, M., Escalera, S.: Bi-directional ConvLSTM U-net with Densley connected convolutions. In: IEEE International Conference on Computer Vision Workshops (2019). https://doi.org/10.1109/ICCVW.2019.00052

18. Song, H., Wang, W., Zhao, S., Shen, J., Lam, K.-M.: Pyramid dilated deeper ConvLSTM for video salient object detection. In: Ferrari, V., Hebert, M., Sminchisescu, C., Weiss, Y. (eds.) ECCV 2018, Part XI. LNCS, vol. 11215, pp. 744–760. Springer, Cham (2018). https://doi.org/10.1007/978-3-030-01252-6_44

19. Iandola, F., Moskewicz, M., Karayev, S., Girshick, R., Darrell, T., Keutzer, K.: DenseNet: Implementing efficient convnet descriptor pyramids. arXiv preprint arXiv:1404.1869 (2014)

20. Shi, X., Chen, Z., Wang, H., Yeung, D.Y., Wong, W.K., Woo, W.C.: Convolutional LSTM network: a machine learning approach for precipitation nowcasting. In: Advances in neural information processing systems, pp. 802–810 (2015). https:// doi.org/10.5555/2969239.2969329
21. Tetteh, G., Rempfler, M., Zimmer, C., Menze, B.H.: Deep-FExt: deep feature extraction for vessel segmentation and centerline prediction. In: Wang, Q., Shi, Y., Suk, H.-I., Suzuki, K. (eds.) MLMI 2017. LNCS, vol. 10541, pp. 344–352. Springer, Cham (2017). https://doi.org/10.1007/978-3-319-67389-9_40
22. Wu, Y., Xia, Y., Song, Y., Zhang, Y., Cai, W.: Multiscale network followed network model for retinal vessel segmentation. In: Frangi, A.F., Schnabel, J.A., Davatzikos, C., Alberola-López, C., Fichtinger, G. (eds.) MICCAI 2018, Part II. LNCS, vol. 11071, pp. 119–126. Springer, Cham (2018). https://doi.org/10.1007/978-3-030-00934-2_14
23. Jin, Q., Meng, Z., Pham, T.D., Chen, Q., Wei, L., Su, R.: DUNet: a deformable network for retinal vessel segmentation. Knowl.-Based Syst. **178**, 149–162 (2019). https://doi.org/10.1016/j.knosys.2019.04.025
24. Liu, B., Gu, L., Lu, F.: Unsupervised ensemble strategy for retinal vessel segmentation. In: Shen, D., et al. (eds.) MICCAI 2019. LNCS, vol. 11764, pp. 111–119. Springer, Cham (2019). https://doi.org/10.1007/978-3-030-32239-7_13
25. Wang, B., Qiu, S., He, H.: Dual encoding U-Net for retinal vessel segmentation. In: Shen, D., et al. (eds.) MICCAI 2019. LNCS, vol. 11764, pp. 84–92. Springer, Cham (2019). https://doi.org/10.1007/978-3-030-32239-7_10
26. Liskowski, P., Krawiec, K.: Segmenting retinal blood vessels with deep neural networks. IEEE Trans. Med. Imaging. **35**(11), 2369–2380 (2016). https://doi.org/10.1109/TMI.2016.2546227
27. Maninis, K.-K., Pont-Tuset, J., Arbeláez, P., Van Gool, L.: Deep retinal image understanding. In: Ourselin, S., Joskowicz, L., Sabuncu, M.R., Unal, G., Wells, W. (eds.) MICCAI 2016, Part II. LNCS, vol. 9901, pp. 140–148. Springer, Cham (2016). https://doi.org/10.1007/978-3-319-46723-8_17
28. Son, J., Park, S.J., Jung, K.H.: Retinal vessel segmentation in fundoscopic images with generative adversarial networks. arXiv preprint arXiv:1706.09318 (2017)

# Choroidal Neovascularization Segmentation Based on 3D CNN with Cross Convolution Module

Xiwei Zhang[1], Mingchao Li[1], Yuhan Zhang[1], Songtao Yuan[2], and Qiang Chen[1(✉)]

[1] School of Computer Science and Engineering,
Nanjing University of Science and Technology, Nanjing, China
chen2qiang@njust.edu.cn
[2] Department of Ophthalmology,
The First Affiliated Hospital with Nanjing Medical University, Nanjing, China

**Abstract.** Choroidal neovascularization (CNV) is a retinal vascular disease that new vessels sprout from the choroid and then grow into retina, which usually appears in the late stage of Age-related macular degeneratoin (AMD). Because of the complex characteristics of CNV, it is time-consuming and laborious to manually segment CNV from spectral-domain optical coherence tomography (SD-OCT) images. In this paper, we propose an improved 3D U-Net model based on anisotropic convolution to segment CNV automatically. First, to adapt the special morphology and differences in the scale of CNV, a cross convolution module (CCM) is designed based on anisotropic convolution instead of general convolution with kernel size 3×3×3 in standard 3D U-Net architecture. Then, the first layer is adjusted to reduce the usage of GPU. In addition, a dilated connection layer (DCL) is proposed to improve the ability to capture multi-scale information while using low-level features. After verifying on a dataset that consists of 376 cubes from 21 patients, we obtain the mean dice of 84.73%. The experimental evaluation shows that our approach is able to achieve effective segmentation of CNV from SD-OCT images.

**Keywords:** Choroidal neovascularization segmentation · Optical coherence tomograph · Anisotopic convolution module · Dilated connection layer · 3D U-Net

## 1 Introduction

Choroidal neovascularization (CNV) generally appears in the late stage of Age-related macular degeneration (AMD) which can lead to permanent visual loss. Specifically, because excessive vascular endothelial growth factor (VEGF) is produced in patients with AMD, new blood vessels grow out of choroid, break through Bruch's membrane (BM), and enter the retina, which forms CNV [1].

This work was supported in part by National Natural Science Foundation of China (61671242), Key R&D Program of Jiangsu Science and Technology Department (BE2018131).

W. Gao et al. (Eds.): FICC 2020, CCIS 1385, pp. 14–21, 2021.
https://doi.org/10.1007/978-981-16-1160-5_2

Optical Coherence Tomography (OCT) that is based on the basic principle of weak coherent light interferometer, is a noninvasive, high resolution 3D volume imaging technique. Clear image of retinal tomographic structure can be captured by OCT equipment, which can show subtle structure and pathological characteristic [2]. As the second generation of OCT technology, spectral-domain OCT (SD-OCT) has the advantages of high speed and superior sensitivity [3]. Most patients with CNV will receive anti-VEGF injection therapy. In this case, the accurate segmentation of CNV can provide clinicians with a lot of information about the lesions, such as volume and shape, which is important to analyze the current condition, and it can also help to evaluate the impact of injected drugs on the disease [4].

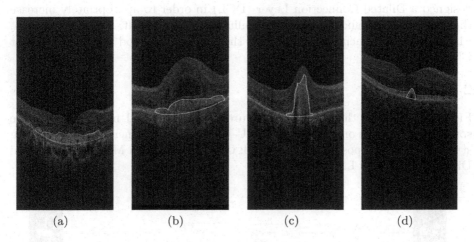

|     |     |     |     |
| (a) | (b) | (c) | (d) |

**Fig. 1.** Pathological characteristics of CNV. The CNV area is surrounded by the red line (Color figure online)

With the growth of CNV, there will be a variety of complications, such as subretinal hemorrhage, fluid exudation, lipid deposition, detachment of the retinal pigment epithelium from the choroid, fibrotic scars, and the combination of these mentioned finding. As shown in Fig. 1, CNV has complex pathological characteristics which are manifested in different scales, blurred boundaries, and non-uniform pixel distribution in SD-OCT images. All these make manual segmentation time-consuming and laborious, and also bring great challenges to automatic segmentation.

In recent years, some algorithms for CNV segmentation in SD-OCT images have been proposed. Li et al. [5] detected CNV by random forest model with 3D-HOG feature in B-scans and obtained CNV segmentation results in fundus images. Zhang et al. [6] proposed a multi-scale parallel branch CNN that combines deep and shallow features and multi-scale information to segment CNV. Su et al. [7] innovatively designed a differential amplification block inspired by the differential amplifier in electronic circuits with the purpose of extracting high-frequency and low-frequency information from SD-OCT images. Although many of currently proposed methods work well, none of them make full use of the three-dimensional characteristics of SD-OCT images to improve the continuity between frames.

Due to the diverse forms of CNV, we proposed a scheme based on anisotropic convolution to capture CNV features more comprehensively. At the same time, in order to improve the continuity between frames, 3D U-Net [8] was chosen as the basic network structure. Another important issue is that the standard 3D U-Net model, which is used to process medical images, consumes a lot of GPU resource. Inspired by Resnet [9], the first layer of the network was adjusted to reduce the size of the generated data. The Cross Convolution Module (CCM) was designed for the complex pathological characteristics of CNV. Additionally, we designed a Dilated Connection Layer (DCL) in order to appropriately increase the receptive field and improve the ability to capture multi-scale information. Finally, the experiments indicated that the proposed method is able to segment CNV efficiently.

## 2    Method

Figure 2 illustrates the network structure of the proposed method. We made some improvements on the basis of 3D U-Net for CNV segmentation. Two main structures were proposed in our method: Cross Convolution Module (CCM) and Dilated Connection Layer (DCL).

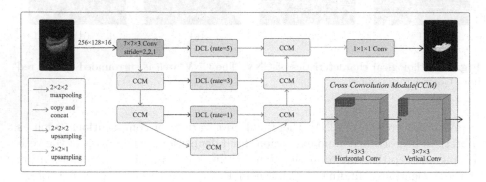

**Fig. 2.** An overview of the proposed method

**Baseline.** For SD-OCT volume, the entire data is composed of many slices, namely B-scans, and each B-scan shows the tomographic structure of the retina. Most of the existing CNV segmentation methods are based on B-scans. However, considering that the data between adjacent B-scans is continuous, we take 3D U-Net as the basic framework of our method. Compared with 2D convolution, 3D convolution has a receptive field in depth, which makes it possible to obtain more information when processing volume data.

**First Stage.** The encoding path of the network is divided into four stages to process data of different resolutions. In particular, we redesign the first stage of the network. A convolution kernel with size of 7×7×3 and a step size of (2,2,1) are adopted to make the network have a larger receptive field, which also enables rapidly reduction of the resolution and reduction of information loss in the longitudinal direction.

**Cross Convolution Module.** As shown in Figs. 1(a)(c), CNV will arch up in the axial direction or extend in the lateral direction, so we designed a Cross Convolution Module (CCM) to capture this morphological characteristic. The remaining stages adopt this structure which specially performs convolution with volumetric kernels having size 7×3×3 and 3×7×3 known as anisotropic convolution [10] instead of size 3×3×3 in basic structure.

**Dilated Connection Layer.** Further more, Dilated Connection Layer (DCL) was designed to extract multi-scale features. We replaced the original identity mapping with dilated convolution on the skip connection, and set appropriate dilation rates [11]. For all three skip connections (skip1, skip2, skip3) in the network from top to bottom, we set the convolution kernel size 3×3×3, and the dilation rate 5, 3, 1. The dilation rate in the longitudinal direction was maintained as 1. Dilated convolution expands the receptive field on the skip connection, which helps to reduce the influence of other lesions on CNV segmentation, and makes our model more robust. Table 1 shows the comparison of the size of the receptive field on the DCL and the direct skip connection.

**Table 1.** Size of receptive field of Dilated Connection Layer (DCL) and direct skip connection (axial×lateral×azimuthal)

| Type | skip1 | skip2 | skip3 |
|---|---|---|---|
| DCL | 27×27×5 | 65×65×16 | 125×125×68 |
| Direct skip connection | 7×7×3 | 41×41×12 | 109×109×30 |

## 3 Experiment

### 3.1 Dataset

The dataset used in our experiment consists of 376 cubes from 21 patients with CNV. Each cube is a 1024×512×128 SD-OCT volume data captured by a Cirrus HD-OCT device, corresponding to a 2×6×6 mm$^3$ retinal area. In order to make each 3D sample contain more adjacent frame information without taking up too much memory, we used bilinear interpolation algorithm to reduce the size of the raw cube. We resized both cube and ground truth from 1024×512×128 to 256×128×128. The dataset were randomly divided into training set, validation set and test set, with 228, 30 and 100 cubes respectively. The remaining 18 cubes from 3 patients were used for patient independent experiments. During the training process, 16 consecutive frames were randomly selected from the entire cube to form a 3D sample as the input of the network.

## 3.2   Implementation Details

We implemented our network based on the Pytorch framework. In all experiments, we fixed some basic parameter settings. Adam optimizer was employed for training, with a fixed learning rate $10^{-4}$, weight decay $10^{-4}$. In addition, the network was trained in parallel on two GPUs with batch size 10, epoch 50. We also used Binary Cross Entropy (BCE) loss function for training.

## 3.3   Evaluation Criterion

In order to better evaluate the segmentation results of our proposed model, five evaluation metrics were selected for quantitative analysis: dice coefficient (Dice), overlap ratio (Overlap), false positive ratio (FPR), false negative ratio (FNR), standard deviation of Dices (Std). The evaluation metrics were calculated based on SD-OCT volumes in the test set, and then the mean value was taken.

The specific calculation formula is shown as follows:

$$Dice = \frac{2\,|Pred \cap Label|}{|Pred|\bigcup|Label|} \tag{1}$$

$$Overlap = \frac{|Pred \cap Label|}{|Pred \bigcup Label|} \tag{2}$$

$$FPR = \frac{FP}{FP+TN} \tag{3}$$

$$FNR = \frac{FN}{FN+TP} \tag{4}$$

where Pred represents the set of pixels predicted by the model, and Label represents the lesion area in ground truth.

## 3.4   Result

To verify the effectiveness of the proposed method, some experiments have been performed for comparison. The four-stage 3D U-Net was selected as the baseline. After modifying the structure of the first stage, we successively experimented with adding CCM and DCL.

Table 2 shows the specific result of each experiment. We compared our method with different settings with 3D U-Net and Z-Net [12]. In order to verify the effectiveness of CCM, CCM was incorporated into our method and compared with the baseline. As we can see, both Dice and Overlap have been improved.

**Table 2.** Performace comparison of different experiments

| Methods | Dice | Overlap | FPR | FNR | Std |
|---|---|---|---|---|---|
| 3D U-Net | 0.8173 | 0.7064 | 0.1369 | 0.1566 | 0.1250 |
| Z-Net | 0.7988 | 0.6825 | 0.2045 | 0.1128 | 0.1370 |
| Ours(CCM) | 0.8228 | 0.7150 | 0.1788 | **0.1061** | 0.1290 |
| Ours(CCM+bottleneck) | 0.8317 | 0.7234 | 0.1851 | 0.1185 | 0.1040 |
| Ours(CCM+DCL) | **0.8473** | **0.7422** | 0.1439 | 0.1138 | **0.0769** |
| Ours(CCM+bottleneck+DCL) | 0.8435 | 0.7383 | **0.1247** | 0.1368 | 0.0880 |

**Table 3.** Performace comparison of patient independent experiments

| Methods | Dice | Overlap | FPR | FNR | Std |
|---|---|---|---|---|---|
| 3D U-Net | 0.5600 | 0.4149 | **0.1801** | 0.2741 | **0.1099** |
| Z-net | 0.5415 | 0.4094 | 0.2397 | 0.3506 | 0.1293 |
| Ours(CCM) | 0.6125 | 0.4685 | 0.2416 | 0.2032 | 0.1315 |
| Ours(CCM+bottleneck) | 0.6164 | 0.4670 | 0.3577 | **0.1751** | 0.1310 |
| Ours(CCM+DCL) | **0.6201** | **0.4754** | 0.2690 | 0.2553 | 0.1216 |
| Ours(CCM+bottleneck+DCL) | 0.6061 | 0.4685 | 0.2349 | 0.2964 | 0.1496 |

Then, the DCL further improve the performance. The Dice and Overlap are 3% and 3.6% higher than the baseline respectively, and it can be seen from the values of Std that our model is able to achieve more stable results.

In addition, in order to further reduce the amount of the parameters and speed up the convergence of the network, we added a bottleneck structure [13] to CCM. The bottleneck structure here is to add a $1\times1\times1$ convolution layer in front of the CCM to reduce the number of channels, and place a $1\times1\times1$ convolution layer at the end to restore the number of channels. From the experiment results, although adding bottleneck structure to our method did not achieve the highest result, it still has a certain improvement to the method with only CCM.

Table 3 shows the patient independent experiments. Because the growth pattern and pathological characteristics of CNV may vary greatly among patients, the performance of patient dependent experiment is better than that of patient independent experiment. However, the comparison of different methods also shows that our method is more effective in this case.

In order to intuitively compare the segmentation results and the ground truth, we show them based on B-scans. Figure 3 shows the segmentation results of different methods in B-scans. By comparison, we can find that the segmentation results of our method are closer to the ground truth. As shown in the first row of Fig. 3, with the addition of DCL, our method performs better in the presence of retinopathy other than CNV. For the case of small CNV, our method can identify the lesion area better, while the baseline method may miss that in some

B-scans. On some samples which are easy to segment, our method can also refine the segmentation results.

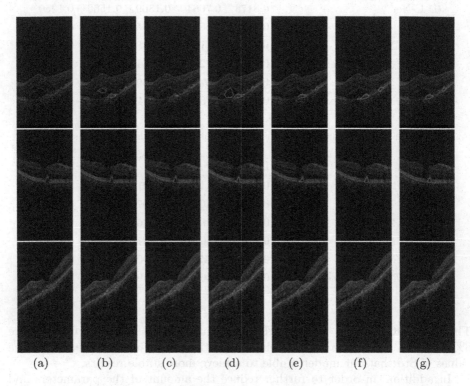

(a)          (b)          (c)          (d)          (e)          (f)          (g)

**Fig. 3.** Comparison of segmentation results between our method and other methods. (a) Original image. (b) 3D U-Net. (c) Z-Net. (d) Ours(CCM). (e) Ours(CCM+bottleneck). (f) Ours(CCM+DCL). (g) Ours(CCM+DCL+bottleneck). The red line represents the ground truth of CNV, and the green line represents the segmentation results (Color figure online)

## 4   Conclusion

In this paper, we propose a method based on 3D U-Net to segment CNV. The Cross Convolution Module (CCM) based on anisotropic convolution is designed to capture features of CNV with different shapes. In the Dilated Connection Layer (DCL), different dilation rates are applied to the convolutional layers of different skip connections to improve the ability to extract multi-scale features. Our method with different settings is compared with other methods, which indicates that our method can effectively segment CNV on SD-OCT data.

# References

1. Grossniklaus, H.E., Green, W.R.: Choroidal neovascularization. Am. J. Ophthalmol. **137**(3), 496–503 (2004)
2. Huang, D., et al.: Optical coherence tomography. Science **254**(5035), 1178–1181 (1991)
3. Yaqoob, Z., Wu, J., Yang, C.: Spectral domain optical coherence tomography: a better oct imaging strategy. Biotechniques **39**(6), S6–S13 (2005)
4. Sulaiman, R.S., et al.: A simple optical coherence tomography quantification method for choroidal neovascularization. J. Ocul. Pharmacol. Ther. **31**(8), 447–454 (2015)
5. Li, Y., Niu, S., Ji, Z., Fan, W., Yuan, S., Chen, Q.: Automated choroidal neovascularization detection for time series SD-OCT images. In: Frangi, A.F., Schnabel, J.A., Davatzikos, C., Alberola-López, C., Fichtinger, G. (eds.) MICCAI 2018, Part II. LNCS, vol. 11071, pp. 381–388. Springer, Cham (2018). https://doi.org/10.1007/978-3-030-00934-2_43
6. Zhang, Y., et al.: MPB-CNN: a multi-scale parallel branch CNN for choroidal neovascularization segmentation in SD-OCT images. OSA Continuum **2**(3), 1011–1027 (2019)
7. Su, J., Chen, X., Ma, Y., Zhu, W., Shi, F.: Segmentation of choroid neovascularization in OCT images based on convolutional neural network with differential amplification blocks. In: Medical Imaging 2020: Image Processing, vol. 11313, p. 1131320. International Society for Optics and Photonics (2020)
8. Çiçek, Ö., Abdulkadir, A., Lienkamp, S.S., Brox, T., Ronneberger, O.: 3D U-Net: learning dense volumetric segmentation from sparse annotation. In: Ourselin, S., Joskowicz, L., Sabuncu, M.R., Unal, G., Wells, W. (eds.) MICCAI 2016, Part II. LNCS, vol. 9901, pp. 424–432. Springer, Cham (2016). https://doi.org/10.1007/978-3-319-46723-8_49
9. He, K., Zhang, X., Ren, S., Sun, J.: Deep residual learning for image recognition. In: Proceedings of the IEEE Conference on Computer Vision and Pattern Recognition, pp. 770–778 (2016)
10. Wang, G., Li, W., Ourselin, S., Vercauteren, T.: Automatic brain tumor segmentation using cascaded anisotropic convolutional neural networks. In: Crimi, A., Bakas, S., Kuijf, H., Menze, B., Reyes, M. (eds.) BrainLes 2017. LNCS, vol. 10670, pp. 178–190. Springer, Cham (2018). https://doi.org/10.1007/978-3-319-75238-9_16
11. Wang, P., et al.: Understanding convolution for semantic segmentation. In: 2018 IEEE Winter Conference on Applications of Computer Vision (WACV), pp. 1451–1460. IEEE (2018)
12. Li, P., Zhou, X.Y., Wang, Z.Y., Yang, G.Z.: Z-net: an anisotropic 3D DCNN for medical CT volume segmentation. arXiv preprint arXiv:1909.07480 (2019)
13. Szegedy, C., et al.: Going deeper with convolutions. In: Proceedings of the IEEE Conference on Computer Vision and Pattern Recognition, pp. 1–9 (2015)

# Task-Free Recovery and Spatial Characterization of a Globally Synchronized Network from Resting-State EEG

Akaysha C. Tang[1,2(✉)], Adam John Privitera[2], Yunqing Hua[2], and Renee Fung[2]

[1] Neural Dialogue Shenzhen Educational Technology, Shenzhen, China
akaysha@mac.com
[2] Neuroscience for Education Group, Faculty of Education, The University of Hong Kong, Hong Kong, China

**Abstract.** Diagnosis, treatment, and prevention of mental illness requires brain-based biomarkers that can serve as effective targets of evaluation. Here we target the neuroanatomically and neurophysiologically well-defined neuromoduatory systems that serve the computational role of generating globally-synchronized neural activity for the purpose of functional integration. By using the second-order blind identification (SOBI) algorithm, which works with temporal information, we show that (1) neuroelectrical signals associated with synchronized global network activity can be extracted from resting state EEG; (2) the SOBI extracted global resting state network (gRSN) can be quantitatively characterized by its spatial configuration, operationally defined as a hits vector; (3) individual differences in the gRSN's spatial configuration can be analyzed and visualized in hits vector defined high-dimensional space. Our streamed-lined process offers a novel enabling technology to support rapid and low-cost assessment of much larger and diverse populations of individuals, addressing several methodological limitations in current investigation of brain function.

**Keywords:** Neuromarker · Biomarker · Resting-state networks · Default mode networks · EEG source imaging · Source localization · gRSN · SOBI · ICA

## 1 Introduction

Three major reasons have been identified for why neurobiology has yet to effectively impact mental health practice [1, 2], which we understand as the following. The first is the nearly exclusive reliance on the research framework of group comparisons, as in the case of comparing diseased populations with healthy controls. The second is the associated lack of sufficient attention to providing critical and meaningful information about the status and outcomes of an individual. The third is the lack of identification of neuromarkers based on neurocomputationally relevant parameters. Here, building upon decades of effort on EEG-based source imaging [3–9], we present an EEG-based source-imaging study of the resting-state brain to show that a neurobiologically-grounded and network configuration-based characterization can be provided for each individual without using

© Springer Nature Singapore Pte Ltd. 2021
W. Gao et al. (Eds.): FICC 2020, CCIS 1385, pp. 22–38, 2021.
https://doi.org/10.1007/978-981-16-1160-5_3

group data, and that cross-individual variability or consistency in the involvement of each brain region in a globally synchronized cortical network can be quantitatively characterized.

In search of neurocomputationally relevant parameters, we capitalize on the extensive psychophysiological and neurobiological literature on the neuromodulatory systems underlying novelty and uncertainty detection and their well-documented expression in the form of the EEG-derived P3 component [10–14]. Ample neurobiological evidence indicates that the most spatially extensive and globally synchronized innervation of the entire cortical mantel is provided by neuromodulatory systems. Cholinergic and noradrenergic-mediated global modulation are among the most extensively studied. These systems consist of projection neurons from subcortical structures to the cerebral cortex, and are thus capable of producing temporally synchronized changes in neuronal excitability and synaptic transmission across the entire cerebral cortex in support of learning and memory. Such global synchrony can manifest itself in scalp recorded electroencephalography signals. The computational challenge is how to measure such synchrony from many other sources of electrical signals, including artifacts, such as ocular artifact associated with eye movement, and neural signals not globally synchronized.

Recent applications of second-order blind identification (SOBI) [15], a blind source separation algorithm, to EEG data offer evidence that global synchronizations in neuroelectrical activity can be separated from these other signals by SOBI. SOBI can recover from different event-related task conditions (oddball as well face perception tasks) the P3 component [16–18], whose scalp projection and subsequent source location all revealed a globally distributed network involving the frontal, temporal, parietal, and occipital lobes [19–22]. The recovery of these P3 components by SOBI is expected because SOBI uses temporal information in the ongoing EEGs provided by the continuous, non-epoched, non-averaged data via minimization of sum squared cross-correlations at multiple temporal delays [23]. Signals originating from different brain structures but with identical time courses will be separated from other functionally distinct signals into a single component. Thus, by definition, the activity of the SOBI-recovered neuronal component will reflect synchronized neuroelectrical activity even if no tasks consisting of repetitive stimuli or responses are involved during EEG data collection.

Because SOBI works by using cross-correlations from the continuous EEG data instead of using event related information, it should be able to recover source signals associated with functionally distinct neural networks without requiring the use of task related information, such as event-related potentials (ERPs). Indeed, an earlier case study partially confirmed this prediction [24]. Among many neuronal components recovered, we found one whose scalp projection pattern closely resembled that of the SOBI-recovered P3 component with a globally distributed scalp projection pattern. This observation motivated us to consider this global resting state network (gRSN) as a potential novel neuromarker to capture synchronized global neuroelectrical activity.

The present study has the following aims: (1) replicating the earlier single-participant finding of the gRSN by investigating a larger sample from 13 individuals; (2) determining whether the underlying source generators can be modeled with equivalent current dipole (ECD) models, and; (3) providing individual-level quantitative characterization of the spatial configuration of the gRSN by introducing a novel measure of a hits vector.

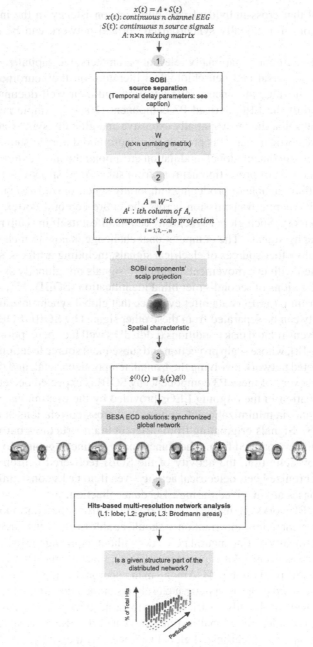

**Fig. 1.** The analysis pipeline used for processing of individual data. (**1**) application of SOBI to continuous EEG data, (**2**) identification of gRSN component using P3 component's spatial characteristics, (**3**) localizing the generators of the gRSN component via equivalent current dipole (ECD) modeling, and (**4**) hits-based analysis of gRSN network configuration.

## 2   Methods

### 2.1   Experimental Procedures

Approval for this study was granted by the Human Research Ethics Committee of the University of Hong Kong. We report data from five minutes of resting-state EEG data collected from thirteen right-handed participants (6 males) between 19–33 years of age (M = 26.50 ± 4.48 years) with no reported neurological conditions. Participants were asked to sit quietly with their eyes closed during data collection. Continuous reference-free EEG data were collected in an unshielded room using an active 128-channel Ag/AgCl electrode cap, ActiCHamp amplifier, and PyCorder data acquisition software (Brain Vision, LLC) with a sampling rate of 1000 Hz (impedance below 10 KΩ, 50 Hz notch filter applied offline). Data were spatially down-sampled to 64-channel to allow for comparison with our previous work [19].

### 2.2   Blind Source Separation of Resting-State EEG Data

SOBI [15] was applied to continuous EEG data, $x(t)$, to decompose the $n$-channel data into $n$-components, $s_i(t)$, i = $1, 2, ... n$, (Fig. 1, **Step 1**). Each SOBI component corresponds to a recovered putative source that contributes to the overall scalp recorded EEG signals. Detailed descriptions of SOBI's usage [23–28], SOBI validation [29, 30], and review of SOBI usage [31–33] can be found elsewhere. Because various underlying sources are summed via volume conduction to give rise to the scalp EEG, each of the $x_i(t)$ is assumed to be an instantaneous linear mixture of $n$ unknown components or sources $s_i(t)$, via an unknown $n \times n$ mixing matrix A,

$$x(t) = As(t)$$

The putative sources, $\hat{s}_i(t)$ are given by

$$\hat{s}_i(t) = Wx(t)$$

Where the unmixing matrix $W = A^{-1}$. SOBI finds the $W$ through an iterative process that minimizes the sum squared cross-correlations between one recovered component at time t and another at time t + τ, across a set of time delays. The following set of delays, τs (in ms), was chosen to cover a reasonably wide interval without extending beyond the support of the autocorrelation function:

τ ∈

{1, 2, 3, 4, 5, 6, 7, 8, 9, 10, 12, 14, 16, 18, 20, 25, 30, 35, 40, 45, 50, 55,

60, 65, 70, 75, 80, 85, 90, 95, 100, 125, 150, 175, 200,

225, 250, 275, 300, 325, 350}

The $i$th component's time course is given by $\hat{s}_i(t)$. The spatial location of the $i$th component is determined by the $i$th column of $A$ (referred to as the component's sensor weights or sensor space projection; Fig. 1, **Step 3**), where $A = W^{-1}$. Note that SOBI is blind to the physical nature of the source signals. Inferring the underlying source locations from the scalp projection is a separate task of source localization, which requires the appropriate physical models for different types of source signals, such as neuro-electrical, neuro-magnetic, etc.

## 2.3  Identification of the SOBI-Recovered gRSN Component

Of the 64 components recovered here, the gRSN component was identified based on its spatial features in the scalp projection pattern of the SOBI component, which was derived from unmixing matrix $W$ (Fig. 1 **Step 2**). The gRSN component was operationally defined as the component with a bilaterally symmetric concentric dipolar field pattern similar to the scalp projection of the well-documented P3 component [19] (for an examples see, Fig. 3B). Such a gRSN cannot be modeled by a small number of ECDs within one or two lobes of neocortex but by globally distributed ECDs throughout 3–4 lobes. This global distribution is central to the interpretation of the gRSN signal as reflecting globally synchronized neuroelectrical activity.

## 2.4  Hypothesis-Driven Source Modeling

Based on the nature of neuromodulatory innervation [10, 13, 14], we anticipated that underlying generators of the gRSN would be broadly-distributed across the neocortex. From previous localization of SOBI-recovered P3 components, we know that such a global network typically consists of $4 \pm 1$ pairs of bilaterally symmetrically placed dipoles. Using this prior knowledge, we began ECD model fitting (BESA Research 6.1, Brain Electrical Source Analysis, MEGIS Software, Munich, Germany) with a template model consisting of four pairs of bilaterally symmetrically placed dipoles at broadly-distributed locations roughly covering all four lobes of the cerebral cortex (Fig. 2). Specifically, the following dipole locations were used as a starting template (X Y Z): $\pm20$ 63 24, $\pm51$ 9 $-28$, $\pm44$ $-3$ 49, and $\pm24$ $-80$ 3 (given in Talairach coordinates [34]), with one pair each located within the occipital (posterior), temporal (lateral), and frontal (anterior) lobes, and one pair located near the border of the frontal and parietal lobes.

## 2.5  Scalp Projection of the gRSN as Input to BESA

Unlike data collected during an oddball task, resting-state EEG data do not afford any ERPs, which are typically projected onto the scalp when localizing P3 components using available software packages. In determining what could serve as alternative signals used as input to BESA, we note the fact that SOBI components' scalp projections are time-invariant (see equation in Fig. 1, **Step 2**). Therefore, the waveforms should not, in principle, affect the localization solution. However, due to numerical error in computation that is inherent to all software implementation of algorithms, fitting at different time points would produce non-identical final solutions. By using BESA's option to fit over a large time window, one can improve stability and avoid an arbitrary decision to fit at one time point over another. Therefore, a single ERP waveform from the P3 component of the same participant (described in [21]) over a time window of 1000 ms (200 ms pre, 800 ms post-stimulus) was used as input to BESA (Fig. 1 **Step 3**).

**Fig. 2.** The starting dipole configurations used for the localization of the gRSN components in BESA. Starting template of the ECD model shown in schematic (A), voltage map (B), and structural MRI views (C).

## 2.6 Iterative ECD Model Fitting Procedure

We performed ECD model fitting iteratively in order to determine the size of the model. This means the model size of four pairs may need to adjusted to be 5 or 3 depending on the fitting results. First, the above defined 4-pair ECD template was used as a starting point. Within each iteration, the Four-Shell Ellipsoidal adult head model (CR = 80) was first used to fit the projected ERP waveforms because it typically converges more quickly and then the Realistic Approximation option, a three-compartment head model (brain/CSF, skull, and scalp) was used to obtain the final model. The resulting model (solution) was increased by one pair of ECDs if (1) the goodness of fit (GoF) value was below 90%, or (2) the variations in voltage values in the residual map were greater than the range of background variations in the P3 component's ERP waveform. Condition 2 is needed because a high GoF value does not always correspond to a good fit between the model, projection, and data.

Occasionally, a participant's best matching gRSN scalp projections were not as global as the typical gRSN. In such a case, in order to still obtain a measure for this participant, we needed to fit the projection with a reduced ECD model consisting of fewer ECDs. Without the reduction, one may face over fitting. To check whether the current model size can be reduced by one pair of ECD, the same evaluation procedure was carried out. Essentially the smallest model that meets both conditions 1 and 2 will be the final solution.

The number of ECD pairs and their distribution in the final solution are the key measures used to validate our claim that the gRSN component reflects synchronized global neuroelectrical activity. If only one pair of ECDs were in the solution, then the SOBI component would reflect only local neuroelectrical activity and would not support the hypothesis that the gRSN captures globally distributed synchrony. If we indeed captured a global network matching the widely spread distribution field of the major neuromodulators, then the final solution would contain a network widely distributed throughout the neocortex.

## 2.7 From ECD Coordinates to Anatomical Structures

BESA solutions of ECD network are given in the form of Talairach coordinates and were used further as inputs to the freeware Talairach Client (TC, version 2.4.3) [35] to identify the corresponding anatomical structures contained within an $11^3$ mm$^3$ volume centered on each pair of ECDs [35]. For example, if a solution consisted *of n* ECD pairs, the total *possible* hits number of a gRSN component would *be n* * 2 * $11^3$. Therefore, the total hits number varies according to the number of ECDs in the final model.

For each set of ECD pairs in the BESA solution, TC determines how many $mm^3$ belong to what anatomical structures within each of the above defined cubes in the form of a table consisting of pairing of names of anatomical structures to their hits numbers. TC offers anatomical labeling at three levels of spatial resolution: the lobe, gyrus, and cell type levels. Thus, our results were also reported for each of the three levels.

Because we were interested in identifying the underlying neural generators of the gRSN, hits associated with all grey matter structures, referred to as the total hits were the focus of our analysis. There are times when subcortical neural structures, such as hippocampus, cerebellum, also have non-zero hits numbers. Their hits numbers are included in the calculation of total hits associated with neural structures. Because of their low frequency of observation and the lack of adequate coverage of these regions by the EEG cap, we do not recommend conclusions to be drawn about non-cortical structures.

## 2.8 Quantitative Characterization of the gRSN's Spatial Configuration

We characterize the involvement of a structure in the gRSN in two ways. The first addresses the question of how reliably an anatomical structure can be observed as part of the gRSN across individuals. Operationally, cross-participant reliability can be measured by its probability of being observed (i.e. % of participants with non-zero hits numbers for that structure regardless of the exact hits value). This is in accordance with the practice in fMRI studies where single voxel activation is included in the analysis [36].

The second addresses the question of how reliably an anatomical structure can be observed *within* an individual relative to other structures within the same gRSN. Such within-participant reliability can be measured by % of total hits contributed by a structure towards that individual's gRSN's total hits number. If an ECD pair falls within the center of a large anatomical structure, thus having a large % of total hits, this structure would be considered more reliably observed in comparison to another structure with only a small number of hits. The introduction of these two novel reliability measures allow quantification of source localization results beyond reporting goodness of fit.

Furthermore, the % of total hits for each of the multiple anatomical structures associated with the gRSN components make up a vector of size of $d$, where $d$ is the number of elements in the union of individual sets of structures observed across a given population. For example, if a hypothetical population consisting of two individuals with one individual having non-zero hits numbers for frontal and temporal lobes (set 1, size $= 2$ structures) and the other having none zero hits numbers for the occipital and temporal lobes (set 2, size $= 2$ structures), then the union of the two sets would have a $d$ of size 3, consisting of three structures (frontal, temporal, and occipital).

Each individual can then be viewed as occupying a location within a thus-defined high-dimensional space. The difference between two individuals or between the same individual observed at different ages, mental states, disease or health conditions, can be calculated as the distance between the two points. For clinical applications, treatment effects or developmental effects can then be measured by changes in spatial configuration of the gRSN. This goes beyond the characterization based on temporal information alone, such as frequency or time domain analysis of the component EEG signals.

## 2.9  Hits Vector-Based Visualization of Individual Differences

Using the concept of hits vector, an individual gRSN's spatial configuration can be visualized with the hits numbers varying across different structures and different individuals. Taking a slice of an individual, one can see how different structures having different hits numbers (the spatial configuration for that individual) and taking a slice of a structure, one can see how individuals vary in their hits numbers for that structure (cross-participant variations).

For clinical populations, treatment effects can then be visualized as a change in the spatial configuration from before to after treatment. Because of the ease of measuring resting state EEG, multiple before- and after- samples can then be used for individual level statistical analysis, making individualized medical treatment well-grounded statistically. Effective treatment effects should be accompanied by a clear separation of clusters of before and after spatial configuration of gRSN. Because no tasks are used, the "practice effects" of repeatedly performing a task are also avoided, though even resting state brain show habituation to repeated observation depending on intervals between repeated measures [37].

## 2.10  Statistical Analysis

For a single pair of bilaterally symmetrically placed ECDs, the hits numbers for a given structure ranges from 0 to 2662, with the hits number being 0 if a structure is not contained in the solution and equal to 2662 (i.e., $2 * 11^3$) if a structure is sufficiently large to fill the entire pair of cubes. Because we are interested in assessing the reliability or consistency of a structure being a part of the gRSN, even if a structure only appears in one participant, we must include this structure because it would be a low reliability structure to be characterized. For this reason, the variable of hits numbers has a large number of 0 values, resulting in a non-normal distribution. Correspondingly, nonparametric statistics were used for all analyses.

Binomial tests were performed on the proportional data and one-sample Wilcoxon signed-rank tests (1-tailed) were performed on the % of total hits for each structure to test for a median significantly greater than 0. To make focused comparison between the contribution of the frontal lobe structure and structures in the other lobes, paired-samples Wilcoxon signed-rank tests (2-tailed) were performed at each of the three levels of analysis (lobe, gyrus, and cell type). Significance levels were adjusted for multiple tests at the lobe level. If a lobe is significantly involved, then each of its constituent structure was tested without further adjustment of $p$ values. Otherwise, $p$ values were adjusted for multiple comparison. All statistical analyses were performed using SPSS (Version 23.0, IBM, Armonk, New York, United States of America).

## 2.11  The "Inverse Problem"

Because a globally-distributed network, such as the gRSN, necessarily consists of multiple underlying generators, this "inverse problem" of finding its generators is considered to have no unique solution. Using SOBI, one effectively reduces the number of sources by virtue of separating the gRSN from various artifacts and other relatively local neural networks, thus significantly lessening, but not solving, the inverse problem. By constraining the solutions for a particular scientific question, in our case, whether the gRSN can reflect globally-distributed neuroelectrical activity, via the use of the template (hypothesis), one can further "lessen" the inverse problem by eliminating solutions incompatible with the hypothesis (such as ECDs clustering locally).

By using a fixed template, reproducible but nevertheless non-unique solutions can be achieved by different human operators on the same computer. The way numerical problems are handled in the implementation of BESA or any other source modeling software can result in different numerical errors. Hence, even though the same procedure is followed, different computers may also produce differences in the exact ECD solutions. How much such potential solutions differ precisely matters less. What is important is whether such variations affect the conclusion or whether your specific solution is tolerant to such numerical errors. In the present case, some movement of the starting ECD locations within a couple of centimeters does result in changes in Talairach coordinates. However, the hits-based analysis remains consistent with the tested hypothesis that the gRSN reflects synchronized neuroelectrical activity across a globally distributed network of neocortical structures.

## 3  Results

### 3.1  Reliable Recovery of gRSN Components from Resting-State EEG

Through visual inspection of all of the scalp projections of SOBI components recovered from individual resting-state EEG data, we were able to identify, in each of the 13 participants, at least one component whose scalp projection patterns mirrored those of the classic P3 components (exact binomial test, $p < 0.001$). Similar to the P3 scalp topographies of the same individuals (Fig. 3**B**, obtained for another study), these resting-state EEG derived gRSN components (Fig. 3**A**), all having bilaterally symmetric and nearly concentric scalp projection patterns, with variations in their centers along the midline, some centered over the central sulcus, and others more anterior or posterior. Note that although the precise topography of these components differs across individuals, the average scalp projections of the two components are rather similar. These results demonstrate that SOBI performs robustly in extracting a GRSN component from each individual's resting-state EEG data.

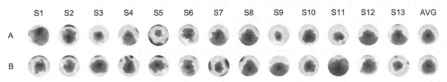

**Fig. 3.**  Reliable identification of gRSN components by SOBI and individual differences in gRSN scalp projection. Compare the scalp projections of the gRSN components (**A**: from resting EEG) with the P3 components (**B**: from EEG during a visual oddball task). Note the similarity between the average scalp projections despite cross-individual difference

## 3.2  Variable Neural Generators Underlying the SOBI Recovered gRSN Component

To determine the underlying neural network that generates the gRSN, the scalp topography of each individual's component was modeled as a network of ECDs, resulting in an average goodness of fit of 97.36 ± 0.4% (n = 13). This is a very high level of goodness of fit, indicating that the ECD model explains the scalp projection pattern very well. The spatial distribution in Fig. 4A shows an example from one individual, for whom the scalp projection of the gRSN (Fig. **4A: Data, shown in voltage**) is well fitted by the similar model scalp projection (Fig. **4A: Model**) with little remaining unexplained variance (Fig. **4A: Residual**). The ECDs of the model are shown against the structural MRI of an average brain provided in BESA (Fig. 4).

Consistent with the variable 2D scalp projection patterns in Fig. 3A, the spatial distribution of the ECDs are also variable cross-individual, as indicated by the overlaid ECD solutions from all participants (Fig. **4B**). Because highly consistent patterns across individuals would result in clustered distribution of ECDs, the lack of clustering indicates a high degree of variation in the spatial configuration of the gRSN. This variability should not be viewed as merely reflecting within group/condition noise, which preventing detecting population level treatment effect. Instead, such variability offers a new window of opportunity for further characterizing the individual.

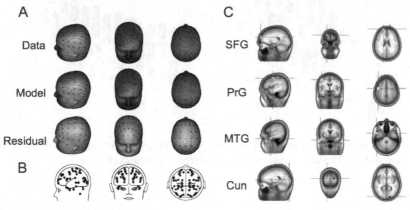

**Fig. 4.** ECD modeling of SOBI-recovered gRSN components. **AB:** example from a single participant. **C.** overlaid ECD solutions from all individuals. **A. Data**: gRSN component's scalp projection; **Model**: the ECD model projection; **Residual**: the difference between data and model, indicating little error. **B.** Overlaid ECDs from all participants' gRSN components, showing no foal clustering, which in turn indicating a high cross participant variability. **C.** the gRSN's ECD solution of an individual, shown against the structural MRI (sMRI), indicating a global distribution of the underlying neuronal sources.

### 3.3  Quantifying Cross-individual Variability in Network Configuration

The probability of a given brain structure being observed as part of the gRSN was used to quantify the cross-individual variability of that structure's involvement in the gRSN. Non-zero hits numbers were found in the frontal lobe in 100% of the participants studied (13 out of 13, exact binomial test, $p < 0.001$), and in the parietal, temporal, and occipital lobes in 77% of the participants (10 out of 13, exact binomial test, $p < 0.05$). At the gyrus level, hits were found in 30 structures out of a possible 55 with the MFG and SFG observed reliably with statistical significance (77%, in 10 of 13 participants, $p < 0.05$). Finally, at the cell type level, while hits were found in 30 out of a possible 71 structures, none of these structures were reliably involved across individuals ($p > .29$) with BA9 and BA18 having the highest probability of being observed in 62% (8 of 13, $p = .29$) of participants. The results at the lobe level support our hypothesis that the gRSN reflects a broadly distributed network across all four lobes of neocortex. The gyrus and cell type levels of analysis with higher spatial resolution further suggests that brain structures may be differentially involved in the gRSN.

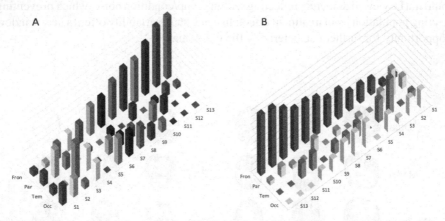

**Fig. 5.**  gRSN network configuration in all individuals. **(AB):** same data are plotted with different colors assigned to different individuals **(A)** and different structures **(B)**.

### 3.4  Quantifying Within-Individual Variations in Network Configuration

The percentage of total hits that each of the constituent structures makes up was used to quantify the relative contribution of each structure towards the gRSN for each individual. Figure **5A** shows how hits numbers vary as a function of both anatomical structure (frontal, temporal, parietal, and occipital) and individual (S1, S2, ...S13). The spatial configuration of an individual is visualized by the profile of hits numbers across all four lobes considered with each individual assigned a unique color. For example, individual S1's gRSN involves the frontal lobe less than S13. This hits-based visualization of the gRSN's spatial configuration offers a quick overview of individual differences in spatial configuration at a glance.

**Table 1.** Analysis of hits numbers associated with structures of the gRSN at the Lobe level. *Z* and *p*: statistics from one-sample Wilcoxon signed-rank test (1-tailed). Total hits refers to all grey matter hits within an individual's gRSN.

| Lobe Level | | | | | | |
|---|---|---|---|---|---|---|
| Structure | Range | *M* (SEM) | Var. | % of Total | Z | p |
| Frontal | 206 - 3155 | 1566 (238) | 733860 | 45% | 3.180 | .001 |
| Parietal | 0 - 1230 | 402 (122) | 192529 | 10% | 2.803 | .003 |
| Temporal | 0 - 2777 | 635 (216) | 603968 | 15% | 2.803 | .003 |
| Occipital | 0 - 1692 | 644 (150) | 293230 | 17% | 2.803 | .003 |

**Table 2.** Analysis of hits numbers associated with structures of the gRSN at the Gyrus level. *See caption for* Table 1.

| Gyrus Level | | | | | | |
|---|---|---|---|---|---|---|
| Structure | Range | *M* (SEM) | Var. | % of Total | Z | p |
| IFG | 0 - 1468 | 341 (152) | 300040 | 11% | 2.023 | .022 |
| MeFG | 0 - 133 | 21 (12) | 1842 | 1% | 2.023 | .022 |
| MFG | 0 - 1147 | 443 (114) | 169645 | 13% | 2.803 | .003 |
| SFG | 0 - 962 | 404 (100) | 130172 | 11% | 2.803 | .003 |
| PrG | 0 - 1034 | 294 (118) | 180477 | 8% | 2.366 | .009 |
| PoG | 0 - 601 | 102 (48) | 29683 | 2% | 2.201 | .014 |
| SPL | 0 - 682 | 94 (58) | 43583 | 2% | 2.023 | .022 |
| ITG | 0 - 496 | 76 (46) | 26983 | 2% | 2.201 | .014 |
| MTG | 0 - 2000 | 349 (163) | 347453 | 8% | 2.201 | .014 |
| STG | 0 - 1367 | 231 (123) | 196361 | 5% | 2.201 | .014 |
| IOG | 0 - 726 | 169 (67) | 58158 | 5% | 2.023 | .022 |
| MOG | 0 - 1209 | 220 (99) | 126384 | 6% | 2.521 | .006 |
| LgG | 0 - 371 | 83 (39) | 20186 | 2% | 1.826 | .034 |
| Cun | 0 - 798 | 115 (65) | 55357 | 3% | 1.826 | .034 |

Tables 1, 2 and 3 shows descriptive statistics as well as $Z$ and associated $p$ values for all of the structures at three levels of analysis respectively. Only those whose % of total hits measures are statistically significantly greater than zero are shown in the tables. Wilcoxon signed rank tests (1-tailed) revealed that 4, 14, 15 structures at the lobe, gyrus, and cell type levels respectively made statistically significant contributions to the gRSN. The highest % of total hits measures at each level of analysis were 45% for the frontal lobe: 13% for MFG (part of the frontal lobe), and 11% for both BA6 and BA8 (also part of the frontal lobe). All of these structures belong to the frontal lobe. The frontal lobe has a significantly greater % of total hits than the parietal ($p = .009$), temporal ($p = .016$), and occipital lobes ($p = .016$). These results indicate a non-uniform involvement of different structures in the gRSN and a potentially unique role of the frontal lobe structures.

**Table 3.** Analysis of hits numbers associated with structures of the gRSN at the Cell Type level. *See caption of* Table 2.

| Structure | Range | M (SEM) | Var. | % of Total | Z | p |
|---|---|---|---|---|---|---|
| BA 10 | 0 - 1057 | 269 (118) | 182117 | 7% | 1.826 | .034 |
| BA 9 | 0 - 525 | 198 (57) | 42125 | 6% | 2.521 | .006 |
| BA 44 | 0 - 896 | 125 (72) | 67785 | 4% | 1.826 | .034 |
| BA 8 | 0 - 1080 | 306 (120) | 187889 | 11% | 2.023 | .022 |
| BA 6 | 0 - 1625 | 380 (169) | 370243 | 11% | 2.201 | .014 |
| BA 4 | 0 - 344 | 87 (32) | 13379 | 3% | 2.201 | .014 |
| BA 3 | 0 - 217 | 23 (17) | 3688 | 1% | 1.826 | .034 |
| BA 7 | 0 - 1230 | 214 (107) | 148325 | 6% | 2.201 | .014 |
| BA 40 | 0 - 909 | 118 (73) | 68642 | 4% | 1.826 | .034 |
| BA 38 | 0 - 1205 | 206 (114) | 169906 | 5% | 2.023 | .022 |
| BA 21 | 0 - 977 | 224 (104) | 140117 | 6% | 2.023 | .022 |
| BA 20 | 0 - 589 | 113 (61) | 47792 | 4% | 1.826 | .034 |
| BA 19 | 0 - 829 | 259 (89) | 102768 | 7% | 2.366 | .009 |
| BA 18 | 0 - 774 | 282 (89) | 102237 | 8% | 2.521 | .006 |
| BA 17 | 0 - 504 | 128 (49) | 31701 | 4% | 2.023 | .022 |

Cell Type Level

Figure **5B** displays the same data but emphasizing the profile of a structure across different individuals by showing each structure with a unique color. Cross-individual variability in the involvement of each structure is further revealed with greater detail in terms of % of total hits than the all-of-none measure used in the probability-based reliability measure (see Sect. 3.3). Thus, the measure of % of total hits offers an additional measure to evaluate the involvement of a given brain structure in the gRSN.

# 4 Conclusion

We presented evidence to support the findings: (1) SOBI can reliably extract the gRSN component from each of the individuals' (n = 13) from five minutes of resting-state EEG; (2) the spatial configuration of the neural network underlying this gRSN component can be modeled by a set of spatially widely distributed ECDs with goodness of fit being 97%; (3) the network structure of the gRSN component can be quantitatively characterized for each individual through a hits vector; (4) individual differences in the neural network underlying gRSN can be quantified and visualized using hits vector.

## 4.1 gRSN: A Spatially Defined High-Dimensional Neural Marker

Frequently used EEG based neural markers are either a quantity measuring amplitude in the time domain or power in the frequency domain. Clinical neuroscientists typically search for brain structures that differ in EEG signal amplitude or power between two disease/health conditions or before/after treatment. Yet, it is possible that individual may differs in the spatial configuration of the globally synchronized neuroelectrical activity, reflecting disease related changes in the global neuromodulatory influences. Here we proposed a novel approach to offer an enriched EEG neural marker—a vector of hits numbers defining the spatial configuration of a globally distributed network. We introduced a work flow for this new approach: (1) extracting synchronized signals from resting state EEG, (2) identify the candidate component with a hypothesized globally distributed underlying network, the gRSN; (3) testing the hypothesis through source localization (point source model here); (4) determining the anatomical structures from ECD coordinates (5) using a hits vector to "mark" an individual's location in the high-dimensional space defined by the gRSN's constituent structures.

## 4.2 Individual Differences in Spatial Configuration of gRSN

The hits-based analysis method offered a novel approach to quantitatively characterize the spatial configuration of a distributed network at the level of individuals. With this method, each individual can be described by a vector, whose length is determined by the number of all observed grey matter brain structures across all participants, and whose elements are defined by each of the structure's hits contribution to the total grey matter hits in that individual. In such a high dimensional space, reliability or variability of each brain structure or each subnetwork in this global network can be observed and quantified. In such a space, individual differences in any of its sub-spaces can be computed as distances between two points and potentially reveal clusters associated with different disease conditions or different cognitive and emotional regulation capacities. The present work lays the methodological foundation for future studies of individual differences in health and disease and in both experimental and natural contexts.

### 4.3 Implications for Medicine

Because the traditional choice of P3 amplitude and latency based neuromarkers are sensitive to stimulus intensity, frequency, inter-trial interval, and past history of experiencing similar stimuli, neuromarkers based on characteristics derived from task-free and resting-stating EEG are, in principle, more likely to capture trait-like individual differences, not "contaminated" by variations associated with these other factors. Similar to the study of the resting-state brain activity using fMRI, here, the quantitative characterization of the gRSN from resting-state EEG data also has all the benefits of not requiring tasks to be performed by the individual under investigation. Different from the resting-state MRI studies, the resting-state EEG studies are substantially less expensive and more convenient for both the study participants and the investigators. This increased feasibility in obtaining the gRSN network parameters as neuromarkers from resting-state EEG also affords a wider range of applications beyond biomedical research and clinical treatment, to include basic research in understanding the brain in natural context as well as in neurotechnology for education.

**Acknowledgements.** This work was supported by a grant from the University of Hong Kong [104004683] and a donation from the Professor Anthony Edward Sweeting Memorial Fund awarded to A.C.T. We thank Drs. R. Sun and E. Tsang, and XY Niu for their assistance.

## References

1. Bullmore, E.: Getting below the surface of behavioral symptoms in psychiatry. Biol. Psychiatry **87**(4), 316–317 (2020)
2. Jollans, L., Whelan, R.: Neuromarkers for mental disorders: harnessing population neuroscience. Front Psychiatry **9**, 242 (2018). https://doi.org/10.3389/fpsyt.2018.00242
3. Babiloni, C., et al.: International Federation of Clinical Neurophysiology (IFCN)–EEG research workgroup: recommendations on frequency and topographic analysis of resting state EEG rhythms. Part 1: applications in clinical research studies. Clin. Neurophysiol. **131**(1), 285–307 (2020)
4. Khanna, A., Pascual-Leone, A., Michel, C.M., Farzan, F.: Microstates in resting-state EEG: current status and future directions. Neurosci. Biobehav. Rev. **49**, 105–113 (2015). https://doi.org/10.1016/j.neubiorev.2014.12.010
5. Phillips, C., Rugg, M.D., Friston, K.J.: Systematic regularization of linear inverse solutions of the EEG source localization problem. Neuroimage **17**(1), 287–301 (2002)
6. Mosher, J.C., Baillet, S., Leahy, R.M.: EEG source localization and imaging using multiple signal classification approaches. J. Clin. Neurophysiol. **16**(3), 225–238 (1999)
7. Valdes-Sosa, P.A., Roebroeck, A., Daunizeau, J., Friston, K.: Effective connectivity: influence, causality and biophysical modeling. Neuroimage **58**(2), 339–361 (2011)
8. Pascual-Marqui, R.D., et al.: Assessing interactions in the brain with exact low-resolution electromagnetic tomography. Philos. Trans. R. Soc. A Math. Phys. Eng. Sci. **369**(1952), 3768–3784 (2011)

9. Pascual-Marqui, R.D., et al.: Assessing direct paths of intracortical causal information flow of oscillatory activity with the isolated effective coherence (iCoh). Front. Hum. Neurosci. **8**, 448 (2014)

10. Ranganath, C., Rainer, G.: Neural mechanisms for detecting and remembering novel events. Nat. Rev. Neurosci. **4**(3), 193–202 (2003). https://doi.org/10.1038/nrn1052

11. Shine, J.M.: Neuromodulatory influences on integration and segregation in the brain. Trends Cogn. Sci. **23**(7), 572–583 (2019). https://doi.org/10.1016/j.tics.2019.04.002

12. Angela, J.Y., Dayan, P.: Uncertainty, neuromodulation, and attention. Neuron **46**(4), 681–692 (2005)

13. Nieuwenhuis, S., Aston-Jones, G., Cohen, J.D.: Decision making, the P3, and the locus coeruleus-norepinephrine system. Psychol. Bull. **131**(4), 510–532 (2005). https://doi.org/10. 1037/0033-2909.131.4.510

14. Nieuwenhuis, S., De Geus, E.J., Aston-Jones, G.: The anatomical and functional relationship between the P3 and autonomic components of the orienting response. Psychophysiology **48**(2), 162–175 (2011). https://doi.org/10.1111/j.1469-8986.2010.01057.x

15. Belouchrani, A., Abed-Meraim, K., Cardoso, J.F., Moulines, E.: A blind source separation technique using second-order statistics. IEEE Trans. Signal Process. **45**(2), 434–444 (1997)

16. Linden, D.E.J.: The P300: where in the brain is it produced and what does it tell us? Neuroscientist **11**(6), 563–576 (2005)

17. Polich, J.: Updating P300: an integrative theory of P3a and P3b. Clin. Neurophysiol. **118**(10), 2128–2148 (2007)

18. Polich, J.: Neuropsychology of P300. In: Oxford Handbook of Event-Related Potential Components, vol. 159, p. 88 (2012)

19. Zhang, Y., Tang, A.C., Zhou, X.: Synchronized network activity as the origin of a P 300 component in a facial attractiveness judgment task. Psychophysiology **51**(3), 285–289 (2014)

20. Randau, M., et al.: Attenuated mismatch negativity in patients with first-episode antipsychotic-naive schizophrenia using a source-resolved method. Neuroimage Clin. **22**, 101760 (2019). https://doi.org/10.1016/j.nicl.2019.101760

21. Privitera, A.J., Tang, A.C.: Reliability and variability of the P3 network configuration revealed by multi-resolution source-space analysis. In: Annual Meeting of the Cognitive Neuroscience Society, vol. G176 (2020)

22. Barry, R.J., Steiner, G.Z., De Blasio, F.M., Fogarty, J.S., Karamacoska, D., MacDonald, B.: Components in the P300: don't forget the novelty P3! Psychophysiology e13371–e13371 (2019). https://doi.org/10.1111/psyp.13371

23. Tang, A.C., Liu, J.-Y., Sutherland, M.T.: Recovery of correlated neuronal sources from EEG: the good and bad ways of using SOBI. Neuroimage **28**(2), 507–519 (2005)

24. Sutherland, M.T., Tang, A.C.: Blind source separation can recover systematically distributed neuronal sources from resting EEG. In: Proceedings of the Second International Symposium on Communications, Control, and Signal Processing (ISCCSP 2006), Marrakech, Morocco, pp. 13–15 (2006)

25. Tang, A.C., Pearlmutter, B.A., Malaszenko, N.A., Phung, D.B.: Independent components of magnetoencephalography: single-trial response onset times. Neuroimage **17**(4), 1773–1789 (2002)

26. Tang, A.C., Pearlmutter, B.A., Malaszenko, N.A., Phung, D.B., Reeb, B.C.: Independent components of magnetoencephalography: localization. Neural Comput. **14**(8), 1827–1858 (2002)

27. Tang, A.C., et al.: Classifying single-trial ERPs from visual and frontal cortex during free viewing. In: The 2006 IEEE International Joint Conference on Neural Network Proceedings, pp. 1376–1383. IEEE (2006)

28. Sutherland, M.T., Tang, A.C.: Reliable detection of bilateral activation in human primary somatosensory cortex by unilateral median nerve stimulation. Neuroimage **33**(4), 1042–1054 (2006)

29. Tang, A.C., Sutherland, M.T., McKinney, C.J.: Validation of SOBI components from high-density EEG. NeuroImage **25**(2), 539–553 (2005)

30. Lio, G., Boulinguez, P.: Greater robustness of second order statistics than higher order statistics algorithms to distortions of the mixing matrix in blind source separation of human EEG: implications for single-subject and group analyses. Neuroimage **67**, 137–152 (2013). https://doi.org/10.1016/j.neuroimage.2012.11.015

31. Tang, A.C., Sutherland, M.T., Yang, Z.: Capturing "trial-to-trial" variations in human brain activity. In: Ding, M., Glanzman, D.L. (eds.) The Dynamic Brain: An Exploration of Neuronal Variability and Its Functional Significance, pp. 183–213. Oxford University Press, New York (2011)

32. Tang, A.: Applications of second order blind identification to high-density EEG-based brain imaging: a review. In: Zhang, L., Lu, B.-L., Kwok, J. (eds.) ISNN 2010. LNCS, vol. 6064, pp. 368–377. Springer, Heidelberg (2010). https://doi.org/10.1007/978-3-642-13318-3_46

33. Urigüen, J.A., Garcia-Zapirain, B.: EEG artifact removal—state-of-the-art and guidelines. J. Neural Eng. **12**(3), 031001 (2015)

34. Talairach, J., Tournoux, P.: Co-planar Stereotaxic Atlas of the Human Brain. Theime, New York (1988)

35. Lancaster, J.L., et al.: Automated Talairach atlas labels for functional brain mapping. Hum. Brain Mapp. **10**(3), 120–131 (2000)

36. Desmond, J.E., Glover, G.H.: Estimating sample size in functional MRI (fMRI) neuroimaging studies: statistical power analyses. J. Neurosci. Methods **118**(2), 115–128 (2002)

37. Tang, A., et al.: Top-down versus bottom-up processing in the human brain: distinct directional influences revealed by integrating SOBI and Granger causality. In: Davies, M.E., James, C.J., Abdallah, S.A., Plumbley, M.D. (eds.) ICA 2007. LNCS, vol. 4666, pp. 802–809. Springer, Heidelberg (2007). https://doi.org/10.1007/978-3-540-74494-8_100

# PRU-net: An U-net Model with Pyramid Pooling and Residual Block for WMH Segmentation

Xin Zhao[1]([✉]), Xin Wang[1], and Hong Kai Wang[2]

[1] School of Information Engineering, Dalian University, Dalian, China
zhaoxin@dlu.edu.cn
[2] School of Biomedical Engineering, Dalian University of Technology, Dalian, China

**Abstract.** Segmentation of white matter hyperintensities (WMHs) from MR images is an essential step in computer-aided diagnosis of brain diseases, especially when considering their effect on cognition or stroke. At present, most of the research for WMH segmentation is based on deep learning methods. Although many deep learning segmentation methods have been proposed, their accuracy of these methods still needs to be improved, especially for discrete and small-sized deep WMHs. To cope with these challenges, and to improve the accuracy of WMH segmentation, an improved 3D U-net model, named PRU-net, was proposed in this paper. PRU-net integrates pyramid pooling and residual convolutional block in bottleneck layer of the U-net architecture. The pyramid pooling block was used to aggregate more context information, and the residual convolutional block was used to deepen the depth of bottleneck layers. Both the two blocks were employed to enhance the feature extraction of U-net. The experiments were based on the MICCAI 2017's WMH Challenge datasets, and the results showed that the Dice similarity coefficient (DSC) of our method was 0.83 and the F1 score was 0.84, which were higher than those of compared methods. Through visual observation of the segmentation results, our method cans not only accurately segment large lesion areas, but also distinguish small lesions which are difficult to segment for conventional U-net models.

**Keywords:** Segmentation · White Matter Hyperintensities · Pyramid pooling · Residual block

## 1 Introduction

White matter hyperintensities (WMH), referred to as leukoaraiosis or white matter lesions, are abnormalities in deep and periventricular white matter areas that exhibit signal hyperintensity on T2-FLAIR magnetic resonance imaging (MRI) sequence. These abnormalities have been commonly found on MRI of clinically healthy elder people. Furthermore, they have been associated with various neurological and geriatric disorders, such as small vessel disease [1], multiple sclerosis [2], Parkinson's disease [3], incident stroke [4], Alzheimer Disease [5], and dementia [6], etc. Studies have reported that periventricular WMHs are associated with a decline in cognitive function, and deep WMHs are of hypoxic/ischemic origin [7, 8].

W. Gao et al. (Eds.): FICC 2020, CCIS 1385, pp. 39–49, 2021.
https://doi.org/10.1007/978-981-16-1160-5_4

The presence, shape, and severity of WMH might provide further insight into healthy aging and the pathophysiology of various disorders. Therefore, the delineation and quantification of WMH areas are of crucial importance for clinical trials and therapy planning. However, accurate segmentation of WMH is not easy because of the heterogeneous intensity and location of the lesions. Hence, it is particularly important to study an automatic and accurate WMH segmentation algorithm.

## 2   Related Work

Manual delineation, as well as machine learning-based segmentation, and deep learning-based segmentation are currently the main segmentation methods of WMH areas on MRI. The manual delineation and quantification of WMHs is a more reliable way to assess WM abnormalities, but the whole process, which is cumbersome and time-consuming for the neuroradiologist, shows high intra-rater and inter-rater variability. To substitute manual delineation, automatic segmentation methods based on convolutional machine learning models have been proposed, but the accuracies of these methods are not high, for the difficulties of WMH features extraction. Therefore, more researchers are inclined to use deep learning methods for the automatic segmentation of WMHs recently.

Convolutional neural networks (CNN) have gained successes in computer vision, but they lack the natural ability to incorporate the anatomical location in their decision-making process. Ghafoorian et al. [9] integrated the anatomical location information into the CNN, in which several deep CNN architectures which consider multi-scale patches or take explicit location features were proposed. Rachmadi et al. [10] proposed a way to incorporate spatial information in the convolution level of CNN for WMH segmentation named global spatial information (GSI), in which four kinds of spatial location information were generated through the original MRI and then used as inputs combined with the original MRI. Although the CNN patch-based methods were better than traditional machine learning algorithms, their limitations and drawbacks were also obvious. For example, repeated computation of the same pixel leads to low computational efficiency, and the size of the pixel block limits the size of the sensing region.

In order to avoid the above situation, researchers began to design WMH segmentation algorithms based on the fully convolutional neural network (FCN) [11], which takes the input of the arbitrary size and produces corresponding-sized output with efficient inference and learning. The U-net proposed by Ronneberger et al. [12] also belongs to the FCN category, this architecture consists of a contracting path to capture context and a symmetric expanding path that enables precise localization. Moreover, U-net can work efficiently even with limited training samples. Because brain MRI is a three-dimensional structure, researchers generally used two methods to segment 3D images. One method is to split the 3D image into multiple 2D slices as input to the network, and the second method is to change the model to 3D models. Using 2D slices will ignore the three-dimensional spatial information of adjacent slices and reduce the segmentation accuracy, so the second method is more effective. Cicek et al. [13] proposed a network that extends the previous U-net architecture from Ronneberger et al. [12], by replacing all 2D operations with their 3D counterparts. Xu et al. [14] utilized transfer learning by using a pre-trained VGG model on Image-Net for natural image classification and a fine-tuned model on WMH data. Wang and his colleagues [15] proposed a two-step segmentation method to segment WMH: In the first step, three different dimensions of patch sampling were performed on brain MR scanning samples and they were input into a separate FCN; in the second step, the segmentation results generated from the FCN were combined using an integrated network. Zhang et al. [16] proposed a post-processing method to improve the white matter hyperintensity segmentation accuracy for randomly-initialized U-net. Wu et al. [17] proposed a novel skip connection U-net (SC U-net) which integrates classical image processing and deep neural network for segmenting WHM. Jeong et al. [18] used saliency U-Net and irregularity map (IAM) to decrease the U-net architectural complexity without performance loss.

Although the above fully convolutional neural networks have provided solutions for the automatic segmentation of brain WMH, they still have the following two main limitations. (1) Due to mild white matter lesions and deep WMHs are small in size, current FCNs are lacking in the ability to sensitively and accurately differentiate small WMHs from artifacts. Therefore, using more deep and complex convolutional operations to extract more useful features are necessary for the improvement of the current FCN model. (2) There are various shapes and scales of WMHs in the brain MRI images, which means multi-scale features is essential for the segmentation methods of WMHs. However, most existing FCNs for WMH segmentation have a limited multi-scale processing ability.

In this work we aimed to address the shortcomings above mentioned and proposed a new U-Net based segmentation approach for WMHs, which is named PRU-net (an U-net model with pyramid pooling and residual block). The main innovations behind our solution are as follows:

- Our PRU-net effectively improved the accuracy, recall, and precision of WMH segmentation by using spatial pyramid pooling and residual convolution block simultaneously in the symmetric U-shaped fully convolutional neural networks.
- We improved the feature extraction capabilities of PRU-net by applying a pyramid pooling block in the architectural bottleneck to adaptively incorporate multi-scale feature information.
- We deepened the convolutional layers of the bottleneck layer by employing residual convolution block, which further enhanced the feature extraction capabilities of U-net and improved the accuracy of WMH segmentation.

The rest of this paper is organized as follows. Section 3 describes in detail the improved U-net model with pyramid pooling and residual convolution block. Section 4 describes the datasets and evaluation metrics on segmentation performance, as well as demonstrates the experimental results. Finally, we concluded our method in Section 5.

## 3   Method

### 3.1   Work Flow

The flow chart of our method is shown in Fig. 1. Firstly, to expand the limited datasets, brain MRI images in the original dataset were preprocessed through data augmentation. Then the expanded datasets were divided into a training dataset, a validation dataset,

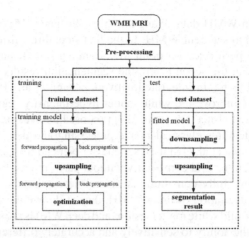

**Fig. 1.** Work flow of training and testing

and a test dataset. Our model was trained on the training dataset and produced a result after down-sampling and up-sampling, which was then compared with the target for optimization. Validation datasets were used for regularization by early stopping (stopping training when the error on the validation dataset increases, as this is a sign of overfitting to the training dataset). After training, the fitted model with down-sampling and up-sampling was used to predict WMH segmentation on the test dataset.

### 3.2 U-net Based Fully Convolutional Neural Network

We proposed a U-net based fully convolutional neural network for WMH segmentation, which consisted of three parts: encoding, bottleneck and decoding. In the encoding part, the feature maps of WMH were extracted by convolutional operation and the receptive field was increased by maximum pooling; In the bottleneck part, we used a pyramid pooling block to learn multi-scale features and used residual connection block to extract high-dimensional features; The feature maps of the encoding part and the decoding part were connected by concatenating, which combine the local information in the shallow layers and the high layers of the U-shaped neural network.

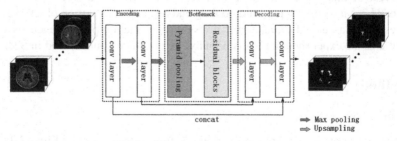

**Fig. 2.** Architecture of the PRU-net

The model we used, as shown in Fig. 2, first performed two consecutive $3 \times 3 \times 3$ convolutions and $2 \times 2 \times 2$ pooling after inputting, these steps could reduce the dimension of inputs and expanding the receptive field of convolutional observation. The resulting feature maps of the encoding part were then inputted into the bottleneck part, which could learn more "useful" information through combined pyramid pooling block and residual connection block. The output of the bottleneck was then subjected to two consecutive up-sampling of $2 \times 2 \times 2$ and convolution of $3 \times 3 \times 3$. The features of low and high layers were concatenated using skip connections, in order to learn contextual information. Finally, the image size was restored, and the segmented result was output after the sigmoid activation.

### 3.3 Pyramid Pooling Block

WMH has a variety of shapes and sizes, so it is difficult to be distinguished from normal brain tissue. Due to the receptive fields of the convolutional filter and pooling filter limits the overall understanding of WMH, information flow in convolutional neural networks is restricted inside local neighborhood regions, which leads to losing spatial information of WMH features. Inspired by the [19], we designed a pyramid pooling block and employed it into the bottleneck layer of the U-net structure. The pyramid pooling block can gain global context information through context aggregation based on different regions.

The pyramid pool block is shown in Fig. 3. It fuses features under three different pyramid scales. Three average pooling operations were executed after feature maps inputting, with pooling rate of 1/2, 1/5, and 1/10 respectively, which output three feature maps of different scales. The output was fed into three parallel convolutional paths to reduce the dimension and then up-sampled to the original input size. Finally, the up-sampled results of the three paths are spliced together to obtain multi-scale fusion features.

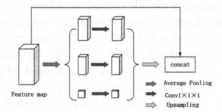

**Fig. 3.** Structure of pyramid pooling block

### 3.4 Residual Connection Block

Deep networks naturally integrate low/mid/high-level features, the "levels" of features can be enriched by the number of stacked layers (depth). The more the convolutional layers, the stronger the feature extraction ability of the network. Therefore, after pyramid pooling, we intend to add more convolutional layers to the bottleneck layer in order to extract more higher-dimensional features. However, simply stacking the number of network layers can easily cause vanishing/exploding gradients. In other words, with the network depth increasing, accuracy gets saturated (which might be unsurprising) and then degrades rapidly. Inspired by Resnet [20], we utilized the residual connection block after the pyramid pooling block in order to deepen the depth of the network and address the degradation phenomena.

The residual block we designed including four sub-residual-blocks, as shown in Fig. 4 . In each sub-block, there is a straight-connected path referred to residual connection and a main path with three consecutive convolution layers. A special note to the sub-block is that the first convolution of $1 \times 1 \times 1$ was used to decrease channel dimension in order to decrease the training load, while the second convolution of $1 \times 1 \times 1$ was used to keep the same dimension of channels as the sub-block input.

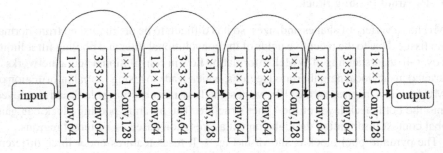

**Fig. 4.** Residual connection block

# 4 Experiment

## 4.1 Datasets and Preprocessing

The dataset used in our experiment was downloaded from the website https://wmh.isi. uu.nl/, which is a public dataset of the MICCAI 2017's WMH Challenge. We only use the FLAIR images of the dataset, which has only 60 images. In order to get enough examples for training, we expanded the dataset by rotating, affine, and perspective transformation the images, and finally reached a number of 480. Considering that the FLAIR image sizes from different scanners are inconsistent, we tailor and fill all FLAIR images and unify all slices to $200 \times 200$. Since intensities can also vary between patients, we normalize the intensities per patient to be within the range of [0, 1], which can also improve the convergence speed of the network. We also divided the 480 preprocessed images into three groups: training dataset, validation dataset, and test dataset, which account for 70%, 15%, and 15% of the total data volume respectively.

## 4.2 Experimental Setup

Our model was trained on the Google drive. The model training used Dice (DSC) loss function and Adam optimization algorithm, in which the Adam parameters learning rate was 0.0001, the batch size was 3, and epoch iterations was 60. When the loss value on the validation dataset keeps increasing in four training consecutive epochs, the training should be terminated.

## 4.3  Evaluation Criteria

To evaluate the accuracy of an automatically segmented WMH region, we compared and analyzed the segmentation results of different models on the evaluation indicators such as recall, precision, F1 score, and DSC score. Recall refers to the proportion of correctly predicted WMH regions to real WMH regions. Precision refers to the proportion of the correctly predicted WMH area to the predicted WMH area in segmentation results. Both precision and recall are expected to be as high as possible, but in fact, precision and recall are contradictory. F1 metric is an evaluation indicator that integrates precision and recall and is used to comprehensively reflect the overall effect of segmentation results. The higher the F1 score, the better the segmentation effect; DSC is statistical information used for comparing the similarity of two sets, the larger the DSC, the closer the segmentation result is to Ground Truth. These evaluation criteria are defined as:

$$Recall = TP/(TP + FN) \tag{1}$$

$$Precision = TP/(TP + FP) \tag{2}$$

$$F1 = 2 \ (Precision * Recall)/(Precision + Recall) \tag{3}$$

$$DSC = 2TP/(FN + FP + 2TP) \tag{4}$$

where TP is the True Positive, FP is the False Positive, FN is the False Negative.

## 4.4  Comparison of Different Models

We compared our method with other U-net-based semantic segmentation methods, all of which use the same training and testing datasets. We used "Unet_5" to represent the U-net model with 5 layers. The Unet_5 based model with pyramid pooling block in the bottleneck layer was named "Unet_5+P", and the Unet_5 based model with the residual convolutional block was named "Unet_5+R". As can be seen from Table 1, all the evaluation indicators of the segmentation results of the model "Unet_5" are the lowest compared with the other three. Compared with the model "Unet_5", the evaluation indicators of model "Unet_5+P" and model "Unet_5+R" were all improved. This means the pyramid pooling block and the residual convolutional blocks can improve the performance of the U-shaped model "Unet_5". We tried to add both the pyramid pooling block and the residual convolutional block at the same time to the bottleneck of the model "Unet_5" and found that the performance of the model was further improved as we expected, see the last row in Table 1.

The segmentation results of the four models listed in Table 1 was shown as Fig. 5. Through visual observation, we can see that the segmentation result of our model is more similar to the ground truth than the other segmentation results of compared models in Fig. 5, and all four models did well segmentation of WMHs with larger areas, but the model of this paper has the best capability of distinguishing small WMH lesions.

**Table 1.** Comparisons between U-net5 based models

| Model | DSC | Recall | Precision | F1 |
|---|---|---|---|---|
| Unet_5 | 0.75 | 0.69 | 0.83 | 0.76 |
| Unet_5+P | 0.76 | 0.69 | 0.87 | 0.77 |
| Unet_5+R | 0.80 | 0.72 | 0.89 | 0.80 |
| Ours | 0.83 | 0.80 | 0.88 | 0.84 |

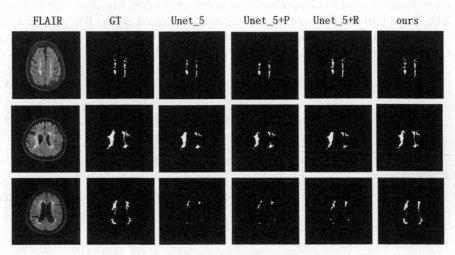

**Fig. 5.** Visual comparison of the WMH segmentation results

### 4.5 Comparison with Existing Approaches

We compared the evaluation indicators of PRU-net with those of several other well-established segmentation methods: Random Forests [21], FCN & Transfer-learning [14], SC-U-net [17], U-net & post-processing [16], Res-U-net [22], Multi-scale U-net [23], as shown in Table 2.

It can be seen from Table 2 that the DSC of the traditional RF method is only 0.5, the recall is 0.27, and the precision is 0.29, all of these are the lowest among compared methods. This shows that compared with the fully convolutional network, the traditional machine learning method does not have advantages in WMH segmentation. The DSC of our model is the highest among compared methods, which means the segmentation results of our model are the closest in similarity to the annotations of human experts. The precision of our method is 0.88, which means that 88% of the segmentation results are correct lesions. The F1 score of our method is 0.84, which is much higher when compared with other methods. This means that our method has relatively fewer false detection areas. However, the Recall of our method is only 0.80, which means our model can predict 80% of the lesion area. It is not the highest compared with other methods. That means our model automatically focused on the precisions more than on the Recall and got less FP but a little more FN in segmentation result.

**Table 2.** Comparison with other existing approaches.

| Author | Segmentation technique | Recall | F1 | DSC |
|---|---|---|---|---|
| Bento et al. [21] | Random Forests | 0.27 | 0.29 | 0.50 |
| Xu et al. [14] | FCN & Transfer-learning | 0.63 | 0.67 | 0.73 |
| Wu et al. [17] | SC-U-net | 0.81 | 0.71 | 0.78 |
| Zhang et al. [16] | U-net & post-processing | – | – | 0.69 |
| Jin et al. [22] | Res-U-net | 0.81 | 0.69 | 0.75 |
| Li et al. [23] | Multi-scale U-net | 0.86 | 0.77 | 0.80 |
| Our | Unet + pyramid pooling + Resnet block | 0.80 | 0.84 | 0.83 |

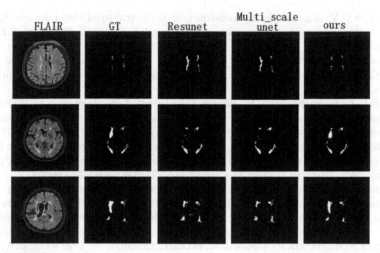

**Fig. 6.** Visual comparison of segmentation results of existing methods

In addition, we visually compared the segmentation results of our method and the two methods involved in the comparison and found that the segmentation results of our method is more accurate and closer to the ground truth as shown in Fig. 6. In the first row of Fig. 6, the two methods involved in the comparison both yielded too many false negatives. While in the second row of Fig. 6, the two methods yielded too many false positives. In the third row of Fig. 6, our segmentation result is closer to the gold standard in shapes, and the segmentation of small lesions is more accurate. In the segmentation results of other slices, there are similar comparisons among the above three situations. Due to space limitations, we only show these three, but there are similar cases to the above three situations in the segmentation results of other slices.

In summary, our method is better than other methods not only visually but also in terms of quantitative evaluation indicators.

## 5 Discussion

Discrete and small-sized deep WMHs are difficult to segment using the current automatic approaches. In order to accurately segment brain white matter lesions, a 3D U-shaped fully convolutional neural network segmentation algorithm combining pyramid pooling and Residual convolutional block was proposed. Experimental results showed that the evaluation criteria of this method are mostly higher than those of other compared methods. Through the visual observation, our model has the best capability of distinguishing small WMH lesions and the segmentation results of our model are more similar to the ground truth than other Unet-5 based model. Future research will focus on improving the recall and accuracy of our model for WMH segmentation.

## References

1. Norden, A.G.V., et al.: Causes and consequences of cerebral small vessel disease. The RUN DMC study: a prospective cohort study. Study rationale and protocol. Bmc Neurol. **11**(1), 29 (2011)
2. Schoonheim, M.M., Vigeveno, R.M., Lopes, F.C.R., et al.: Sex-specific extent and severity of white matter damage in multiplesclerosis: implications for cognitivedecline. Hum. Brain Mapp. **35**(5), 2348–2358 (2014)
3. Marshall, G.A., Shchelchkov, E., Kaufer, D.I., Ivanco, L.S., Bohnen, N.I.: White matter hyperintensities and cortical acetylcholinesterase activity in parkinsonian dementia. Acta Neurol. Scand. **113**(2), 87–91 (2006). https://doi.org/10.1111/j.1600-0404.2005.00553.x
4. Weinstein, G., Beiser, A.S., DeCarli, C., Rhoda, A., Wolf, P.A., Seshadri, S.: Brain imaging and cognitive predictors of stroke and alzheimer disease in the framingham heart study. Stroke **44**(10), 2787–2794 (2013). https://doi.org/10.1161/STROKEAHA.113.000947
5. Hirono, N., Kitagaki, H., Kazui, H., et al.: Impact of white matter changes on clini-cal manifestation of Alzheimer's disease: a quantitative study. Stroke **31**(9), 2182–2188 (2000)
6. Smith, C.D., Snowdon, D.A., Wang, H., et al.: White matter volumes and periventricular white matter hyperintensities in aging and dementia. Neurology **54**(4), 838–842 (2000)
7. Caligiuri, M.E., Perrotta, P., Augimeri, A., Rocca, F., Quattrone, A., Cherubini, A.: Automatic detection of white matter hyperintensities in healthy aging and pathology using magnetic resonance imaging: a review. Neuroinformatics **13**(3), 261–276 (2015). https://doi.org/10.1007/s12021-015-9260-y
8. Hinton, G.E., Osindero, S., Teh, Y.W.: A fastlearning algorithm for deep belief nets. Neural Comput. **18**(7), 1527–1554 (2006)
9. Ghafoorian, M., Karssemeijer, N., Heskes, T., et al.: Location sensitive deep convolutional neural networks for segmentation of white matter hyperintensities. Scientific Reports **7**(1), 1–12 (2017)
10. Rachmadi, M.F., del Maria, C., Valdés-Hernández, M.L., Agan, F., Di Perri, C., Komura, T.: Segmentation of white matter hyperintensities using convolutional neural networks with global spatial information in routine clinical brain MRI with none or mild vascular pathology. Comput. Med. Imaging Graph. **66**, 28–43 (2018). https://doi.org/10.1016/j.compmedimag.2018.02.002
11. Long, J., Shelhamer, E., Darrell, T.: Fully convolutional networks for semantic segmentation. IEEE Trans. Pattern Anal. Mach. Intell. **39**(4), 640–651 (2014)
12. Ronneberger, O., Fischer, P., Brox, T.: U-Net: convolutional networks for biomedical image segmentation. In: Navab, N., Hornegger, J., Wells, W.M., Frangi, A.F. (eds.) MICCAI 2015. LNCS, vol. 9351, pp. 234–241. Springer, Cham (2015). https://doi.org/10.1007/978-3-319-24574-4_28

13. Çiçek, Ö., Abdulkadir, A., Lienkamp, S.S., Brox, T., Ronneberger, O.: 3D U-Net: Learning Dense Volumetric Segmentation from Sparse Annotation. In: Ourselin, S., Joskowicz, L., Sabuncu, M.R., Unal, G., Wells, W. (eds.) Medical Image Computing and Computer-Assisted Intervention – MICCAI 2016, pp. 424–432. Springer International Publishing, Cham (2016). https://doi.org/10.1007/978-3-319-46723-8_49

14. Xu, Y., Géraud, T., Puybareau, É., Bloch, I., Chazalon, J.: White matter hyperintensities segmentation in a few seconds using fully convolutional network and transfer learning. In: Crimi, A., Bakas, S., Kuijf, H., Menze, B., Reyes, M. (eds.) BrainLes 2017. LNCS, vol. 10670, pp. 501–514. Springer, Cham (2018). https://doi.org/10.1007/978-3-319-75238-9_42

15. Wang, Z., Smith, C.D., Liu, J.: Ensemble of multi-sized FCNs to improve white matter lesion segmentation. In: Shi, Y., Suk, H.-I., Liu, M. (eds.) MLMI 2018. LNCS, vol. 11046, pp. 223–232. Springer, Cham (2018). https://doi.org/10.1007/978-3-030-00919-9_26

16. Zhang, Y., Chen, W., Chen, Y., Tang, X.: A postprocessing method to improve the white matter hyperintensity segmentation accuracy for randomly-initialized U-net. In: 2018 IEEE 23rd International Confere-nce on Digital Signal Processing (DSP). pp. 1–5. IEEE (2018)

17. Wu, J., Zhang, Y., Wang, K., et al.: Skip connection U-Net for white matter hyperi-ntensities segmentation from MRI. IEEE Access **7**, 155194–155202 (2019)

18. Jeong, Y., Rachmadi, M.F., Valdés Hernández, M.D.C., et al.: Dilated saliency u-net for white matter hyperintensities segmentation using irregularity age map. Frontiers in aging neuroscience **11**, 150 (2019)

19. Zhao, H., Shi, J., Qi, X., et al.: Pyramid scene parsing network. In: Proceedings of the IEEE Conference on Computer Vision and Pattern Recognition, pp. 2881–2890 (2017)

20. He, K., Zhang, X., Ren, S., et al.: Deep residual learning for image recognition. In: 2016 IEEE Conference on Computer Vision and Pattern Recognition (CVPR). IEEE Computer Society, pp. 770–778 (2016)

21. Bento, M., de Souza, R., Lotufo, R., Frayne, R., Rittner, L.: WMH segmentation chal-lenge: a texture-based classification approach. In: Crimi, A., Bakas, S., Kuijf, H., Menze, B., Reyes, M. (eds.) Brainlesion: Glioma, Multiple Sclerosis, Stroke and Traumatic Brain Injuries, pp. 489–500. Springer International Publishing, Cham (2018). https://doi.org/10.1007/978-3-319-75238-9_41

22. Jin, D., Xu, Z., Harrison, A.P., et al.: White matter hyperintensity segmentation from T1 and FLAIR images using fully convolutional neural networks enhanced with residual connections. In: 2018 IEEE 15th International Symposium on Biomedi-cal Imaging (ISBI), pp. 1060–1064 IEEE (2018)

23. Li, H., Zhang, J., Muehlau, M., et al.: Multi-scale convolutional stack aggregation for robust white matter hyper intensities segmentation. In: Crimi A., Bakas S., Kuijf H., Keyvan F., Reyes M., van Walsum T. (eds.) Brainlesion: Glioma, Multiple Sclerosis, Stroke and Traumatic Brain Injuries. BrainLes 2018. Lecture Notes in Computer Science, vol 11383. Springer, Cham.

# Two-Way Perceived Color Difference Saliency Algorithm for Image Segmentation of Port Wine Stains

Wenrui Kang, Xu Wang, Jixia Zhang, Xiaoming Hu[✉], and Qin Li

School of Life Science, Key Laboratory of Convergence Medical Engineering System and Healthcare Technology, The Ministry of Industry and Information Technology, Beijing Institute of Technology, Beijing 100081, China
coverycovery@163.com

**Abstract.** The image segmentation of port wine stains (PWS) lesions is of great significance to assess PDT treatment outcomes. However, it mainly depends on the manual division of doctors at present, which is time-consuming and laborious. Therefore, it is urgent and necessary to explore an efficient and accurate automatic extraction method for PWS lesion images. A two-way perceived color difference saliency algorithm (TPCS) for PWS lesion extraction is proposed to improve the efficiency and accuracy, and is compared with other image segmentation algorithms. The proposed algorithm shows the best performance with 88.91% accuracy and 96.36% sensitivity over 34 test images of PWS lesions.

**Keywords:** Port wine stains · Image segmentation · Saliency algorithm

## 1 Introduction

### 1.1 A Subsection Sample

PWS is a congenital cutaneous vascular disease with an incidence of 3‰–5‰ in newborns [1]. The skin of patients with PWS presents varying degrees of erythema, which won't fade spontaneously, and may deteriorate with age if patients don't receive effective treatments [2]. Photodynamic therapy (PDT) is the mainstream method for clinical treatment of PWS for its high effective clearance rate, low recurrence rate, few side effects, and high patient satisfaction [3]. However, PDT also shows some defects including low complete clearance rate, no effects on partial patients, long treatment cycle, and so on [4]. Numerous experimental results show that single PDT outcomes are related to the classification of preoperative lesions. Therefore, it is necessary to extract PWS lesions accurately for establishing the stable and effective relation between PDT treatment parameters and the outcomes of PWS treatment, which will conclude an individual therapeutic regimen for each patient. At present, the PWS lesions are generally extracted by the manual division of doctors. Although the manual division is very accurate, it's also time-consuming and laborious.

© Springer Nature Singapore Pte Ltd. 2021
W. Gao et al. (Eds.): FICC 2020, CCIS 1385, pp. 50–60, 2021.
https://doi.org/10.1007/978-981-16-1160-5_5

At present, many image segmentation algorithms have been applied in medical images, such as threshold segmentation, clustering segmentation, region growing, region split and merge, segmentation based on graph cuts, segmentation based on active contour model, segmentation based on deep learning, segmentation based on visual saliency. Tumor images from computed tomography (CT) and magnetic resonance imaging (MRI) help doctors to develop the therapeutic regimen doctors plan before surgery to a great extent. Randike [5] tests the performance of mean-shift clustering, K-means clustering (K-means), Fuzzy c-means clustering (FCM), and OTSU in extracting brain tumors from MRI images and finds the OTSU has an accuracy much better than the others. Mubarak proposes a cloud model computing (CMC) theory to realize automatic and adaptive segmentation threshold selection in Region growing algorithm (RG), and it shows strong robustness in the segmentation of bone X-rays and brain MRI images [6]. The segmentation of dental X-ray images is one key issue in dental based human identification. Lu proposes an algorithm based on full threshold segmentation, which is able to improve the accuracy of Grabcut algorithm by generating a proper mask image. The results show the proposed algorithm can effectively overcome the problems of uneven grayscale distribution and adhesion of adjacent crowns in dental X-ray images [7].

Besides, some researchers have paid attention to image segmentation of skin lesions. Rebia proposes an intelligent method by implementing the histogram decision to judge whether the contrast of input image needs to be enhanced, and the results show it is helpful to avoid time complexity [8]. Vesal builds a multi-task convolutional neural network (CNN) for skin lesion detection and segmentation, which utilizes a faster region-based CNN to detect the lesion bounded by a box, and then further segments the rectangular region by SkinNet, a modified version of U-Net. This framework shows very high accuracy (96.8%) and sensitivity (97.1%) [9]. In order to overcome the problem of low discrimination caused by low contrast between lesions and normal skin, Alan proposes an automated saliency algorithm based on skin lesion segmentation, which shows an accuracy of 91.66% in 160 dermoscopic images from PH2 public dataset [10].

However, to our knowledge, only a few researchers have developed image segmentation algorithms for PWS. Tang designs an image segmentation algorithm combining the threshold segmentation and connected component labeling to extract lesions accurately in the PWS treatment process through binocular surveillance system [11]. Although the algorithm shows good performance, it's only suitable for the ongoing process of PDT.

Therefore, the two-way perceived color difference saliency algorithm (TPCS) aiming at extracting the PWS lesion accurately is proposed. The principle and processes of TPCS are introduced in Sect. 2, and its accuracy and precision are compared with other typical algorithms, such as OTSU [12], K-means [13], FCM [14], RG [15], Grabcut [16] and saliency algorithms including ITTI [17], HC [18], LC [19], FT [20], MR [21], MC [22], RBD [23] and PCA [24], and the results of proposed algorithm show the best performance in test algorithms, in Sect. 4.

## 2 Principle of Proposed Algorithm

Pixel saliency, the kernel of the proposed algorithm, includes foreground saliency and background saliency, which represents the possibility of each pixel belonging to foreground (lesion) or background (normal skin) respectively. By comparing the saliency

of each pixel with foreground and background saliency thresholds varying with iteration number, each pixel can be classified into three types: foreground, background, and
undefinition, and then each pixel in the undefinition group will be reclassified continuously until it is classified into foreground or background. The elaborate processes of
the proposed algorithm are shown in Fig. 1, including three main steps: pre-processing,
image segmentation, and post-processing.

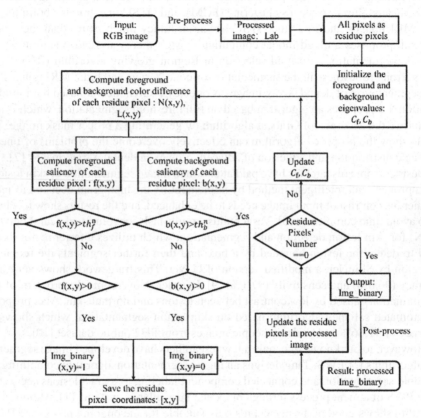

**Fig. 1.** The process of TPCS algorithm.

## 2.1  Pre-processing

Two pre-processing operations are implemented on each image to be segmented. Firstly,
the color space of the original image is converted from RGB to CIE-L*a*b to adjust
color channels and the brightness channel separately. In color space conversion, original
values of RGB are corrected according to Eq. (1–3), where R, G, B is the original three-
channel data, r, g, b represents the corrected three-channel values, and the values of
$\alpha 1$, $\alpha 2$, $\gamma_c$ are 0.055, 1.055 and 2.4 [25]. Then the color space of the corrected image

converts from RGB to CIE-XYZ, and further converts to CIE-L*a*b through the ITU-R BT.709 standards.

$$r = \left(\frac{R+\alpha 1}{\alpha 2}\right)^{\gamma_c} \tag{1}$$

$$g = \left(\frac{G+\alpha 1}{\alpha 2}\right)^{\gamma_c} \tag{2}$$

$$b = \left(\frac{B+\alpha 1}{\alpha 2}\right)^{\gamma_c} \tag{3}$$

Secondly, the brightness of the image is normalized to attenuate the effect of brightness to overcome the problems that often occur in PWS images, such as partial regional reflection caused by skin grease, and uneven illumination.

## 2.2 Image Segmentation

Image segmentation is implemented by comparing the saliency of each pixel with the foreground and background saliency thresholds, which indicates proper saliency construction and saliency threshold selection are two kernels of the proposed algorithm.

Pixel saliency is computed according to Eq. (4–5), where $f(x, y)$ and $b(x, y)$ represent the foreground and background saliency of the pixel with coordinates [x y], and $\Delta N(x, y)$ and $\Delta L(x, y)$ represent its color differences between its pixel value with the background and foreground eigenvalues. In this paper, color differences are computed by CIEDE2000 color difference formula for its capability of detecting the minor color difference [26].

$$f(x, y) = 255 \frac{\Delta N(x, y)}{\Delta N(x, y) + \Delta L(x, y)} Heaviside(\Delta N(x, y) - \Delta L(x, y)) \tag{4}$$

$$b(x, y) = 255 \frac{\Delta L(x, y)}{\Delta N(x, y) + \Delta L(x, y)} Heaviside(\Delta L(x, y) - \Delta N(x, y)) \tag{5}$$

Obviously, both eigenvalues need to be initialized and updated before the initialization and update of pixel saliency. For initializing the foreground and background eigenvalues, two rectangular areas are drawn manually in the foreground and background respectively, and then the average values in three color channels of foreground and background rectangular areas are calculated as the eigenvalues of foreground and background. But for updating the eigenvalues, the foreground and background eigenvalues are computed Eq. (6) and (7), where $C_f^n$ and $C_b^n$ represent the foreground and background eigenvalues in $n$th iteration, and $M_f^{n-1}$ and $M_b^{n-1}$ represent the average pixel values of foreground and background.

$$C_f^n = (2C_f^{n-1} + M_f^{n-1})/3 \tag{6}$$

$$C_b^n = (2C_b^{n-1} + M_b^{n-1})/3 \tag{7}$$

The saliency thresholds are able to classify pixels into the pixels with high foreground saliency, the pixels with high background saliency, and the pixels with low saliency, which correspond to the foreground, background, and undefinition. And then the pixels in the undefinition group are extracted to the next classification. Specifically, in this paper, the foreground and background thresholds of the first cycle are set by Eq. (8) and (9), where $th_f^1$ and $th_b^1$ represent the saliency thresholds of foreground and background; $\bar{p}_f$ and $P_{f\_min}$ represent the average and minimal foreground saliency of pixels whose foreground saliency is over 127; and $\bar{p}_b$ and $P_{b\_min}$ represent the average and minimal background saliency of pixels whose background saliency is over 127. Then both thresholds in the second cycle are set to 127, directly resulting in the ending of the cycle.

$$th_f^1 = (\bar{p}_f + 2 * P_{f\_min})/3 \qquad (8)$$

$$th_b^1 = (\bar{p}_b + 2 * P_{b\_min})/3 \qquad (9)$$

### 2.3 Post-processing

In order to further optimize the segmentation results, the original image generated in section B was processed by morphological processing. Firstly, the median filter is used to eliminate small burrs and noise. Secondly, morphological opening operation with a 7 × 7 size square operator is used to smooth the contour of the object, disconnect the fine connection, and remove the fine protrusion. Finally, morphological closing operation with a 7 × 7 size square operator is implemented to smooth the contour of the object and fill long and thin grooves, and the holes smaller than the operator size.

## 3   Data Sources

34 images tested in this study are from PWS patients receiving the Phase III clinical trial of HMME photodynamic therapy. This clinical trial led by Peking University First Hospital began in 2009. The interval between two treatments is 8–12 weeks and the follow-ups last 2–3 cycles. The PWS lesion images are acquired by SONY DSC_W1 from positive, lateral, and 45° oblique lateral position before each course of treatment. The ground truth images used to evaluate the performance of different image segmentation algorithms for PWS lesions are established by LabelMe software with the help of dermatologists.

## 4   Results

Four evaluation parameters: accuracy rate (A), dice coefficient (D), sensitivity rate (S), and recall coefficient (R), which represent the accuracy of segmenting lesions and normal skins, the consistency between the results and true values, the accuracy of lesions extraction, and the integrity of lesions extraction respectively. The parameters calculation methods are shown in Fig. 2.

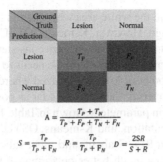

$$A = \frac{T_P + T_N}{T_P + F_P + T_N + F_N}$$

$$S = \frac{T_P}{T_P + F_N} \quad R = \frac{T_P}{T_P + F_N} \quad D = \frac{2SR}{S + R}$$

**Fig. 2.** Calculation of evaluation parameters.

The experimental results of the TPCS algorithm and other typical algorithms including OTSU, K-means, FCM, RG, Grabcut, and saliency algorithms including LC, HC, MR, MC, PCA, ITTI are also compared. The details of the quantitative evaluation results of TPCS and other typical algorithms are shown in Table 1. It's worth noting that saliency algorithms can only obtain the grayscale saliency maps instead of the binary images, and that all saliency maps need to further binarization. Therefore, the deviation between saliency algorithms results and ground truth images are from the deviation of generating saliency maps and further binarization. In this paper, OTSU is selected to binarize the saliency maps.

**Table 1.** Evaluation scores of different image segmentation algorithms for PWS

| Type | A | D | S | R |
|---|---|---|---|---|
| LC | 57.65 | 66.62 | 61.83 | 77.13 |
| HC | 60.49 | 63.52 | 65.87 | 65.92 |
| FT | 60.90 | 70.82 | 64.39 | 84.01 |
| MC | 67.22 | 69.07 | 88.42 | 57.82 |
| PCA | 60.03 | 56.49 | 74.80 | 49.50 |
| RBD | 61.03 | 54.69 | 77.25 | 44.68 |
| ITTI | 72.64 | 80.10 | 82.29 | 80.37 |
| MR | 76.94 | 80.46 | 92.26 | 72.42 |
| OTSU | 87.96 | 92.59 | 96.04 | 89.68 |
| K-means | 87.92 | 92.55 | 96.15 | 89.51 |
| FCM | 86.45 | 90.29 | 93.55 | 87.52 |
| RG | 84.70 | 89.76 | 95.45 | 85.35 |
| Grabcut | 80.45 | 86.30 | 91.20 | 82.81 |
| TPCS | 88.91 | 93.21 | 96.36 | 90.53 |

For quantitative evaluation, the white area was taken as lesions and the black area as normal skin in the binary image. However, K-means, OTSU, FCM, and all test saliency algorithms expect TPCS can't ensure the results are consistent with the above definition, thus each image whether the gray value of all pixels needs to be inversed simultaneously. If the segmentation result is too poor to be recognized where the lesions are, the scheme that keeps more white area covering the lesions as far as possible was adopted.

According to the evaluation parameters shown in Table 1 , TPCS has the best performance in PWS lesions segmentation, and K-means, OSTU, and FCM have similar and much better performance than RG, Grabcut, and other saliency algorithms. In saliency algorithms, MR and ITTI show much better performance than others, but are still far from satisfactory.

To show the segmentation performance differences between non-saliency algorithms clearly, some typical PWS lesion images and their segmentation results by non-saliency algorithms are shown in Fig. 3, which includes various extreme situations. Specifically, Img1 and Img2 represent the cases of scattered lesions and single lesion in good contrast between lesions and normal skin, while Img3 represents the case of low contrast between lesions and normal skins; Img4, Img5, and Img6 represent the cases of overexposure, uneven illumination, existing shadow respectively; and Img7 and Img8 represent the cases of multi-color lesions, and existing eye and hair respectively. As shown in Fig. 3, OTSU, K-means, FCM, and RG can't handle the situations of existing multi-color lesions (Img2), and overexposure (Img5). Grabcut has the least small empties in extracted lesions, but obtain good and complete segmentation results only in Img2, which satisfy the conditions of good contrast, single lesion, and simple shape lesion at the same time.

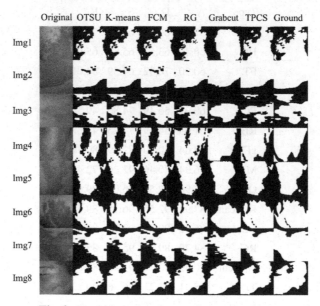

**Fig. 3.** Typical results of non-saliency algorithms.

Figure 4 shows some typical images and their saliency maps of saliency algorithms. As shown in Fig. 4, MR shows good performance, except the situation of low contrast (Img3) between lesions and normal skin. ITTI is able to distinguish foreground and background, except the situations of few background areas (Img2), uneven illumination (Img5), and dark illumination (Img11). Meanwhile, the contrast of pixels saliency is very low and lacks clear boundaries, which affects final binary segmentation results seriously. The saliency maps of MC show clear boundaries between the foreground and background, and it is beneficial for further binarization. However, it is obvious that the area of high saliency is smaller, which is consistent with the low recall rate of MC. Moreover, the saliency maps of LC and FT are similar and both show low contrast, dark brightness, and show good performance only when the difference between the proportion of foreground and background is large. The saliency maps of HC show a much better than LC and FT from observation, however it highlights the small areas with rare values excessively. It's not surprising that the above three algorithms show low accuracy and sensitivity. Contrary to LC and RC, the saliency maps of RBD show high contrast like MR, however, it just shows good performance in the situation of low connections between lesions and image margins (Img10, Img11). Moreover, PCA tends to highlight the areas with drastic changes like the boundaries between lesions and normal skins, which is consistent with the very low accuracy and recall.

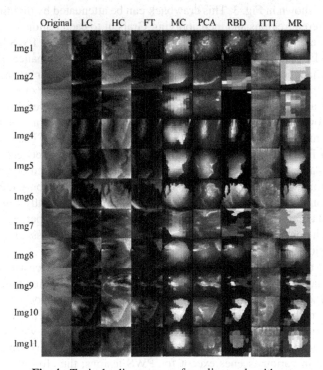

**Fig. 4.** Typical saliency maps for saliency algorithms.

Comparing with other algorithms, TPCS can segment lesions accurately and precisely in each extreme situation, meanwhile, there are very few empties in extracted lesions interior.

## 5 Discussion

As shown in the results, OTSU, K-means, and FCM show good performance in test images. However, with the increase of background complexity, the possibility of misclassification increases. For K-means and FCM, increasing cluster categories may improve the performance, if the category numbers can be automatically chosen.

As we know, the results of RG are related to the initial seed point location and growth threshold. Indeed, in the process of attempting the best results through selecting the initial seed point and the growth threshold manually, it is found that even a slight change in growth threshold or seed point location might lead to different results, especially in the situation of scattered lesions, lesions with multi-grades, and low contrast between the lesions and normal skins. This doesn't mean that RG is unsuited to extract the PWS lesions. On the contrary, it points out the valuable directions such as developing multi-seed points and self-adaption high-accuracy RG algorithms. Grabcut is a segmentation algorithm based on graph cuts that utilize max-flow min-cut theorem, so it can only get single lesion as shown in Fig. 3. This drawback can be attenuated by dividing the whole image into different regions and then applying Grabcut to each region.

All test saliency algorithms show unsatisfactory performance in PWS lesion extraction, in which LC, HC, FT, PCA, and RBD show very bad performance, and ITTI and MR show reluctant performance. This phenomenon is related to their intrinsic attributes. The pixel saliency of LC and HC is calculated by the sum of gray or color difference between itself and all other pixels, so only pixels with rare gray level or color in the image will show high saliency. This also explains why the contrast of HC between lesions and normal skin is higher than LC. Similar to LC and HC, FT tends to highlight the areas with little proportion for defining the color difference between a pixel with average color value in CIE-L*a*b as the pixel saliency. RBD is based on the hypothesis that the foreground has fewer connections with image margin, so only the lesions that are isolated from the margin or connected with image margins slightly can be extracted accurately and completely. Although the results of PCA in this paper are not good, its principle which considers the pattern distinctness, color distinctness, and organization prior is surprising.

ITTI shows reluctant performance because it includes a comprehensive integration of various information in color lightness and orientation. However, the numerous parameters in the processes are constant, which may influence the performance of ITTI in PWS lesion extraction. Therefore, developing more suitable parameters for PWS lesion extraction through machine learning will be undertaken in our next works. Meanwhile, numerous segmentation algorithms based on machine learning have been proposed and widely applied in various scenes except for PWS lesions, therefore the effectiveness of segmentation algorithms based on machine learning is worth investing. MR, which applies graph-based manifold ranking to the differences between superpixels with foreground and background information is an outstanding algorithm in recent years. In this paper, it shows the best performance on PWS lesion extraction in non-saliency algorithms except TPCS, however, the accuracy and recall are still low, even compared with Grabcut, the algorithm shows the worst performance in non-saliency algorithms. Therefore, more novel non-saliency algorithms need to be refined on their performance for PWS lesion extraction.

Similar to MR, the proposed algorithm considers the foreground and background saliency simultaneously and show excellent performance for PWS lesions extraction. However, the proposed algorithm still has great potential to further improve its performance by optimizing the pixel saliency calculation formula, the saliency threshold selection, and the eigenvalue selection. At the same time, refining the categories of pixels whose values are far from the foreground and background eigenvalues is also beneficial for TPCS to meet more complex situations such as including hair, eyes, and lips.

## 6 Conclusion

In this paper, a two-way perceived color difference saliency algorithm for improving the accuracy and precision in PWS lesions extraction is proposed, and its accuracy and precision are compared with other typical algorithms. The results show that the non-saliency algorithms have better segmentation performance than the saliency algorithms and the proposed algorithm has the best performance in test algorithms, indicating the proposed algorithm is a promising image segmentation algorithm for PWS.

**Acknowledgments.** This research is supported by the National Natural Science Foundation of China (81773349).

## References

1. Chen, J.K., Ghasri, P., Aguilar, G., Drooge, A.M.V., Wolkerstorfer, A., Kelly, K.M.: An overview of clinical and experimental treatment modalities for port wine stains. J. Am. Acad. Dermatol. **67**(2), 289–304 (2012)
2. Arnstadt, B., Ayvaz, A., Weingart, V., Wallner, J., Allescher, H.D., Bühren, V.: Port wine stain laser treatments and novel approaches. Facial Plast. Surg. **28**(6), 611–620 (2012)
3. Zhang, Y., Zou, X., Chen, H., Yang, Y., Lin, H., Guo, X.: Clinical study on clinical operation and post-treatment reactions of HMME-PDT in treatment of PWS. Photodiagn. Photodyn. Ther. **20**(9), 253–256 (2017)
4. Yuan, K.H., Gao, J.H., Huang, Z.: Adverse effects associated with photodynamic therapy (PDT) of port wine stain (PWS) birthmarks. Photodiagn. Photodyn. Ther. **9**(4), 332–336 (2012)
5. Gajanayake, G.M.N.R., Yapa, R.D., Hewawithana, B.: Comparison of standard image segmentation methods for segmentation of brain tumors from 2D MR images. In: International Conference on Industrial & Information Systems, pp. 301–305. IEEE (2010)
6. Mubarak, D.M.N.: A hybrid region growing algorithm for medical image segmentation. Int. J. Comput. Inf. Technol. **4**(3), 61–70 (2012)
7. Mao, J., Wang, K., Hu, Y., Sheng, W., Feng, Q.: GrabCut algorithm for dental X-ray images based on full threshold segmentation. IET Image Proc. **12**(12), 2330–2335 (2018)

8. Javed, R., Saba, T., Shafry, M., Rahim, M.: An intelligent saliency segmentation technique and classification of low contrast skin lesion dermoscopic images based on histogram decision. In: 12th International Conference on Developments in eSystems Engineering, pp. 164–169. IEEE (2019)

9. Vesal, S., Malakarjun Patil, S., Ravikumar, N., Maier, A.K.: A multi-task framework for skin lesion detection and segmentation. In: Stoyanov, D., et al. (eds.) CARE/CLIP/OR 2.0/ISIC -2018. LNCS, vol. 11041, pp. 285–293. Springer, Cham (2018). https://doi.org/10.1007/978-3-030-01201-4_31

10. Ahn, E., Lei, B., Youn, H.J., Jinman, K.: Automated saliency-based lesion segmentation in dermoscopic images. In: 37th Annual International Conference of the IEEE Engineering in Medicine and Biology Society, pp. 3009–3012. IEEE (2015)

11. Tang, X., Fan, X., Ying, L., Liu, W., Han, X.: Algorithm design of image segmentation in port wine stains photodynamic therapy binocular surveillance system. In: 3rd International Conference on Bioinformatics and Biomedical Engineering, pp. 1–4. IEEE (2009)

12. Otsu, N.: A threshold selection method from gray-level histograms. IEEE Trans. Syst. Man Cybern. 9(1), 62–66 (1979)

13. Hartigan, J.A., Wong, M.A.: A K-means clustering algorithm. Appl. Stats. 28(1), 100–108 (1979)

14. Bezdek, J.C., Ehrlich, R., Full, W.: FCM: The fuzzy C-means clustering algorithm. Comput. Geoences 10(2–3), 191–203 (1984)

15. Steven, W.Z.: Region growing: childhood and adolescence. Comput. Graph. Image Process. 5(3), 382–399 (1976)

16. Rother, C.: GrabCut: interactive foreground extraction using iterated graph Cut. ACM Trans. Graph. 23(3), 309–314 (2004)

17. Itti, L.: A model of saliency-based visual attention for rapid scene analysis. IEEE Trans. Pattern Anal. Mach. Intell. 20(11), 1254–1259 (1998)

18. Cheng, M., Zhang, G., Mitra, N.J., Huang, X., Hu, S.: Global contrast based salient region detection. In: IEEE Conference on Computer Vision and Pattern Recognition, pp. 409–416. IEEE (2011)

19. Zhai, Y., Shah, M.: Visual attention detection in video sequences using spatiotemporal cues. In: 14th ACM International Conference on Multimedia, New York, pp. 815–824. ACM (2006)

20. Achanta, R., Hemami, S., Estrada, F., Susstrunk, S.: Frequency-tuned salient region detection. In: IEEE Conference on Computer Vision and Pattern Recognition, New York, pp. 1597–1604. IEEE (2009)

21. Yang, C., Zhang, L., Lu, H., Ruan, X., Yang, M.: Saliency detection via graph-based manifold ranking. In: IEEE Conference on Computer Vision and Pattern Recognition, New York, pp. 3166–3173. IEEE (2013)

22. Jiang, B., Zhang, L., Lu, H., Yang, C., Yang, M.: Saliency detection via absorbing markov chain. In: IEEE International Conference on Computer Vision, New York, pp. 1665–1672. IEEE (2013)

23. Zhu, W., Liang, S., Wei, Y., Sun, J.: Saliency optimization from robust background detection. In: IEEE Conference on Computer Vision and Pattern Recognition, New York, pp. 2814–2821. IEEE (2014)

24. Margolin, R., Tal, A., Zelnik, M.L.: What makes a patch distinct?. In: 2013 IEEE Conference on Computer Vision and Pattern Recognition, New York, pp. 1139–1146. IEEE (2013)

25. Fairchild, M.D., Berns, R.S.: Image color-appearance specification through extension of CIELAB. Color Res. Appl. 18(3), 178–190 (1993)

26. Lee, Y.K., Powers, J.M.: Comparisons of CIE lab, CIEDE 2000 and DIN 99 color differences between various shades of resin composites. Int. J. Prosthodont. 18(2), 150–155 (2005)

# A New Pathway to Explore Reliable Biomarkers by Detecting Typical Patients with Mental Disorders

Ying Xing and Yuhui Du[✉]

School of Computer and Information Technology, Shanxi University, Taiyuan, China
sxxying@126.com, duyuhui@sxu.edu.cn

**Abstract.** Identifying neuroimaging-based biomarkers is greatly needed to boost the progress of mental disorder diagnosis. However, it has been well acknowledged that inaccurate diagnosis on mental disorders may in turn raise unreliable biomarkers. In this paper, we propose a new method that can detect typical patients with specific mental disorders, which is beneficial to further biomarker identification. In our method, we extend an advanced sample noise detection technology based on random forest to identify typical patients, and apply it to identify typical subjects from schizophrenia (SZ) and bipolar disorder (BP) patients with neuroimaging features estimated from resting fMRI data. To evaluate the capacity of our method, we investigate the typical subjects and whole subjects with respect to group differences, classification accuracy, clustering, and projection performance based on the identified typical subjects. Our results supported that the typical subjects showed greater group differences between SZ and BP, higher classification accuracy, more compact clusters in both clustering and projection. In short, our work presents a novel method to explore discriminative and typical subjects for different mental disorders, which is promising for identifying reliable biomarkers.

**Keywords:** Biomarker · fMRI · Sample selection · Schizophrenia · Bipolar disorder

## 1 Introduction

Diagnosis of mental illness that is defined by behavior/symptoms depends on experience and subjective judgment of psychiatrists, thus likely leading to inaccurate clinical diagnosis [1–4]. There are no existing gold standards that can be used to define different mental disorders, especially for those sharing similar symptoms, such as schizophrenia (SZ) and bipolar disorder (BP). Therefore, there is an urgent need to explore biomarkers that can help understand the substrates of mental disorders and assist in diagnosis [5]. Researchers have been working on the investigation of biological abnormalities associated with mental illness based on neuroimaging techniques, such as functional magnetic resonance imaging (fMRI) [6–8].

Supported by National Natural Science Foundation of China (Grant No. 62076157 and 61703253 to YHD).

Schizophrenia and bipolar disorder share significant overlap in clinical symptom and risk genes, which raise a great difficulty in separating them in clinical practice [2,9,10]. It has been found that both disorders have similar and unique disruptions of the normal-range operation of brain functions [2,6]. In recent years, many efforts have been made in using fMRI data to discover biomarkers for distinguishing the two disorders [11–13]. In general, statistical analysis and classification approaches are applied for the goal. For identifying significant biomarkers that can differentiate the two disorders, Calhoun et al. discovered a key role of the default mode in distinguishing SZ and BP based on independent component analysis (ICA) [6]. Rashid et al. also utilized ICA on fMRI data and presented group differences between SZ and BP in patterns of functional connectivity involving the frontal and frontal-parietal regions [13]. Argyelan et al. claimed that SZ had significantly lower global connectivity than that of BP [7]. Birur reported that the aberrant connectivity of SZ and BP were different in default mode network [8].

All the above-mentioned works used statistical analysis and supervised machine learning techniques to explore biomarkers, which greatly depend on the diagnosis label. It is apparent that misdiagnosed patients can influence the identified biomarkers, and probably result in poor effectiveness in dealing with new coming patients [2,14,15]. Therefore, there has been increasing interest in developing neuroimaging-based biotypes by clustering patients with mental disorders based on neuroimaging measures [14–16]. Unfortunately, the progress of biotype development has been slow and limited due to the difficulty in detecting subjects fitting in cluster patterns using high-dimensional features via an unsupervised way.

Detecting typical patients who present more consistent brain changes may help solve the problem in exploring reliable biomarkers and biotypes. In this paper, we propose a new method to identify typical patients with mental disorders using neuroimaging measures, and apply our method to 113 SZ and 113 BP patients with resting fMRI data for a comprehensive investigation. Brain functional connectivity is used as input features. We extend the Complete Random Forest (CRF) model [17] to divide the subjects into typical and atypical samples. Considering that the typical subjects should have: significant difference between groups, high similarity within groups, and clear separability between groups, we design three studies to validate whether the performances of the selected typical subjects are better than all subjects. Study 1 focuses on if the differences between groups of the selected typical subjects are more significant than the whole subjects. In study 2, we compare the classification performance of the typical subjects and whole subjects based on four popular classifiers. In study 3, we compare the clustering performance of the selected typical subjects and whole subjects. Finally, we apply t-SNE technology [18] to project the typical subjects and all subjects into 2D planes, respectively, in order to show the separation of the two subject sets intuitively.

## 2    Methods

### 2.1    Data and Neuroimaging Measures

In this work, we analyze fMRI data of 226 subjects including 113 SZ and 113 BP from the multi-site Bipolar and Schizophrenia Network on Intermediate Phenotypes (BSNIP-1) study. For each subject, the functional connectivity (FC) is estimated using resting-state fMRI across the entire brain in 116 predefined regions of interest (ROIs) from the automated anatomical labeling (AAL) template [5]. First, the averaged blood-oxygen-level dependent (BOLD) time-series over the entire scan time are computed for each ROI. Then by computing the Pearson correlation coefficients between pairwise BOLD time-series of ROIs, a symmetrical FC matrix (size: $116 \times 116$) is obtained. The FC matrix is converted to a vector containing only upper triangular 6670 elements (reflection of FC's strength) as the input features. Based on each computed FC, we further explore the difference between any pair of groups using a two-tailed two-sample t-test. Then the FCs with $p\ value$ less than 0.01 are regarded as important features that are used for subsequent analysis.

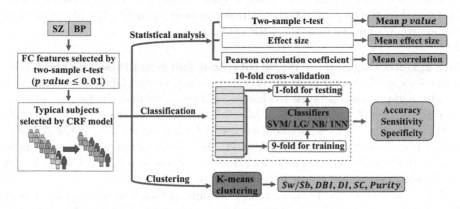

**Fig. 1.** The flowchart of the proposed pipeline. SVM, LG, NB, and 1NN represent four classifiers, i.e. support vector machine based on radial basis function, logistic regression, naïve Bayes classifier, and 1-nearest neighborhood classifier. $Sw/Sb$, $DBI$, $DI$, $SC$, and $Purity$ are clustering measures, in which $Sw/Sb$, $DBI$, $DI$, $SC$ are used to measure compactness within a group and the separability between groups, $Purity$ is used to reflect the overlap degree between the clustering results and diagnosis labels. The above measures will be introduced in detail in Sect. 2.4.

### 2.2    Overview of Our Method

Figure 1 shows the flowchart of our proposed pipeline which mainly consists of three steps. Using the functional connectivity, the significant features are selected by two-sample t-test between SZ and BP firstly. Secondly, the typical subjects are identified based on the advanced sample noise detection model which is called CRF. Thirdly, we evaluate the selected subjects based on mainstream tasks in the field of brain imaging analysis, including statistical analysis, classification analysis, clustering, and projection analysis.

### 2.3   Detection of Typical Subjects

In the existing related literature, sample noise is described as incorrect label observation, mainly due to insufficient information and subjective label errors caused by experts. Here, the atypical patients that cause label errors can be regarded as sample noises. In this section, we present an advanced sample noise detection method using CRF model [17] to filter the atypical patients. The CRF model consists of the following two parts to identify the sample noises.

**Part 1: Using All Subjects to Construct the CRF with $N$ Complete Random Decision Trees (CRDT).** A CRDT is constructed according to the following rules firstly. The root node of each tree contains all subjects. The non-leaf node of each tree contains two child nodes, and the division rule is based on randomly selecting features and feature split values. The label of each node is determined by the majority labels of the subjects in the node. The above process is repeated $N$ times to form the CRF of $N$ trees.

**Part 2: Detecting Sample Noise (atypical Subjects) Based on the CRF.** A previous work from Xia et al. [17] supposes that the label of a node that contains sample noise is prone to change, and after the first change, it can keep a certain degree of stability, which is called the noise intensity ($NI$). For each CRDT in the forest, we first calculate the $NI$ for each subject. If the $NI$ of a subject is no less than the given threshold, the subject is considered as a sample noise in the tree. If more than 50% of trees in the forest identify a subject as a sample noise, then the subject is considered as a sample noise.

After the above two steps, we remove sample noises from the whole SZ and BP subjects and keep the non-noise subjects, i.e. typical SZ and BP patients, for the subsequent analysis. The evaluations of the selected typical subjects are introduced in the following section.

### 2.4   Evaluation

To validate the selected typical subjects, we design the following three studies. In study 1, the statistical analyses are employed to verify whether the difference between SZ and BP using the selected typical patients is more significant, with lower $p$ $value$, greater effect size, and lower correlation between samples from different groups. Study 2 focuses on investigating whether the selected typical patients are more distinguishable (with higher classification accuracy) through classifying SZ and BP. Study 3 aims to examine whether the selected typical subjects are more separable between different groups based on clustering and projection analyses.

**Study 1: Investigating the Group Differences Between the Selected Typical Subjects Using Statistical Analysis.** Exploring neurological differences between healthy and diseased populations or between different mental disorder groups is a pivotal step for understanding the internal mechanisms of brain diseases [2,3,19,20]. Therefore, in this study, we employ three statistical analysis technologies to reflect differences between groups, including two-sample t-test, effect size, and Pearson correlation coefficient.

By using all subjects or only the selected typical patients, we first perform a two-tailed two-sample t-test on each functional connectivity measure between two groups. As we know, smaller $p$ $value$ means a more significant group difference. We expect that using the selected typical subjects can yield a smaller mean $p$ $value$ than using all subjects. So, we compare the mean $p$ $value$ across all connectivity measures, the smallest $p$ $value$ and the biggest $p$ $value$ between using all subjects and only the typical patients.

However, it is clear that the $p$ $value$ is related to the number of subjects since more samples (i.e. patients) tend to generate a smaller $p$ $value$. That is to say, using all subjects is supposed to result in a smaller $p$ $value$ than using part of subjects (e.g. typical patients). In order to evaluate fairly, we also assess the mean effect size, the smallest effect size, and the biggest effect size between different groups from all subjects and typical subjects, respectively. Furthermore, we calculate the correlation coefficients between patients in different groups based on Pearson correlation coefficient, and then compare the mean correlations between using whole subjects and using only typical patients. We hope that the mean correlation between different groups is lower in using typical subjects than using all subjects, which supports the stronger group differences between two typical disorder groups.

**Study 2: Investigating the Distinguishing Ability of the Selected Typical Subjects Based on Classification Task.** Another important task of neuroimaging analysis is to train a classification model with the guidance of labeled subjects to predict new subjects [2,21]. Here, we perform the classification task to verify that the typical patients in BP and SZ selected by our method can be easily classified, compared to classifying all patients.

In this study, we apply an unbiased 10-fold cross-validation framework to classify subjects as shown in Fig. 1. Nine of ten folds are used as the training data to build a classifier, and then the remaining fold is used as the testing data to evaluate the model. We use four popular classifiers, including support vector machine based on radial basis function (SVM-RBF), logistic regression (LR), naïve Bayes classifier (NB), and 1-nearest neighborhood classifier (1NN) to test. In our experiments, we compare the classification performance, including the average accuracy, average sensitivity, and average specificity between using the selected typical subjects and using all subjects in the classification.

**Study 3: Investigating the Separation of the Selected Typical Subjects Based on Clustering and Projection Analysis.** Heterogeneity is one of the challenges in exploring the mechanism of brain diseases [22]. Previous neuroimaging studies have shown that clustering could help identify biotypes of psychiatric disorders using features including neuroimaging measures [16,23].

In this study, we apply K-means clustering to the selected typical subjects and all subjects, respectively, and compare the clustering performance of the two subject sets. Here, we set the cluster number to 2 for K-means. Next, we use five cluster evaluation measures to compare the performance between the two different subject sets. Four measures (1–4) reflect the compactness within a group and the separability between groups without labels, and another one (5) reflects the consistency of the clustering results regarding diagnostic labels.

(1) $Sw/Sb$ is the ratio of $Sw$ and $Sb$, where $Sw$ is the average distance between pairwise samples within a cluster, and $Sb$ is the average distance between pairwise samples in different clusters. A smaller $Sw/Sb$ value means better performance. Suppose there is a set of samples $\{x_1, x_2, ..., x_n\}$, which are divided into two clusters $C_1$ and $C_2$ by K-means algorithm, $\mu_1$ and $\mu_2$ are the centers of $C_1$ and $C_2$, respectively, $d(x_i, x_j)$ is a distance function. The calculation formulas of $Sw$ and $Sb$ are presented as follows:

$$Sw = \frac{1}{2} * \sum_{k=1}^{2} \frac{2}{|C_k| * (|C_k| - 1)} \sum_{x_i, x_j \in C_k} d(x_i, x_j), \tag{1}$$

$$Sb = \sum_{x_i \in C_k, x_j \in C_l} d(x_i, x_j). \tag{2}$$

(2) Davies-Bouldin Index ($DBI$) is the ratio of the sum of average distances of pairwise of samples in a cluster to the distance between the center points of the two clusters. A smaller $DBI$ value means better performance [24].

(3) Dunn Index ($DI$) is the ratio of the shortest distance between two clusters to the maximum distance of pairwise samples in any cluster. A bigger $DI$ value means better performance [25].

(4) Silhouette Coefficient($SC$) is the ratio of average compactness within a cluster to the separation between clusters. A bigger $SC$ value means better performance [26].

(5) *Purity* means the consistency of obtained labels of the clustering process and the ground-truth cluster labels within clusters. The bigger *Purity* value reflects higher consistency [27].

In addition, in order to see if typical patients have better and clear cluster patterns than all subjects, we also visualize the distribution of all subjects and the selected typical subjects separately, using a famous projection method, t-SNE [18]. We expect that the typical subjects are more separable than all subjects.

# 3 Results

Based on the CRF model, we detected 134 typical subjects (79 SZ and 55 BP patients) from 226 original subjects (113 SZ and 113 BP patients). As mentioned, we performed different studies to evaluate the selected typical subjects and show the results as below.

## 3.1 Results of Study 1: Typical Patients Show Significant Group Differences Using Statistical Analyses

Table 1 shows three metrics that reflect the group differences in all subjects and the selected typical subjects, respectively. It can be seen the mean $p$ value of using all subjects ($p = 0.0045$) is just a little lower than using typical patients ($p = 0.0057$). That is acceptable and understandable, because more samples are liable to generate lower $p$ value. In fact, the minimum $p$ value of the typical subjects is much lower than that of all subjects.

Besides, the mean effect size between different groups from typical subjects is greater than the mean effect size of all subjects (0.6258 vs. 0.0289). The minimum and maximum effect sizes between different groups of typical subjects are bigger than the minimum and maximum effect sizes between different groups of all subjects. These results indicate that the typical subjects show more pronounced group differences compare to all subjects.

In addition, measured by the between-group correlations, the mean correlation is 0.3632 for using all subjects, which is higher than the mean correlation of 0.2758 for only using selected typical subjects. Figure 2 presents the correlation matrix of all subjects and the selected typical subjects. It is observed that the typical patients selected by our method have smaller intra-group difference and bigger inter-group differences, which is in line with our expectations.

**Table 1.** Group differences in all subjects and the selected typical subjects

| | $P$ value of two-sample t-test | | | Effect size | | | Mean PCC |
|---|---|---|---|---|---|---|---|
| | Minimum | Maximum | Mean | Minimum | Maximum | Mean | |
| All subjects | 2.71e−06 | 0.0100 | 0.0045 | 1.79e−6 | 0.1388 | 0.0289 | 0.3632 |
| Typical subjects | 2.83e−19 | 0.4711 | 0.0057 | 0.0975 | 1.2355 | 0.6258 | 0.2758 |

Remarks: PCC is Pearson correlation coefficient between subjects in different groups.

## 3.2 Results of Study 2: Typical Patients Are More Distinguishable Than Whole Subjects Based on Classification Task

Table 2 presents the averaged classification results of 10-fold cross-validation for using all subjects and only using the typical subjects. The results from each type of classifier are summarized. For the selected typical patients, the three

**(a) Correlation matrix of all subjects    (b) Correlation matrix of typical subjects**

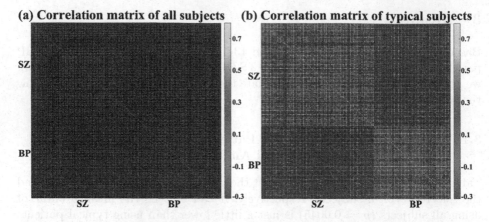

**Fig. 2.** Pearson correlation coefficient matrix between the pairwise subjects for (a) all subjects and (b) the selected typical subjects.

evaluations of each classifier are all greater than 90%, in which LR and SVM-RBF classifiers achieve the highest accuracy, with an average accuracy of 99.23%, an average sensitivity of 98.75%, and an average specificity of 100%. Regarding using all subjects, the NB classifier achieves better performance among the four classifiers, with an average accuracy of 69.35%, an average sensitivity of 67.53%, and an average specificity of 71.17%. The results support that the typical patients are relatively more separable in classification.

**Table 2.** Classification performance from 10-fold cross-validation

| Classifier | Using all subjects in classification | | | Using typical subjects in classification | | |
|---|---|---|---|---|---|---|
| | Accuracy | Sensitivity | Specificity | Accuracy | Sensitivity | Specificity |
| SVM-RBF | 60.06% | 61.17% | 58.96% | 99.23% | 98.75% | 100.00% |
| LR | 63.70% | 61.17% | 66.23% | 99.23% | 98.75% | 100.00% |
| NB | 69.35% | 67.53% | 71.17% | 96.92% | 94.64% | 100.00% |
| 1NN | 65.97% | 62.79% | 69.16% | 94.80% | 93.87% | 96.57% |

### 3.3   Results of Study 3: Typical Patients Show More Compactness Within Groups and Significant Separation Between Groups Using Clustering and Projection Analyses

The comparisons of five evaluations for clustering performance are presented in Table 3. We can see that the results from using the typical subjects are better in each evaluation. The values of $Sw/Sb$ and $DBI$ of all subjects are greater than that of the typical subjects, which are 0.9579 vs. 0.8134 and 7.9330 vs. 6.1414, respectively. The values of $DI$, $SC$, and $Purity$ of all subjects are smaller than

**Table 3.** Clustering performance of all subjects and the selected typical subjects

|  | $Sw/Sb$ ↓ | $DBI$ ↓ | $DI$ ↑ | $SC$ ↑ | $Purity$ ↑ |
|---|---|---|---|---|---|
| All subjects | 0.9579 | 7.9330 | 0.4900 | 0.1024 | 0.6770 |
| Typical subjects | 0.8134 | 6.1414 | 0.5804 | 0.1585 | 0.9747 |

Remarks: "↓" means that the smaller the evaluation value, the better the result; "↑" means that the greater the evaluation value, the better the result.

that of the typical subjects, which are 0.4900 vs. 0.5804, 0.1024 vs. 0.1585, and 0.6770 vs. 0.9747, respectively. The results of $Sw/Sb$, $DBI$, $DI$, and $SC$ indicate that the distances within groups of the selected typical subjects are closer and the distances between groups are larger, relative to that of all subjects. $Purity$ measure supports that the selected typical subjects were grouped to clusters that align their original diagnosis categories, while the whole subjects did not. In short, the clustering performance of the selected typical subjects outperforms that of all subjects.

Figure 3 shows the result of visualization based on t-SNE technology. We can find that most of the selected typical subjects are grouped clearly, while t-SNE does not work well using all subjects. In other words, the typical subjects selected are more separable.

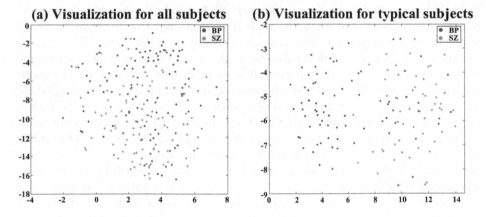

**(a) Visualization for all subjects**        **(b) Visualization for typical subjects**

**Fig. 3.** Visualization of subject distribution for using (a) all subjects, (b) the typical subjects.

## 4   Conclusion

There are many studies using fMRI data to identify biomarkers and help diagnose mental disorders. Most of the previous fMRI studies used statistical analysis or supervised learning technology to identify biomarkers for mental illness [2, 14].

However, using traditional statistical analysis and classification methods could result in biased biomarkers due to the fact that the diagnosis label is often inaccurate. To explore more reliable biomarkers, we provided a new pathway to identify typical patients with mental disorders based on the sample noise detection model using neuroimaging measures.

To validate the reliability of the selected subjects, we evaluated the typical subjects based on the three mainstream tasks. Study 1 demonstrated that the typical subjects had more significant group differences than that of all subjects based on statistical analysis. Study 2 proved that the classification of the selected typical BP and SZ subjects is easier than the classification of the original all subjects, which indicated that those typical patients may be promising to provide more reliable biomarkers for new coming subjects. Compare to the existing classification studies of SZ and BP, we have achieved better classification performance. For example, in our previous work, the overall classification accuracy of SZ and BP based on the recursive feature elimination model was 82% [5]. Calhoun applied multivariate analysis and independent component analysis methods and found that utilizing the temporal lobe and the default mode network as biomarkers, the classification accuracies of SZ and BP were 92% and 83%, respectively [21]. Besides, using the estimated neural responses to verbal fluency for participants, Costafreda reported that SZ and BP patients were correctly distinguished with an average accuracy of 86% [19]. The classification accuracies of the above studies were inferior to ours, probably due to the inaccurate diagnostic labels. Therefore, it is necessary to select reliable typical subjects before exploring biomarkers. In study 3, we verified that the five clustering results of using the typical subjects were better than using all subjects. Specifically, clustering measures including $Sw/Sb$, $DBI$, $DI$, and $SC$ demonstrated that the typical subjects in the same cluster showed more compact correlations, while the typical subjects in distinct clusters showed more salient differences. The $Purity$ results proved that the obtained clustering labels for typical subjects were more consistent with the diagnostic labels. The t-SNE projection results also supported that the selected typical subjects were more separable than that of all subjects. In summary, the selected typical SZ and BP subjects have prominent differences between groups, great correlation within the same group, and significant separability between groups, which is promising for identifying reliable biomarkers and provide insights in exploring biotypes.

This is the first attempt to explore typical subjects with mental diseases based on the biologically meaning measures. Our work lays the foundation for the future exploration of biomarkers. In the future, we will apply different modality data to explore more reliable typical subjects and biomarkers for mental disorders diagnosis.

# References

1. Whalley, H.C., Papmeyer, M., Sprooten, E., Lawrie, S.M., Sussmann, J.E., McIntosh, A.M.: Review of functional magnetic resonance imaging studies comparing bipolar disorder and schizophrenia. Bipolar Disord. **14**(4), 411–431 (2012)
2. Du, Y., Fu, Z., Calhoun, V.D.: Classification and prediction of brain disorders using functional connectivity: promising but challenging. Front. Neurosci. **12**, 525 (2018)
3. Du, Y., et al.: Neuromark: an automated and adaptive ICA based pipeline to identify reproducible fMRI markers of brain disorders. NeuroImage: Clin. **28**, 102375 (2020)
4. Steardo Jr., L.: Application of support vector machine on fMRI data as biomarkers in schizophrenia diagnosis: a systematic review. Front. Psychiatry **11**, 588 (2020)
5. Du, Y., Hao, H., Wang, S., Pearlson, G.D., Calhoun, V.D.: Identifying commonality and specificity across psychosis sub-groups via classification based on features from dynamic connectivity analysis. NeuroImage: Clin. **27**, 102284 (2020)
6. Calhoun, V.D., Sui, J., Kiehl, K., Turner, J.A., Allen, E.A., Pearlson, G.: Exploring the psychosis functional connectome: aberrant intrinsic networks in schizophrenia and bipolar disorder. Front. Psychiatry **2**, 75 (2012)
7. Argyelan, M., et al.: Resting-state fMRI connectivity impairment in schizophrenia and bipolar disorder. Schizophr. Bull. **40**(1), 100–110 (2014)
8. Birur, B., Kraguljac, N.V., Shelton, R.C., Lahti, A.C.: Brain structure, function, and neurochemistry in schizophrenia and bipolar disorder—a systematic review of the magnetic resonance neuroimaging literature. NPJ Schizophr. **3**(1), 1–15 (2017)
9. Mukherjee, S., Shukla, S., Woodle, J., Rosen, A.M., Olarte, S.: Misdiagnosis of schizophrenia in bipolar patients: a multiethnic comparison. Am. J. Psychiatry **140**, 1571–1574 (1983)
10. Shen, H., Zhang, L., Chuchen, X., Zhu, J., Chen, M., Fang, Y.: Analysis of misdiagnosis of bipolar disorder in an outpatient setting. Shanghai Arch. Psychiatry **30**(2), 93 (2018)
11. Alan, A., et al.: Characterizing thalamo-cortical disturbances in schizophrenia and bipolar illness. Cereb. Cortex **24**(12), 3116–3130 (2014)
12. Sui, J., et al.: Discriminating schizophrenia and bipolar disorder by fusing fMRI and DTI in a multimodal CCA+ joint ICA model. Neuroimage **57**(3), 839–855 (2011)
13. Rashid, B., Damaraju, E., Pearlson, G.D., Calhoun, V.D.: Dynamic connectivity states estimated from resting fmri identify differences among schizophrenia, bipolar disorder, and healthy control subjects. Front. Hum. Neurosci. **8**, 897 (2014)
14. Insel, T.R., Cuthbert, B.N.: Brain disorders? precisely. Science **348**(6234), 499–500 (2015)
15. Dwyer, D.B., et al.: Brain subtyping enhances the neuroanatomical discrimination of schizophrenia. Schizophr. Bull. **44**(5), 1060–1069 (2018)
16. Marquand, A.F., Wolfers, T., Mennes, M., Buitelaar, J., Beckmann, C.F.: Beyond lumping and splitting: a review of computational approaches for stratifying psychiatric disorders. Biol. Psychiatry Cogn. Neurosci. Neuroimaging **1**(5), 433–447 (2016)
17. Xia, S., Wang, G., Chen, Z., Duan, Y., et al.: Complete random forest based class noise filtering learning for improving the generalizability of classifiers. IEEE Trans. Knowl. Data Eng. **31**(11), 2063–2078 (2018)

18. van der Maaten, L., Hinton, G.: Visualizing data using t-SNE. J. Mach. Learn. Res. **9**(Nov), 2579–2605 (2008)
19. Costafreda, S.G., et al.: Pattern of neural responses to verbal fluency shows diagnostic specificity for schizophrenia and bipolar disorder. BMC Psychiatry **11**(1), 18 (2011)
20. Rashid, B., Calhoun, V.: Towards a brain-based predictome of mental illness. Hum. Brain Mapp. **41**(12), 3468–3535 (2020)
21. Calhoun, V.D., Maciejewski, P.K., Pearlson, G.D., Kiehl, K.A.: Temporal lobe and "default" hemodynamic brain modes discriminate between schizophrenia and bipolar disorder. Hum. Brain Mapp. **29**(11), 1265–1275 (2008)
22. Varol, E., Sotiras, A., Davatzikos, C., Initiative, A.D.N., et al.: Hydra: revealing heterogeneity of imaging and genetic patterns through a multiple max-margin discriminative analysis framework. Neuroimage **145**, 346–364 (2017)
23. Zhang, T., Koutsouleris, N., Meisenzahl, E., Davatzikos, C.: Heterogeneity of structural brain changes in subtypes of schizophrenia revealed using magnetic resonance imaging pattern analysis. Schizophr. Bull. **41**(1), 74–84 (2015)
24. Davies, D.L., Bouldin, D.W.: A cluster separation measure. IEEE Trans. Pattern Anal. Mach. Intell. **2**, 224–227 (1979)
25. Dunn, J.C.: Well-separated clusters and optimal fuzzy partitions. J. Cybern. **4**(1), 95–104 (1974)
26. Rousseeuw, P.T.: Silhouettes: a graphical aid to the interpretation and validation of cluster analysis. J. Comput. Appl. Math. **20**, 53–65 (1987)
27. Parmar, D., Teresa, W., Blackhurst, J.: MMR: an algorithm for clustering categorical data using rough set theory. Data Knowl. Eng. **63**(3), 879–893 (2007)

# Activities Prediction of Drug Molecules by Using Automated Model Building with Descriptor Selection

Yue Liu[1,2,3], Wenjie Tian[1], and Hao Zhang[4(✉)]

[1] School of Computer Engineering and Science, Shanghai University, Shanghai, China
{yueliu,twenjie}@shu.edu.cn
[2] Shanghai Institute for Advanced Communication and Data Science,
Shanghai, China
[3] Shanghai Engineering Research Center of Intelligent Computing System,
Shanghai 200444, China
[4] Shanghai Chempartner Co., Ltd., Shanghai 201203, China
haozhang@chempartner.com

**Abstract.** Machine learning is a powerful tool for simulating the quantitative structure activity relationship in drug discovery (QSAR). However, descriptor selection and model optimization remain two of most challenging tasks for domain experts to construct high-quality QSAR model. Therefore, we propose a QSAR-special automated machine learning method incorporating Automated Descriptor Selection with Automated Model Building (ADSMB) to efficiently and automatically build high-quality QSAR model. Automated Descriptor Selection provides a QSAR-special molecular descriptor selection mechanism to automatically obtain the descriptors without unique value, redundancy and low importance in QSAR dataset. Based on these QSAR-special descriptors, Automated Model Building constructs high-quality ensemble model of molecular descriptors and target activities under Bayesian optimization through Auto-Sklearn. Finally, we conduct experimental evaluation for our proposed method on Mutagenicity dataset. The results show ADSMB can obtain better and stable performance than the competing methods.

**Keywords:** Automated machine learning · Feature selection · Drug discovery

## 1 Introduction

Quantitative Structure Activity Relationship (QSAR) has been widely applied in pharmaceutical industry to predict biological activities of chemical compounds through analyzing quantitative characteristics of structure features [1]. It is a cost-effective technology to substantially reduce the workload and time needed for traditional approach of drug discovery [2]. Machine learning approach has been a prevailing computational tool in QSAR to guide rational drug discovery

W. Gao et al. (Eds.): FICC 2020, CCIS 1385, pp. 73–84, 2021.
https://doi.org/10.1007/978-981-16-1160-5_7

over the past few decades [3], which are based on molecular structures and target activities, such as physicochemical properties and therapeutic activities [4]. The commonly applied machine learning models in QSAR include Support Vector Machine (SVM) [5,6], Decision Tree [7], Random Forest [8], k-Neighbors Neighbor (kNN) [9] and Artificial Neural Networks [2,9,10]. For examples, Poorinmohammad N, et al. incorporated SVM with pseudo amino acid composition descriptors to predict active anti-HIV peptides, which obtained a prediction accuracy of 96.76% [5]. Weidlich I E, et al. used kNN with a simulated annealing method and RF for 679 drug-like molecules to improve the prediction accuracy of drug discovery [9].

Despite the progress of machine learning in QSAR for drug discovery, there are still some practical constraints on the use of machine learning in QSAR, such as (1) The datasets of QSAR may involve numerous descriptors. They are sparse and few of them are nonzero [2]. Moreover, there exists strong correlation between different descriptors, which is useless for building QSAR model. (2) Algorithm selection and hyperparameter optimization remain two of most challenging tasks in QSAR model construction. The existing QSAR models are human-elaborate, and require extensive and iterative fine-tuning with trial and error to build high-performance QSAR models. Furthermore, the increasing number of machine learning algorithms with their sensitive hyperparameters are put forward. It is very difficult and practically infeasible for domain experts to quickly and efficiently apply machine learning to help with activity prediction of drug molecules.

Automated Machine Learning (AutoML) aims to automatically build appropriate model without human intervention to enable widespread use of machine learning by non-experts [11]. Previous works have demonstrated the success of AutoML at every stage of machine learning process, such as feature engineering [12], hyperparameter optimization [13], model selection [14,15] and neural architecture search [16], and many domain-special applications. For example, Olson R. S. et al. developed an AutoML system to optimize tree-based machine learning pipeline through genetic programming, which was successfully applied in biomedical data [15]. In our previous work [17], we proposed automated feature selection with multiple layers incorporating expert knowledge toward modeling materials with targeted properties successfully. Therefore, it is feasible to incorporate AutoML with QSAR to automatically build high-quality QSAR-special machine learning model for efficient drug discovery.

In this paper, we propose an QSAR-special automated machine learning method incorporating Automated Descriptor Selection with Automated Model Building (ADSMB) to efficiently build high-quality QSAR model. Specially, we highlight the following contributions of our work.

– In order to select more suitable descriptors for QSAR, we propose Automated Descriptor Selection method, which provides a QSAR-special descriptor selection mechanism to automatically eliminate the descriptors with unique value, redundancy and low importance.

- In order to effectively and efficiently construct prediction model for QSAR, we propose Automated Model Building method, which is based on QSAR-special descriptors and constructs high-quality ensemble model under Bayesian optimization through the classical AutoML system, Auto-Sklearn [14].
- Experimental evaluations for our proposed method have been conducted on a drug molecular activity dataset. The results show ADSMB can obtain better and more stable performance than the competing method.

The rest of this paper is organized as follows. In Sect. 2, details of the proposed novel algorithm are described. Experiments are performed on a drug molecular activity dataset in Sect. 3. Finally, Sect. 4 concludes this paper.

## 2 Towards Automated Activities Prediction of Drug Molecules

In order to alleviate the dilemma between tedious descriptor selection and model building in activities prediction of drug molecules, we propose an automated machine learning method incorporating Automated Descriptor Selection with Automated Model Building (ADSMB) to automatically construct the high-quality QSAR model without human intervention. Figure 1 illustrates the process of ADSM B. ADSMB firstly obtains the QSAR-special descriptors through Multi-Layer Descriptor Selection incorporating with expert knowledge. It progressively considers the uniqueness, redundancy and importance in molecular structures descriptors. Based on QSAR-special descriptors, ADSMB employs Auto-Sklearn to automatically construct the high-quality machine learning model to predict the relationship between molecular structures and target activities. It performs meta-learning to find promising machine learning framework to warm-start Bayesian optimization (BO) and construct ensemble model with the individuals searched by BO. Herein, we efficiently construct the high-quality QSAR model for activity prediction of drug molecules instead of extensive and iterative fine-tuning with trial and error.

### 2.1 Automated Descriptor Selection

In order to utilize more valuable descriptors from high-dimension QSAR dataset and improve the efficiency and performance of Auto-Sklearn, we develop QSAR-special Automated Descriptor Selection to obtain the most valuable subset of descriptors based on the characteristic of QSAR dataset. It involves three processing layers: uniqueness evaluation, redundancy evaluation and importance evaluation, which analyze the availability for activities prediction from different perspectives depending on the characteristics of the dataset. Thus, the trigger conditions of Multi-Layer Descriptor Selection are defined as Eq. (1).

$$\text{Multi\_Layer}(X) = \begin{cases} Layer_1(X) & \text{if } \exists x_i \in X \text{ and } x_i \text{ has uniquevalue } \varepsilon \\ Layer_2(X) & \text{if } \exists x_i \in X \text{ and } x_i.\text{redundany} \geq \gamma \\ Layer_3(X) & \text{if } \exists x_i \in X \text{ and } x_i.\text{importance} \leq \lambda \end{cases} \tag{1}$$

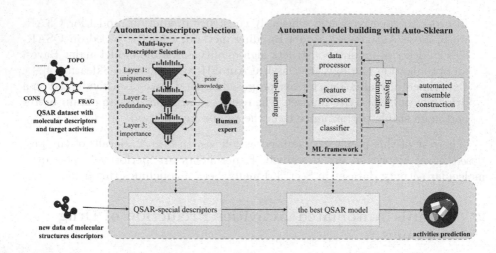

**Fig. 1.** The process of ADSMB

where $X$ indicates the input descriptors and $x_i$ indicates the $x_i$ descriptor. $\varepsilon$, $\gamma$ and $\lambda$ represent the unique value in descriptors, redundancy threshold and importance threshold respectively.

The first layer (i.e. $Layer_1(X)$) is defined as uniqueness evaluation. When the descriptor only has a single unique value $\varepsilon$, it will be removed from the input descriptors, since it is useless for the predictive performance of QSAR model. After the unique descriptors are removed, the second layer ($Layer_2(X)$) is to eliminate redundancy descriptors. i.e. the descriptors strongly correlated with the other descriptors. We employ Pearson Correlation Coefficient to be defined as Eq. (2) to measure the redundancy among descriptors.

$$\text{PCC}(x_i, y) = \frac{\text{cov}(x_i, y)}{\sigma_{x_i}\sigma_y} \tag{2}$$

where $cov(x_i, y)$ represents the covariance of descriptor $x_i$ and $y$. $\sigma_{x_i}$ and $\sigma_y$ represent the standard deviation of $x_i$ and $y$ respectively. Herein, $Layer_2(X)$ is defined as Eq. (3).

$$\text{Layer}_2(X) = |\text{PCC}(x_i, y)| \tag{3}$$

If the absolute PCC value is greater than redundancy threshold $\gamma$, one of the two descriptors $x_i$ and $y$ will be removed. The redundancy descriptors are eliminated in $Layer_2(X)$, however, in the subset of descriptors obtained, some of the descriptors may still be unimportant to the target activities. Therefore, in the third layer ($Layer_3(X)$), we evaluate the importance of each descriptor to activities prediction. The $Layer_3(X)$ is defined as Eq. (4).

$$\text{Layer}_3(X) = \text{LightGBM}(X, ta) = \text{info\_gain}(x_i) \tag{4}$$

where $ta$ indicates the target activities. Herein, $Layer_3(X)$ computes the total information gain used in LightGBM model to measure the importance of each descriptor. If the information gain of a descriptor is lower than importance threshold $\lambda$, it will be removed from the input descriptors.

Purely data-driven Multi-Layer Descriptor Selection may be biased in practice, since it can remove the descriptors that human experts consider important to QSAR problem. Therefore, it is necessary to introduce expert knowledge to Multi-Layer Descriptor Selection. The number of descriptors in QSAR dataset is numerous, which renders that it is infeasible to score and weight for each descriptor according to our previous work [17]. In our Automated Descriptor Selection, we simply describe the expert knowledge as Eq. (5).

$$ek\left(x_i\right) = \begin{cases} 0 \\ 1 \end{cases} \tag{5}$$

where $ek$ indicates whether a descriptor $x_i$ is important and necessary to QSAR problem. Then, the selected descriptors $ADS\left(X\right)$ from input descriptors $X$ are define as Eq. (6).

$$\text{ADS}\left(X\right) = ek\left(X\right) \cdot \text{Multi\_Layer}\left(X\right) \tag{6}$$

Herein, we eliminate the unique, redundancy and unimportant descriptors for QSAR model building, which renders Auto-Sklearn concentrates on the more valuable descriptors obtained by Automated Descriptor Selection.

## 2.2   Automated Model Building

Although QSAR-special descriptors have been obtained by Automated Descriptor Selection, human experts still face the challenge of how to select algorithm and optimize hyperparameter to build high-quality QSAR model. In order to efficiently construct QSAR model, we employ Auto-Sklearn to automatically build predictive model for activities prediction of drug molecules based on the selected QSAR-special descriptors. Auto-Sklearn is the most classical AutoML system to automatically build appropriate machine learning model for given datasets. It uses meta-learning to warm-start Bayesian optimization process and performs automated ensemble construction following the CASH framework (combining algorithm selection and hyperparameter optimization to execute them simultaneously). Therefore, the Automated Model Building process for QSAR dataset is briefly described as follows.

**Step 1:** Define the search space of Auto-Sklearn for QSAR dataset as a collection of preprocessing and classification algorithms. Enable the preprocessing algorithms to be composed of *extreml. rand. trees prepr.*, *fast ICA*, *feature agglomeration*, *kernel PCA*, *rand. kitchen sinks*, *linear SVM prepr.*, *no preprocessing*, *nystroem sampler*, *PCA*, *polynomial*, *random trees embedding*, *select percentile*, *select rates*, *one-hot encoding*, *imputation*, *balancing* and *rescaling* and classification algorithms to be composed of *adaboost*, *Bernoulli naïve Bayes*, *decision trees*, *Gaussian naïve Bayes*, *gradient boosting*, *kNN*, *LDA*, *linear SVM*, *kernel SVM*, *multinomial naïve Bayes*, *passive aggressive*, *QDA*, *random forest* and *SGD*.

**Step 2:** Select promising instantiations of machine learning framework (including data processor, feature processor and classifier) that are likely to perform well on QSAR dataset by using meta-learning embedded in Auto-Sklearn, which learns how different configuration performs across various datasets from prior performance. For QSAR dataset, Auto-Sklearn computes its meta-features (characteristics of the dataset), ranks all prior datasets by their L1 distance to QSAR dataset in meta-feature space and selects the promising machine learning framework instantiations of 25 nearest datasets.

**Step 3:** Start Bayesian optimization process with these promising instantiations for evaluation under prediction accuracy metric as objective function. In this process, Auto-Sklearn stores the models performing almost as well as the best.

**Step 4:** Enable automated ensemble construction in Auto-Sklearn. It uses an efficient post-processing method to automatically construct a weighted ensemble model out of the model stored by Step 3.

Automated Model Building process provides an automated and efficient mode l building strategy to dealing with QSAR problem. It can output high-quality ensemble model for activities prediction of drug molecules without human involvement.

## 3    Experiments

### 3.1    Dataset

The Mutagenicity (available on http://www.niss.org) dataset is used. Totally 1863 samples are included in this dataset and there are 618 structure parameters including 47 CONS descriptors, 260 Topological Indexes, 64 BUCT descriptors and 247 FRAG descriptors. We set the activity label as output.

### 3.2    Experimental Setup

In our experiments, 75% samples are used as the training set, while the remaining 25% are used as the test set. For Automated Descriptor Selection, we set the unique value $\varepsilon$ as set of real numbers and the redundancy threshold $\lambda$ as 0.95. Following the expert knowledge, we select 15 most important descriptors from layer 2. For Auto-Sklearn, we set the time limits of 3600 s and 360 s for the search of appropriate models and a single call to machine learning model, respectively. Moreover, the maximum number of models added to the ensemble is set as 50. We ran all procedures on Windows server with Intel(R) Xeon(R) CPU E5-2609 v4 @1.79GHZ processor, and the process was limited to a single CPU core.

## 3.3   Experimental Analysis

Automated Descriptor Selection eliminates the unique, redundancy and unimportant descriptors for Auto-Sklearn. 58 descriptors with a single unique value 0 and 225 descriptors with strong correlation which is greater than 0.95 are removed from Mutagenicity dataset. In the remaining 335 descriptors, we evaluate the importance to target activities through LightGBM. The results are shown in Fig. 2 and Fig. 3. We normalize the importance of each descriptor, which renders the total normalized importance as 1. From Fig. 2, we can observe that there exist numerous unimportant descriptors since the cumulative normalized descriptor importance exceeds 0.9 on 95 descriptors. Therefore, it is necessary to analyze the importance of descriptors and filter the unimportant descriptors for model building. In our experiments, we select 15 descriptors with the highest descriptor importance for Auto-Sklearn. Figure 3 shows these important descriptors for activities prediction of drug molecules and corresponding importance. From top to bottom, they are *AAC, PSA, MLOGP, MAXDN, BEHv1, N-078, MAXDN, nR2CHX, MAXDP, Jhetv, Hy, AMW, PCR, SIC2, BEHm1, 0–061* respectively.

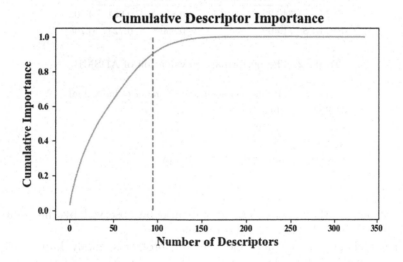

**Fig. 2.** The cumulative descriptor importance

To verify the effectiveness of Automated Descriptor Selection, we achieve Support Vector Machine (SVM) as machine learning classification model based on original descriptors and selected descriptors respectively. The comparison is shown in Table 1. SVM and SVM-FS are the short for SVM with original descriptors and selected descriptors respectively. We can observe that the prediction accuracy of SVM-FS is 16.74% higher than SVM. Moreover, most predictive outputs of SVM with original descriptors on test set are 0, which indicates SVM

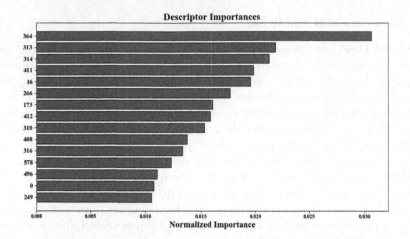

**Fig. 3.** The 15 descriptors with highest normalized importance

**Table 1.** The comparison of different descriptor set.

| Model | Prediction accuracy | Precision | Recall | F1 |
|---|---|---|---|---|
| SVM | 54.72% | 0.857 | 0.054 | 0.102 |
| SVM-FS | 71.46% | 0.690 | 0.724 | 0.706 |

**Table 2.** The performance evaluation of ADSMB.

| Model | Prediction accuracy | Number of individuals |
|---|---|---|
| RBFNN | 49.89% | – |
| SVC | 54.72% | – |
| ORNNEUD | 66.24% | 17 |
| ADSMB | **82.19%** | 19 |

faces a strong overfitting problem. It can also be observed by the minuscule Recall (0.054) and F1 (0.102) score in Table 1.

When tackling specific machine learning problems, many human experts require the off-the-shelf model. In order to choose a right classification algorithm or hyperparameter configuration, they tend to machine learning algorithms with high reputation such as SVC (SVM classifier) or neural network (NN) with simple hyperparameter optimization. Our proposed ADSMB can efficiently build high-quality machine learning model in an automatic manner without human intervention. It allows human experts to focus more on the dataset itself, which also facilitates better scientific discovery and analysis. Therefore, we compare the performance of ADSMB with the most used SVC, NN model and elaborate model by experts, ORNNEUD, for activities prediction. ORNNEUD [10] is an optimal neural network ensemble method with uniform design and selects 10 descriptors as input. The result is shown in Table 2. SVC and RBFNN are

**Table 3.** The top 5 weight individuals provided by our method.

| No | Individuals | Weight |
|---|---|---|
| 1 | balancing:strategy: weighting<br>classifier: adaboost<br>data_preprocessing:imputation: median<br>data_preprocessing:rescaling: minmax<br>adaboost:algorithm: SAMME<br>adaboost:learning_rate: 1.1224843045217607<br>adaboost:max_depth: 7<br>adaboost:n_estimators: 261 | 0.12 |
| 2 | balancing:strategy: weighting<br>classifier: adaboost<br>data_preprocessing:imputation: most_frequent<br>data_preprocessing:rescaling: quantile_transformer<br>quantile_transformer:n_quantiles: 1131<br>quantile_transformer:output_distribution: normal<br>adaboost:algorithm: SAMME.R<br>adaboost:learning_rate: 0.18265490638193532<br>adaboost:max_depth: 1<br>adaboost:n_estimators: 96 | 0.12 |
| 3 | balancing:strategy: none<br>classifier: random_forest<br>data_preprocessing:imputation: most_frequent<br>data_preprocessing:rescaling: normalize<br>feature_preprocessor: polynomial<br>polynomial:degree: 2<br>polynomial:include_bias: False<br>polynomial:interaction_only: False<br>random_forest:bootstrap: True<br>random_forest:criterion: gini<br>random_forest:max_depth: None<br>random_forest:max_features: 0.48772464140872207<br>random_forest:max_leaf_nodes: None<br>random_forest:min_impurity_decrease: 0<br>random_forest:min_samples_leaf: 1<br>random_forest:min_samples_split: 16<br>random_forest:min_weight_fraction_leaf: 0 | 0.1 |
| 4 | balancing:strategy: weighting<br>classifier: random_forest<br>data_preprocessing:imputation: median<br>data_preprocessing:rescaling: standardize<br>feature_preprocessor: select_rates<br>select_rates:alpha: 0.061500733991527654<br>select_rates:mode: fdr<br>select_rates:score_func: f_classif<br>random_forest:bootstrap: False<br>random_forest:criterion: gini<br>random_forest:max_depth: None<br>random_forest:max_features: 0.5804208006044023<br>random_forest:max_leaf_nodes: None<br>random_forest:min_impurity_decrease: 0<br>random_forest:min_samples_leaf: 5<br>random_forest:min_samples_split: 2<br>random_forest:min_weight_fraction_leaf: 0 | 0.1 |
| 5 | balancing:strategy: none<br>classifier: adaboost<br>data_preprocessing:imputation: median<br>data_preprocessing:rescaling: standardize<br>feature_preprocessor: select_percentile_classification<br>select_percentile_classification:percentile: 50.28741992371421<br>select_percentile_classification:score_func: f_classif<br>adaboost:algorithm: SAMME<br>adaboost:learning_rate: 0.12476367665786196<br>adaboost:max_depth: 2<br>adaboost:n_estimators: 459 | 0.1 |

built on original descriptors and the results of RBFNN and ORNNEUD are from
Ref. [10]. We can observe that it is difficult for a single model without descriptor
selection to obtain good performance for activities prediction of drug molecules.
Incorporating with Table 1, SVC-FS (71.46%) has higher prediction accuracy
than ORNNEUD (66.24%), which demonstrates that appropriate descriptor
selection is also important for this problem. Therefore, ADSMB, equipping
Auto-Sklearn with Automated Descriptor Selection, provides drastic improvement in prediction accuracy which is at least 15.95% higher than the competing
models. For number of individuals, our proposed method only increases 2 individuals in final ensemble model, which also demonstrates the feasibility and
effectiveness of ADSMB. Table 3 shows the top 5 weight individuals outputted
by our method. It can be observed that the activities prediction problem is more
inclined to select the machine learning algorithms based on tree as individuals.

In order to verify the stability of ADSMB, we repeatly test our proposed
method, and this validation process is repeated 10 times. Figure 4 shows the
result. We observe that ADSMB can maintain stable performance during 10
times of test and consistently achieve at least 13% higher prediction accuracy
than ORNNEUD.

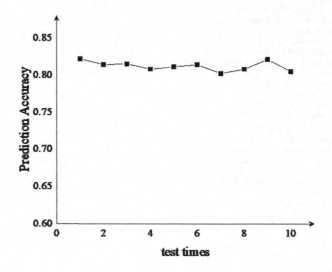

**Fig. 4.** The prediction accuracy of 10 test times

## 4    Conclusions

Machine learning is widely applied in QSAR for drug discovery. However, it is
difficult and time-consuming for domain experts to build a high-quality QSAR
model, as it requires extensive optimization with trial and error. Therefore, in
this paper, we propose a QSAR-special automated machine learning method

incorporating Automated Descriptor Selection with Automated Model Building (ADSMB) to build high-quality QSAR model in an efficient manner. Automated Descriptor Selection considers QSAR-special descriptor selection to filter the descriptors with unique value, high redundancy and low importance. Based on remaining descriptors, Automated Model Building automatically constructs high-quality ensemble model under Bayesian optimization through Auto-Sklearn. Experiments evaluations have been conducted on Mutagenicity dataset. The results show ADSMB can obtain at least 15.95% improvement in prediction accuracy and more stable performance than the competing methods. Our future work includes evaluating ADSMB on more QSAR datasets to demonstrate its generation ability and improving the embedded expert knowledge in Automated Descriptor Selection to obtain more appropriate descriptor subsets for QSAR model.

## Statement

This manuscript is original research that has not been previously published and is not currently in press, under review, or being considered for publication elsewhere, in whole or in part.

**Acknowledgement.** This work is supported by the National Key Research and Development Plan of China (Grant No.2016YFB1000600 and 2016YFB1000601) and the State Key Program of National Nature Science Foundation of China (Grant No. 61936001).

## References

1. Peter, S.C., Dhanjal, J.K., Malik, V., et al.: Quantitative Structure-Activity Relationship (QSAR), Modeling Approaches to Biological Applications (2019)
2. Ma, J., Sheridan, R.P., Liaw, A., et al.: Deep neural nets as a method for quantitative structure-activity relationships. J. Chem. Inf. Model. **55**(2), 263–274 (2015)
3. Zhang, L., Tan, J., Han, D., et al.: From machine learning to deep learning: progress in machine intelligence for rational drug discovery. Drug Discovery Today **22**(11), 1680–1685 (2017)
4. Lavecchia, A.: Machine-learning approaches in drug discovery: methods and applications. Drug Discovery Today **20**(3), 318–331 (2015)
5. Poorinmohammad, N., Mohabatkar, H., Behbahani, M., et al.: Computational prediction of anti HIV-1 peptides and in vitro evaluation of anti HIV-1 activity of HIV-1 P24-derived peptides. J. Pept. Sci. **21**(1), 10–16 (2015)
6. Jain, N., Gupta, S., Sapre, N., et al.: In silico de novo design of novel NNRTIs: a bio-molecular modelling approach. RSC Adv. **5**(19), 14814–14827 (2015)
7. Gupta, S., Basant, N., Singh, K.P.: Estimating sensory irritation potency of volatile organic chemicals using QSARs based on decision tree methods for regulatory purpose. Ecotoxicology **24**(4), 873–886 (2015). https://doi.org/10.1007/s10646-015-1431-y
8. Kumari, P., Nath, A., Chaube, R.: Identification of human drug targets using machine-learning algorithms. Comput. Biol. Med. **56**, 175–181 (2015)

9. Weidlich, I.E., Filippov, I.V., Brown, J., et al.: Inhibitors for the hepatitis C virus RNA polymerase explored by SAR with advanced machine learning methods. Bioorg. Med. Chem. **21**(11), 3127–3 137 (2013)

10. Liu, Y., Yin, Y., Teng, Z., Wu, Q., Li, G.: Activities prediction of drug molecules by using the optimal ensemble based on uniform design. In: Huang, D.-S., Wunsch, D.C., Levine, D.S., Jo, K.-H. (eds.) ICIC 2008. LNCS, vol. 5226, pp. 106–113. Springer, Heidelberg (2008). https://doi.org/10.1007/978-3-540-87442-3_15

11. Yao, Q., Wang, M., Chen, Y., et al.: Taking human out of learning applications: a survey on automated machine learning. arXiv preprint arXiv:1810.13306 (2018)

12. Katz, G., Shin, E.C.R., Song, D.: ExploreKit: automatic descriptor generation and selection. In: 16th International Conference on Data Mining, pp. 979–984. IEEE (2016)

13. Klein, A., Falkner, S., Bartels, S., et al.: Fast Bayesian optimization of machine learning hyperparameters on large datasets. In: Artificial Intelligence and Statistics, pp. 528–536 (2017)

14. Feurer, M., Klein, A., Eggensperger, K., et al.: Efficient and robust automated machine learning. In: Advances in Neural Information Processing Systems, pp. 2962–2970 (2015)

15. Olson, R.S., Urbanowicz, R.J., Andrews, P.C., Lavender, N.A., Kidd, L.C., Moore, J.H.: Automating biomedical data science through tree-based pipeline optimization. In: Squillero, G., Burelli, P. (eds.) EvoApplications 2016, Part I. LNCS, vol. 9597, pp. 123–137. Springer, Cham (2016). https://doi.org/10.1007/978-3-319-31204-0_9

16. Zoph, B., Le, Q.V.: Neural architecture search with reinforcement learning. arXiv preprint arXiv:1611.01578 (2016)

17. Liu, Y., Wu, J., Avdeev, M., Shi, S.: Multi-layer descriptor selection incorporating weighted score-based expert knowledge toward modelling materials with targeted properties. Adv. Theory Simul. **3**(2), 1900215 (2020)

# Survival Prediction of Glioma Tumors Using Feature Selection and Linear Regression

Jiewei Wu[1,2], Yue Zhang[2], Weikai Huang[2], Li Lin[1,2], Kai Wang[1(✉)], and Xiaoying Tang[2(✉)]

[1] School of Electronics and Information Technology, Sun Yat -sen University, Guangzhou, China
wangkai23@mail.sysu.edu.cn
[2] Department of Electrical and Electronic Engineering, Southern University of Science and Technology, Shenzhen, China
tangxy@sustech.edu.cn

**Abstract.** Early diagnosis of brain tumor is crucial for treatment planning. Quantitative analyses of segmentation can provide information for tumor survival prediction. The effectiveness of convolutional neural network (CNN) has been validated in medical image segmentation. In this study, we apply a widely-employed CNN namely UNet to automatically segment out glioma sub-regions, and then extract their volumes and surface areas. A sophisticated machine learning scheme, consisting of mutual information feature selection and multivariate linear regression, is then used to predict individual survival time. The proposed method achieves an accuracy of 0.475 on 369 training data based on leave-one-out cross-validation. Compared with using all features, using features obtained from the employed feature selection technology can enhance the survival prediction performance.

**Keywords:** Brain tumor segmentation · Feature selection · Survival prediction

## 1 Introduction

Glioma tumor is one of the most common brain malignancies, the median survival of which is 15 months for high-grade ones [1]. It is of great importance to automatically predict tumor survival in advance, for which magnetic resonance imaging (MRI) is a very useful tool [2]. MRI-based quantitative analyses of brain tumors can provide vital information for overall survival prediction, typically based on segmentations of brain tumors and their sub-regions.

Recently, CNN has proven its effectiveness in natural and medical image segmentation tasks when given a large amount of data. For brain tumor segmentation, there exists a publicly-available dataset of pre-operative multimodal MR images and their corresponding manual annotations provided by the multimodal Brain Tumor Segmentation (BraTS) challenge [3–5]. This dataset makes it feasible to build a fully-automated tumor segmentation model. Isensee et al.

© Springer Nature Singapore Pte Ltd. 2021
W. Gao et al. (Eds.): FICC 2020, CCIS 1385, pp. 85–92, 2021.
https://doi.org/10.1007/978-981-16-1160-5_8

demonstrated that a well trained UNet [6] can generate state-of-the-art tumor segmentations [7]. Inspired by Isensee's work, we apply UNet to segment out brain tumors and their sub-regions.

As a subsequent task of tumor segmentation, survival prediction has a very high clinical value in tumor prognosis and clinical decision-making [8,9]. In the BraTS challenge, additional data including age and resection status for a subset of the dataset has also been provided for the survival prediction task. Xue et al. proposed a linear regression model utilizing features extracted from tumor segmentations and this simple model outperformed other methods in the test phase of the BraTS 2018 challenge [9]. Building basis on top of Xue's work, we extract two more features and perform effective feature selection to improve the survival prediction accuracy.

## 2   Materials and Methods

### 2.1   Dataset

The dataset used in this study are obtained from the BraTS 2020 challenge [3–5]. For the tumor segmentation task, there are 369 training cases and 125 validation cases. For each case, there are MR images of four different modalities, i.e., T1-weighted (T1), contract enhanced T1-weighted (T1c), T2-weighted (T2), and Fluid Attenuation Inversion Recovery (FLAIR) images. All cases have been segmented into three regions, i.e., necrotic core and non-enhancing tumor (NCR/NET–label 1), edema (ED–label 2), and enhancing tumor (ET–label 4). The whole tumor (WT) is a combination of label 1, label 2, and label 4, and the tumor core (TC) is comprised of label 1 and label 4. The Dice scores of WT, TC, and ET are used as metrics to evaluate the performance of the segmentation model. Five fold cross-validation experiments are conducted on the training data.

For the survival prediction task, there are 236 training cases and 29 validation cases. Each case is classified into one of three classes, namely short-survivors ($< 10$ months), mid-survivors (between 10 and 15 months), and long-survivors ($> 15$ months). The age and resection status of each train case are also provided. In terms of resection status, some cases are labeled as NA, suggesting that their resection states are not available. For all other cases, the resection status is either Gross Total Resection (GTR) or Subtotal Resection (STR). The survival is given in days. Accuracy and mean error are used as a metric to evaluate the prediction performance. To get reliable and robust results, we conduct leave-one-out cross-validation (LOOCV) analysis on the training data rather than the validation data because the sample size of the validation data is too small.

### 2.2   Tumor Segmentation and Feature Extraction

In this study, a multi-channels UNet is utilized to segment out glioma sub-regions. MR images with four modalities (i.e. T1, T1c, T2 and FLAIR) are jointly inputted to our segmentation model (namely UNet), and their corresponding human annotations are treated as the expected output. The flowchart of the segmentation procedure is shown in Fig. 1.

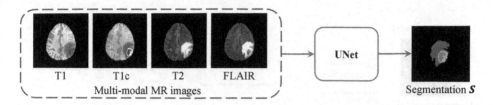

**Fig. 1.** Flowchart of the segmentation procedure.

After obtaining an automated segmentation $S$, we extract certain imaging and non-imaging features. Xue et al. [9] calculated the volumes and surface areas of several regions (i.e. NCR/NET, ED, and ET). Age and resection status were also used as two additional features and inputted to a subsequent regression model in Xue's work. In addition to Xue's work, we also calculate the volume and surface area of WT and input to the subsequent regression model. Different from that in Xue et al. [9] which used a two-dimensional feature vector to represent the resection status, we use 1, 2, and 3 to respectively represent NA, GTR, and STR.

## 2.3 Feature Seletion and Regression

Feature selection is effective in preparing data for feature-based machine learning problems, and it can be used to automatically remove irrelevant features to improve the overall performance and a model's generalization ability [10].

Univariate feature selection strategy is employed in our experiments. We estimate the weight of each feature, sort features according to their weights, and then identify the top $K$ features to be our input features. Two scoring methods are employed in this study, namely univariate linear regression and mutual information based [11–13]. Univariate linear regression calculates an F-value between the labels and each feature. In our experiments, we use *f_regression* function from Scikit-learn [14] to obtain the F-value between the survival time and each feature, and use them as the features's weight. For the second scoring method, the mutual information between the labels and each feature is calculated and used as the feature's weights. We use *mutual_info_regression* function from Scikit-learn [14] to estimate the mutual information between each feature and the survival time.

In predicting individual survival, we simply apply a multivariate linear regression. As shown in Fig. 2, the baseline model proposed by Xue et al. [9] consists of tumor segmentation, feature extraction, and linear regression. In addition to that baseline model, we perform the aforementioned feature selection before inputting features into the linear regression model and obtain the newly predicted survival $D^*$. In another experiment setting, the linear regression is replaced with decision tree regression as we want to investigate whether other regression methods can improve the prediction performance.

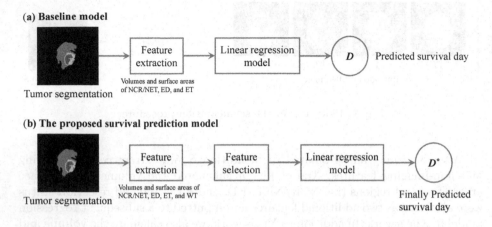

**Fig. 2.** The entire procedure of the proposed survival prediction pipeline.

## 2.4   Implementation Details

All experiments are implemented in python. The segmentation model builds its basis on nnUNet[1] [15]. The proposed survival prediction model is based on Xue's work[2] [9]. The feature selection strategy is conducted using Scikit-learn [14].

## 3   Results

For brain tumor segmentation, we obtain mean Dice scores of 91.42%, 86.75%, and 78.76% for WT, TC, and ET on the training data. Figure 3 shows a representative segmentation result of the training data. Obviously, our automated segmentation results are highly consistent with the manually annotated ones.

In our experiments, the baseline features are the features extracted by Xue et al. [9], and the WT features refer to the volume and surface area of WT. Table 1 shows the accuracies and mean errors of different experiment settings. When using linear regression as the regression model, compared with using baseline features proposed by Xue et al. [9], adding the volume and surface area of WT improves the prediction accuracy from 0.445 to 0.458, indicating that the WT features can provide complementary information for survival prediction. Moreover, when incorporating feature selection strategy, the accuracy is further promoted from 0.458 to 0.475. However, decision tree regression does not improve

---

[1]  https://github.com/MIC-DKFZ/nnUNet.
[2]  https://github.com/xf4j/brats18.

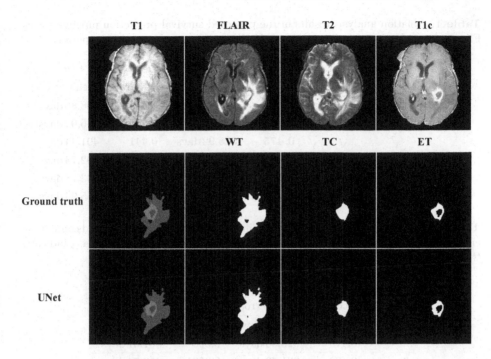

**Fig. 3.** A representative example of segmentation results of the training data.

the prediction performance, especially in mean error, when the other settings are the same. In addition, similar papttern is also observable when directly using groudtruth as the segmentation results.

Table 2 shows the results of different feature selection methods when considering all ten features and using linear regression as the regression method. The best performance in terms of each metric is highlighted. As shown in Table 2, the proposed pipeline, with mutual information being the scoring method and the number of selected features ($K$) being eight or night, achieves a prediction accuracy of 0.475. Compared with using all features, the accuracy is increased from 0.458 to 0.475, indicating feature selection is effective for improving the survival prediction performance. The selected features of the proposed pipeline are shown in Table 3.

There are several other interesting findings that are worthy of mentioning. Firstly, when using decision tree regression as the regression model, the accuracy does not be increased. Secondly, when using univariate regression as the scoring method, the feature selection strategy does not enhance the prediction accuracy. Thirdly, when using mutual information as the scoring method, the feature selection strategy yields the highest accuracy when the value of $K$ is eight or night. When using a $K$ value less than eight, the prediction performance is worse than using all ten features. This is sort of reasonable given that ten features are indeed not very redundant and thus not many features to abandon.

**Table 1.** Ablation analysis results of the proposed survival prediction pipeline. Keys: FS - feature selection. ↓ indicates that a smaller value represents a better performance.

|  | Baseline | WT | FS | Linear regression | | Decision tree regression | |
|---|---|---|---|---|---|---|---|
|  |  |  |  | Accuracy | Mean error ↓ | Accuracy | Mean error ↓ |
| UNet | √ |  |  | 0.445 | 327.10 days | 0.419 | 508.86 days |
|  | √ | √ |  | 0.458 | **325.20** days | 0.377 | 536.92 days |
|  | √ | √ | √ | **0.475** | 325.92 days | 0.441 | 491.44 days |
| Groudtruth | √ |  |  | 0.436 | 323.75 days | 0.424 | 492.73 days |
|  | √ | √ |  | 0.441 | 324.09 days | 0.407 | 533.78 days |
|  | √ | √ | √ | **0.483** | **323.27** days | 0.407 | 485.99 days |

**Table 2.** Survival prediction results, with different feature selection methods and different numbers of selected features (number), obtained from the training data. ↓ indicates that a smaller value represents a better performance.

| Number | Univariate regression | | Mutual information | |
|---|---|---|---|---|
|  | Accuracy | Mean error ↓ | Accuracy | Mean error ↓ |
| 1 | 0.390 | 335.02 days | 0.322 | 359.14 days |
| 2 | 0.415 | 328.00 days | 0.356 | 351.07 days |
| 3 | 0.403 | 331.49 days | 0.360 | 352.74 days |
| 4 | 0.411 | 333.20 days | 0.377 | 350.43 days |
| 5 | 0.407 | 332.67 days | 0.407 | 347.61 days |
| 6 | 0.407 | 334.01 days | 0.407 | 344.28 days |
| 7 | 0.403 | 337.83 days | 0.415 | 345.49 days |
| 8 | 0.441 | 325.31 days | **0.475** | 325.92 days |
| 9 | 0.458 | 325.20 days | **0.475** | 325.92 days |
| 10 (All) | 0.458 | 325.20 days | 0.458 | **325.20** days |

**Table 3.** The selected features of the proposed survival prediction pipeline, sorted by the weights of features from largest to smallest.

| Index ↓ | Feature |
|---|---|
| 1 | Surface area of ED |
| 2 | Volume of WT |
| 3 | Age |
| 4 | Volume of NCR/NET |
| 5 | Surface area of ET |
| 6 | Surface area of WT |
| 7 | Volume of ED |
| 8 | Surface area of NCR/NET |
| 9 | Volume of ET |

# 4  Discussion and Conclusion

In this study, we proposed and validated a tumor segmentation-based survival prediction pipeline. For the survival prediction task, we extracted eight imaging features from brain tumor segmentation and combined them with two non-imaging features (i.e. age and resection status). Multivariate linear regression with feature selection was used to predict the survival time. The experiment results show that a combination of feature selection with the scoring method being mutual information and the number of finally selected features being eight or night yields the best prediction performance.

In our experiments, we found that using decision tree regression decreases the prediction performance compared with using linear regression. It might imply that the relationship between the features and their corresponding survival time is more like linear rather than non-linear. We also noticed that adding the volume and surface area of WT can enhance the prediction accuracy when using a linear regression model. However, the prediction performance decreases when using decision tree regression. It was probably because WT has linear correspondences with survival time. Moreover, we found that feature selection with the scoring method being mutual information can further improve the prediction accuracy when using linear regression. And our results suggested that feature selection with the scoring method being mutual information is superior to that with the scoring method being multivariate regression.

Recently, Yang et al. ensembled CNN-based regression and random forest regression, and yielded a prediction accuracy of 0.475 [16]. Gates et al. extracted image features from brain tumor segmentations and applied a cox model to achieve an accuracy of 0.445 for predicting survival days [17]. Compared with the above methods, our proposed pipeline achieved comparable results.

There are two potential limitations of this work. Firstly, we only utilized the volumes and surface areas of NCR/NET, ED, ET, and WT in our experiments. It is possible to further enhance the survival prediction performance by extracting new features from brain tumor segmentations and adding them to the proposed pipeline. Secondly, we only considered two types of univariate feature selection techniques in our experiments. In the future, we aim to perform other feature selection methods to enhance our proposed pipeline.

**Acknowledgement.** This work was supported by the Shenzhen Basic Research Program (JCYJ20190809120205578), the National Key R&D Program of China (2017YFC0112404), and the National Natural Science Foundation of China (81501546).

# References

1. Stupp, R., Mason, W.P., et al.: Radiotherapy plus concomitant and adjuvant temo-zolomide for glioblastoma. N. Engl. J. Med. **352**(10), 987–996 (2005)
2. Kumar, V., Gu, Y., Basu, S., Berglund, A., et al.: Radiomics: the process and the challenges. Magn. Reson. Imaging **30**(9), 1234–1248 (2012)

3. Bakas, S., et al.: Advancing the cancer genome atlas glioma MRI collections with expert segmentation labels and radiomic features. Sci. Data **4**, 170117 (2017)
4. Menze, B.H., et al.: The multimodal brain tumor image segmentation benchmark (BraTS). IEEE Trans. Med. Imaging **34**(10), 1993–2024 (2015)
5. Bakas S., Reyes M., Jakab A., et al.: Identifying the best machine learning algorithms for brain tumor segmentation, progression assessment, and overall survival prediction in the BRATS challenge. arXiv preprint arXiv:1811.02629 (2018)
6. Ronneberger, O., Fischer, P., Brox, T.: U-net: Convolutional networks for biomedical image segmentation. In: Navab, N., Hornegger, J., Wells, W.M., Frangi, A.F. (eds.) MICCAI 2015. LNCS, vol. 9351, pp. 234–241. Springer, Cham (2015). https://doi.org/10.1007/978-3-319-24574-4_28
7. Isensee, F., Jger, P. F., Kohl, S. A., et al.: Automated design of deep learning methods for biomedical image segmentation. arXiv preprint arXiv:1904.08128 (2020)
8. Sun, L., Zhang, S., Luo, L.: Tumor segmentation and survival prediction in Glioma with deep learning. In: Crimi, A., Bakas, S., Kuijf, H., Keyvan, F., Reyes, M., van Walsum, T. (eds.) BrainLes 2018. LNCS, vol. 11384, pp. 83–93. Springer, Cham (2019). https://doi.org/10.1007/978-3-030-11726-9_8
9. Feng, X., Tustison, J., Patel, H., et al.: Brain tumor segmentation using an ensemble of 3D U-nets and overall survival prediction using radiomic features. Front. Comput. Neurosci. **14**, 25 (2020)
10. Li, J., Cheng, K., Wang, S., et al.: Feature selection: a data perspective. ACM Comput. Surv. (CSUR) **50**(6), 1–45 (2017)
11. Kozachenko, L.F., Leonenko, N.N.: Sample estimate of the entropy of a random vector. Problemy Peredachi Informatsii **23**(2), 9–16 (1987)
12. Kraskov, A., Stögbauer, H., Grassberger, P.: Estimating mutual information. Phys. Rev. E **69**(6), 066138 (2004)
13. Ross, B.C.: Mutual information between discrete and continuous data sets. PloS one **9**(2), e87357 (2014)
14. Pedregosa, F., Varoquaux, G., Gramfort, A., et al.: Scikit-learn: machine learning in Python. J. Mach. Learn. Res. **12**, 2825–2830 (2011)
15. Isensee, F., Kickingereder, P., Wick, W., Bendszus, M., Maier-Hein, K.H.: No new-net. In: Crimi, A., Bakas, S., Kuijf, H., Keyvan, F., Reyes, M., van Walsum, T. (eds.) BrainLes 2018. LNCS, vol. 11384, pp. 234–244. Springer, Cham (2019). https://doi.org/10.1007/978-3-030-11726-9_21
16. Yang, H.-Y., Yang, J.: Automatic brain tumor segmentation with contour aware residual network and adversarial training. In: Crimi, A., Bakas, S., Kuijf, H., Keyvan, F., Reyes, M., van Walsum, T. (eds.) BrainLes 2018. LNCS, vol. 11384, pp. 267–278. Springer, Cham (2019). https://doi.org/10.1007/978-3-030-11726-9_24
17. Gates, E., Pauloski, J.G., Schellingerhout, D., Fuentes, D.: Glioma segmentation and a simple accurate model for overall survival prediction. In: Crimi, A., Bakas, S., Kuijf, H., Keyvan, F., Reyes, M., van Walsum, T. (eds.) BrainLes 2018. LNCS, vol. 11384, pp. 476–484. Springer, Cham (2019). https://doi.org/10.1007/978-3-030-11726-9_42

# AI and Big Data

# Root Cause Localization from Performance Monitoring Metrics Data with Multidimensional Attributes

Bo Zhou[✉], Ping Zhang, and Runlin Zhou

National Computer Network Emergency Response Technical Team/Coordination Center
of China, Xinjiang Branch, Beijing, Xinjiang, China
{Zhoubo,zhourunlin}@cert.org.cn, lucia619@163.com

**Abstract.** Time series data with multidimensional attributes (such as network
flow, page view and advertising revenue) are common and important performance
monitoring metrics in large-scale Internet services. When the performance monitoring metrics data deliver abnormal patterns, it is of critical importance to timely
locate and diagnose the root cause. However, this task remains as a challenge
due to tens of thousands of attribute combinations in search space. Moreover,
faults will propagate from one attribute combination to another, resulting in a subtle mathematical relationship between the different attribute combinations. Only
after spotting the root cause from the huge search space, one can take appropriate
actions to mitigate the problem and keep the system running uninterrupted. In
this paper, we propose a novel root cause localization algorithm to identify the
attribute combinations most likely to blame. Tests on the real-world data from an
Internet company show that this algorithm achieves an averaged F-score over 0.95
with a localization time less than 30 s.

**Keywords:** Root cause localization · Fault diagnosis · Multidimensional
attributes · Heuristic search · Discrete optimization

## 1 Introduction

During the daily management of operations and maintenance in large Internet companies, huge amounts of performance monitoring data on various levels are collected and
stored. The historical and real-time data empower big data analytics related tasks such as
fault detection and localization, performance management [1, 2]. Particularly, in order
to ensure the stable and efficient operation of information systems and meet the strict
requirements on service quality, system operators must monitor a massive of key performance indicators (KPIs) on a regular basis [3, 4]. Time series with multidimensional
attributes is a common and important type of such KPI. These records usually possess
more than one attribute. At every sampling time, the total KPI can be decomposed into
separate groups with different attribute combinations and the attribute is characterized
by a range of discrete values in each dimension.

© Springer Nature Singapore Pte Ltd. 2021
W. Gao et al. (Eds.): FICC 2020, CCIS 1385, pp. 95–106, 2021.
https://doi.org/10.1007/978-981-16-1160-5_9

Here, an example is given to illustrate the corresponding terminologies involved in this paper. Page View (PV) of a website (the number of accesses) is a total KPI that has a set of associated attributes, such as Province, Internet Service Provider, Data Center, User Agent and so on, as depicted in Table 1. Each attribute has many possible values and describes one certain aspect of the KPI value. For example, the attribute Province specifies the geographical location of visitor's IP and its values can be one of the provinces of China, e.g., Beijing, Shanghai, Guangzhou, etc. The Data Center depicts the cluster where that request is served (e.g., DC1, DC2, DC3, etc.).

When the total KPI suffers from a sudden change, it is the system operators' duty to diagnose which combination of attribute values has the most potential to contribute to the increase or decrease of the total KPI [5]. Such analyses help the system operators to take proactive actions before anomalies cause costly consequences in the system [6]. The challenges in root cause analysis come from two aspects. On the one hand, the root cause can be combinations of attribute values in any attribute dimensions. Moreover, in each attribute dimension, the root cause may consist of several distinct attribute values. Thus, exhaustive enumeration of all the possibilities is computationally prohibited. On the other hand, because of the additive property, the KPI value of the attribute combinations that include or are included in the true root cause will also be affected, although the former is more coarse-grained and the latter is more fine-grained.

**Table 1.** Terms and examples.

| Term | Definition | Example |
| --- | --- | --- |
| Attributes | The categorical information of each KPI record | Province, Internet Service Provider, Data Center, User Agent |
| Attribute values | The possible values for each attribute | IE7.0, Firefox30.0, Chrome67.0 for Client Agent |
| Dimension combination | The attribute dimensions | (Province, *, *, *) for the first dimension: (Province, Internet Service Provider, *, *) for the first two dimensions |
| Attribute combinations | A set of attribute values in some dimension combinations | {(Beijing, *, DC1, *); (Shanghai, *, DC2, *)} for dimension combination (Province, *, Data Center, *) |
| KPI record | An original record of the time series data with timestamp, attributes and KPI value | 08:00:00; Beijing, Mobile, DC1, IE7.0; 103.46 |
| KPI value | The time series value of one attribute combination | The KPI value of (Beijing, *, DC1, *) is the sum of KPI values of KPI records whose Province attribute is Beijing and Data Center attribute is DC1 |

In the real world, the root cause set can be rather complicated. However, exiting works can only cope with simple cases. Idice [7] is tailored to the situation where the root cause set contains less than two attribute combinations. Adtributor [8] can only deal with root causes of anomalous cases in a single attribute dimension. Hotspot [9] is another anomaly localization framework proposed for additive KPIs with multidimensional attributes, but both the object function and the search algorithm can be further revised to improve its accuracy and robustness.

To handle the above challenges, we firstly propose a novel indicator that quantitatively characterizes the potential score of a root cause set. This indicator captures the intrinsic system behaviors of the multidimensional additive KPI affected by root causes. Based on this new indictor, a heuristic search framework consisted of breadth-first search algorithm and genetic algorithm is utilized to uncover the most possible root cause for the KPI change. The main contributions of this paper can be summarized as follows. (1) A new objective function is developed, which can measure the potential of a set of attribute combinations to be the true root cause. The physical explanation behind the objective function is given and analyzed. (2) A heuristic search framework combining breadth-first search algorithm, genetic algorithm and pruning strategy is used to reduce the search space effectively. (3) Tests on real-world data from an Internet company show that F-measure values of our proposed approach are above 0.95 with an averaged run time less than 30 s.

## 2    Problem Formulation

The KPI data involved in this paper have two characteristics. Firstly, they are time series with temporal order, which means each sample possesses a timestamp attribute [10]. Secondly, they are multidimensional data where each record has many multidimensional other attributes. Such a combination of time series data is commonly seen in many Internet companies.

Most of the time, the total KPI values as well as the sums of KPI values in the combined attributes follow normal patterns. However, sometimes the KPI values in some attribute combinations may deliver a sudden decrease (or increase). This could be due to a faulty configuration updated on multiple data centers, or the network transmission failure in a province. In this case, site reliability engineers are supposed to identify the root cause of abnormalities as soon as possible, i.e., figure out which attribute combinations contribute most to the abnormalities and provide useful information to isolate the problem.

### 2.1    Problem Statement

The problem of root cause localization from time series data with multidimensional attributes is to identify the effective attribute combinations that can explain the anomalous change of total KPI value. In this paper, attribute combinations in a root cause set are assumed to be in the same dimension combination, since other cases are rare in reality.

In Table 2, a simple example of a two-dimensional KPI with attributes of Province and User Agent is adopted to clarify the problem. In this case, there exist three dimension combinations: (Province, *), (*, User Agent) for one dimension and (Province, User

Agent) for two dimensions. Each dimension combination contains a set of attribute combinations, e.g., (Province, *) contains {(Beijing, *), (Shanghai, *)}; {*,User Agent} contains {(*, IE7.0), (*, Firefox30.0)}; (Province, User Agent) contains {(Beijing, IE7.0), (Beijing, Firefox30.0), (Shanghai, IE7.0), (Shanghai, Firefox30.0)}.

At time 08:00, the KPI value is under normal condition. One minute later at 08:01, the KPI value of (Beijing, IE7.0) changes from 10 to 100 and that of (Beijing, Firefox30.0) changes from 20 to 200. As one can note, the KPI values of other attribute combinations (Beijing, *), (*, IE7.0) and (*, Firefox30.0) are also impacted. The total KPI value suffers from an anomalous change from 75 (10 + 20 + 30 + 15 = 75) to 345 (100 + 200 + 30 + 10 = 345), which is above the monitoring threshold. Thus, Operators need to look through the entire KPI records, compare the KPI values at different timestamps and determine which attribute combinations are the most possible root cause for this anomaly.

**Table 2.** Example of root cause localization.

| Timestamp | Province | User agent | KPI value |
|-----------|----------|------------|-----------|
| 08:00 | Beijing | IE7.0 | 10 |
| 08:00 | Beijing | Firefox30.0 | 20 |
| 08:00 | Shanghai | IE7.0 | 30 |
| 08:00 | Shanghai | Firefox30.0 | 15 |
| 08:01 | Beijing | IE7.0 | **100** |
| 08:01 | Beijing | Firefox30.0 | **200** |
| 08:01 | Shanghai | IE7.0 | 30 |
| 08:01 | Shanghai | Firefox30.0 | 15 |

## 2.2 Challenge

There are two challenges in identifying the root cause set. The first challenge is efficiency. The number of possible attribute combinations could be explosively huge if there are many attributes while each attribute has many different values. For example, if there are 100 different attribute combinations in one dimension combination, then the number of all the possible sets of attribute combinations will be $2^{100} - 1$. Investigating all the possible sets one by one is infeasible. The second challenge is effectiveness. The entire set of dimension combinations forms a layer structure, as depicted in Fig. 1. Each node denotes a dimension combination, each edge denotes a parent-and-child relationship. Attribute combinations in different dimension combinations may share the same set of KPI records, e.g., (Beijing, IE7.0) belongs to both (Province, *) and (*, User Agent). As a result, the faults may propagate from one attribute combination to another, i.e., if the KPI values of KPI records where Province is Beijing increase, then the KPI values of KPI records where User Agent is IE7.0 will also increase due to the fact that (Beijing, IE7.0) belongs to both (Province, *) and (*, User Agent). Thus, it is not an easy task to distinguish which attribute combinations are the true root cause.

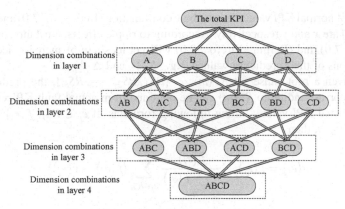

**Fig. 1.** Structure of dimension combinations.

## 3 Proposed Approach

To tackle the two challenges discussed in Sect. 2.2, we propose an approach to effectively identify possible attribute combinations as the root cause automatically. Two core strategies have been designed to achieve this goal: (1) A new objective function is proposed to measure the possibility and potential of attribute combinations to be the real root cause. This function serves as a guide for the search process. (2) A heuristic search based on genetic algorithm is adopted to significantly reduce the search space and avoid enumerating all the possible combinations. In the following subsection, we will present these two strategies in detail.

### 3.1 Definition of Objective Function

In the definition of objective function, three types of KPI values are indispensable: the actual fault KPI values $v \in R^n$, the predicted normal KPI values $f \in R^n$, and the predicted fault KPI values $a \in R^n$, where $n$ is the number of the KPI records at each timestamp.

When an anomaly in the total KPI is detected at time $t_i$ (We assume the fault detection algorithms are already available and interested readers may refer to references [11, 12] for further information), the actual fault KPI values $v$ at time $t_i$ denote the original KPI data recorded by the monitoring instrument. The forecast KPI values $f$ are intermediate values calculated under the assumption that the system is still under normal condition and no anomaly occurs to the KPI data at time $t_i$. Many time series predicting algorithm are well-qualified for this task, e.g., AR, MA, ARMA models. In this paper, we simply take the means of historical KPI values of several adjacent KPI records before time as the predicted values. The predicted fault KPI values $a$ are calculated under the assumption that the root causes are given and the rules of anomaly propagation follow "ripple effect" [9].

Ripple effect describes how the change of the KPI values in one attribute combination causes the change of KPI values of subsets of this attribute combination. For example, if the sum of KPI values for attribute combination (Beijing, *) increases by $\Delta$ and the predictions of normal KPI values for attribute combination (Beijing, IE 7.0) and (Beijing, Firefox30.0) are $x$ and $y$ respectively. According to ripple effects, attribute combination (Beijing, IE 7.0) and (Beijing, Firefox30.0) will get its share of increase according to the proportions of their prediction values, e.g., $\Delta \frac{x}{x+y}$ and $\Delta \frac{y}{x+y}$.

Thus, given a root cause set $RS(m) = \{RS_1, RS_2, \cdots, RS_m\}$, the predicted fault KPI values corresponding to $RS(m)$ can be obtained from actual fault KPI value $v$ and predicted normal KPI value $f$ in the following formulas. If $e_j \in RS_i$, $f(e_j) \neq 0$, $1 \leq j \leq n$, $1 \leq i \leq m$, then

$$a(e_j) = f(e_j) - \frac{f(e_j)}{\sum\limits_{e_j \in RS_i} f(e_j)} \sum\limits_{e_j \in RS_i} (f(e_j) - v(e_j)) \tag{1}$$

If $e_j \in RS_i, f(e_j) = 0$, $1 \leq j \leq n$, $1 \leq i \leq m$, then

$$a(e_j) = v(e_j) \tag{2}$$

For those $e_j \notin RS_i$, $1 \leq j \leq n$, $1 \leq i \leq m$,

$$a(e_j) = f(e_j) \tag{3}$$

The notation $e_j \in RS_i$ means that the attribute combinations in KPI record $e_j$ are subsets of attribute combination $RS_i$. Equation (1) can be reduced to $a(e_j) = \frac{f(e_j)}{\sum\limits_{e_j \in RS_i} f(e_j)} \sum\limits_{e_j \in RS_i} v(e_j)$, where one can note how the real fault KPI values $\sum\limits_{e_j \in RS_i} v(e_j)$ propagate to the predicted KPI values $a(e_j)$ with its proportional share of $\sum\limits_{e_j \in RS_i} f(e_j)$. It is worthy of note that the predicted KPI value $a$ will be updated $m$ times with Eq. (1) for the $m$ attribute combinations in the root cause set $RS(m)$.

Now we can define an objective function for the root cause set $RS(m) = \{RS_1, RS_2, \cdots, RS_m\}$ based on predicted normal KPI value $f$, predicted fault KPI value $a$ and actual fault KPI value $v$ as

$$score(RS(m)) = \frac{score1 \times score2}{m^c} \tag{4}$$

where $score1 = \frac{\sum\limits_{e_j \in RS(m)} |v(e_j) - f(e_j)|}{\|v - f\|_1}$, $score2 = 1 - \frac{\|v - a\|_1}{\|v - f\|_1}$, $m$ is the number of the attribute combinations and $c > 0$ is a free coefficient that can be tuned for different data set.

The objective function defined in Eq. (4) has the following properties: (1) If a new attribute combination $RS_{m+1}$ that is not a member of the true root cause set is added to $RS(m)$, resulting in $RS(m+1) = \{RS_1, \cdots, RS_m, RS_{m+1}\}$, then $score(RS(m)) > score(RS(m+1))$; (2) if an existing attribute combination $RS_i$, $1 \leq i \leq m$ that is a member of the true cause set is removed from $RS(m)$, resulting in $RS(m-1) = \{RS_1, \cdots, RS_{i-1}, RS_{i+1}, \cdots, RS_m\}$, then $score(RS(m)) > score(RS(m-1))$. In the following subsection, we will explain the psychical meanings behind the objective function and why it has the above two properties.

## 3.2  Explanations of Objective Function

The terms $v(e_j)$ and $f(e_j)$ in $score1$ correspond to real fault KPI value and predicted normal KPI value, thus the difference between them $|v(e_j) - f(e_j)|$ measures the amplitude of the fault and its contribution to the change of the total KPI value. Therefore, $score1$ of a set $RS(m)$ is a normalized metric ranging from 0 to 1. If a root cause set has a higher $score1$, then it will be considered to cover more KPI records that are affected by anomalies.

$score2$ can be further rewritten as

$$score2 = \frac{\left( \sum_{e_j \in RS(m)} (|v(e_j) - f(e_j)| - |v(e_j) - a(e_j)|) + \sum_{e_j \notin RS(m)} (|v(e_j) - f(e_j)| - |v(e_j) - a(e_j)|) \right)}{\|v - f\|_1} \quad (5)$$

If the KPI value prediction based on ripple effect is completely accurate, i.e., $a(e_j) = v(e_j), \forall e_j \in RS(m), a(e_j) = f(e_j), \forall e_j \notin RS(m)$ then $score2$ and the objective function can be reduced as

$$score2 = \frac{\sum_{e_j \in RS(m)} |v(e_j) - f(e_j)|}{\|v - f\|_1} \quad (6)$$

$$score(RS(m)) = \frac{\left( \sum_{e_j \in RS(m)} |v(e_j) - f(e_j)| \right)^2}{\|v - f\|_1^2 m^c} \quad (7)$$

where $score2$ score is exactly the same as $score1$. Then it is clear that the term $\sum_{e_j \in RS(m)} (|v(e_j) - f(e_j)|)$ in Eq. (5) measures the contribution of root cause set $RS(m)$ to amplitude of anomalies (same as $score1$) and the term $\sum_{e_j \in RS(m)} (|v(e_j) - a(e_j)|)$ in Eq. (5) measures the degree to which the KPI values of KPI records in root cause set $RS(m)$ match the ripple effect assumption.

In the objective function, there exists a penalty term $m^c$ for attribute combination number. Comparing $score(RS(m)), RS(m) = \{RS_1, \cdots, RS_i, \cdots, RS_m\}$ with $score(RS(m-1)), RS(m) = \{RS_1, \cdots, RS_{i-1}, RS_{i+1}, \cdots, RS_m\}$ as

$$\frac{score(RS(m))}{score(RS(m-1))}$$

$$= \left( \frac{\sum_{e_j \in RS(m)} (|v(e_j) - f(e_j)|)}{\sum_{e_j \in RS(m-1)} (|v(e_j) - f(e_j)|)} \right)^2 \left( \frac{m-1}{m} \right)^c, \text{it can be concluded}$$

$$= \left( \frac{\sum_{e_j \in RS(m-1)} (|v(e_j) - f(e_j)|) + \sum_{e_j \in RS_i} (|v(e_j) - f(e_j)|)}{\sum_{e_j \in RS(m-1)} (|v(e_j) - f(e_j)|)} \right)^2 \left( \frac{m-1}{m} \right)^c$$

that $score(RS(m)) > score(RS(m-1))$ holds if and only if

$$\sum_{e_j \in RS_i} |v(e_j) - f(e_j)| > \left( \left( \frac{m-1}{m} \right)^{\frac{c}{2}} - 1 \right) \sum_{e_j \in RS(m-1)} ||v(e_j) - f(e_j)|| \tag{8}$$

That is, the necessary and sufficient condition of adding attribute combination $RS_i$ to the candidate root cause set is that the change of KPI values in $RS_i$ is lager then a fraction of the change of KPI values in $RS(m-1)$, which equals to $\left( \frac{m-1}{m} \right)^{\frac{c}{2}} - 1$.

### 3.3 Heuristic Search Framework

The search process consists of three parts.

a) A breadth-first search from lower lay of dimension combinations to higher layer of dimension combinations. The dimension combinations in the time series follow a multi-layer hierarchy structure, as depicted in Fig. 1. The lower layer with less attribute dimensions will be searched first. In each layer, we search the attribute combinations in each dimension combination and locate the root cause set of attribute combinations with the best objective function. The best objective function as well as the corresponding set will be stored and then updated after searching each dimension combination.

b) Selection of the *TOPN* attribute combinations in each dimension combination. The number of attribute combinations in one dimension combination can be more than tens of thousands. To cope with the huge amount of searching space, the *TOPN* attribute combination individuals $RS_i^{TOP}$ ($1 \leq i \leq N$) with the highest *score*1 will be selected out. In this step, the unlikely elements with a relatively low *score*1 will be eliminated, as the KPI values of these attribute combination remain almost unchanged, meaning that these attribute combinations are not affected by the fault.

**Remark 1.** In this step, one can prune some dimension combinations to avoid further search in step (c).If every attribute combination in one dimension combination has a very low *score*1, then it is unlikely that the true cause set is in this dimension combination which can be removed from search space.

c) A genetic-algorithm-based search from the *TOPN* attribute combinations. A brute force search from *TOPN* attribute combinations is still inefficient, recalling that the search space of the $N$ attribute combination candidates is $2^N - 1$.In order to further reduce the computation load and improve its online performance, the genetic algorithm is utilized. The genetic algorithm is a solution of discrete optimization problems based on the process of natural selection process that mimics biological evolution. It performs well on large and complex data sets through an iterating process of a population of individual solutions [13].

In this paper, each individual of the population is encoded as a binary bit string of length $N$ where the $i$-th bit denotes whether the $i$-th attribute combination in the *TOPN* arrange is chosen into the root cause set: "1" means that the attribute combination is included and "0" means that it is excluded, i.e., the compositions of the root cause set are mathematically represented by a binary string as

$$s = s_1 \cdots s_i \cdots s_N, s_i \in \{0, 1\}, s_i = 1 \ iff \ RS_i^{TOP} \in RS, 1 \leq i \leq N \qquad (9)$$

The fitness function of an individual $s$ is the objective function defined in Eq. (4) with the root cause set $RS$ characterized by string $s$ in Eq. (9).

Once the population of individual solutions is initialized, the genetic algorithm leverages the process of genetic crossover and mutation to optimize the process of evolution. It is a selection process that the individual solution of a higher fitness will tend to survive, i.e., the best objective function in the generation will tend to increase. When the number of generations reaches a predetermined threshold, we terminate the iteration process, and then the root cause set represented by the binary string with the highest objective function value is the best root cause set we can obtain.

## 4 Evaluation

In this section, we describe the experimental evaluation of the proposed approach on real-world data. The efficiency and effectiveness of the algorithm under different parameter configurations are analyzed.

### 4.1 Date Set

The data set is collected from an Internet company for two weeks. The KPI record is stored as "1536940800000, i01, e01, c1, p01, l1, 12.0" where the first column denotes timestamp whose granularity is millisecond and the last column denotes the KPI value. Each record has five attributes $i$, $e$, $c$, $p$, $l$ and the attribute values in each attribute dimension are 147, 13, 9, 37 and 5 respectively. 200 anomaly cases with different root cause sets are injected into the data set. The number of attribute combinations in each case varies from 1 to 5 and the number of dimensions varies from 1 to 5. Figure 2 shows the KPI value in one example attribute combination with two anomalies injected.

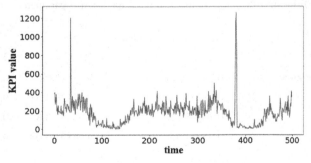

**Fig. 2.** The KPI values with the anomalies.

## 4.2 The Effectiveness and Efficiency of Algorithm

The effectiveness of the algorithm is measured by *F-score*, with the definition $F-score = \frac{2\times Precision\times recall}{Precision+recall}$, where $Precision = \frac{TP}{TP+FP}$ and $Recall = \frac{TP}{TP+FN}$. TP (true positive) is the number of attribute combinations correctly reported. FP (false positive) is the number of attribute combinations wrongly reported. FN (false negative) is the number of attribute combinations that are not reported. The efficiency of the algorithm is evaluated by the averaged execution time (in seconds) of each case on a PC with (Inter(R) Core(TM) i5-3210 CPU @ 2.5 GHz with 4G RAM).

Table 3 shows the results for root cause detection evaluation under different $c$ in Eq. (4), where $N$ in the *TOPN* is set as 10, the number of the individuals is set as 10 and the evolution generation is set as 20. Generally, coefficient $c$ can be chosen in a range from 0 to 1 according to the characteristics of the underlying data. The closer $c$ approaches 1, the more succinct the results will be, i.e., the number of the attribute combinations will tend to be less. The total number of attribute combinations in the 200 anomaly cases is 362. When the coefficient $c$ varies from 0.2 to 1, the number of attribute combinations recognized by the algorithm decreases from 419 to 354. One can note that F-score reaches the highest in the five configurations when the parameter $c$ is 0.4, and the corresponding attribute combination number is 369. The averaged F-score in Table 3 is 0.9514, which illustrates the effectiveness and robustness of the proposed approach.

**Table 3.** Effectiveness comparison between different parameters.

| $c$ | Attribute combination number | F-score | Time (s) |
|-----|------------------------------|---------|----------|
| 0.2 | 419 | 0.9014 | 21.34 |
| 0.4 | 369 | 0.9684 | 19.61 |
| 0.6 | 358 | 0.9667 | 18.78 |
| 0.8 | 355 | 0.9623 | 17.97 |
| 1   | 354 | 0.9581 | 17.37 |

**Table 4.** Efficiency comparison between different parameters.

| F-score/Time (s) | | Number of individuals | | |
|------------------|----|-----------------|-------|-------|
| | | 5 | 10 | 20 |
| Number of generations | 5 | 0.7332/5.93 | 0.8382/9.55 | 0.8579/12.31 |
| | 10 | 0.9125/7.95 | 0.9332/14.53 | 0.9685/20.21 |
| | 20 | 0.9685/10.94 | 0.9685/23.83 | 0.9685/35.92 |
| | 40 | 0.9685/18.43 | 0.9685/43.08 | 0.9685/67.16 |

Table 4 compares the F-scores and the execution time of the proposed approach under different parameters of the genetic-algorithm-based search. $c$ is set as 0.4 and

$N$ in the *TOPN* is set as 10. The search space of the *TOP10* attribute combinations is $2^{10} - 1 = 1,023$. But in our algorithm when the individual number is 5 and generation number is 20, only $5 * 20 = 100$ cases have been explored, resulting in a F-score of 0.9685, which is very close to the optimal value 0.9699 obtained by the method of exhaustion of the 1023 cases. Thus, the generic-algorithm-based search achieves an approximate optimal objective function by exploring only 9.8% of the total solution space. The above results prove the efficiency of our proposed approach.

The accuracy (i.e., F-score) of the algorithm can be further improved by evaluation with different predicted values. The forecast normal values in this paper are calculated based on the means of the historical KPI values. Therefore, in the framework described in this paper, we can parallelize our algorithm by comparing the F-scores under different means of the historical KPI values and choosing the highest F-score and corresponding attribute combinations as the final results. In Table 3, the forecast values are the means of 3 adjacent historical KPI values. We compare the results with the means of 3, 4, 5 and 6 adjacent historical KPI values and obtain a F-score of 0.9876, which is higher than 0.9685.

## 5  Conclusion

When the overall KPI value suffers from an abnormal change, it remains as a challenge to look through tens of thousands of KPI records with multi-dimensional attributes to localize the root cause. In this paper, we propose a novel mechanism to transform the root localization to a search problem. A new objective function is developed as a potential score to guide the search process. A breadth-first search combined with a genetic-algorithm-based heuristic is utilized to reduce the search space efficiently. Experiments on the real-world data show that our proposed approach achieves an averaged F-score of 0.95 with a localization time less than 30 s.

## References

1. O'Donovan, P., Leahy, K., Bruton, K., O'Sullivan, D.T.J.: An industrial big data pipeline for data-driven analytics maintenance applications in large-scale smart manufacturing facilities. J. Big Data 2(1), 1–26 (2015). https://doi.org/10.1186/s40537-015-0034-z
2. Shah, S.Y., Yuan, Z., Lu, S., Zerfos, P.: Dependency analysis of cloud applications for performance monitoring using recurrent neural networks. In: IEEE International Conference on Big Data, pp. 1534–1543 (2017)
3. Chen, M.Y., Kiciman, E., Fratkin, E., Fox, A., Brewer, E.: Pinpoint: problem determination in large, dynamic internet services. In: International Conference on Dependable Systems and Networks, DSN 2002, Proceedings, pp. 595–604. IEEE (2002)
4. Lin, J., Zhang, Q., Bannazadeh, H., Leon-Garcia, A.: Automated anomaly detection and root cause analysis in virtualized cloud infrastructures. In: NOMS 2016 - 2016 IEEE/IFIP Network Operations and Management Symposium, April 2016, pp. 550–556 (2016)
5. Jayathilaka, H., Krintz, C., Wolski, R.: Performance monitoring and root cause analysis for cloud-hosted web applications. In: Proceedings of the 26th International Conference on World Wide Web, Series WWW 2017, pp. 469–478. Republic and Canton of Geneva, International World Wide Web Conferences Steering Committee, Switzerland (2017)

6. Shah, S.Y., Dang, X., Zerfos, P.: Root cause detection using dynamic dependency graphs from time series data. In: 2018 IEEE International Conference on Big Data (Big Data), Seattle, WA, USA, pp. 1998–2003 (2018)
7. Lin, Q., Lou, J., Zhang, H., Zhang, D.: iDice: problem identification for emerging issues. In: Proceedings of the 38th International Conference on Software Engineering, pp. 214–224. ACM (2016)
8. Bhagwan, R., et al.: Adtributor: revenue debugging in advertising systems. In: 11th USENIX Symposium on Networked Systems Design and Implementation (NSDI 2014), pp. 43–55 (2014)
9. Sun, Y., et al.: HotSpot: anomaly localization for additive KPIs with multi-dimensional attributes. IEEE Access 6, 10909–10923 (2018)
10. Eichler, M.: Granger causality and path diagrams for multivariate timeseries. J. Econometrics 137(2), 334–353 (2007)
11. Liu, D., Zhao, Y., Xu, H., Sun, Y., Pei, D., et al.: Opprentice: towards practical and automatic anomaly detection through machine learning. In: Proceedings of IMC, pp. 211–224. ACM (2015)
12. Aminikhanghahi, S., Cook, D.J.: A survey of methods for time series change point detection. Knowl. Inf. Syst. 51(2), 339–367 (2016). https://doi.org/10.1007/s10115-016-0987-z
13. Srinivas, M., Patnaik, L.M.: Genetic algorithms: a survey. Computer 27(6), 17–26 (1994)

# A Performance Benchmark for Stream Data Storage Systems

Siqi Kang[1,2]($\boxtimes$), Guangzhong Yao[1,2], Sijie Guo[3], and Jin Xiong[1,2]($\boxtimes$)

[1] SKL Computer Architecture, ICT, CAS, Beijing, China
{kangsiqi,yaoguangzhong,xiongjin}@ict.ac.cn
[2] University of Chinese Academy of Sciences, Beijing, China
[3] StreamNative, Beijing, China
guosijie@gmail.com

**Abstract.** Modern business intelligence relies on efficient processing on very large amount of stream data, such as various event logging and data collected by sensors. To meet the great demand for stream processing, many stream data storage systems have been implemented and widely deployed, such as Kafka, Pulsar and DistributedLog. These systems differ in many aspects including design objectives, target application scenarios, access semantics, user API, and implementation technologies. Each system use a dedicated tool to evaluate its performance. And different systems measure different performance metrics using different loads. For infrastructure architects, it is important to compare the performances of different systems under diverse loads using the same benchmark. Moreover, for system designers and developers, it is critical to study how different implementation technologies affect their performance. However, there is no such a benchmark tool yet which can evaluate the performances of different systems. Due to the wide diversities of different systems, it is challenging to design such a benchmark tool. In this paper, we present SSBench, a benchmark tool designed for stream data storage systems. SSBench abstracts the data and operations in different systems as "data streams" and "reads/writes" to data streams. By translating stream read/write operations into the specific operations of each system using its own APIs, SSBench can evaluate different systems using the same loads. In addition to measure simple read/write performance, SSBench also provides several specific performance measurements for stream data, including end-to-end read latency, performance under imbalanced loads and performance of transactional loads. This paper also presents the performance evaluation of four typical systems, Kafka, Pulsar, DistributedLog and ZStream, using SSBench, and discussion of the causes for their performance differences from the perspective of their implementation techniques.

**Keywords:** Stream data storage system · Performance benchmark · Performance evaluation

© Springer Nature Singapore Pte Ltd. 2021
W. Gao et al. (Eds.): FICC 2020, CCIS 1385, pp. 107–122, 2021.
https://doi.org/10.1007/978-981-16-1160-5_10

# 1   Introduction

Nowadays Internet companies are generating tremendous amount of data everyday. They keep recording various events all the time. For example, social media platforms keep recording state changes of their users; online shopping platforms keep recording transactions as well as clicking events of their users; data centers keep recording machine states, including CPU, memory, network and disk. These data could contain great value. Through efficient processing of these data, the companies can recommend preferences for users, formulate more accurate advertising strategies for businesses, or propose more efficient solutions for data center resource scheduling. Therefore, stream processing platforms are widely used to analyze real-time data quickly and efficiently [24–27]. The characteristics of stream processing are real-time, scalability and openness [1]. The processing objects of stream processing are data streams. Writers can continuously append data to the data stream, and the data stream can simultaneously accept multiple readers to start reading data from a certain position in the data stream. In general, stream data has the following characteristics [2]:

- Real-time: Stream data are generated in real time, and each piece of data describes in detail what happened.
- Continuous flow: Stream data are generated ceaselessly, that is, the applications always generate new data without stopping.
- Append-only: New data are written into the data stream in an append mode, and no modification operation will be performed on the existing data.
- Sequentiality: Stream data are generally appended to the data stream in chronological order.

Since the continuously generated stream data contains rich value, it is important to persist them for later processing using streaming or batch queries. To support various stream processing scenarios, stream data storage systems should provide high read and write performance. Many such storage systems are developed either in academia or industry to store stream data. However, each of them only satisfies some of the characteristics of stream data. Moreover, these systems have different interfaces, functions, technologies, and performance.

According to their technologies, existing stream data storage systems can be divided into two categories: *pub-sub systems* and *shared log systems*. Typical pub-sub systems include Kafka [3], Pulsar [4], RocketMQ [5], etc. They abstract their data as a *topic*, and provide a publish/subscribe interface to the topic. They can provide fast read and write to stream data by using technologies including partitioning, sharding, multi-layer architecture and replication. Distributed log systems, such as Corfu [6], vCorfu [7], Tango [29] and BookKeeper [8], abstract their data as *logs*, and provide read and write interfaces to logs. They emphasis on data consistency guarantees by using the transaction technology and a global sequencer.

Due to the wide diversities of stream data storage systems, each system only uses its own performance measuring tool for evaluation. Currently, it is impossible to evaluate different systems using the same benchmark tool. Therefore, it is difficult for users to choose the best-performing system according to their needs. However, to design a benchmark tool for performance evaluation of different systems is non trivial, because they have different design goals, access interfaces, semantics, and technologies. There is no such benchmark tool yet which can evaluate these systems.

In this paper, we present SSBench, a performance benchmark tool designed for evaluate stream data storage systems. SSBench abstracts the data and operations in different systems as "data streams" and "reads/writes" to data streams. By translating stream read/write operations into the specific operations of each system using its own APIs, SSBench can evaluate different systems using the same loads. In addition to measure simple read/write performance, SSBench also provides several specific performance measurements for stream data, including end-to-end read latency, performance under imbalanced loads and performance of transactional loads. Moreover, as a case study, we use SSBench to evaluate four typical systems, Kafka, Pulsar, DistributedLog [14] and ZStream [28], and analyze the causes for their performance differences from the perspective of their implementation techniques.

The paper is organized as follows. Section 2 provides an overview of stream data storage systems, including the differences between typical systems. Section 3 describes SSBench, the performance benchmark tool we designed and implemented. Section 4 presents using SSBench to systematically compare the performance of different systems. Section 5 discusses related work, and Sect. 6 concludes the paper.

## 2  Overview of Stream Data Storage Systems

In this section, we first discuss typical application scenarios and their requirements for stream data storage systems. Then, we summarize the critical technologies of existing stream data storage systems. And finally, we briefly survey several typical stream data storage systems, highlighting the differences between them.

### 2.1  Typical Application Scenarios

Several typical application scenarios of stream processing are listed as follows.

- **Website activity tracking.** In the LinkedIn social platform [9], it is necessary to record the information that each user searched for and followed. By using machine learning models, it can predict social connections, match users with suitable positions, and optimize advertising. By analyzing the events of each person clicking on the company and position, we can show users some similar job opportunities. This requires the data platform to quickly analyze and process the user's click event data in real time, and quickly make recommendations for users in a short time.

- **Monitoring data.** In the traffic management platform [10], The platform evaluates and monitors real-time congestion and accidents by monitoring GPS and SCATS sensor data. In addition, it also needs to monitor the content posted by local social platform to determine traffic conditions. In the network flow monitoring scenario [11], network monitoring data (such as TCP data packets sent and received within a certain period of time) are usually generated in the form of streams, which need to be continuously analyzed and monitored.
- **Social network platform.** In the Twitter open social platform scenario [12], users can send tweets with a certain hashtag, and users can click on a certain hashtag to view all the tweets under that topic. The real-time tweets sent by the user are appended to the data stream. The data stream is partitioned and partitioning rule is to hash the hashtags. This calculation method makes the tweets of the same hashtag be sent to the same partition. When some hot events occur, hot topics will appear. There will be a large number of read and write requests for a certain topic within a certain period of time. In a scenario where topics are used as the partition standard, hot partitions will be caused. Some partitions have less access, and some partitions have more access, which will cause data skew and access skew.
- **Commit log.** The system log needs to record all the operations of the system. It is a very important data source and contains many values. In LinkedIn [13] and Twitter [14], the stream data storage system is used as the write ahead commit log of the Espresso [30] and Manhattan databases [31] to record the modification operations on the database. Batch processing, stream processing, or the operation of data warehouses can analyze submission logs and extract important values.
- **Distributed transaction.** In the scenario of bank transfer [15], using the above-mentioned stream data storage system as the commit log of the balance database needs to ensure the correct execution of distributed transactions. There cannot be an intermediate situation where only one operation succeeds and another operation fails.

## 2.2  Requirements

The various scenarios involved in stream processing put forward several requirements for the storage of stream data:

- **Fast reading and writing.** It can quickly store a large amount of detailed data generated in real time, and can support fast reading for real-time data, so that the data can be quickly analyzed.
- **Efficient analysis.** Stream data has no semantics, and the stream processing system cannot only extract key fields for query analysis. If the data storage itself can be semantically aware, and the processing system can quickly extract important fields, query efficiency can be greatly improved.
- **Load balancing.** Stream data storage systems usually use partitions to increase concurrency. Partitions will inevitably cause hot spots. Stream data storage systems should load balance hotspot partitions to reduce the writing and reading of hot partitions and servers.

– **Distributed transactions.** For certain scenarios of stream data storage systems, it is required to provide support for distributed transactions and support the atomic execution of multiple write operations.

## 2.3   Critical Technologies

Based on the requirements of stream data storage, the key issues that the stream data storage system mainly solves are as follows:

– **Fast reading and writing.** Stream data records the detailed data that occurred at the time of recording. Storing the stream data is essential for data analysis and processing. Stream processing requires that stream data can be read and written quickly to support efficient stream processing analysis. Therefore, fast storage of stream data and provision of fast access are the most critical technologies for stream data storage systems.
– **Efficient analysis query support.** Stream data is required to support low-latency stream processing requirements as well as efficient query and analysis requirements. Therefore, the data transfer from non-semantic-aware stream data into column data stream data and the performance guarantee of low latency and high bandwidth have become challenges for stream data.
– **Automatic load balancing.** Load balancing needs to determine the partitions and nodes with heavier loads accurately in order to migrate the heavier loads. Therefore, determining the time point of load balancing trigger and the location of load migration is the focus of load balancing work. And it is also very important to ensure the correctness of load migration and correctness after failure recovery in the process of load balancing.
– **Efficient distributed transaction support.** Stream data storage systems should address the need for atomic completion of multiple write operations in storage. Distributed transactions need to solve the problem of isolation of uncommitted transaction messages. The characteristics of data streams cause the stream data storage system to only support append-only writes. For reads, users can read stream continuously from a certain position. Therefore, if we mix ordinary data, committed transaction data, and uncommitted transaction together, it is difficult to guarantee the isolation and transactional reading and writing will interfere with the reading and writing of ordinary events, which is a key point in distributed transactions.

## 2.4   Typical Systems

**Pub-Sub Systems.** Kafka is the most classic publish-subscribe messaging system. Kafka increases read and write throughput through partitioning in parallel [3]. In terms of read, sendfile technology is used to reduce multiple copies of data between kernel mode and user mode. Since the data in Kafka is appended to the topic in a non-semantic sense, the data cannot be quickly analyzed. For the load balancing of hot data, load migration will cause the migration of old data and waste cluster resources. The advanced version of Kafka has supported distributed transactions. The mixed storage of ordinary messages and transaction messages can cause some problem. Short transactions must wait for long transactions to be submitted before their data can be seen by users. Therefore, this transaction isolation method is less efficient.

Pulsar is a relatively new publish-subscribe messaging system. Its data abstraction and interface are similar to Kafka. Pulsar also uses partitioning to increase the throughput of reads and writes [4]. Similarly, Pulsar does not provide a semantic-aware storage format and cannot perform efficient analysis and processing of data. In Pulsar, load balancing is also considered, but Pulsar's load balancing stays in the service layer. In addition, Pulsar does not yet support the distributed transaction feature, which does not meet all our requirements for stream data storage systems.

**Shared-Log Systems.** In Corfu [6] and vCorfu [7], the global log is responsible for ensuring consistency, global order, and transactionality, and the virtual materialized flow increases scalability. But in essence, every write must firstly go through the global log before being written to the partitioned stream. This shared log uses a centralized sequencer, and its throughput is limited by the speed at which the sequencer allocates address space. FuzzyLog [16] is based on the shortcomings of Corfu/vCorfu. At the cost of strict order, it supports the causal order of each additional operation rather than a global order by establishing a directed acyclic graph of events. But FuzzyLog is not orderly for each partition. In some strictly orderly scenarios, there may be errors.

The distributed log storage system BookKeeper [8] is a log stream service that provides persistent storage. It uses striping and read-write separation mechanism to provide high performance. However, BookKeeper does not support semantic awareness for storing data, and cannot provide cross-ledger transactions, which does not meet all our requirements for stream data storage systems. DistributedLog [14] is a distributed log warehouse. It uses log stream to represent the continuous data stream. It uses BookKeeper as a storage component. So DistributedLog has the same problem as BookKeeper.

ZStream [28] is a stream data storage system that divides a stream into more fine-grained partitions, called ranges. ZStream asynchronously converts row stream data into column stream data, and supports efficient columnar reading performance. By dynamically splitting and merging the partitions, the imbalanced load of the partitions is solved. And the transaction buffer technology solves the problem of distributed transaction isolation and the interference of transactional requests to ordinary requests.

## 3    Design of SSBench

In order to compare the performance of systems, a general performance evaluation tool is needed. The results of the evaluation of each system under the same dimension, unified environment and unified test conditions can be comparable. As far as we know, there is still a lack of such a performance evaluation tool for stream data storage systems. So we designed and implemented a unified stream data storage system benchmark tool called SSBench, which abstracts based on the common characteristics of each system. SSBench aims at evaluating the performance of different systems. It encapsulates the read and write operations of stream data that is separate from the specific system, and supports multiple read-write mode test scenarios. In addition, we also considered the scalability of SSBench itself to facilitate the evaluation of adding new systems.

## 3.1   Architecture and Functions

**Architecture.** The system architecture of SSBench is shown in Fig. 1. The top layer is the user interface. By providing users with a simple and easy-to-use operation interface, users can input simple test parameters and specific system commands to define test scenarios, start test programs, and obtain test results.

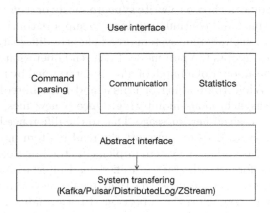

**Fig. 1.** Architecture.

The command parsing module determines the number of read and write client processes to start, the specific settings of each read and write task, and creates client processes on each test node by parsing user commands. The communication module is responsible for the communication between client processes. It includes the distribution of read and write tasks on multiple test machines, the maintenance of client life status, and the unified collection of test data. The statistical data module is responsible for collecting the performance of each client process in the process of performing read and write tasks. When the performance information generated, results will be output to the console and files.

In addition, SSBench also provides a more flexible evaluation method. Users can start a single client process through commands, and use scripts to start multiple client processes. Therefore, in multi-client read and write tasks, different parameters can be set for different client processes to achieve more flexible evaluation.

The abstract interface layer provides a unified read and write interface (send/read) for the stream data storage system, with the functions of sending and receiving messages.

**Unified Interface.** For the two major types of stream storage systems, the data abstraction and writing unit are different. In general, the message system data is abstracted as a topic, and the distributed log is a log stream. So we abstract these operating objects as streams. Reader reads streams and writer writes to streams respectively. Users of SSBench only need to specify how to read and write and does not need to care about how the specific system reads and writes.

For write operations of different stream data storage systems, the client can write one message at a time and return the result asynchronously. In addition, some systems can also write synchronously, that is, when writing a message, the writing process is blocked, waiting for the returned result before writing the next message. Therefore, we abstract a unified client-side write operation interface. For each system, you can choose to write synchronously or asynchronously, write one message at a time, and continuously loop to support continuous writing.

For read operations, different systems read slightly differently. For example, the Kafka consumer adopts the pull mode. Each read operation will pull a batch of data to read. The amount of a batch of data is determined by time. The default setting is 100ms. While the reading of Pulsar and DistributedLog needs to be obtained through asynchronous monitoring. Once a new message is available, then it returns the received new message. Due to the difference between the two methods, our framework does not support the read-by-item interface similar to the write interface. Instead, each system implements the function of continuously reading data. Therefore, we abstract a unified read interface, and each system reads data in its own way, returns the read results asynchronously, and the unified read interface completes the collection of performance results.

Therefore, for a specific stream data storage system, you only need to implement the above abstract interface, and you can use this framework for testing. So SSBench can support a unified test platform and compare the differences between systems horizontally.

**Functions.** For the design, we refer to the framework structure of YCSB (Yahoo! Cloud Serving Benchmark) [18]. In order to test the performance of different databases, YCSB abstracts the addition, deletion, modification, and query interfaces. Specific databases need to implement these interfaces, while the test scenarios are implemented by YCSB, such as database performance under different mixing ratio operations. In SSBench, we use the following functions to fairly test different systems:

1) Given a list of available client nodes, users of SSBench can specify the number of clients, that is, how many write processes and how many read processes. Then the test framework will evenly start all write (or read) processes on given nodes. Therefore, the write (or read) process can run on different machines, so SSBench is a distributed performance evaluation tool.

2) For the writing and reading process, the status of writing or reading, such as throughput and latency, is displayed in the form of a window (fixed time) and the running time can also be configured.

3) For the writing process, users of SSBench can specify the data format to be sent, such as fixed size, random content, etc. They can also specify the sending rate, such as fixed rate, unlimited rate, gradual rate, and random rate. In particular, some stream data storage systems support distributed transactions, so for the write process, users can specify whether it is a transaction write request, as well as transaction-related sending interval, number of sent and other parameters.

4) For the reading process, the user of SSBench can specify where to start reading, such as reading from the beginning or reading from the end. The former is to read from the oldest data, and the latter is to read from the latest data.

## 3.2   Common Read/Write Performance

For ordinary reading and writing, different systems use different technologies to ensure fast reading and writing. For example, for the replication mechanism, Kafka adopts the leader-follower model for multi-copy replication. First, a partition is sent to a leader node and then the followers pull data from the leader. Compared with Pulsar and DistributedLog, BookKeeper provides quorum replication mechanism. Data will be sent to three Bookies at the same time, and it only needs to receive most Bookie's confirmation so that the message can be known as persisted. Based on different copy strategies, we hope to understand the differences in write performance of different systems in their respective modes.

Therefore, for ordinary read and write performance evaluation, this paper evaluates the difference in write performance of different systems in the synchronous write mode, and summarizes the impact of different replication mechanisms based on the read and write performance of stream data storage systems.

## 3.3   Column Read/Write Performance

For column read and write evaluation, this paper mainly evaluates the difference in read and write performance of different systems in a multi-field scenario. Users can customize the number of event fields, field types, the number of fields of each type, and the number of read and write client processes to evaluate the impact of different field numbers on the performance of different stream data storage systems.

## 3.4   Imbalanced Load Performance

SSBench can simulate unbalanced load scenarios. User-defined evaluation of the number of streams and the ratio of write speed in each stream can verify read and write performance under load balanced and load imbalanced conditions. For ZStream, due to its own implementation of load balancing, this paper uses SSBench to adjust ZStream to a better performance state by setting different load balancing parameters of ZStream.

### 3.5   Transactional Load Performance

Distributed transactions simulate distributed transaction scenarios in stream data storage. Users can set the number of streams involved in each transaction, the number of data written in each stream, and the interval between transaction sending to evaluate the performance of distributed transactions. For long transactions, users can set the duration of long transactions and the proportion of long transactions in all transaction requests to evaluate the impact of long transaction requests on performance.

## 4   Performance Evaluation of Typical Systems

For the evaluation of reading and writing of ordinary messages, Sect. 4.1 uses the synchronous writing method to evaluate the write throughput of Kafka, Pulsar and DistributedLog when writing multiple copies. For column storage characteristics, Sect. 4.2 uses multi-event data sets to compare and evaluate ZStream and Kafka. For the load balancing feature, the evaluation results of ZStream, Kafka and Pulsar are compared in the case of balanced and imbalanced loads in Sect. 4.3. For distributed transactions, Sect. 4.4 evaluates the comparison results of ZStream and Kafka.

This paper uses a cluster platform with five nodes interconnected, each of which is configured with two Intel Xeon E5645 CPUs (each CPU has 12 hyper-threaded cores). The nodes are connected by 10 Gigabit Ethernet. The stream data storage system cluster uses 4 nodes, one of which serves as the master node to start the NameNode (for HDFS) or ZooKeeper (for ZStream, Kafka, and Pulsar) services. The other four are used as servers or clients. Storage nodes use SSDs as data storage.

### 4.1   Common Read/Write Performance

In this experiment, SSBench creates a write process, which creates a stream, and writes messages to this stream synchronously at a rate of 100000 msg/s (each message size is 1KB).

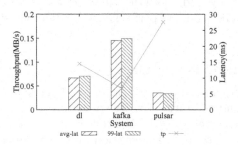

**Fig. 2.** Write performance in sync mode.

The results are shown in Fig. 2. It can be seen that Pulsar has the lowest and most stable write latency, followed by DistributedLog, and Kafka is the worst, which are about 2 times and 3 times that of DistributedLog and Pulsar, respectively. Kafka also has the worst write throughput, reaching only 30% of Pulsar and 50% of DistributedLog. The reason is Kafka is a two-step serial replication. First, a message is sent to the leader node, and the two follower nodes pull data from the leader node. While Pulsar and DistributedLog are one-step parallel replication, they send a message to all replica nodes. DistributedLog and Pulsar use BookKeeper to store data persistently, and the message will be sent to multiple Bookie nodes at the same time. They only need to receive the confirmation returned by most Bookie to consider that the message has been persisted in all Bookie nodes. Since the Bookie node uses SSD as the log disk, data can be written sequentially at high speed, so the write latency of DistributedLog and Pulsar is low.

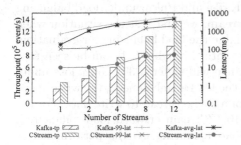

**Fig. 3.** Write performance. Each event has 6 fields and totally 38 bytes.

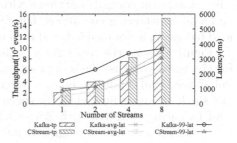

**Fig. 4.** Read performance. Each event has 6 fields and totally 38 bytes.

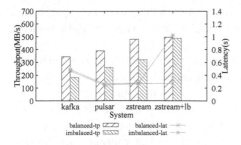

**Fig. 5.** Write throughput under balanced and imbalanced load.

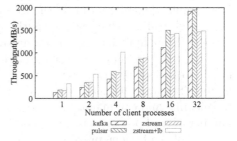

**Fig. 6.** Read throughput under imbalanced load.

### 4.2 Column Stream

In this experiment, SSBcnch creates a stream, and creates a read process and a write process. The writing process is responsible for writing multi-field data to the image system. Each event includes six fields, including types Int, Long, Float, Double, Boolean, and String. The total length of each event is 38 bytes, to verify the difference in read and write performance.

Figure 3 shows the write performance. It can be seen from the figure that the write performance of ZStream is significantly better than that of Kafka. The difference in write performance of Kafka is mainly due to its leader-follower replication strategy. ZStream uses BookKeeper's Quorum mechanism, which leads to low latency of writing. Figure 4 shows the read performance. When the number of streams increases, the end-to-end latency of both ZStream and Kafka changes. Because of its lower write latency, the end-to-end latency of ZStream is also lower than that of Kafka. In addition, due to its high network utilization, it also shows higher throughput.

### 4.3   Load Balance

This experiment compares ZStream with Kafka and Pulsar under load balancing and load imbalancing conditions. For ZStream, we first test the situation when load balancing is not enabled, and then enable the load balancing for evaluation.

Figure 5 shows the comparison of the write throughput. It can be seen that for the three systems, the write performance will suffer a 30%–50% loss under imbalanced load condition. Because when the load is imbalanced, the write speed of a single stream is limited by the speed of a single thread to write data. For the load balancing situation with ZStream+lb (ZStream enabled load balancing), the partition will not be triggered for load balancing, so the performance is consistent with the original ZStream load balancing situation. In the case of load imbalancing, the partition with large traffic can be divided into two sub-partitions, so writing increases the degree of parallelism, it can show higher write performance. Pulsar also supports load balancing, but Pulsar only migrates the partitions with large traffic to the lighter load machine, and does not split the large traffic. Therefore, during the experiment, there will always be a partition with large traffic, which is constantly migrating between the two data nodes, but the throughput has not been improved.

Figure 6 shows how the read throughput of the three systems varies with the number of client processes when the load is imbalanced. As can be seen in the figure, ZStream with load balancing enabled increases in concurrent reads due to splitting, and can maintain a performance advantage of 30% to 50% when the number of read processes is small, and is limited to the ZStream reader when the number of read client processes is large. Performance disadvantages are shown without aggregation processing and multi-process sharing.

### 4.4   Distributed Transactions

In this experiment, a write client process continuously sends write requests and transaction requests with four read client processes read data in real time. Transaction requests are sent every 100 ms.

Figure 7 and 8 show the end-to-end latency with the write time of each long transaction and the proportion of long transactions. The end-to-end latency changes relatively smoothly with the long transaction execution time and the proportion of long transactions. In Kafka, as the long transaction request time and the proportion of long transactions increase, the end-to-end latency is on the rise. For Kafka, reading of ordinary messages has to be blocked by uncommitted long transactions. Especially when the long transaction takes a heavier part and the execution time of the long transaction is long, most of the real-time message reading will be in a blocking state, so Kafka shows poor end-to-end latency. ZStream uses transaction buffers. When long transactions are not committed, ordinary messages and short transaction messages can still be read in real time. Only long transaction data has a long end-to-end latency, and the overall performance is 70% to 90% better than Kafka.

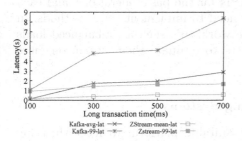

**Fig. 7.** End-to-end latency of different long transaction time settings.

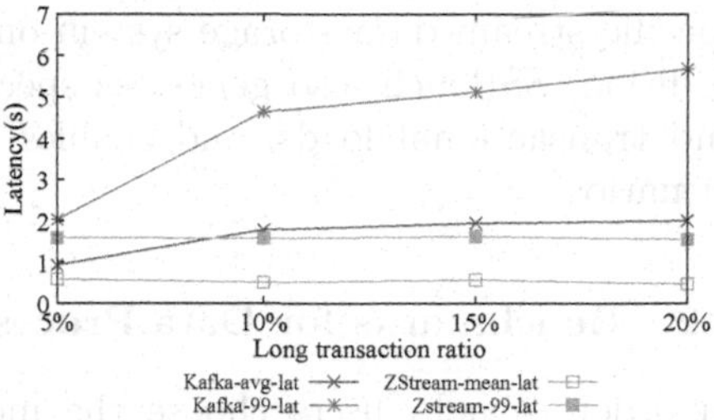

**Fig. 8.** End-to-end latency of different long transaction ratio settings.

## 5    Related Works

### 5.1    Benchmarks for Storage Systems

fio is an IO test tool that can run on a variety of systems [19]. It can test the performance of local disks, network storage, etc. fio simulates different loads by specifying the type of I/O to be tested, configuring I/O size, block size, engine, I/O depth and other parameters to simulate different loads to verify storage performance under different loads. Filebench [20] is an automated testing tool for file system performance. It tests the performance of the file system by quickly simulating the load of a real application server. It can not only simulate micro-operations (such as copyfiles, createfiles, randomread, randomwrite), but also simulate complex applications (such as varmail, fileserver, oltp, dss, webserver, webproxy). Filebench is more suitable for testing file server performance, but it is also an automatic load generation tool, and can also be used for file system performance.

The goal of the YCSB project is to develop a frame work and a set of common workloads to evaluate the performance of different key value stores and cloud service storage systems [18]. Users can install different database systems in the same hardware environment, use YCSB to generate the same workload, and compare and evaluate different systems. YSCB is to test the performance of different databases through abstract addition, deletion, modification, and query methods. Specific databases need to implement these methods, while the test scenarios are implemented by YCSB, such as database performance under different mixing ratio operations.

YCSB++ is an extension of YCSB to improve the testing of advanced database functions [21]. YCSB++ includes multi-tester coordination for load increase and eventual consistency measurement, multi-stage workloads for quantifying the consequences of work delays and the benefits of expected configuration optimization (such as B-tree pre-splitting or batch loading), and high-level features in abstract API benchmarks for explicit consolidation.

We learns a lot from YCSB's method. Similar to YCSB, SSBench abstracts the basic interfaces such as read/write APIs for the basic operations, and each specific stream data storage system only need to implement these methods. In addition, SSBench also generates specific workloads such as imbalanced loads and transactional loads, and enables users to evaluate their system on more scenarios.

## 5.2   Benchmarks for Data Processing Systems

In order to help users choose the most suitable platform to meet their big data real-time stream processing needs, Yahoo has designed and implemented a stream processing benchmark program based on real-world scenarios and released it as an open source [22]. In this benchmark test, Kafka and Redis were introduced for data extraction and storage to build a complete data pipeline to more closely simulate actual production scenarios. The benchmark test platform scenario for stream processing is different from the storage scenario, and does not meet the needs of the storage system evaluation in the paper.

OLTP-Bench is a scalable DBMS benchmark test platform [23]. The main contribution of OLTP-Bench is its ease of use and scalability, support for transaction mixing, strict control of request rate and access allocation, and the ability to support all major DBMS platforms. In addition, it also bundles 15 workloads with varying complexity and system requirements, including 4 comprehensive workloads, 8 workloads from popular benchmarks, and 3 jobs from real applications load. The OLTP evaluation evaluates the transaction characteristics of the database management system. It has a single function and does not meet the multiple requirements of the storage system in the stream processing scenario.

## 6   Conclusion

This paper implements a benchmark tool based on stream data storage system called SSBench, and uses it to evaluate and compare the performance of multiple stream data storage systems. Our experimental results show that for common read and write operations, different replication mechanisms and caching strategies can bring great performance differences. For column data read and write, unbalanced load read and write, and distributed transaction scenarios, ZStream has 30% to 90% advantages over the other systems.

# References

1. Shahrivari, S.: Beyond batch processing: towards real-time and streaming big data. Computers **3**(4), 117–129 (2014)
2. The log: What every software engineer should know about real-time data's unifying abstraction. https://engineering.linkedin.com/distributed-systems/log-what-every-software-engineer-should-know-about-real-time-datas-unifying. Accessed 16 Dec 2013
3. Kreps, J., Narkhede, N., Rao, J., et al.: KAFKA: a distributed messaging system for log processing. In: Proceedings of the NetDB, pp. 1–7. IEEE (2011)
4. Francis, J., Merli, M.: Open-sourcing pulsar, pub-sub messaging at scale, 8, September 2016. https://www.richardnelson.it/post/150096199100/open-sourcing-pulsar-pub-sub-messaging-at-scale
5. ALIBABA: How to support more queues in rocketmq? 23 September 2016. https://rocketmq.apache.org/rocketmq/how-to-support-more-queues-in-rocketmq/
6. Balakrishnan, M., Malkhi, D., Prabhakaran, V., et al.: CORFU: a shared log design for flash clusters. Proceeding of the 9th USENIX Symposium on Networked Systems Design and Implementation (NSDI 2012), pp. 1–14, USENIX (2012)
7. Wei, M., Tai, A., Rossbach, C.J., et al.: vCorfu: a cloud-scale object store on a shared log. In: Proceedings of the 14th USENIX Symposium on Networked Systems Design and Implementation (NSDI17), pp. 35–49 (2017)
8. Junqueira, F.P., Kelly, I., Reed, B.: Durability with bookkeeper. ACM SIGOPS Oper. Syst. Rev. **47**(1), 9–15 (2013)
9. Goodhope, K., Koshy, J., Kreps, J., et al.: Building Linkedin's real-time activity data pipeline. IEEE Data Eng. Bull. **35**(2), 33–45 (2012)
10. Panagiotou, N., et al.: Intelligent urban data monitoring for smart cities. In: Berendt, B., et al. (eds.) ECML PKDD 2016. LNCS (LNAI), vol. 9853, pp. 177–192. Springer, Cham (2016). https://doi.org/10.1007/978-3-319-46131-1_23
11. Bär, A., Finamore, A., Casas, P., et al.: Large-scale network traffic monitoring with DBstream, a system for rolling big data analysis. In: 2014 IEEE International Conference on BigData (Big Data), pp. 165–170. IEEE 2014
12. McCreadie, R., Macdonald, C., Ounis, I., et al.: Scalable distributed event detection for Twitter. In: IEEE International Conference on Big Data, pp. 543–549. IEEE (2013)
13. Wang, G., Koshy, J., Subramanian, S., et al.: Building a replicated logging system with Apache Kafka. Proc. VLDB Endowment **8**(12), 1654–1655 (2015)
14. Sijie, G., Robin, D., Leigh, S.: DistributedLog: a high performance replicated log service. In: IEEE 33rd International Conference on Data Engineering. ICDE, pp. 1183–1194 (2017)
15. Mehta, A., Gustafson, J.: Transactions in Apache Kafka[EB/OL], 17 July 2017. https://www.confluent.io/blog/transactions-apache-kafka/
16. Lockerman, J., Faleiro, J.M., Kim, J., et al.: The FuzzyLog: a partially ordered shared log. In: 13th USENIX Symposium on Operating Systems Design and Implementation (OSDI 2018), pp. 357–372 (2018)

17. Kreps, J.: Benchmarking Apache Kafka: 2 million writes per second (on three cheap machines), April 2014. https://engineering.linkedin.com/Kafka/benchmarking-apache-Kafka-2-million-writes-second-three-cheap-machines
18. Cooper, B.F., Silberstein, A., Tam, E., et al.: Benchmarking cloud serving systems with YCSB. In: Proceedings of the 1st ACM symposium on Cloud computing, pp. 143–154. ACM (2010)
19. fio. http://freshmeat.sourceforge.net/projects/fio
20. Tarasov, V., Zadok, E., Shepler, S.: Filebench: a flexible framework for file system benchmarking. USENIX Login **41**(1), 6–12 (2016)
21. PatilS, P.M.: YCSB++: benchmarking and performance debugging advanced features in scalable tablestores. In: Proceedings of the 2nd ACM Symposium on Cloud Computing, Cascais, Portugal, vol. 9 (2011)
22. Chintapalli, S., Dagit, D., Evans, B., et al.: Benchmarking streaming computation engines: storm, flink and spark streaming. In: IEEE International Parallel and Distributed Processing Symposium Workshops (IPDPSW), pp. 1789–1792. IEEE (2016)
23. Difallah, D.E., Pavlo, A., Curino, C., et al.: OLTP-bench: an extensible testbed for benchmarking relational databases. Proc. VLDB Endowment **7**(4), 277–288 (2013)
24. Toshniwal, A., Taneja, S., Shukla, A., Ramasamy, K.: Storm@twitter. In: Proceedings of the 2014 ACM SIGMOD International Conference on Management of Data, pp. 147–156. ACM (2014)
25. Kulkarni, S., Bhagat, N., Fu, M., et al.: Twitter heron: stream processing at scale. In: Proceedings of the 2015 ACM SIGMOD International Conference on Management of Data, pp. 239–250. ACM (2015)
26. Carbon, P., Katsifodimos, A., Ewen, S., et al.: Apache flink: stream and batch processing in a single engine. IEEE Data Eng. Bull. **38**(4), 28–38 (2015)
27. Venkataraman, S., Panda, A., Ousterhout, K., et al.: Drizzle: fast and adaptable stream processing at scale. In: Proceedings of the 26th Symposium on Operating Systems Principles, pp. 374–389. ACM (2017)
28. Shen, Y., Yao, G., Guo, S., et al.: A unified storage system for whole-time-range data analytics over unbounded data. In: Proceedings of 2019 IEEE Intl Conference on Parallel and Distributed Processing with Applications, Big Data and Cloud Computing, Sustainable Computing and Communications, Social Computing and Networking, pp. 967–974. IEEE (2019)
29. Balakrishnan, M., Malkhi, D., Wobber, T., et al.: Tango: distributed data structures over a shared log. In: Proceedings of the 24th ACM Symposium on Operating Systems Principles, pp. 325–340. ACM (2013)
30. Qiao, L., Surlaker, K., Das, S., et al.: On brewing fresh espresso: LinkedIn's distributed data serving platform. In: Proceedings of the 2013 ACM SIGMOD International Conference on Management of Data, pp. 1135–1146. ACM (2013)
31. Manhattan, our real-time, multi-tenant distributed database for Twitter scale, 2 April 2014. https://blog.twitter.com/engineering/en_us/a/2014/manhattan-our-real-time-multi-tenant-distributed-database-for-twitter-scale.html

# Failure Characterization Based on LSTM Networks for Bluegene/L System Logs

Rui Ren[1(✉)], JieChao Cheng[2], Hao Shi[1,3], Lei Lei[1,3], Wuming Zhou[1], Guofeng Wang[1,3], and Congwu Li[1,3]

[1] China Electronics Technology Research Institute of Cyberspace Security Co., LTD., Beijing, China
renrui871@qq.com, hshi@nscslab.net, llchina@139.com, zhouwm1983@sina.com, wangguofeng@bupt.edu.cn, 502227039@qq.com

[2] International School of Software, Wuhan University, Wuhan, China
jetrobert19@gmail.com

[3] China Academy of Electronics and Information Technology, Beijing, China

**Abstract.** As the scales of cluster systems increase, failures become normal and make reliability be a major concern. System logs contain rich semantic information of all components, and they are always used to detect abnormal behaviors and failures by mining methods.

In this paper, we perform a failure characterization study of the Bluegene/L system logs. First, based on the event sequences parsed by LogMaster, we take advantage of a prediction method based on Long Short-Term Memory Network (LSTM), and train the N-ary event sequence patterns in Many-to-One scenario; Second, we extract the failure rules from the minimal event sequence patterns, which identify the key events (failure signs) that correlate to system failures in the large-scale cluster; At last, we evaluate our experiments on the publicly available production Bluegene/L dataset, and obtain some interesting rules and correlations of failure events.

**Keywords:** LSTM network · N-ary event sequence patterns · Failure rules

## 1 Introduction

Large cluster systems are common platforms for both cloud computing and high performance. As the scales of cluster systems increase, failures become normal [1] and make reliability be a major concern [2]. In addition, due to the increase of software complexity and the diversity of workload behaviors, failure diagnosis and fault management are becoming more and more challenging. System logs record the execution trajectory of systems and exist in all components of systems, so they contain rich semantic information, and are always used to detect abnormal behaviors and failure events of systems by mining a large number of logs [3].

Based on system logs, there are many analysis methods for failure characterization. For instance, the rule-based approaches are common methods, which

© Springer Nature Singapore Pte Ltd. 2021
W. Gao et al. (Eds.): FICC 2020, CCIS 1385, pp. 123–133, 2021.
https://doi.org/10.1007/978-981-16-1160-5_11

require expert knowledge, even though more precise. Traditional statistical-based methods are generally suitable for analyzing small data, and it is challenging when dealing with large-scale data. Moreover, machine learning methods are often used for mining and analyzing the large-scale system logs, especially deep neural networks (e.g., CNNs [4], RNNs [5], LSTM [6], etc.) are increasingly used in log sequence analysis. And LSTM is able to capture the long-range dependency across sequences, therefore outperforms traditional supervised learning methods in our application domain.

In this paper, we perform a failure characterization study of the Bluegene/L system logs by using LSTM networks [6]. First, we take advantage of a prediction method based on Long Short-Term Memory Network (LSTM) to train the N-ary event sequence patterns; Second, we extract the failure rules from the minimal event sequence patterns, which identify the key events (failure signs) that correlate to system failures in the large-scale cluster; At last, we evaluate our experiments on the publicly Bluegene/L system logs, and obtain some interesting rules, which can be used to reveal deeper failure event correlations.

The rest of this paper is organized as follows. Section 2 presents the methodology of failure characterization. Section 3 gives the experiment results and evaluations on the Bluegene/L logs. Section 4 describes the related work. At last, Sect. 5 draws a conclusion and discuss the future work (Fig. 1).

## 2   Methodology

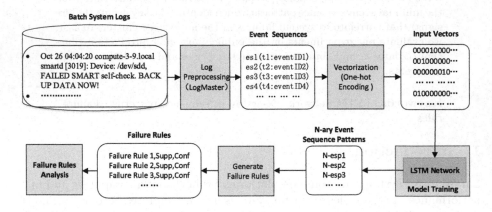

**Fig. 1.** The methodology.

### 2.1   Log Preprocessing

In log preprocessing, we make use of LogMaster [7] to parse logs with different log formats into a sequence of events. After formatting event logs, LogMaster also remove repeated events and periodic events [7]. Here, an *event sequence*

$es = <e_1, e_2, ..., e_n>$ is an ordered sequence of different log events in chronological order, here, $e_i$ ($1 \le i \le n$, and $n$ indicates the number of events in the logs.) is the event element in $es$, and event $e_{i-1}$ occurs before $e_i$. And $es$ is a sequence of $eventIDs$, which is represented by $eventID$ in the form of natural number. Specifically, $eventID$ is described by 2-tuples $(severity\ degree, event\ type)$, $severity\ degree$ indicates the priority or importance of events, which includes: Info, Warning, Error, Failure, Fatal, Severe, etc., and $event\ type$ indicates the type of events, which includes: Hardware, System, Filesystem, Network, and so on.

## 2.2   Vectorization

We use the *event sequence es* which has been defined in LogMaster [7] as the input data for model training. Then, We use an $K$-dimensional vector to represent a $eventID$, here, $K$ is defined as the total number of $eventIDs$. To employ the LSTM on event sequence $es$, we encode the $eventID$ by using the one-hot encoding method. Here, one hot encoding refers to splitting the column which contains numerical categorical data to many columns depending on the number of categories present in that column. Each column contains "0" or "1" corresponding to which column it has been placed. That is, each $eventID$ is encoded with a binary vector of $K$ bits with one of them is hot (i.e. 1) while others are 0. Specifically, we sort the $eventIDs$ in ascending order, and set the location of a certain $eventID$ as "1" while others are 0. For instance, $eventID\ 1 = [1, 0, 0, 0, ..., 0]$, $eventID\ 3 = [0, 0, 1, 0, ..., 0]$, and so on.

## 2.3   Model Training

Long Short Term Memory network [6] (usually just called "LSTM") is a special kind of RNN, capable of learning long-term dependencies. It is introduced by Hochreiter and Schmidhuber, and has recently been successfully applied in a variety of sequence modeling tasks, including handwriting recognition, character generation and sentiment analysis.

LSTM introduces long-term memory into recurrent neural networks. It mitigates the vanishing gradient problem, which is where the neural network stops learning because the updates to the various weights within a given neural network become smaller and smaller. Specifically, the LSTM model introduces an intermediate type of storage via the memory cell. A memory cell is a composite unit, built from simpler nodes in a specific connectivity pattern, with the novel inclusion of multiplicative nodes. And it uses a series of "gates", and "gates" are a distinctive feature of the LSTM approach. A gate is a sigmoidal unit that, like the input node, takes activation from the current data point as well as from the hidden layer at the previous time step. A gate is so-called because its value is used to multiply the value of another node. It is a gate in the sense that if its value is zero, then flow from the other node is cut off. If the value of the gate is one, all flow is passed through [5]. There are three types of gates within a unit:

- *Input Gate*: The value of the input gate multiplies the value of the input node. And the input gates are scales input to cell (write).
- *Forget Gate*: These gates were introduced by Gers et al. [8]. They provide a method by which the network can learn to flush the contents of the internal state. This is especially useful in continuously running networks. And the forget gates are scales old cell value (reset).
- *Output Gate*: The value ultimately produced by a memory cell is the value of the internal state multiplied by the value of the output gate . It is customary that the internal state first be run through a tanh activation function, as this gives the output of each cell the same dynamic range as an ordinary tanh hidden unit. And the output gates are scales output to cell (read).

Each gate is like a switch that controls the read/write, thus incorporating the long-term memory function into the model (Fig. 2).

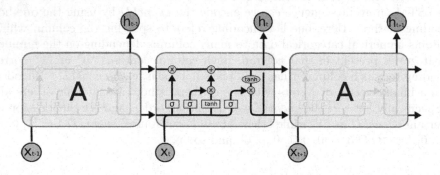

**Fig. 2.** The cell of LSTM network.

Based on LSTM network, we could use it to solve sequence predict problems. And the sequence predict problems can be broadly categorized into the following categories [9]:

- One-to-One: There is one input and one output. Typical example of a one-to-one sequence that through one event to predict a single event, such as $A_i \rightarrow B_j$ $(1 \leq i \leq n, 1 \leq j \leq n)^1$.
- One-to-Many: In one-to-many sequence predict scenario, we have a single event as input and give a sequence of outputs, such as $A_i \rightarrow B_j, B_{j+1}, ...$ $(1 \leq i \leq n, 1 \leq j < j + 1 \leq n)$.
- Many-to-One: In many-to-one sequence predict scenario, we have a event sequence as input and predict a single output, such as $A_i, A_{i+1}, ... \rightarrow B_j$ $(1 \leq i < i + 1 \leq n, 1 \leq j \leq n)$.

---

[1] In order to distinguish the antecedent and consequent of event sequence, the *eventID* of antecedent is represented by $A$, and the *eventID* of subsequent is represented by $B$.

– Many-to-Many: Many-to-many sequence problems involve a sequence input and a sequence output, such as $A_i, A_{i+1}, \dots \rightarrow B_j, B_{j+1}, \dots$ ($1 \leq i < i+1 \leq n, 1 \leq j < j+1 \leq n$) (Fig. 3).

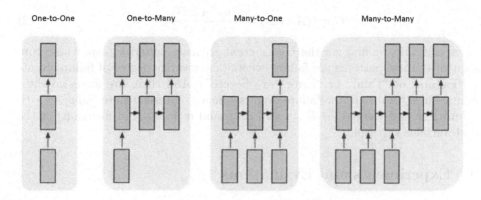

**Fig. 3.** The categories of sequence predict problems.

In this work, we build the Many-to-One sequence predict scenario based on the LSTM training network, and generate the *N-ary event sequence pattern N-esp*, which is an implication like $A_i, A_{i+1}, \dots \rightarrow B_j$, and $A_i, A_{i+1}, \dots$ is the antecedent, $B_j$ is the consequent.

## 2.4  Failure Rules Mining

Because failure rules record a series of events that lead to machine failures, and can be used to understand the failure behaviors of machine cluster, and so on. So we extract the failure rules from the *minimal event sequence patterns $esp_{min}$*, which is the event sequence pattern that are generated by compressing the adjacent events in antecedent, for example, the minimal event sequence pattern of (a,a,a,b,e) → f is (a,b,e) → f, here, *a, b , e, f* indicates the specific events.

Moreover, we call the series of events that lead to failures as *failure signs*, and the *failure rules* are actually the association rules between failure signs and failures. Faced with the mined numerous failure rules, we also need to measure which failure rules are more valuable. General, the *support Supp* and *confidence Conf* are the commonly used objective measurements of event rules' interestingness [10].

If A → B is a minimal event sequence pattern, the support *Supp(A → B)* denotes the probability that *A* and *B* appears simultaneously, which indicates the practicality of event rules, and is calculated according to the formula (1). Here, *A* may be a N-ary sequence.

The confidence *Conf(A → B)* denotes the conditional probability of the consequent actually occurring after the antecedent events occurred, which indicates whether useful association rules can be derived from the discovered patterns, and is calculated according to the formula (2).

$$Supp(A \rightarrow B) = Supp(A \cup B) \tag{1}$$

$$Conf(A \rightarrow B) = \frac{Sup(A \cup B)}{Sup(A)}) \tag{2}$$

Specifically, we first get the *failure event sequence patterns fesp*, whose consequent of these patterns are failure events (the severity degree of failure events is "Failure" or "Fatal" or "Error" or "Severe"). And then, We focus on these failure rules, which are the failure event sequence patterns whose *Supp(A → B)* ≥ *min_sup* and *Conf(A → B)* ≥ *min_conf*, and *min_sup* and *min_conf* is the customizable threshold.

## 3  Experiments and Evaluations

We evaluate our experiments on the publicly available production Bluegene/L dataset. Based on the pre-processing results of LogMaster, the total number of *eventID* is 74. So the 74-dimensional vector by using the one-hot encoding method is fed to a LSTM network. Our LSTM network has one hidden layer of 64 LSTM cells. We train the network using mini-batch stochastic gradient descent with batch size 64 and exponential decay method with base learning rate 10 and decay factor 0.1. In order to get a relation optimum solution quickly, we first set a larger learning rate and then gradually reduce the learning rate through iteration, which could make the model more stable in the later stage of training.

We also use 80% of this data for training the LSTM network and test it on the remaining 20% of data. RMSE is used to calculate accuracy of the LSTM network for both training and testing phase. And the Accuracy, Precision, Recall, Fscore of LSTM Model is shown in Table 1.

**Table 1.** The Accuracy, Precision, Recall, Fscore of LSTM Model.

| Accuracy | Precision | Recall | Fscore |
|---|---|---|---|
| 96.8611% | 96.9236% | 96.8611% | 96.8615% |

After training of LSTM network, it generates 5-ary event sequence patterns 5-esp, such as (49, 69, 69, 69) → 66, and the total number of 5-esp is 442523; And then, we extract the minimal event sequence patterns $esp_{min}$, and the total number of $esp_{min}$ is 4867; Based on the severity degree of consequent of minimal event sequence patterns, we get 2682 failure event sequence patterns, which are listed in Table 2.

**Table 2.** The 5-ary event sequence patterns and minimal event sequence patterns after training of LSTM network.

| Patterns | Formats | Number |
|---|---|---|
| 5-esp | (49, 69, 69, 69) → 66 | 442523 |
| $esp_{min}$ | (49, 69) → 66 | 4867 |
| fesp | 44 → 59 | 2682 |

In addition, the consequent of failure event sequence patterns have different event types, so we classify the failure event sequence patterns based on the event type of consequent. And the classification of failure event sequence patterns is listed in Table 3.

**Table 3.** The classification of failure event sequence patterns.

| Consequent | Event type | Desc of event | Num of fesp |
|---|---|---|---|
| 1 | Hardware | $R\_DDR\_STR$ Error | 38 |
| 2 | Hardware | $R\_DDR\_EXC$ Failure | 23 |
| 3 | Network | Error | 10 |
| 40 | Application | MONILL Failure | 20 |
| 44 | Application | APPCIOD Fatal | 853 |
| 45 | Application | APPDCR Fatal | 21 |
| 47 | Application | MASFAIL Failure | 12 |
| 50 | Application | MMCS Error | 48 |
| 51 | System | KILLJOB Fatal | 70 |
| 54 | Hardware | DIS Error | 326 |
| 55 | Hardware | DIS Severe | 329 |
| 57 | Hardware | HARDSEVE Severe | 86 |
| 58 | Hardware | MONTEMP Failure | 69 |
| 59 | Network | MONLINK Failure | 40 |
| 60 | Hardware | MONHARD Failure | 22 |
| 61 | Hardware | MONFANM Failure | 16 |
| 62 | Hardware | MONFANS Failure | 11 |
| 63 | Network | MONPGOOD Failure | 9 |
| 64 | Hardware | MONHARDS Severe | 14 |
| 71 | System | KERNFDDR Fatal | 46 |
| 72 | Application | KERNFPI Fatal | 382 |
| 73 | Filesystem | KERNFPLB Fatal | 59 |
| 74 | I/O | KERNFEII Fatal | 178 |

We sorted the failure event sequence patterns according to *supp*, and then selected the top 10 failure rules with *conf* > 0.1, which are shown in Table 4. From the Table 4, we try to conclude that: 1) One type of failure events are likely to cause the same type of failure events to occur, such as 44 → 44, and 44 is an failure type of application; 2) Some types of events always occur simultaneously, and form the event clusters. For example, (53, 54, 55) is an event cluster dominated by DIS events, here, 53 is the event type of DIS warining, 54 is the event type of DIS error, and 55 is the event type of DIS severe.

**Table 4.** The top 10 failure rules.

| Patterns | Supp | Conf | Desc |
|---|---|---|---|
| 44 → 44 | 53674 | 0.35 | APPCIOD Fatal → APPCIOD Fatal |
| 69 → 44 | 40542 | 0.36 | KERNICIOD Info → APPCIOD Fatal |
| 44 → 72 | 20556 | 0.13 | APPCIOD Fatal → KERNFPI Fatal |
| 72 → 44 | 19054 | 0.36 | KERNFPI Fatal → APPCIOD Fatal |
| 69 → 72 | 13928 | 0.12 | KERNICIOD Info → KERNFPI Fatal |
| 66 → 44 | 8620 | 0.36 | KERNIPI Info → APPCIOD Fatal |
| (53, 55, 54, 53) → 44 | 4316 | 0.39 | DIS Warning, DIS Severe, DIS Error, DIS Warning → APPCIOD Fatal |
| (54, 53, 55, 54) → 44 | 4252 | 0.39 | DIS Error, DIS Warining, DIS Severe, DIS Error → APPCIOD Fatal |
| (55, 54, 53, 55) → 44 | 4107 | 0.38 | DIS Severe, DIS Error, DIS Warning, DIS Severe → APPCIOD Fatal |
| 66 → 72 | 4107 | 0.38 | KERNIPI Info → KERNFPI Fatal |

## 4   Related Work

Since System logs of large-scale clusters track system behaviors by accurately recording detailed data about the system's changing states, they are the primary resources for implementing dependability and failure management [11]. Based on system logs, numerous log mining tools have been designed for different systems [12].

Many use rule-based approaches [13,14], which are limited to specific application scenarios and also require domain expertise. M. Cinque et al. [13] analyzed the limitations of current logging mechanisms and proposed a rule-based approach to make logs effective to analyze software failures. R. Ren et al. [14] proposed a hybrid approach that combines prior rules and machine learning algorithms to detect performance anomalies, such as, the detection methods of straggler tasks, task assignment imbalance and data skew are rule-based.

Other generic methods that use system logs for anomaly detection and failure management are statistical-based approaches [15,16]. Liang et al. [15] investigated the statistical characteristics of failure events, as well as the statistics about the correlation between occurrence of non-fatal and fatal events. Gujrati et al. [16] presented a meta-learning method based on statistical analysis and standard association rule algorithm, but it only focused on some specific failure patterns and predicted one of them would happen without further information.

In addition, there are many methods that using machine learning algorithms to mine the failure rules and significant patterns. For example, LogMaster [7] proposed an improved Apriori-LIS and Apriori-semiLIS algorithms to mine rules, which can improve sequence mining efficiency. Zhou et al. [17] presented an online log analysis algorithm Apriori-SO, and an online event prediction method that can predict diversities of failure events with the great detail. Z. Zheng et al. [18] mined the relationships between fatal and non-fatal events to predict failures that including location and lead time. A. Gainaru et al. [2] used a dynamic window strategy to find the frequent sequences, by utilizing an Apriori-modified algorithm. Y. Watanabe et al. [19] proposed an online failure prediction method by real-time message pattern learning in cloud datacenter, which identifies the relationship between the messages and failures.

Recently, several researchers have used deep neural networks in system log analysis. Zhang et al. [20] presented a log-driven failure prediction system for complex IT systems, which uses clustering techniques on the raw text from multiple log sources to generate feature sequences fed to an LSTM for hardware and software failure predictions. Du et al. [12] employ customized parsing methods on the raw text of system logs to generate sequences for LSTM Denial of Service attack detection. Brown et al. [21] presented recurrent neural network (RNN) language models augmented with attention for anomaly detection in system logs. Ren et al. [22] proposed a log classification system based on deep convolutional neural network for event category classification on distributed cluster systems.

# 5   Conclusion and Future Work

In this paper, we utilize the Long Short Term Memory network to train and mine the event sequence patterns of Bluegene/L logs. On the basis of event sequence patterns, we further mine the correlations between failure signs and failure events, and generate the valuable failure rules with higher support and confidence. In addition, we evaluate our experiments on the publicly Bluegene/L system logs. From the obtained interesting rules, we find that one type of failure events are likely to cause the same type of failure events to occur, and some types of events always occur simultaneously to form the event clusters.

**Acknowledgment.** This project is supported by the National Key Research and Development Program of China under Grant No. 2017YFB1001602. The authors appreciate the valuable comments provided by the anonymous reviewers. Authors thank the experimental platform provided by Center for Advanced Computer Systems, Institute of Computing Technology, Chinese Academy of Science.

# References

1. Liang, Y.: Filtering failure logs for a bluegene/l prototype. In: Proceedings of DSN (2005)
2. Gainaru, A., Cappello, F., Fullop, S., Trausan-Matu, J., Kramer, W.: Adaptive event prediction strategy with dynamic time window for large-scale hpc systems. In: Managing Large-scale Systems via the Analysis of System Logs and the Application of Machine Learning Techniques(SLAML) (2011)
3. Wang, M., Xu, L., Guo, L.: Anomaly detection of system logs based on natural language processing and deep learning. In: The 4th International Conference on Frontiers of Signal Processing (2018)
4. Krizhevsky, A., Sutskever, I., Hinton, G.: Imagenet classification with deep convolutional neural networks. In: Proceedings of the 25th International Conference on Neural Information Processing Systems (NIPS), pp. 1097–1105 (2012)
5. Lipton, Z.C., Berkowitz, J., Elkan, C.: A critical review of recurrent neural networks for sequence learning. In: https://arxiv.org/abs/1506.00019 (2015)
6. Hochreiter, S., Schmidhuber, J.: Long short-term memory. Neural Comput. **9**(8), 1735–1780 (1997)
7. Fu, X., Ren, R., Zhan, J., Zhou, W., Jia, Z., Lu, G.: Logmaster: mining event correlations in logs of large-scale cluster systems. In: IEEE 31st Symposium on Reliable Distributed Systems (SRDS) (2012)
8. Gers, F.A., Schmidhuber, J., Cummins, F.: Long short-term memory. Neural Comput. **12**(10), 2451–2471 (2000)
9. Malik, U.: Solving sequence problems with lstm in keras. https://stackabuse.com/solving-sequence-problems-with-lstm-in-keras/
10. Geng, L., Hamilton, H.J.: Interestingness measures for data mining: a survey. ACM Comput. Surv. **38**, 9–es (2006)
11. Fu, X., Ren, R., McKeez, S.A., Zhan, J., Sun, N.: Digging deeper into cluster system logs for failure prediction and root cause diagnosis. In: IEEE International Conference on Cluster Computing(Cluster), pp. 103–112 (2014)
12. Du, M., Li, F., Zheng, G., Srikumar, V.: Deeplog: anomaly detection and diagnosis from system logs through deep learning. In: Proceedings of the 2017 ACM SIGSAC Conference on Computer and Communications Security (2017)
13. Cinque, M., Cotroneo, D.o., Pecchia, A.: Event logs for the analysis of software failures: a rule-based approach. IEEE Trans. Softw. Eng. **39**(6), 806–821 (2013)
14. Ren, R., et al.: Hybridtune: spatio-temporal performance data correlation for performance diagnosis of big data systems. J. Comput. Sci. Technol. **34**, 1167–1184 (2019)
15. Liang, Y.: Bluegene/l failure analysis and prediction models. In: DSN, pp. 426–435 (2006)
16. Gujrati, P.: A meta-learning failure predictor for blue gene/l systems. In: Proceedings of ICPP (2007)
17. Zhou, W., Zhan, J., Meng, D., Zhang, Z.: Online event correlations analysis in system logs of large-scale cluster systems. In: Network and Parallel Computing (NPC) (2010)
18. Zheng, Z., Lan, Z., Gupta, R., Coghlan, S., Beckman, P.: A practical failure prediction with location and lead time for bluegene/p. In: Proceedings of DSN-W (2010)

19. Watanabe, Y., Otsuka, H., Sonoda, M., Kikuchi, S., Matsumoto, Y.: Online failure prediction in cloud datacenters by real-time message pattern learning. In: International Conference on Cloud Computing Technology and Science, pp. 504–511 (2012)
20. Zhang, K., Xu, J., Min, M.R., Jiang, G., Pelechrinis, K., Zhang, H.: Automated it system failure prediction: a deep learning approach. In: 2016 IEEE International Conference on Big Data (Big Data) (2016)
21. Brown, A., Tuor, A., Hutchinson, B., Nichols, N.: Recurrent neural network attention mechanisms for interpretable system log anomaly detection. In: arXiv:1803.04967v1 [cs.LG] 13 Mar 2018 (2018)
22. Ren, R., et al.: Deep convolutional neural networks for log event classification on distributed cluster systems. In: IEEE International Conference on Big Data (2018)

# Traffic Crowd Congested Scene Recognition Based on Dilated Convolution Network

Xinlei Wei[1]([✉]), Yingji Liu[1], Wei Zhou[1], Haiying Xia[1], Daxin Tian[2], and Ruifen Cheng[3]

[1] Research Institute of Highway Ministry of Transport, Beijing, China
weixinlei-2-2-2@126.com
[2] School of Transportation Science and Engineering, Beihang University, Beijing, China
dtian@buaa.edu.cn
[3] School of Management, Zhengzhou University of Industrial Technology, Zhengzhou, China
1067772794@qq.com

**Abstract.** With the development of the city, the traffic crowd congested scene is increasing frequency. And the traffic crowd congested may bring disaster. It is important for city traffic management to recognize traffic crowd congested scene. However, the traffic crowd scene is dynamically and the visual scales are varied. Due to the multi-scale problem, it is hard to distinguish the congested traffic crowd scene. To solve the multiple scales problem in traffic crowd congested scene recognition, in this paper, a traffic crowd congested scene recognition method based on dilated convolution network is proposed, which combines the dilated convolution and VGG16 network for traffic crowd congested scene recognition. To verify the proposed method, the experiments are implemented on two crowd datasets including the CUHK Crowd dataset and Normal-abnormal Crowd dataset. And the experimental results are compared with three states of the art methods. The experimental results demonstrate that the performance of the proposed method is more effective in congested traffic crowd scene recognition. Compared with the three state of the art methods, the average accuracy value, and the average AUC values of the proposed method are improved by 15.87% and 11.58% respectively.

**Keywords:** Traffic crowd · Congested scene · Dilated convolution · Deep learning · Two-stream convolution

---

This work was supported in part by the Central Public Research Institutes Special Basic Research Foundation No. 2020–9004, No. 2020–9065, in part by the National Natural Science Foundation of China under Grant Nos. 61672082 and 61822101, in part by Beijing Municipal Natural Science Foundation No. 4181002.

W. Gao et al. (Eds.): FICC 2020, CCIS 1385, pp. 134–146, 2021.
https://doi.org/10.1007/978-981-16-1160-5_12

# 1   Introduction

With the development of city and tourism, crowd often appear in scenic spots, train station and bustling urban areas, especially during holidays. Public safety events often occur when people are crowded, such as trample and so on. If congestion is detected in advance, it is possible to prevent accidents. It is important to detect crowd congestion for public safety.

At present, most methods of crowd congestion detection employ two kinds of data, which include crowdsourcing data and video data. The crowdsourcing data often brings much noise and it is not intuitive to detect crowd congestion scene. Thus the performance of crowd congestion prediction based on it is not significantly improved. The advantages of video data are intuitive and effective to detect crowd congestion. Thus this paper concentrates on crowd congestion detection based on video data.

While the traffic crowd congestion detection based on video are facing with two challenges at present. The first challenge is background independence of crowd congestion scene detection. Most crowd congestion detection models are trained in a certain context. It does not cross background to detect crowd congestion scene.

Another challenge is multi-scale crowd congestion scene detection. The most methods of crowd congestion detection based on global crowd density estimation and crowd counting estimation. However, in most situations, due to limitation of visual scale of video, the global crowd density and crowd counting do not reflect the real crowd congestion level. Such as in outdoor scenic spot, though there are many people in global, people do not feel congestion. But, in narrow entryway, there are few people, but the velocity is slow and most people feel crowd congestion. The reason of crowd congestion is not only related to the crowd counting and density, but also related to the crowd movement and around environment.

In this paper, the main tasks are to solve the back ground independence problem and multi-scale problem in crowd congestion scene recognition.

To solve the background independence problem of crowd congestion scene recognition, the motion map [1] is used. The most of cross-scene crowd counting methods [2] are based on the density map [3]. To generate the density map, in training stage, the local position pixels need to be found by hand at first and the Gaussian distributions of these corresponding pixels need to be calculated by 2D Gaussian kernel. It costs a lot of manual work to label the position of person in the crowd scene. However, in crowd congestion scene, too many people are occupied to mark the position of people.

Normally, in crowd congestion scene, the motion pattern is analogous and background independent. And the motion feature of crowd congestion scene is a key factor to detect crowd congestion scene. When the velocity of movement of crowd is fast, it means that the crowd is not congested. While the velocity of movement of crowd is slowly, it means that the congestion is arises in the case of a high probability.

The motion features have various formats, such as optical flow, dynamic texture and so on. But the motion map is achieved by calculating the KLT trajectory of crowd movement. So it is able to reflect the movement pattern of crowd and background independence. The motion map includes three scene-independent motion descriptors, which are collectiveness descriptor, stability descriptor and conflict descriptor. In the static channel, the values of RGB image imply the static features of environment around the group and static crowd. In the motion channel, the motion map are extracted from crowd trajectories imply the motion feature of congestion crowd. The target of these features are to represent the crowd scene. At present, it mainly faces two problems including large variety scale of traffic crowd scene and traffic crowd scene feature represent.

To solve the multi-scale problem of crowd congestion scene recognition, the dilated convolution is employed. Dilated convolution has good performance in multi-scale scene analysis, such as crowd counting [4] and image segmentation [5]. But it is not able to distinguish the crowd congestion scene. However the motion map contains the motion information and the dilated convolution is also used in the motion map to make a stand against multi-scale problem of crowd scene.

In general, the depth is deeper of convolutional layers and it is more robust to recognize object and scene [6]. To extract effective feature of the motion map and static map, the very deep learning network architecture VGG16 model is employed. And it has been proved that the VGG16 model is robust and effective in crowd congestion scene [4].

The end-to-end two-column deep learning architecture is effective for recognizing the congestion scene. The two columns include the deep static feature column and deep motion feature column, where the inputs are RGB image and motion map respectively. The RGB image and the corresponding motion map are pairs. In fact, the corresponding motion map reflects the temporal relation, which is constructed by the clustering of three crowd motion trajectories [1]. However, the motion map is just superficial characterization of crowd motion and there are too many objects in RGB map. Hence, it is difficult to represent the intrinsic quality of congestion crowd by shallow network. The very deep convolution network [7] is effective to distill the intrinsic feature. In this paper, the very deep convolutional layers are employed in the two-column deep learning architecture and it can improve the performance of the two-column deep learning method. Thus, a two-column very deep learning method is proposed to recognize crowd congestion scene.

Our objective in this paper is to construct an effective traffic crowd congestion scene recognition model in video. For that, in this paper, the main works are summarized as follows:

(1) A two-column dilated convolutional network is proposed to recognize crowd congestion scene;
(2) An effective crowd congestion scene recognition model based on very deep learning is constructed;

(3) Verifying proposed method on two public dataset Normal-abnormal dataset and CUHK Crowd dataset and the experimental results demonstrate that proposed method is effective to recognition crowd congestion scene in video.

We structure the remainder of this paper as follows. In Sect. 2, the related works are introduced. The proposed two-column very deep learning based crowd congestion detection model is introduced in Sect. 3. Performance studies on two public crowd datasets, Normal-abnormal dataset and CUHK Crowd dataset, are given in Sect. 4. Finally, we conclude the paper in Sect. 5.

## 2   Related Work

### 2.1   Congested Scene Recognition Based on Sensed Data

At present, two kinds of data sources are used to detect congested crowd, which are sensed data and video data. In sensed data field, Elhamshary, et al. [8] leverage sensed data from smart phones to predict congestion levels. In this method, the smart phone data is not easy to obtain and has amount noise. Pattanaik, Vishwajeet, et al. [9] propose a real-time congestion avoidance method, which utilize K-Means Clustering of Global Positioning System data from mobile phone to predict shortest route. Nguyen, Hoang, et al. [10] propose an algorithm which constructs a causality trees from congestion information and estimates the propagation probabilities based on temporal and spatial information. The method is validated by experiments on a travel time data set recorded from an urban road network. Scott Workman et al. [11] use overhead imagery to learn the dynamic model of traffic, which is based on the computer vision method. The method is not able to detect the crowd congestion scene.

The shortcoming of sensed data is that the sensed data contains a lot of noise and is not intuitive enough. However, the video data is intuitive for crowd congestion scene detection.

Recently, video transmission technology is able to real-time translate the high quality video [3], hence, most crowd congestion scene detection methods based on surveillance video data are able to recognition the crowd congestion scene with real-time. Huang, Lida, et al. [12] utilize the velocity entropy to detect the congestion of pedestrian. In this method, the velocity entropy is computed by optical flow. Thus it is easily upset by illumination noise. Yuan Yuan et al. [13] propose congested scene classification via unsupervised feature learning and density estimation. In this method, it is not reasonable for identification crowd congestion by the global density estimation. The crowd congestion is related to the local density and the crowd motion. Bek, Sebastian et al. [14] utilize track density to estimate the congestion level. This method takes into account crowd motion factors by the local inertia which is based on track density, but this method is effective to motion crowd. However, many people are almost immovable in crowd congestion scene. This method is not able to estimate the static crowd.

## 2.2    Dilated Convolutional Neural Networks

The Dilated Convolutional Neural Networks (CSRNet) are proposed by Yuhong Li et al. [4] to estimate the crowd counting and crowd density of congested scene for understanding congestion scene. The key of this method is that a dilated convolution layer based on dilated kernel is embedded into back-end of the network to replace the pooling layers. In fact, the main performance of this method is to estimate crowd density and crowd counting. It does not directly detect the crowd congestion scene. The crowd congestion scene is not a static scene. In some situation, though there are few people, the crowd congestion still happen, such as in bottleneck place. Thus the motion factor needs to be considered in crowd congestion detection. Shuai Bai et al. [15] propose the adaptive dilated network for crowd counting to meat the variation of scales in different scenes. This method utilize the adaptive dilated convolution to solve the multi-scale problems in crowd scenes. The dilate convolution is also used in image aesthetics assessment [16]. In summary, it can be demonstrated that the dilate convolution is worked in scene recognition.

## 2.3    Limitations of the State-of-the-art Approaches

Most recently, Yuhong Li et al. [4] propose the dilated convolution based network CSRNet, which is used to estimate the number of people in congested scene. In this method, three dilated rates are used and they are non-adaptation with the variation of object scales in different scenes. Yingying Zhang et al. [17] propose multi-column network to estimate the crowd density map for counting the number of people. For these methods, to archive the task of estimating crowd density map, the locations of heads in the crowd congestion scene are marked in training stage. However it's unpractical for raw videos/images of crowd congestion scene. They are not directly detection crowd congestion and it needs to dependents on manual judgment for crowd congestion recognition. Meanwhile, the motion feature is not employed.

# 3    Crowd Congestion Scene Detection Based on Two-Column Very Deep Learning

In this section, proposed crowd congestion scene detection method is introduced. The proposed two-column very deep dilated convolution network architecture for crowd congestion scene detection and the corresponding piplin are introduced in Subsects. 3.1, 3.2, 3.3, and 3.4.

## 3.1    Dilated Convolution on Two-Column Network

In the traffic scenes, there is large scale variation in different scenes. Even in the same scene, the scale still changes dramatically due to perspective phenomenon [15]. The dilated convolution is effective to larege scale variation. In the proposed

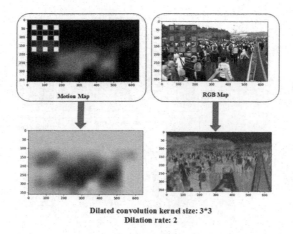

Fig. 1. 3*3 convolution kernels with dilation rate as 2 on RGB map and motion map respectively. (Color figure online)

two columns network, the dilated convolution layers are employed. In specially, the dilated convolution is able to reduce the loss of information in motion-channel column. The 2-D dilated convolution of motion map can be defined as follow:

$$y(h,w) = \sum_{i=1}^{H} \sum_{j=1}^{W} map(h + r \times i, w + r \times j)k(i,j) \qquad (1)$$

where $y(h,w)$ is the output of dilated convolution layer. $map(h,w)$ is the motion map and a filter $k(i,j)$ with the high $H$ and width $W$ respectively. $r$ is the dilation rate. In this paper, the $3*3$ Sobel kernel [4] is used in both operations while the dilation rate is 2. It is shown in Fig. 1.

Dilated convolution is demonstrated in congestion crowd counting task based on RGB image and estimating crowd density map task with significant improvement of accuracy. In the proposed two-column network, the dilated convolution is used in the two columns, which are RGB column and motion column respectively. The network architecture is shown in Fig. 2.

### 3.2 Proposed Crowd Congestion Detection Framework

The proposed architecture takes into account the motion factor and static factor. Meanwhile, the very deep convolutional layers are used to improve the robust. The pipline is shown in Fig. 1. At first, the trajectories are calculated by KLT method [18] from frame sequence. Thus the motion map of frame is calculated by the trajectory. The very deep learning is used to extract very deep convolutional features from key frame and corresponding motion map respectively. In Fig. 1 the blue cubes represent the convolutional layers. The pink cubes represent the pooling layers. The Dilated Conv represents the dilated convolution. In Fig. 1, the Dilated Convolutional layers parameters are denoted as Dilated

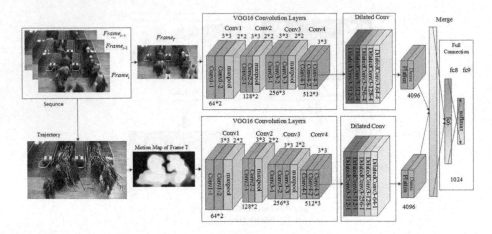

**Fig. 2.** The framework of Two-column very deep dilated convolution network architecture for crowd congestion detection (Color figure online)

Conv-(kernel size)-(number of filters)-(dilated rate). The orange cube represents the merge layer, which is used to fuse the feature which is from key frame and feature which is from motion map. The green rectangles represent the full connection layers. The sigmoid function is selected to be as final outputs.

The pipeline of proposed method has eight steps, which are shown in Algorithm 1. In Algorithm 1 the TCDC-Net is the short name of Two-column very deep Dilated Convolution Network(TCDC-Net).

### 3.3  Constructing Two-Column Dilated Convolutional Network Architecture

The proposed two-column very deep learning architecture is two channel deep learning architectures and each channel is very deep convolutional layers which are same as the convolutional layers of VGG16 model. After the two-column convolutional layers, there is the merge layer, which is used to fuse the two channel features. In follow, they are full connectional layers, which are used to optimize the fuse feature. The activation function of output layer is softmax. The details of the network architecture are shown in Fig. 1. The neuron nodes values are calculated by Eq. (2), which are in full connection layer. And exponential linear unit (ELU) [19] is selected as the active function which is shown by Eq. (3) and Eq. (4), where $\alpha$ is the hyperparameter.

$$y_i^{(l+1)} = f'\left(w_i^{(l+1)} \bullet y_i^{(l)} + b_i^{(l+1)}\right) \tag{2}$$

$$f(x) = \begin{cases} x & if\ x \geq 0 \\ \alpha(exp(x) - 1) & if\ x < 0 \end{cases} \tag{3}$$

$$f'(x) = \begin{cases} 1 & if \ x \geq 0 \\ f(x) + \alpha \ if \ x < 0 \end{cases} \tag{4}$$

The convolutional layers of proposed two-column very deep learning network is VGG16 model convolutional layers and the parameters are also from the VGG16 model trained on imagenet dataset.

---

**Algorithm 1.** Traffic Crowd Congested Scene Recognition Based on Dilated Convolution Network Algorithm

---

**Input:** Frame sequence *Seq*
**Output:** Crowd congestion label *Y*
  Initialize: Set the interval of frame in the sequence *Seq*
  Step1: Calculate the KLT trajectory *T*
  Step2: Extract key frame every *S* frames
  Step3: Calculate motion map of corresponding key frame
  Step4: Train the proposed TCDC-Net network
  Step5: Get the proposed two-column network model TCDC-Net
  Step6: Input the test frame into the trained network model TCDC-Net
  Step7: Output the crowd congestion label *Y*

---

### 3.4 Learning Crowd Congestion Scene Recognition Model Based on Two-Column Dilated Convolution Network

Learning crowd congestion scene recognition model is iterative optimization process, where gradient descent method is employed. It is shown in Algorithm 2. In optimization process, the object function is cross entropy loss function. It is shown in Eq. (5).

$$L(y, o) = -\frac{1}{N} \sum_{i=1}^{N} (y_i \log(o_i) + (1 - y_i)log(1 - o_i)) \tag{5}$$

Where, $y_i \in y, i = 1, 2, ..., N$ is the ground truth label and $o_i \in o, i = 1, ..., N$ is the predicted likelihood value. $N$ is the number of training samples.

## 4   Experiment

This section introduces the implementation of experiment. It includes dataset and metrics, experimental results and analysis.

---

**Algorithm 2.** TCDC-Net Training Algorithm

---

**Input:** Input training set $trainX_{rgb}$, $trainX_{motion}$, $trainY$, iterations number $iter = c$, batch size $bs$,

**Output:** Crowd congestion recognition model $W$

    Initialize very deep convolutional layers parameters $W_{vgg}^{(0)} = W_{vgg}$ by VGG16 model; and full connection layers weight paramters $W_f^{(0)} = random()$ by random, batch size $bs$

    **while** i<iter **do**

        $l = 0$

        **for** j=1:bs **do**

            $f_{motion}^{vgg} = VGG(trainX_{motion}(j), W_{motion-vgg}^{(i)})$

            $f_{rgb}^{Dilated} = DilatedConv(f_{rgb}^{vgg})$

            $f_{rgb}^{vgg} = VGG16(trainX_{rgb}(j), W_{rgb-vgg}^{(i)})$

            $f_{motion}^{Dilated} = DilatedConv(f_{motion}^{vgg})$

            $output_f = fcl([f_{rgb}^{Dilated}, f_{motion}^{Dilated}], W_f^{(i)})$

            $y^{'} = sigmoid(output_f)$

            $l^{(j+1)} = L(y^T, y)$

        **end for**

        $delt^{(i)} = SGD(l^{(j)}, W_f^{(i)}, W_{rgb-vgg}^{(i)}, W_{motion-vgg}^{(i)})$

        $W_{rgb-vgg}^{(i+1)} \leftarrow Update(delt^{(i)}, W_{rgb-vgg}^{(i)})$

        $W_{motion-vgg}^{(i+1)} \leftarrow Update(delt^{(i)}, W_{motion-vgg}^{(i)})$

        $W_f^{(i+1)} \leftarrow Update(delt^{(i)}, W_f^{(i)})$

    **end while**

    $W = [W_{rgb-vgg}^{(i+1)}, W_{motion-vgg}^{(i+1)}, W_f^{(i+1)}]$

    **return** $W$

---

### 4.1 Dataset and Metrics

To evaluate the proposed method, two crowd datasets are selected to test proposed method, which are CUHK Crowd dataset [1] and Normal-abnormal Crowd dataset. The Normal-abnormal Crowd dataset is constructed by collecting 300 crowd videos, where there are 120 normal crowd videos and 80 abnormal crowd videos. In the Normal-abnormal Crowd dataset, the most frames video has 800 frames and the lest frames video has 230 frames. Meanwhile, these videos are splitted by every 48 frames. At final, these videos are splitted into 5836 simples. In this paper, 70% samples are selected as training dataset and 30% samples are selected as test dataset. The details information of the two dataset is shown in Table 1. The crowd types of each clips is distinguished and two widely used metrics, mean Area Under ROC Curve (AUC) [1] and Mean Accuracy [20] are used to evaluate performance of methods.

### 4.2 Training Details

In the training process, there are two parameters are able to influence the performance of trained model, which are iterations number and batch size of network

**Table 1.** The crowd types video dataset

| Dataset | CUHK crowd dataset | Normal-abnormal crowd dataset |
|---|---|---|
| Videos number | 320 | 300 |
| Samples image | 4997 | 5836 |
| Scenes | City | City and station |
| Indoor/Outdoor | Outdoor and indoor | Outdoor and indoor |

input. In order to more objective comparison, the iterations numbers of all deep learning based algorithms are set as 1000. Meanwhile, the batch sizes of network inputs are set as 16. To improve the training efficiency of the proposed two-stream model, the exponential attenuation method is used to achieve the decayed learning rate $dlr$, which is updated by every iteration.

$$dlr = lr \bullet dr^{gs/ds} \qquad (6)$$

where, $lr$ is the initial learning rate and set as $0.8.dr$ is decay rate and set as $0.99$. $gs$ is the number of current iterating rounds. $ds$ is decay steps $ds = \frac{bN}{bS}$. $bN$ is the number of batch and $bS$ is the batch size.

### 4.3   Comparison Results

There are three methods are selected as the comparison methods, that include three deep learning based method and three low-level features based method. The experimental results of two datasets are shown in Tables 2 and 3 respectively.

It can be observed that the accuracy value of the proposed TCDC-Net method is higher than accuracy values of other methods from the Table 3. The reasons are that the proposed TCDC-Net method utilized the very deep convolution layers based on VGG16 architecture, which can extract more abstracted image feature, meanwhile, the two columns architecture including the static image channel and the motion map channel are employed. The two columns architecture can extract motion information and scene information. And the dilated convolution layers are added after VGG16 convolution layers to resist multi-scale.

From Tables 2 and 3, it can be observed that the experimental results of deep learning based method are supor than the experimental results of STL method. The MCNN method is multi-columns convolution neural network, but there is only one channel which is the static channel e.g. RGB image to be used in the method for dynamic crowd scenes. It is not enough power to represent the crowd motion characteristics. In our works, the motion channel is used, which can represent the motion of crowd scene. The parts of experimental results are shown in Fig. 3. There are six different scens. It can be observed that the proposed method is able to recognition the most congestion scene. These scene are about traffice scenes including station scenes, road scenes and car scenes.

**Table 2.** The experimental results on normal-abnormal dataset

| Method | Accuracy | AUC |
|---|---|---|
| CSRNet_C [4] | 0.501 | 0.5 |
| TCDC-Net (proposed) | 0.7305 | 0.5 |
| STL [18] | 0.5623 | 0.5241 |
| MCNN [17] | 0.6752 | 0.5 |

**Table 3.** The experimental result on UCHK Crowd dataset

| Method | Accuracy | AUC |
|---|---|---|
| CSRNet_C [4] | 0.5724 | 0.5 |
| TCDC-Net (proposed) | 0.7325 | 0.568 |
| STL [18] | 0.470476 | 0.4625 |
| MCNN [17] | 0.6552 | 0.5 |

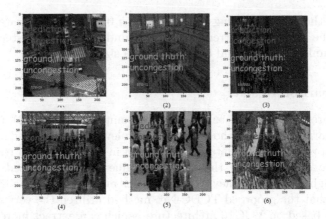

**Fig. 3.** The recognition results of congestion by proposed TCDC-Net method and the ground truth.

## 5    Conclusion

To solve traffic crowd congestion scene recognition, the dilated convolution and VGG16 network are combined to construct the traffic crowd congested scene recognition method is proposed. To solve the method, the experiment is implemented on two datasets. The experimental results demonstrate that the performance of proposed method is effective. In future work, the semantic objects of crowd scene are considered for traffic crowd congestion scene recognition.

**Acknowledgements.** This work was supported in part by the Central Public Research Institutes Special Basic Research Foundation No. 2020–9004 and No. 2020–9065, in part by the National Natural Science Foundation of China under Grant Nos. 61672082 and 61822101, in part by Beijing Municipal Natural Science Foundation No. 4181002.

# References

1. Shao, J., Kang, K., Loy, C.C., Wang, X.: Deeply learned attributes for crowded scene understanding, pp. 4657–4666 (2015)
2. Zhang, C., Li, H., Wang, X., Yang, X.: Cross-scene crowd counting via deep convolutional neural networks, pp. 833–841 (2015)
3. Guo, J., Gong, X., Liang, J., Wang, W., Que, X.: An optimized hybrid unicast/multicast adaptive video streaming scheme over MBMS-enabled wireless networks. IEEE Trans. Broadcast. **64**(4), 791–802 (2018)
4. Li, Y., Zhang, X., Chen, D.: CSRNet: dilated convolutional neural networks for understanding the highly congested scenes, pp. 1091–1100 (2018)
5. Yu, F., Koltun, V.: Multi-scale context aggregation by dilated convolutions. Vision and Pattern Recognition. arXiv: Computer (2015)
6. Lempitsky, V., Zisserman, A.: Learning to count objects in images, pp. 1324–1332 (2010)
7. Simonyan, K., Zisserman, A.: Very deep convolutional networks for large-scale image recognition (2014)
8. Elhamshary, M., Youssef, M., Uchiyama, A., Yamaguchi, H., Higashino, T.: Crowd-Meter: congestion level estimation in railway stations using smartphones, pp. 1–12 (2018)
9. Pattanaik, V., Singh, M., Gupta, P.K., Singh, S.K.: Smart real-time traffic congestion estimation and clustering technique for urban vehicular roads, pp. 3420–3423 (2016)
10. Nguyen, H., Liu, W., Chen, F.: Discovering congestion propagation patterns in spatio-temporal traffic data. IEEE Trans. Big Data **3**(2), 169–180 (2017)
11. Workman, S., Jacobs, N.: Dynamic traffic modeling from overhead imagery. In: IEEE Conference on Computer Vision and Pattern Recognition (CVPR) (2020)
12. Huang, L., Chen, T., Wang, Y., Yuan, H.: Congestion detection of pedestrians using the velocity entropy: a case study of love parade 2010 disaster. Physica A-Stat. Mech. Appl. **440**, 200–209 (2015)
13. Yuan, Y., Wan, J., Wang, Q.: Congested scene classification via efficient unsupervised feature learning and density estimation. Pattern Recogn. **56**, 159–169 (2016)
14. Bek, S., Monari, E.: The crowd congestion level a new measure for risk assessment in video-based crowd monitoring, pp. 1212–1217 (2016)
15. Bai, S., He, Z., Qiao, Y., Hu, H., Wu, W., Yan, J.: Adaptive dilated network with self-correction supervision for counting. In: CVPR2020 (2020)
16. Chen, Q., et al.: Adaptive fractional dilated convolution network for image aesthetics assessment, April 2020
17. Zhang, Y., Zhou, D., Chen, S., Gao, S., Ma, Y.: Single-image crowd counting via multi-column convolutional neural network, pp. 589–597 (2016)
18. Zhang, Y., Qin, L., Ji, R., Zhao, S., Huang, Q., Luo, J.: Exploring coherent motion patterns via structured trajectory learning for crowd mood modeling. IEEE Trans. Circuits Syst. Video Technol. **27**(3), 635–648 (2017)

19. Clevert, D., Unterthiner, T., Hochreiter, S.: Fast and accurate deep network learning by exponential linear units (ELUs). arXiv: Learning (2015)
20. Luo, Y., Wen, Y., Tao, D.: On combining side information and unlabeled data for heterogeneous multi-task metric learning, pp. 1809–1815 (2016)

# Failure Prediction for Large-Scale Clusters Logs via Mining Frequent Patterns

Rui Ren[1](✉), Jinheng Li[2], Yan Yin[3], and Shuai Tian[4]

[1] China Electronics Technology Research Institute of Cyberspace Security Co., LTD,
Beijing, China
renrui871@qq.com
[2] School of Computer and Software, Sichuan University, Chengdu, China
leeebucks@gmail.com
[3] Institute of Computing Technology, Chinese Academy of Sciences, Beijing, China
yinyan512@foxmail.com
[4] Sinotrans, Beijing, China
simontiancn@gmail.com

**Abstract.** As the scales of cluster systems increase, failures become normal and have made reliability management be a major concern for system administrators. Failure prediction is a proactive measure through mining failure patterns and predicting when the systems will fail. In general, it is helpful to improve the accuracy of failure prediction by mining true failure patterns. And currently, the statistical and data mining driven methods are often used for mining failure patterns. However, since the overwhelming volumes and complicated interleaving of logs, the efficient and accurate failure pattern detection and automated failure prediction are still challenging.

In this paper, we utilize the FP-Growth algorithm based on Spark (Spark-FPGrowth) to mine the correlations among different events, which can obtain the long-tail frequent event sequences effectively. Since the preprocessed event transactions are not suitable for using frequent pattern mining algorithms, we propose an adaptive sliding window division method based on event density with/without overlapping to construct event sequence transactions. At last, we analyze the log characteristics and predict failures for three large-scale production systems, and the evaluation results show that the average accuracy rates have higher accuracy and efficiency in CMRI-Hadoop, LANL-HPC and Bluegene/L logs respectively.

**Keywords:** Failure prediction · Spark FP-Growth · Adaptive sliding window division

© Springer Nature Singapore Pte Ltd. 2021
W. Gao et al. (Eds.): FICC 2020, CCIS 1385, pp. 147–165, 2021.
https://doi.org/10.1007/978-981-16-1160-5_13

# 1    Introduction

Large cluster systems have become the common platforms for both high performance computing and cloud computing. As the scales of cluster systems increases, failures become normal [1] and it makes reliability be a major concern [2]. Once failures occur in such distributed environment, they may spread rapidly and widely, and affect the related components for the complicated inter-dependencies between components [3]. It makes troubleshooting and management more difficult and time-consuming for administrators, so proactive fault management technique is essential.

Failure prediction is a proactive measure for preventing disastrous failures and minimizing system crash, which catches signs of system failures and speculates when a failure is going to happen [3,4]. Based on data type, there are two failure prediction methods: metric-based method and log-based method [5]. Since logs record the important information of failures, and they are the first place for system administrators checking, so log-based method is more suitable for failure prediction in large-scale systems [6]. Currently, there are some existing works [2–4,7–9] for automated failure prediction by analyzing logs, which using statistical or data mining methods. For instance, some works proposed Apriori-like algorithms, such as, LogMaster [7]. And P. Gujrati [8] proposed a meta-learning failure predictor without providing details or rare events. Yan *et al.* [10] mined some frequent closed subsequences only, which cannot consider the entire logs. Since the overwhelming volumes, wide variations in formats, and complicated interleaving relationship, it is still a great challenge for efficient rule mining and accurate failure prediction. Especially, as the log size increases, the efficiency of rule mining algorithms may be very low.

In order to improve the efficiency of frequent sequence mining for large-scale logs, we utilize the frequent pattern mining algorithm based on Spark (Spark-FPGrowth) for mining correlations among different events. On the basis of event correlations, we further mine the correlations between failure signs and failure events, which are used for predicting system failures. Since the preprocessed event transactions are not suitable for using frequent pattern mining algorithms, we design an adaptive sliding window division method based on event density with/without overlapping, which is used to generate event sequence transactions. Meanwhile, to obtain the long-tail frequent event sequences and get a higher recall rate, we also set a low threshold $min\_sup$. At last, we analyze the characteristics and predict system failures for three production large-scale systems, a production Hadoop-based cloud computing system at Research Institution of China Mobile, and a production HPC cluster system at Los Alamos National Lab (LANL), Bluegene/L systems respectively. And the average accuracy rates are 78.18%, 64.63% and 67.29%, respectively.

The rest of this paper is organized as follows. Section 2 explains the terminologies. Section 3 presents the methodology and design. Experiment results and evaluations on the large-scale cluster logs are summarized in Sect. 4. In Sect. 5, we describe the related work. We draw a conclusion in Sect. 6.

# 2    Terminology

In this section, we first define some notations that are listed as follows:

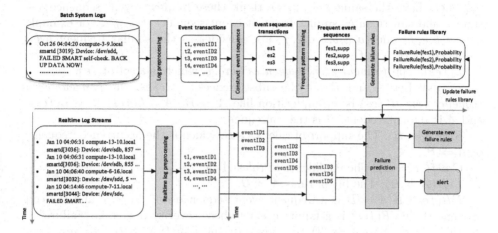

**Fig. 1.** The overview of failure prediction framework.

$e$: The event that defined in LogMaster [7], which is represented by $eventID$ in the form of natural number. Here, $eventID$ is described by 2-tuples (*severity degree, event type*). *severity degree* indicates the priority or importance of events, which includes: Info, Warning, Error, Failure, Fatal, etc., *event type* indicates the type of events, which includes: Hardware, System, Filesystem, Network, and so on.

$Trans(e)$: The *event transactions* is composed of multiple events and timestamps of events, and it is expressed as $Trans(e) = (t_i, e_i)(1 \leq i \leq n)$, here, $t_i$ is the timestamp of event $e_i$.

$es$: An *event sequence* $es = < e_1, e_2, ..., e_n >$ is an ordered sequence of different log events in chronological order, here, $e_i(1 \leq i \leq n)$ is the event element in $es$, and event $e_{i-1}$ occurs before $e_i$. In this paper, $es$ is a sequence of $eventIDs$.

$Trans(es)$: The *event sequence transactions* with time constraints, which is composed of multiple $es_j(1 \leq j \leq m)$ in different periods, and it is expressed as $Trans(es) = (t_j, es_j)(1 \leq j \leq m)$, here, $t_j$ is the initial time of *event sequence* $es_j$.

$sup(es_j)$: The support count of *event sequence* $es_j$, which is the occurrence frequency of $es_j$ in *event sequence transactions* $Trans(es)$.

$sup\_rate(es_j)$: The support rate of *event sequence* $es_j$, which is the proportion of the support count $sup(es_j)$ in the total number of event sequences in $Trans(es)$.

$fes$: The *frequent event sequence* in $Trans(es)$. Here, given a positive integer $min\_sup$ that represents the minimum support count threshold, if $Sup(es_j) \geq min\_sup$, we consider the event sequence $es_j$ is a frequent event sequence $fes$ in $Trans(es)$.

$fes_{min}$: The *minimal frequent event sequence*, which is the event sequences that are generated by compressing the adjacent events in a event sequence, for example, the minimal frequent event sequence of $(a, a, a, b, e, f)$ is $(a, b, e, f)$.

$fes_{simi}$: The *similar frequent event sequence*, if two or more frequent event sequences have the same $fes_{min}$, we think these frequent event sequences are similar and call them as similar frequent event sequence $fes_{simi}$. For example, $(a, a, a, b, e, f)$ and $(a, a, b, b, e, f, f)$ are consider as the similar frequent event sequences of $(a, b, e, f)$.

$L(fes)$: The *frequent event sequences library*, which is a set of $fes$.

$R(fes)$: Let $A$ and $B$ are the subsequences of a $fes$, the *frequent event sequence rule* $R(fes)$ is an implication like $A \Rightarrow B$ $(A \in L(fes), B \in L(fes))$, and $A$ is the antecedent, $B$ is the consequent.

$Sup(A \Rightarrow B)$: The support of rule $A \Rightarrow B$, which is the occurrence frequency of $A \Rightarrow B$.

$Conf(A \Rightarrow B)$: The confidence of rules $A \Rightarrow B$.

$Lift(A \Rightarrow B)$: The lift of rules $A \Rightarrow B$.

$FR(fes)$: If $A \Rightarrow B$ is a frequent event sequence rule $R(fes)$, and the consequent of this $R(fes)$ is a failure event (whose severity degree is "Failure" or "Fatal"), when $Sup(A \Rightarrow B) >= min\_sup$ and $Conf(A \Rightarrow B) >= min\_conf$ and $Lift(A \Rightarrow B) > 1$, we call this rule as a *failure rule* $FR(fes)$.

$L(fr)$: The *failure rules library*, which is a set of $FR(fes)$.

## 3    Methodology

### 3.1    Failure Prediction Framework

The failure prediction framework includes two major phases: *Offline batch training* and *Online failure prediction*. As shown in Fig. 1, in the offline training phase, the failure prediction framework analyzes the whole batch logs to generate failure rules. Specifically, it first leverages LogMaster [7] to execute logs preprocessing and construct event transactions, then builds event sequence transactions by dividing sliding windows, and mines frequent event sequences from the event sequence transactions, at last, generates the failure rules and builds the failure rules library for online failure prediction. In the online failure prediction phase, the failure prediction framework also leverages the similar preprocessing method for realtime log streams, then uses the failure rules library to predict system failures. If there is a failure forecast in a prediction valid duration, the failure prediction framework will give an alert.

### 3.2    Construct Event Transactions

In previous LogMaster, the system log messages with multiple fields are parsed into the nine-tuples (*timestamp, logID, nodeID, eventID, severity degree, event type, application name, processID, userID*) [7], and filtered by removing the repeated *logID*s and periodic *logID*s. Then LogMaster utilizes the improved

Apriori algorithms to discover the association rules between *LogID*s, and the recall rate of prediction is relatively low. In experiments, we also mine the frequent sequences of *LogID*s by using the frequent pattern mining algorithms, and we find that the frequent sequences of *LogID*s with low support count account for a large proportion, which indicates that many *LogID*s sequences appear occasionally. Originally, the *LogID* is composed by *NodeID* and *EventID*, that is, the same event on different nodes will be assigned as different *LogID*, even if they occur in the same period. In addition, since there are different and changeable node topologies in different applications, the correlations of *LogID*s also change with the changes of node topologies, so it is hard to find the particularly strong correlations between different *LogID*s.

Different from discovering the association rules between *LogID*s in LogMaster [7], we decide to build the event transactions $Trans(e)$ through *eventID*s. And the reason is that, *eventID*s abstracts the event types and the events' priority, and the correlations between events are also universal.

### 3.3  Construct Event Sequence Transactions

Before mining frequent event sequences, we first need to divide the event windows for cluster logs. Generally, there are two types of sliding windows: 1) time based sliding window, that is, when setting sliding window size as a time interval $T_{sliding}$, the sliding window is $[T_{start}, T_{start} + T_{sliding}]$ (here, $T_{start}$ is the start timestamp); 2) event density based sliding window, that is, a certain number of events are saving in this sliding window. In this section, we combine these two sliding windows for log data division.

**Adaptive Sliding Window Division Based on Event Density.** In this section, we define *event density*, which is the number of events in a unit sliding window in log data streams. In order to check *event density*, we assume that the size of unit window is 60 min, and then sequentially store the events in sliding window $[T_{start}, T_{start} + 60\ min]$, calculate the event numbers in each sliding window. For example, through the histogram statistics information in Fig. 5, we find that the number of events in unit sliding window is almost 0–10 for three used cluster logs. In addition, we also find that there may be a great amount of burst events in a short period.

In order to construct event sequence transactions, we try to divide the event windows based on event density. Although A. Gainaru [2] has proposed the dynamic time window division method by time interval between the same event type, but this method may cause large sliding window, when the time interval between a certain event type is large. And the large sliding window will bring about greater complexity of association knowledge mining and decrease of mining efficiency. So we propose an *adaptive sliding window division based on event density*, that is, the size of sliding window can be changed automatically according to the event density in log streams. The basic idea of this division algorithm is: making the events in the sliding window $[T_{start}, T_{start} + T_{sliding}]$

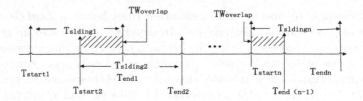

**Fig. 2.** The adaptive sliding window with overlapping $(0 < TW_{overlap} <= T_{slidingi})$.

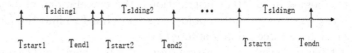

**Fig. 3.** The adaptive sliding window without overlapping $(TW_{overlap} = 0)$.

as the event sequences, and counting the events number in the sliding window; if the events number in the sliding window is greater than the defined threshold $max\_Count_{sw}$, the event sequence just contains the previous $max\_Count_{sw}$ events, and the later events are placed in the next sliding window.

**Adaptive Sliding Window Division Based on Event Density With-/without Overlapping.** Due to artificial dividing operations, some events that occur almost simultaneously may be divided into two sliding windows. In order to prevent this case, we design the overlapping windows based on adaptive density-based sliding window. Specifically, we define $T_{endi}(= T_{starti} + T_{slidingi}$, $1 <= i <= n)$ as the end timestamp of $i$-th sliding window, $TW_{overlap}$ as the size of overlapping window, here, $0 <= TW_{overlap} <= T_{slidingi}$.

The adaptive sliding window with overlapping is shown in Fig. 2, and the adaptive sliding window without overlapping is shown in Fig. 3. And Algorithm 1 describes the process of adaptive sliding window division based on event density with/without overlapping. This algorithm includes three steps: first, construct the first event sequence; second, move the sliding window backward according to $TW_{overlap}$; second, according to step 2 until the sliding time window moves to the last event of $Trans(e)$.

## 3.4  Frequent Event Sequences Mining

Association rule mining [7,11] has often been proposed to discover certain interesting correlation relationships between the events of cluster logs. Indeed, frequent itemsets mining is an effective step in the process of association rule mining, such as, some well-known conventional algorithms (Apriori [12], FP-Growth [13] and Prefixspan [14], etc.) working on a single computer, have shown good performance in dealing with small amount of data. Nevertheless, conventional approaches come across significant challenges when computing power and memory space are limited in big data era. So some practices and attempts have

**Algorithm 1.** Adaptive sliding window division based on event density with-/without overlapping

---

**Input:**
  The event transactions $Trans(e)$
  The size of sliding window $T_{sliding}$
  The size of Overlapping window $TW_{overlap}$
  The threshold of event count in sliding window $max\_Count_{sw}$

**Output:**
  Event Sequence Transactions $Trans(es)$

1: (1) Construct the first event sequence:
2: Gets the timestamp of the first event in $Trans(e)$, which is record as $T_{start1}$;
3: Divide the first sliding window $SW_1$: $T_{end1} = T_{start1} + T_{sliding}$;
4: Count the number of events in $SW_1$: $Count_{sw_1}$;
5: **if** $Count_{sw_1} <= max\_Count_{sw}$ **then**
6:    Generate the first event sequence $ES_1$, which is composed of all events in $SW_1$;
7: **else**
8:    **if** $Count_{sw_1} > max\_Count_{sw}$ **then**
9:       Generate the first event sequence $ES_1$, which is composed of top $max\_Count_{sw}$ events in $SW_1$;
10:       Update $T_{end1}=$ the timestamp of $max\_Count_{sw}$-th event;
11:    **end if**
12: **end if**
13: (2) Move sliding windows backward according to $TW_{overlap}$:
14: **if** $TW_{overlap}=0$ **then**
15:    Determines $T_{start2}$ of the next sliding window $SW_2$: $T_{start2}=$the timestamp of the next event after $T_{end1}$
16:    Construct the event sequence according to step 1;
17: **else**
18:    **if** $0 < TW_{overlap} < T_{sliding}$ **then**
19:       Determines the start time of the next sliding window $SW_2$: $T_{start2}=T_{end1}$-$TW_{overlap}$;
20:       Construct the event sequence according to step 1;
21:    **end if**
22: **end if**
23: (3) Move the time window backward according to step 2, until the sliding time window moves to the last event of $Trans(e)$.

---

been made to mine frequent itemsets from massive data by using parallel computing technologies. For example, some researches apply message passing interface (MPI) [15], MapReduce [16] and Spark [17] to mine frequent itemsets.

In this section, we utilize the FP-Growth algorithm based on Spark.mllib to mine frequent event sequences. The FP-Growth algorithm is described in the paper [13], and it mines frequent patterns without candidate generation, where FP stands for a frequent pattern. The FP-Growth algorithm mines the frequent itemsets by using a divide-and-conquer strategy and has two phases: 1) Construction of the FP-Tree, that is, it compresses the database representing frequent itemsets into a frequent pattern tree (FP-Tree), which retains the itemset association information as well; 2) Discovery of frequent patterns in the FP-Tree, specifically, the step is to divide a compressed database into a set of conditional databases (a special kind of projected database), and each associated with one frequent item; and then mine each such database separately [18].

The FP-Growth algorithm based on Spark.mllib is a parallel FP-Growth (PFP) [19], and PFP distributes the work of growing FP-trees based on the size of transactions, and is more scalable than a single-machine implementation. When executing this algorithm, there are two predefined parameters: $min\_sup\_rate$ and $num\_partitions$. $min\_sup\_rate$ is the minimum support rate for an event sequence to be identified as frequent, for example, if an event sequence $es_j$ appears 3 out of 5 transactions, it has a support rate $sup\_rate(es_j)$ of $3/5 = 0.6$ according to the Formula (1). And $num\_partitions$ is the number of partitions used to distribute the work.

$$sup\_rate(es_j) = \frac{sup(es_j)}{\# \ of \ event \ sequences \ in \ Trans(es)} \qquad (1)$$

Moreover, the support threshold value $min\_sup\_rate$ plays an important role in FP-Growth. The larger the $min\_sup\_rate$ is, the fewer result patterns are returned, and it also consumes the lower cost of computation and storage. Usually, for a large scale transactions, $min\_sup\_rate$ has to be set large enough, otherwise the FP-Tree would overflow the storage. For our mining tasks, we typically set $min\_sup\_rate$ to be very low to obtain the long-tail event sequences, even though this low setting may require unacceptable computational time [19]. Afterwards, we will get the frequent event sequences library $L(fes)$.

## 3.5   Building Failure Rules Library

In this section, we extract the failure rules $FR(fes)$ from the frequent event sequences library $L(fes)$, because failure rules record a series of events that lead to system failures, which can be used to understand the failure behaviors and predict failures. Here, we call the series of events that lead to failures in $FR(fes)$ as *failure signs*. And the failure rules are actually the association rules between failure signs and failures.

Faced with the mined numerous failure rules, we also need to measure which failure rules are more valuable. The *support, confidence* and *lift* are the commonly used objective measurements of association rules' interestingness [20].

If $A \Rightarrow B$ is a frequent event sequence and $B$ is a failure event, the support $Sup(A \Rightarrow B)$ denotes the probability that $A$ and $B$ appears simultaneously, which indicates the practicality of failure rules, and is calculated according to the Formula (2).

The confidence $Conf(A \Rightarrow B)$ denotes the conditional probability of the consequent actually occurring after the antecedent events occurred, which indicates whether useful association rules can be derived from the discovered patterns, and is calculated according to the Formula (3).

The lift represents the degree to which one event or event sequence occurrence predicts another event or event sequence occurrence, and using lift can

---

**Algorithm 2.** Failure rules extraction

---

**Input:**
    Frequent Event sequences Library $L(fes)$
**Output:**
    Failure Rules Library $L(fr)$
1: Find the antecedent $A_i$ and consequent $B_j$ in $L(fes)$, here, the consequent $B_j$ is a
    meta event and a failure event;
2: **for** each $A_i$ and consequent $B_j$ **do**
3:     Generate the minimal frequent event sequence $A_{i\_min}$:
        Delete the adjacent same *event ID* in $A_i$ and leave just one *event ID*;
4:     Find similar frequent event sequences $A_{i\_simi}$ of $A_i$ according to $A_{i\_min}$;
5:     Calculate the support count of $A_i$, which is the number of $A_{i\_simi}$ in $L(fes)$:
        $Sup(A_i) = Sup(A_{i\_simi})$
6:     Calculate the support count of $A_i \Rightarrow B_j$ in $L(fes)$:
        $Sup(A_i \Rightarrow B_j) = Sup(A_{i\_simi} \Rightarrow B_j)$
7:     Calculate the confidence of $Conf(A_i \Rightarrow B_j)$:
        $Conf(A_i \Rightarrow B_j) = \frac{Sup(A_i \Rightarrow B_j)}{Sup(A_i)}$
8:     Calculate the $Sup(B_j)$, the total number $Count_{L(fes)}$ and the lift of $Lift(A_i \Rightarrow$
        $B_j)$:
        $Lift(A_i \Rightarrow B_j) = \frac{Conf(A_i \Rightarrow B_j)}{\frac{Sup(B_j)}{Count_{L(fes)}}}$
9:     **if** $Sup(A_i \Rightarrow B_j) > min\_sup$ & $Conf(A_i \Rightarrow B_j) >= min\_conf$ & $Lift(A_i \Rightarrow$
        $B_j) > 1$ **then**
10:        the *fes* $A_i \Rightarrow B_j$ is a failure rule, and save it into $L(fr)$.
11:    **end if**
12: **end for**

---

help to filter out some unpleasant association rules. The lift $Lift(A \Rightarrow B)$ is calculated according to the Formula (4). Here, $p(A)$, $p(B)$ represents the occurrence probability of the event sequence $A$ or $B$ in $L(fes)$, $P(A \cup B)$ represents the occurrence probability of $B$ occurs at least once in a period after event sequence $A$ occurs.

$$Sup(A \Rightarrow B) = Sup(A \cup B) \tag{2}$$

$$Conf(A \Rightarrow B) = \frac{Sup(A \cup B)}{Sup(A)}) \tag{3}$$

$$Lift(A \Rightarrow B) = \frac{P(A \cup B)}{P(A)P(B)} = \frac{Conf(A \Rightarrow B)}{\frac{Sup(B)}{Count_{L(fes)}}} \tag{4}$$

In addition, we define the *fes* $A \Rightarrow B$ as the failure rule, when $Sup(A \Rightarrow B) > min\_sup$ and $Conf(A \Rightarrow B) >= min\_conf$ and $Lift(A \Rightarrow B) > 1$. And Algorithm 2 describes the detailed process of extracting failure rules from frequent event sequences library.

## 3.6    Online Failure Prediction

In this section, we use the failure rules to predict system failures, and the process of prediction system is shown in Fig. 4.

For each prediction, there are three important timing points: *predicting point*, *predicted point*, and *expiration point*, *ch13LogMaster*. The prediction system begins predicting events at the timing of the *predicting point*. The *predicted point* is the occurrence timing of the predicted event. The *expiration point* refers to the expiration time of a prediction, which means this prediction is not valid if the actual occurrence timing of the event passed the expiration point [7].

In addition, there are two important derived properties for each prediction: *prediction time*, and *prediction valid duration*. The *prediction time* is the time difference between the predicting point and the predicted point, which is the time span left for system administrators to respond with the possible upcoming failures. The *prediction valid duration* is the time difference between the predicting point and the expiration point [7]. We analyze the events occur in the prediction time window to produce the failure prediction.

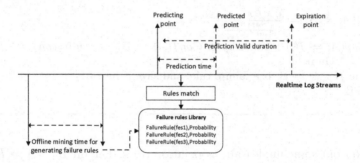

**Fig. 4.** The time relations in our event correlation mining and failure prediction systems.

However, in the process of real-time failure prediction, the prediction framework selects the failure rule with the greatest confidence from failure rule library, and update the failure rules library periodically simultaneously. First, we need set an update period; second, we use the previous frequent pattern mining algorithm to dig out the frequent patterns and failure rules, and then update the support, conf and lift of failure rules library by using the support counts of new frequent patterns and failure rules.

## 4    Experiments and Evaluations

### 4.1    Experiment Settings

In the experiments, we use our approach to analyze three real cluster logs generated by a production Hadoop system (not publicly available), a HPC system in Los Alamos National Lab (LANL) and a BlueGene/L system.

The experiment platform is a Spark cluster (consists of 1 master and 5 slaves) deployed upon the Hadoop Yarn framework, which have 32 GB memory per server and use Intel(R) Xeon(R) CPU E5645@2.40 GHz as the processer. Specifically, we set spark configuration parameters as spark.executor.core = 24, spark.executor.memory = 32G, spark.driver.maxResultSize = 20G.

We also use the holdout method to divide the system log into two parts: the previous part (about 2/3) is used for offline training and the latter part (about 1/3) is used for prediction and evaluations. Table 1 gives the summary of these cluster logs.

**Table 1.** The summary of cluster logs

|  | CMRI-Hadoop | LANL-HPC | BlueGene/L |
|---|---|---|---|
| Days | 67 | 1005 | 215 |
| Start date | 2008-10-26 | 2003-07-31 | 2005-06-03 |
| End date | 2008-12-31 | 2006-04-40 | 2006-01-03 |
| Log size | 130MB | 31.5MB | 118MB |
| # of Raw records | 977858 | 433490 | 4399503 |
| # of Events | 26538 | 132650 | 422554 |
| # of Event ID | 402 | 36 | 74 |
| # of Training Set | 18090 | 90808 | 284255 |
| # of Test Set | 8257 | 42879 | 138299 |

## 4.2  Log Characteristics Analysis

Different logs have different distribution characteristics, and these distribution characteristics have a certain impact on our follow-up experiments. In order to guide the window division and frequent event sequences mining experiments, we analyze two characteristics of these cluster logs: 1) the event density distribution, and 2) the time interval distribution between adjacent events.

**Event Density Distribution.** In order to describe the range of event density easily, we define $R\_density[x_i, x_j]$ according to Formula (5):

$$R\_density[x_i, x_j] = \frac{\# \ of \ event \ density \ is \ [x_i, x_j]}{\# \ of \ all \ unit \ sliding \ windows} \tag{5}$$

$$(x_i >= 0 \ and \ x_i < x_j)$$

We plotted the event density distribution histograms of the three logs, which are shown in Figs. 5a, b and c. From the figures we see that, the number of events in a unit sliding window is almost 0–10 for three cluster logs. However, there may be a great amount of burst events in a short period and it results in burst windows. For example, the $R\_density[100, \infty]$ of CMRI-Hadoop, LANL-HPC and Bluegene/L logs respectively are 2.0%, 5.7% and 17.6%. More specifically, we find that the frequency of CMRI-Hadoop logs' event density which exceeds 500 is 2, such as, the number of events is 1464 in the time window after 2008-11-12 07:07:39, and the number of events is 814 in the time window after 2008-12-09 18:10:26. The frequency of LANL-HPC logs' event density which exceeds 500 is 39, such as, the number of events is 1352 in the time window after 2004-03-18 21:03:00, and the number of events is 1322 in the time window after 2004-11-18 21:11:00. The frequency of Bluegene/L logs' event density which exceeds 500 is 101, such as, the number of events is 55297 in the time window after 2005-06-14 10:10:32, and the number of events is 22640 in the time window after 2005-09-20 17:55:01.

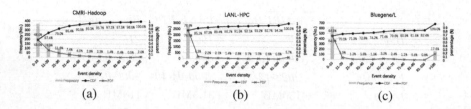

(a)                          (b)                          (c)

**Fig. 5.** The event density distribution histogram of (a) CMRI-Hadoop, (b) LANL-HPC and (c) Bluegene/L logs.

Without considering the burst windows, we find that the $R\_density[0, 20]$ of LANL-HPC and Bluegene/L logs are 85.1% and 70.1%, the $R\_density[0, 40]$ of CMRI-Hadoop logs is 86.4%, and the ratio of each event density in the above range is greater than 5%. Meanwhile, the ratio of other event densities is relatively low.

**Time Interval Distribution Between Adjacent Events.** In order to determine the size of the overlapping window, we also analyzed the distribution of time interval between adjacent events. So we define $R\_interval[x_i, x_j)$ to describe the ratio of time interval between adjacent events, which is expressed in Formula (6) and (7).

$$R\_interval[0] = \frac{\# \ of \ adjacent \ interval \ is \ 0}{\# \ of \ Training \ Set} \quad (6)$$

$$R\_interval[x_i, x_j) = \frac{\# \ of \ adjacent \ interval \ is \ [x_i, x_j)}{\# \ of \ adjacent \ interval \ is \ not \ 0} \quad (7)$$

$$(x_i > 0 \ and \ x_i <= x_j)$$

First, we calculated the $R\_interval[0]$ of the three logs, the $R\_interval[0]$ of CMRI-Hadoop, LANL-HPC and Bluegene/L are respectively 28.13%, 81.17%, 88.18%. We see that the latter two logs record a lot of simultaneous events.

Then we analyzed the ratio of non-zero adjacent time intervals, which are shown in Figs. 6a, b and c. For example, the $R\_interval(0, 1 \ min)$ of CMRI-Hadoop and Bluegene/L logs are 52.3% and 88.8%. The time interval distribution of LANL-HPC logs shows a leaping-type distribution, the $R\_interval(0, 1\ min)$ is 75.2%, and the remaining adjacent intervals are distributed in the range of $[49\,\text{min}, 61\,\text{min})$ and $[119\,\text{min}, 206103\,\text{min})$.

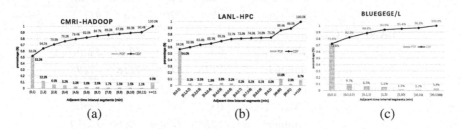

**Fig. 6.** The ratio of non-zero adjacent time intervals in (a) CMRI-Hadoop, (b) LANL-HPC and (c) Bluegene/L logs.

### 4.3 Event Sequence Transactions

In this section, we use adaptive sliding window division based on event density to construct the event sequence transactions for each cluster log. In our experiments, there are two parameters $max\_Count_{sw}$ and $TW_{overlap}$ needed to determine.

According to the ratio of event densities which are shown in Sect. 4.2, we set the $max\_Count_{sw}$ as 20 for LANL-HPC and Bluegene/L logs, and set the $max\_Count_{sw}$ as 40 for CMRI-Hadoop logs, which are shown in Table 2.

After setting the parameter $max\_Count_{sw}$, we set different $TW_{overlap}$ for these three logs. Based on the above observations in time interval distributions, we could set the range of overlapping window size as $(0, 20\,\text{s}, 40\,\text{s}, 1\,\text{min})$ for LANL-HPC and Bluegene/L logs, and the range of overlapping window size as $(0, 1\,\text{min}, 5\,\text{min}, 10\,\text{min})$ for CMRI-Hadoop logs. Table 2 illustrates the number of event sequences (that is, the number of sliding windows) for three logs. From Table 2 we see that, because events are unevenly distributed in logs, as the overlapping windows become larger, there is no obvious trend for the number of divided sliding windows. Here, we tend to choose the parameter $TW_{overlap}$ as 1 min for subsequent analysis, it can prevent multiple events that occur almost simultaneously to be divided into different sliding windows, and ensure the number of window events not be too much.

**Table 2.** The number of event sequences in event sequences transaction with different overlapping windows.

| CMRI-Hadoop | $max\_Count_{sw}$ | 40 | | | |
|---|---|---|---|---|---|
| | $TW_{overlap}$ | $0\,min$ | $1\,min$ | $5\,min$ | $10\,min$ |
| | number | $939$ | $970$ | $948$ | $941$ |
| LANL-HPC | $max\_Count_{sw}$ | 20 | | | |
| | $TW_{overlap}$ | $0\,s$ | $20\,s$ | $40\,s$ | $1\,min$ |
| | number | $7118$ | $7118$ | $7118$ | $7749$ |
| BlueGene/L | $max\_Count_{sw}$ | 20 | | | |
| | $TW_{overlap}$ | $0\,s$ | $20\,s$ | $40\,s$ | $1\,min$ |
| | number | $12960$ | $12960$ | $12960$ | $12870$ |

### 4.4  Rules Mining Results

Although a low $min\_sup\_rate$ will result in inefficient FP-Trees or huge storage overheads, but in order to obtain the long-tail frequent event sequences and ensure the recall rate, we still decide to set $min\_sup\_rate = 2$.

In the experiment, we utilize the FP-Growth algorithm based on Spark (Spark-FPGrowgh) to mine the frequent event sequences, and the mining results are shown in Table 3. In addition, we also analyze and evaluate the execution efficiency of different mining algorithms, by using the average mining time $MineTime_{avg}$ in Formula 8. By comparing with other algorithms Apriori-LIS [7] and Apriori-semiLIS [7], we see that the efficiency of Spark-FPGrowgh is higher than others from Table 3.

$$MineTime_{avg} = \frac{MineTime_{all}}{Num_{fes}} \tag{8}$$

However, we find that the mined $fes$ of CMRI-Hadoop are significantly more than that of other two logs: the $Num_{fes}$ of CMRI-Hadoop is 5244915, and the $Num_{fes}$ of LANL-HPC, Bluegene/L are respectively 4884, 2965. The reason is that *we perform the mining algorithms on the basis of eventID, and the number of eventID that after pre-processing in LANL-HPC and Bluegene/L logs are few.* From the number of failure rules $Num_{FR(fes)}$, we also see that the $Num_{FR(fes)}$ of CMRI-Hadoop mined by Spark-FPGrowth is obviously more than the failure rules mined by other two methods.

### 4.5  Evaluation of Failure Predication

On the basis of failure rules obtained, we predict the failures in the latter 1/3 part of these three logs. We evaluate the effectiveness of failure prediction through three indicators: Precision, Recall and Accuracy [21]. *Precision* indicates the exactness of the prediction, and *Recall* indicates the completeness. Furthermore, the higher the *Precision* is, the lower the false alarm rate is; and the higher

**Table 3.** The average mining time comparison with different mining algorithms.

| Log Type | | CMRI-Hadoop | LANL-HPC | Bluegene/L |
|---|---|---|---|---|
| Spark-FPGrowth (min_sup = 2) | $MineTime_{avg}$ | 20.41 ms | 0.11 ms | 0.03 ms |
| | $Num_{fes}$ | 5244915 | 5988 | 2965 |
| | $Num_{FR(fes)}$ | 271064 | 170 | 748 |
| Apriori-LIS (min_sup = 5) | $MineTime_{avg}$ | 213.34 ms | 313.34 ms | 285.27 ms |
| | $Num_{fes}$ | 2413 | 4726 | 1492 |
| | $Num_{FR(fes)}$ | 463 | 123 | 655 |
| Apriori-semiLIS (min_sup = 5) | $MineTime_{avg}$ | 55.23 ms | 75.23 ms | 63.58 ms |
| | $Num_{fes}$ | 1520 | 3990 | 1158 |
| | $Num_{FR(fes)}$ | 390 | 117 | 633 |

the *Recall* is, the lower the false negative rate is. However, neither *Precision* nor *Recall* alone can judge the goodness of failure prediction. So we introduce *Accuracy*, the harmonic mean of *Precision* and *Recall*.

$$Precision = \frac{\#\ of\ successful\ detections}{\#\ of\ total\ alarms} \tag{9}$$

$$Recall = \frac{\#\ of\ successful\ detections}{\#\ of\ total\ failures} \tag{10}$$

$$Accuracy = \frac{2 * Precision * Recall}{Precision + Recall} \tag{11}$$

In our experiments, we set the *prediction valid duration* as 60 min. The Precision, Recall and Accuracy of predicting failures in CMRI-Hadoop logs, LANL-HPC logs and BlueGene/L logs are shown in Fig. 7. We notice that the average accuracy rates of Spark-FPGrowth are 78.18%, 64.63% and 67.29% for CMRI-Hadoop, LANL-HPC and Bluegene/L logs respectively, and the accuracy rates are equivalent to the other two algorithms (Apriori-LIS and Apriori-semiLIS). However, the spark-based frequent pattern mining algorithm Spark-FPGrowth has higher efficiency.

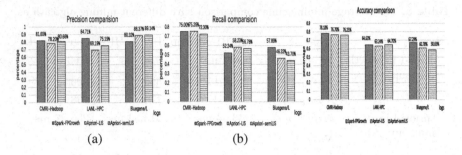

(a)                                (b)

**Fig. 7.** The Precision, Recall and Accuracy comparison of three rules mining algorithms in (a) CMRI-Hadoop, (b) LANL-HPC and (c) Bluegene/L logs.

## 5   Related Work

In the past decades, failure prediction has been proved as an effective method to achieve proactive fault management. Based on the type of used data, the failure prediction method can be divided into *metric-based method* and *log-based method*, [5]. The metric-based method is using system performance metrics to analyze system status or analyzing periodically system variables [22]. However, the log-based method is more suitable in large-scale systems, which has been proved in paper [6]. Of course, there are more recent works that combining these two methods together [23], but they are used for fault diagnosis and root cause analysis generally. This paper focuses on failure detection and prediction, which generally utilizes the system logs, for these logs directly or indirectly record important information of failures.

Based on the method that failure prediction used, various failure detection and prediction methods have been proposed, by using both *statistical* and *data mining driven* methods for analysing large system log files [2]. On one hand, based on the statistical analysis approach, Liang et al. [24] investigated the statistical characteristics of failure events, as well as the statistics about the correlation between occurrence of non-fatal and fatal events; A. Gainaru et al. [25] proposed a novel way of characterizing the normal and faulty behaviors of system by using signal analysis, and implemented a filtering algorithm and short-term fault prediction methodology based on the extracted models. Gujrati et al. [8] presented a meta-learning method based on statistical analysis and standard association rule algorithm, but it only focused on some specific failure patterns and predicted one of them would happen without further information. On the other hand, many methods are using data mining algorithms to extract sequences of events that lead to failures. For example, LogMaster [7] proposed an improved Apriori-LIS and Apriori-semiLIS algorithms to mine rules, which can improve sequence mining efficiency; Zhou et al. [26] presented an online log analysis algorithm Apriori-SO, and an online event prediction method that can predict diversities of failure events with the great details. Z. Zheng et al. [27] mined the relationships between fatal and non-fatal events, and predicted failures that including location and lead time. A. Gainaru et al. [2] used a dynamic window strategy to find the frequent sequences, by utilizing an Apriori-modified algorithm. Y. Watanabe et al. [3] proposed an online failure prediction method by real-time message pattern learning in cloud datacenter, which identifies the relationship between the messages and failures.

Nevertheless, conventional data mining approaches have come across significant challenges when computing power and memory space are limited in big data era. For example, due to scanning the database several times, the efficiency of general association rule mining algorithms are always not high. So some practices and attempts have been made to mine rules from massive data by using parallel computing technologies, such as applying message passing interface (MPI) [15], MapReduce [16] and Spark [17], and so on. However, in spite of certain advantages in MPI's iterative computation, the disadvantages are its high communication loads, which due to data exchanges between different computer nodes and the lacking of fault tolerance. And MapReduce framework is not appropriate for iterative computation, because the repeated read/write operations to Hadoop distributed file system (HDFS) would lead to high I/O load and time cost. Moreover, Spark is more suitable for processing iterative jobs for it using resilient distributed datasets (RDDs). And it also offers an open-source distributed machine learning library MLlib, which helps with efficient iterative learning. In this paper, we use Spark FP-Growth algorithm to mine a large number of long tail rules between $eventIDs$ quickly and efficiently, which helps to improve the rules mining efficiency.

## 6   Conclusion and Future Works

In this paper, we utilize the frequent pattern mining algorithm based on Spark, and mine the correlations among different events that generated by large-scale cluster logs. On the basis of event correlations, we further mine the correlations between failure signs and failure events, which are used for predicting system failures. Since the preprocessed event transactions are not suitable for using frequent pattern mining algorithms, we design an adaptive sliding window division method based on event density with/without overlapping, to construct event sequence transactions. Meanwhile, in order to obtain the long-tail frequent event sequences and ensure a higher recall rate, we set a low $min\_sup\_rate$ in our frequent pattern mining algorithm. At last, we analyze the log characteristics and predict the system failures for three large-scale production clusters, and the average prediction accuracy are 78.18%, 64.63% and 67.29% in CMRI-Hadoop, LANL-HPC and Bluegene/L logs, respectively.

**Acknowledgment.** This project is supported by the National Key Research and Development Program of China under Grant No.2017YFB1001602. The authors appreciate the valuable comments provided by the anonymous reviewers. Authors thank the experimental platform provided by Center for Advanced Computer Systems, Institute of Computing Technology, Chinese Academy of Science.

# References

1. Liang, Y.: Filtering failure logs for a Bluegene/l prototype. In: Proceedings of DSN (2005)
2. Gainaru, A., Cappello, F., Fullop, J., Trausan-Matu, S., Kramer, W.: Adaptive event prediction strategy with dynamic time window for large-scale HPC systems. In: Managing Large-scale Systems via the Analysis of System Logs and the Application of Machine Learning Techniques (SLAML) (2011)
3. Watanabe, Y., Otsuka, H., Sonoda, M., Kikuchi, S., Matsumoto, Y.: Online failure prediction in cloud datacenters by real-time message pattern learning. In: International Conference on Cloud Computing Technology and Science, pp. 504–511 (2012)
4. Navarro, J.M., Parada, G.H.A., Dueñas, J.C.: System failure prediction through rare-events elastic-net logistic regression. In: International Conference on Artificial Intelligence, Modelling and Simulation (AIMS) (2014)
5. Salfner, F., Lenk, M., Malek, M.: A survey of online failure prediction methods. ACM Comput. Surv. (CSUR) **42**(3), 1–42 (2010)
6. Yu, L., Zheng, Z., Lan, Z., Coghlan, S.: Practical online failure prediction for blue gene/p: Periodbased vs event-driven. In: Dependable Systems and Networks Workshops (DSN-W) (2011)
7. Fu, X., Ren, R., Zhan, J., Zhou, W., Jia, Z., Lu., G.: LogMaster: mining event correlations in logs of large-scale cluster systems. In: IEEE 31st Symposium on Reliable Distributed Systems (SRDS) (2012)
8. Gujrati, P.: A meta-learning failure predictor for blue gene/l systems. In: Proceedings of ICPP (2007)
9. Zheng, Z.: System log pre-processing to improve failure prediction. In: Proceedings of DSN (2009)
10. Yan, X., Han, J., Afshar, R.: CloSpan: mining closed sequential patterns in large datasets. In: SIAM International Conference on Data Mining (2003)
11. Zhou, W.: Research on failure correlation analysis in large-scale sluster systems. J. Parallel Distrib. Comput. (2010)
12. Han, J., Kamber, M.: Morgan Kaufmann Publishers (2000)
13. Han, J., Pei, J., Yin, Y.: Mining frequent patterns without candidate generation. In: International Conference on Management of Data (SIGMOD), pp. 205–216 (1996)
14. Pei, J., et al.: PrefixSpan: mining sequential patterns efficiently by prefix-projected pattern growth. In: International Conference on Data Engineering (2001)
15. Otey, M.E., Wang, C., Parthasaratlty, S., Veloso, A., Meira, W.: Mining frequent itemsets in distributed and dynamic databases. In: Proceedings of ICDM (2003)
16. Dean, J., Ghemawat, S.: MapReduce: simplified data processing on large clusters. In: USENIX Symposium on Operating Systems Design and Implementation (OSDI) (2004)
17. Gui, F., et al.: A distributed frequent itemset mining algorithm based on spark (2015)
18. Said, A.M., Dominic, P., Abdullah, A.B.: A comparative study of FP-growth variations. Int. J. Comput. Sci. Netw. Secur. **9**, 266–272 (2009)
19. Li, H., Wang, Y., Zhang, D., Zhang, M., Chang, E.Y.: PFP: parallel FP-growth for query recommendation. In: ACM Conference on Recommender Systems, pp. 107–114 (2008)

20. Geng, L., Hamilton, H.J.: Interestingness measures for data mining: a survey. ACM Comput. Surv. **38** (2006)
21. Wang, C., Talwar, V., Schwan, K., Ranganathan, P.: Online detection of utility cloud anomalies using metric distributions. In: IEEE Network Operations and Management Symposium (NOMS), pp. 96C–103 (2010)
22. Sahoo, R.K.: Critical event prediction for proactive management in large-scale computer clusters. In: International Conference on Knowledge Discovery and Data Mining, pp. 426–435 (2003)
23. Gurumdimma, N., Jhumka, A., Liakata, M., Chuah, E., Browne, J.C.: Crude: combining resource usage data and error logs for accurate error detection in large-scale distributed systems. In: Symposium on Reliable Distributed Systems (SRDS) (2016)
24. Liang, Y.: Bluegene/l failure analysis and prediction models. In: DSN, pp. 426–435 (2006)
25. Gainaru, A., Cappello, F., Kramer, W.: Taming of the shrew: modeling the normal and faulty behaviour of large-scale HPC systems. In: International Parallel and Distributed Processing Symposium (IPDPS), pp. 1168–1179 (2012)
26. Zhou, W., Zhan, J., Meng, D., Zhang, Z.: Online event correlations analysis in system logs of large-scale cluster systems. In: Network and Parallel Computing (NPC) (2010)
27. Zheng, Z., Lan, Z., Gupta, R., Coghlan, S., Beckman, P.: A practical failure prediction with location and lead time for Bluegene/p. In: Proceedings of DSN-W (2010)

# FLBench: A Benchmark Suite
# for Federated Learning

Yuan Liang[1,3,4], Yange Guo[1,3], Yanxia Gong[1,3], Chunjie Luo[2], Jianfeng Zhan[2],
and Yunyou Huang[1,3(✉)]

[1] Guangxi Key Lab of Multi -Source Information Mining & Security, Department
of Computer Science, Guangxi Normal University, Guilin 541004, China
liangyuan@gxnu.edu.cn, {gyx19980201,gyg2019010376}@stu.gxnu.edu.cn
[2] Institute of Computing Technology, Chinese Academy of Sciences,
Beijing 100190, China
{luochunjie,zhanjianfeng}@ict.ac.cn
[3] BenchCouncil R&D Lab - Guilin, Guilin, China
huangyunyou@gxnu.edu.cn
[4] Guangxi Key Laboratory of Trusted Software, Guilin University of Electronic
Technology, Guilin 541004, China

**Abstract.** Federated learning is a new machine learning paradigm. The
goal is to build a machine learning model from the data sets distributed
on multiple devices–so-called an isolated data island–while keeping their
data secure and private. Most existing federated learning benchmarks
work manually splits commonly-used public datasets into partitions to
simulate real-world isolated data island scenarios. Still, this simulation
fails to capture real-world isolated data island's intrinsic characteristics. This paper presents a federated learning (FL) benchmark suite
named FLBench. FLBench contains three domains: medical, financial,
and AIoT. By configuring various domains, FLBench is qualified to evaluate federated learning systems and algorithms' essential aspects, like
communication, scenario transformation, privacy-preserving, data distribution heterogeneity, and cooperation strategy. Hence, it becomes a
promising platform for developing novel federated learning algorithms.
Currently, FLBench is open-sourced and in fast-evolution. We package
it as an automated deployment tool. The benchmark suite is available
from https://www.benchcouncil.org/flbench.html.

## 1 Introduction

Google recently proposed the concept of Federated Learning (FL). The main idea
is to build a machine learning model from the data sets distributed on multiple
devices–so-called an isolated data island–while preventing data leakage [11,16,
24]. FL has become a hot research topic in both industry and academia [15,
18,39]. Unfortunately, most existing FL benchmarks work [2,20,25,29,36,38,
40] manually splits commonly-used public data sets into partitions to simulate
isolated data island scenarios [5]; however, they fail to capture the intrinsic

W. Gao et al. (Eds.): FICC 2020, CCIS 1385, pp. 166–176, 2021.
https://doi.org/10.1007/978-981-16-1160-5_14

characteristics of real-world scenarios. We call an isolated data island scenario a scenario for abbreviation in the rest of this paper.

On the one hand, the statistical data characterization of simulated scenario, which are manually split, are different from those of the real-world one. For example, in centralized training, data can be assumed to be independent and identically distributed (IID). this assumption is unlikely to hold in federated learning settings. For example, Chandra et al. [32] use the MNIST and CIFAR-10 datasets to simulate a scenario and assume that the datasets are random, disjoint, and evenly distributed between clients. However, in the real-world scenario, the medical image data will involve the magnetic field intensity issue. For example, the magnetic field intensity at 1.5 T and the magnetic field intensity at 3 T of the same image will show different lesions. The more details are shown in Table 1 and Table 2.

Second, without considering the data characterstics, the FL algorithms developed based on the simulated scenario cannot be migrated to a real-world one. For example, in an Alzheimer's diagnosis scenario, it is a CT image–a 3d black and white image, so the FL algorithms developed on the MNIST and CAFAR-10 data sets [40] are hard to migrate to an Alzheimer's diagnosis.

We observe that there are several previous FL benchmarking researches. Due to the massive gap between the simulated scenario and the real-world one, FedML [8] and OARF [10] fail to achieve the goal some extent. FedML [8] focuses on deploying the benchmark to distributed training, mobile device training, and stand-alone simulation. However, the authors fail to justify their methodology for choosing data sets and workloads. The common datasets (CIFAR-10, CINIC-10) are partitioned for each participant with statistical methods. Hu et al. [10] shows that the OARF benchmark suite is diverse in data size, distribution, feature distribution, and learning task complexity. Still, it focuses on benchmarking systems instead of algorithms. In addition, Luo et al. [22] and Hsu et al. [9] presume independent and identical distribution, this data type is rare in real life.

**Table 1.** The Summary of the Latest FL Publications.

| Venue | Datasets | Simulation approaches | Isolated data island[a] | Task | Consistent or not[b] | real-world scenario[c] |
|---|---|---|---|---|---|---|
| ICLR [38] | MNIST; CIFAR-10; Tiny-imagenet | Training: 80%; Testing: 20%; [d] | No | Image classification and loan status prediction | No | No |
| ICML [25] | Adult dataset, Cornell movie dataset, Penn TreeBank (PTB) dataset, Fashion MNIST | Adult data and Fashion MNIST[e] | No | Census income forecast, language modeling and image classification | Yes | Yes |
| ICLR [19] | Public datasets and Synthetic dataset[e] | – | No | Image classification, emotion analysis, language modeling, vehicle prediction | No | No |
| ICLR [20] | MNIST | MNIST balanced and MNIST un-balanced [g] | No | Image classification | No | No |

a. Data exists in the form of isolated islands in real scenarios.

b. Whether the features of the simulated scenario are consistent with those of the real one.

c. Whether it is used in the real scenario or migrated to the real scenario

d. Training: 80%; Testing: 20%; The training set is equally divided into 100 participants.

**Table 2.** The Summary of the Latest FL Publications.

| Venue | Datasets | Simulation approaches | Isolated data island[a] | Task | Consistent or not[b] | real-world scenario[c] |
|---|---|---|---|---|---|---|
| ICLR [27] | MNIST, Amazon Review, DomainNet, Office-Caltech10 | 10 Titan-Xp GPU cluster and simulate the federated system on a single machine | No | Image classification, target recognition, DomainNet, emotion analysis | Yes | Yes |
| ICML [2] | Fashion-MNIST, UCI Adult census dataset | Fashion MNIST and UCI Adult census dataset[h] | No | Image classification and census income forecast | No | No |
| ICML [40] | MNIST, CAFAR-10 | Randomly divided into J batches[i] | No | Image classification | No | No |
| ICML [29] | CIFAR-10/100; FEMNIST; PersonaChat | CIFAR and FEMNIST [j] | No | Image classification, dialogue prediction for personality | No | No |
| ICML [36] | MNIST; CIFAR-10; shakespeare dataset | One centralized node in the distributed cluster is regarded as the data center, and the other nodes are regarded as local clients.[k] | No | Image classification, language modeling | Image classification: No; Language modeling: Yes | No |

e. Adult data: divide the dataset into two domains with and without doctorates; Fashion MNIST: We extract three categories of data subsets: T-shirts, pullovers, and shirts, and then divide this subset into three areas, each containing a garment.

f. Public datasets: vehicle dataset; text data built from the complete works of William Shakespeare; Omniglot; tweet data curated from Sentiment 140.

g. The former is balanced so that the number of samples on each device is the same, while the latter is highly unbalanced. The number of samples between devices follows the power law.

h. Fashion MNIST: a 3-layer convolution neural networks (CNN)-based an offline model is used. UCI adult census dataset: uses fully connected neural networks. Set the number of agents K to 10 and 100. When $k = 10$, all agents are selected in each iteration, while when $k = 100$, one-tenth of agents are randomly selected in each iteration.

i. These datasets are randomly divided into J batches. Two partitioning strategies are of interest: (a) uniform partitioning, in which each class in each batch has approximately equal proportions and (b) miscellaneous new partitions with unbalanced batch size and class proportions.

j. CIFAR: uses 50000 training data points and 10000 validated standard training/test splits. The dataset is divided into 10000 (CIFAR-10) and 50000 (CIFAR-100) clients. Each client has five (CIFAR-10) and one (CIFAR-100) data point from a single target class. In each round, 1% of the clients participated, resulting in a total batch size of 500 (100 clients of CIFAR-10 have 5 data points, and 500 clients of CIFAR-100 have 1 data point). Federated EMNIST(FEMNIST)62 classes(upper- and lower-case lettersplus digits)which is formed by partitioning the EMNIST dataset such that each client in FEMNIST contains characters written by a single person.

k. For the CIFAR-10 dataset, data enhancement (random clipping and flipping) is used, and each image is normalized. We propose two data partitioning strategies to simulate the joint learning scheme. Homogeneous partitioning: the proportion of each local client is approximately equal in each class. Heterogeneous partitioning: the number of data points and the proportion of classes are unbalanced. For Shakespeare's dataset, we treat each speaking role as a client, resulting in naturally heterogeneous partitions. We preprocess the Shakespeare dataset by filtering out clients with less than 10k data points and sampling a random subset of J = 66 clients. We allocate 80% of the data for training and merge the rest into the global test set.

Therefore, this paper calls attention to building an FL benchmark suite to provide various scenarios. We propose a benchmark framework with flexible customization and configuration. Covering the three most concerning domains: medical, financial, and AIoT, FLBench provides three scenario benchmarks [5] for developing novel FL systems and algorithms as the real-world scenario is

**Table 3.** Comparison with existing federated learning benchmarks.

| | Scenario configuration | Given fixed scenario | Customized scenario | Medicine | Finance | AIoT | Evaluation metrics | Automated deployment tool |
|---|---|---|---|---|---|---|---|---|
| FedML [8] | X | ✓ | X | X | X | ✓ | X | ✓ |
| OARF [10] | X | X | X | X | X | X | ✓ | ✓ |
| IDFL [9] | X | X | ✓ | X | X | ✓ | ✓ | ✓ |
| FVC [22] | X | X | X | X | X | ✓ | X | ✓ |
| FLBench | ✓ | ✓ | ✓ | ✓ | ✓ | ✓ | ✓ | ✓ |

unavailable for most of the researchers. Each scenario benchmark models the critical paths of a real-world application scenario as a permutation of essential modules [5]. Our suite can evaluate FL system and algorithms' various aspects, including communication, scenario transformation, privacy-preserving, data distribution heterogeneity, and cooperation strategy. Table 2 summarizes the critical differences between FLBench and existing FL libraries and benchmarks. Our key contributions are:

1) We propose a configurable FL benchmark suite–FLBench, covering the three most concerning domains: medical, financial, and AIoT. FLBench can be used to evaluate FL systems and algorithms' different aspects, including communication, scenario transformation, privacy-preserving, data distribution heterogeneity, and cooperation strategy.
2) FLBench provides various customized scenarios for developing novel FL algorithms.
3) FLBench is packaged as an automated deployment tool and can be deployed in mobile, distributed, and standalone manners.

## 2 Related Work

### 2.1 Federated Learning

Federated Learning is to build a machine learning model from the data sets distributed on multiple devices while preventing data leakage [11,16,24]. According to data distribution characteristics, FL is mainly divided into horizontal federal learning, vertical federal learning, and federal transfer learning [8,39]. FL benchmarking should consider both systems and algorithms' innovation and performance, so we design a new benchmark suite by evaluating the performance of the FL systems and algorithms from different perspectives. Specifically, we consider the following aspects:

1) **Communication.** In the federated network, communication is a key bottleneck. Also, due to the privacy problem of sending original data, the data generated on each device must be kept local. A federated network may consist of many devices, such as millions of smartphones. The speed of communication in the network may be many orders of magnitude slower than local

computing. To match the model with the data generated by the devices in the federated network, it is necessary to develop a communication-efficient method to send small messages or model updates iteratively as a part of the training process [3, 11].

2) **Scenario Transformation**. Federated learning systems are significantly different from the traditional distributed environment. The main idea of scenario transformation is to transfer a local machine learning model to federated learning settings. Scenario transformation enables people to explore statistical training models on remote devices to suit different scenarios, achieving the purpose of data privacy protection [1, 27].

3) **Privacy-preserving**. Privacy is often a major concern in federated learning applications. Federated learning takes a step towards protecting the data generated on each device by sharing model updates (such as gradient information) rather than raw data. However, during the whole training process, the model updating communication can still disclose sensitive information to the third party or central server. Although current methods aim to enhance federated learning's privacy by using secure multiparty computation or differential privacy, these methods usually provide privacy at the cost of reducing model performance or system efficiency [17, 33, 34, 37]. There is a big gap between the theoretical results and the real results.

4) **Data Distribution Heterogeneity**. Devices often generate and collect data on the network in a non-IID manner. For example, mobile phone users use different languages in the context of the next word prediction task. Also, the number of data points across devices may vary greatly, and there may be an underlying structure that captures the relationships between devices and their related distributions. This data generation paradigm violates the i.i.d. assumption that is often used in distributed optimization, increases the likelihood of stragglers and may increase the complexity of modeling, analysis, and evaluation [20, 27, 29, 36]. Data heterogeneity also includes the characteristic heterogeneity of other data, such as small sample data of intelligent terminal, which cannot form a stable distribution. The characteristic heterogeneity is a an issue that has not been considered in FL algorithms benchmarking.

5) **Cooperation Strategy**. To fully commercialize federal learning between different organizations, it is necessary to develop a fair platform and cooperation strategy. After establishing the model, the model's performance will be reflected in practical application and recorded in the permanent data recording cooperation strategy, such as blockchain-based ones. The model's effectiveness depends on the organizations' contribution to providing high-quality data; the high-quality model relies upon the federation mechanism distributed to all parties. The high-quality model continues to motivate more organizations to join the data federation [30, 33, 35], and vice versa.

## 2.2    Benchmarks

In recent years, deep learning and machine learning benchmarks have played an important role in the machine learning area. Typical benchmarks include

AIBench [5–7,13,21,31], DAWNBench [4], and MLPerf [12,23]. These benchmarks have provided various metrics and results for machine learning training and inference. For example, AIBench is a comprehensive AI benchmark suite, distilling real-world application scenarios into AI Scenario [5], Training [6,31], Inference, and Micro Benchmarks across Datacenter, HPC [13], IoT [21], and Edge [7]. AIBench Scenario benchmarks are proxies to industry-scale real-world applications scenarios [31]. However, today's AI faces two major challenges. In most industries, data exists in isolated islands; second, data privacy and security concerns matter. However, in many cases, we are forbidden to collect, fuse and use data in different AI processing places. How to solve the problem of data fragmentation and isolation legally is a major challenge for AI researchers and practitioners. Therefore, many researchers have proposed a possible solution: safe federated learning [11,16,24].

We notice that there have been some researches on FL benchmarking, i.e., FedML [8] and OARF [10]. Due to the massive gap between simulated and real-world scenarios, it has not been solved well. FedML [8] conducted experiments in different system environments, but they did not detail the benchmarking methodology. Hu et al. [10] shows their OARF benchmark suite is diverse in data size, distribution, feature distribution, and learning task complexity, but they lack algorithm-level benchmarking. Luo et al. [22], and Hsu et al. [9] seldom consider essential aspects like communication, scenario transformation, privacy-preserving, data distribution heterogeneity, and cooperation strategy. They mainly focus on the independent and identical distribution of data. Also, they fail to cover federated learning's mainstream scenarios like medical, financial, and AIoT.

## 3  FLBench Methodology and Design

This section presents FLBench methodology, decisions, and implementation.

### 3.1  FLBench Methodology

1) We investigate the most concerning scenarios. The candidate scenarios involve many domains such as medicine and electricity. However, for a benchmark suite, it is impossible and unnecessary to provide all scenarios since they. Besides, providing many scenarios is very costly. Thus, the first step to construct an FL benchmark is selecting several kinds of scenarios to cover FL's different fundamental aspects.

2) According to the output from Step 1), the data generated by several kinds of real-world scenarios need to be collected for constructing the scenarios. Meanwhile, we perform complex data pre-processing, which requires professional domain knowledge, in this step.

3) According to the output from Step 2), we propose configurable scenarios to evaluate the FL systems and algorithms. For example, the primary concerns about algorithm evaluation are fairness and algorithm robustness. It requires

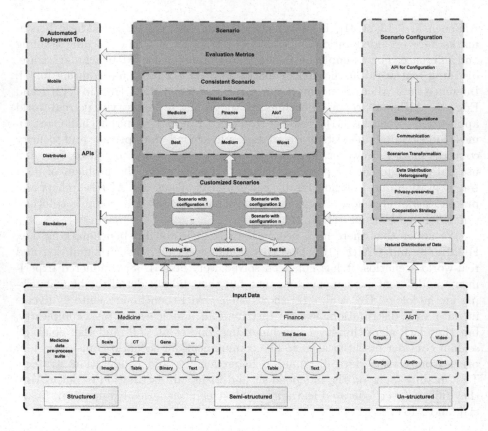

**Fig. 1.** The FLBench framework.

the FL benchmark can provide various scenarios according to the specific researches for every domain. However, it is very costly to construct scenarios for every potential detailed research. Thus, in this step, we designed the user-oriented configurable scenario.

4) According to the output from Step 3), we construct two main scenarios for evaluating FL algorithms. For an FL benchmark, two main functions are necessary: a)provide a fixed scenario, which refers to a limited set of scenarios provided for all algorithm evaluations in a research direction of an application domain for the fair comparison of FL. b) give an easily-customized scenario for the development of a novel FL algorithm. Besides, we propose specific evaluation metrics for every scenario.

5) based on the above outputs, we design and implement an automated deployment tool to deploy the scenario on different platforms.

## 3.2  FLBench Design

Today's artificial intelligence still faces two significant challenges. One is that, in most industries, data exists in the form of isolated islands. The other is data privacy and security issue. As analyzed in the FLBench requirements, the FLBench framework (shown in Fig. 1) including four parts.

**Input Data**: Most of the current researches on FL are carried out on the simulation scenario, which is constructed by commonly used dataset such as CIFAR-10. However, there is a vast difference between the commonly used dataset data and the real-world scenario data in data type and data mode. This considerable difference leads to that FL algorithms developed based on simulated scenarios cannot be migrated to real-world scenarios. We collect data from the three most concerning scenarios to solve this issue, including medicine, finance, and AIoT. Besides, a particular data pre-process tool is necessary for medicine data since medicine data need special processing.

**Scenario Configuration**: To achieve the robustness and multi-faceted evaluation of the FL algorithms, we propose a scenario configuration function. First, we analyze the current innovation of FL researches and then classify the innovation directions of FL into the following categories: communication, scenario transformation, privacy-preserving, data distribution heterogeneity, cooperation strategy. Second, for each innovation direction on each domain, we provide a basic configuration according to the natural distribution of data and an API to modify the configuration to simulate various scenarios according to requirements.

**Scenarios**: Benchmark has two functions: first, it provides an open and fair comparison; second, it will provide a research basis for later researchers to develop more advanced algorithms and determine the selection of some important parameters. Thus, we construct two kinds of scenarios: fixed scenarios and customized scenarios. We modify the basic configuration mentioned above to achieve customized scenarios.

**Automated Deployment Tool**: We will update FLBench step by step to make it adapt to future development needs. Besides, we continue to expand the benchmark and provide more scenarios and related APIs. We hope that more people will join our benchmark research, which will make our benchmarks suite more perfect and comprehensive.

## 3.3  FLBench Implementation

Currently, FLBench contains: four datasets (medicine: ADNI [28], MIMIC-III [14]; finance: Adult dataset [26]; AIoT: iNaturalist-User-120k [9]), one basic configuration file (Alzheimer's diagnosis scenario configuration). The Alzheimer's diagnosis scenario configuration can provide various scenarios for NO-IID (data distribution heterogeneity) research in the medicine domain.

FLBench is a fully open and evolving benchmark; next, we will provide $3 * 3 = 9$ datasets for three domains(medicine, finance, and AIoT), and $3 * 3 * 5 = 45$ basic configuration files on different research aspects, including communication, scenario transformation, privacy-preserving, data distribution heterogeneity, and cooperation strategy. Each configuration file can provide various scenarios according to the requirements of the specific research.

## 4   Conclusion

This paper presents a federated learning benchmark suite named FLBench. FLBench contains three domains: medical, financial, and AIoT. By configuring various domains, FLBench is qualified to evaluate federated learning systems and algorithms' essential aspects, like communication, scenario transformation, privacy-preserving, data distribution heterogeneity, and cooperation strategy. We design and implement a configurable benchmark framework, which can deploy on different platforms and provide simple APIs to users. Hence, it becomes a promising platform for developing novel federated learning algorithms. Currently, FLBench is open-sourced and in fast-evolution. We package it as an automated deployment tool. The benchmark suite is available from https://www.benchcouncil.org/flbench.html.

**Acknowledgment.** We are grateful to anonymous reviewers for their constructive comments. Yuan Liang, Yange Guo, Yanxia Gong, and Yunyou Huang's works are partially supported by Guangxi Key Laboratory of Trusted Software (No. KX202037), the Project of Guangxi Science and Technology (Nos. GuiKeAD20297004 and GuiKeAD20297054) and Guangxi Natural Science Foundation Project (2020GXNSFBA297108). Yunyou Huang is the corresponding author of this paper.

## References

1. Bagdasaryan, E., Veit, A., Hua, Y., Estrin, D., Shmatikov, V.: How to backdoor federated learning. In: International Conference on Artificial Intelligence and Statistics, pp. 2938–2948. PMLR (2020)
2. Bhagoji, A.N., Chakraborty, S., Mittal, P., Calo, S.: Analyzing federated learning through an adversarial lens. In: International Conference on Machine Learning, pp. 634–643 (2019)
3. Chen, M., Yang, Z., Saad, W., Yin, C., Poor, H.V., Cui, S.: A joint learning and communications framework for federated learning over wireless networks. arXiv: Networking and Internet Architecture (2019)
4. Coleman, C., et al.: Dawnbench: an end-to-end deep learning benchmark and competition. Training **100**(101), 102 (2017)
5. Gao, W., et al.: Aibench scenario: Scenario-distilling AI benchmarking (2020)
6. Zheng, C., Zhan, J. (eds.): AIBench: towards scalable and comprehensive datacenter AI benchmarking. Bench 2018. LNCS, vol. 11459, pp. 3–9. Springer, Cham (2019). https://doi.org/10.1007/978-3-030-32813-9_1
7. Hao, T., et al.: Edge AIBench: towards comprehensive end-to-end edge computing benchmarking. In: Zheng, C., Zhan, J. (eds.) Bench 2018. LNCS, vol. 11459, pp. 23–30. Springer, Cham (2019). https://doi.org/10.1007/978-3-030-32813-9_3
8. He, C., et al.: Fedml: A research library and benchmark for federated machine learning. arXiv preprint arXiv:2007.13518 (2020)

9. Hsu, T.M.H., Qi, H., Brown, M.: Federated visual classification with real-world data distribution. arXiv preprint arXiv:2003.08082 (2020)
10. Hu, S., Li, Y., Liu, X., Li, Q., Wu, Z., He, B.: The oarf benchmark suite: Characterization and implications for federated learning systems. arXiv preprint arXiv:2006.07856 (2020)
11. Konen, J., McMahan, H., Yu, F., Richtrik, P., Suresh, A., Bacon, D.: Federated learning: Strategies for improving communication efficiency (2016)
12. Janapa Reddi, V., et al.: Mlperf inference benchmark, pp. 446–459 (2020). https://doi.org/10.1109/ISCA45697.2020.00045
13. Jiang, Z., et al.: HPC AI500: a benchmark suite for HPC AI systems. In: Zheng, C., Zhan, J. (eds.) Bench 2018. LNCS, vol. 11459, pp. 10–22. Springer, Cham (2019). https://doi.org/10.1007/978-3-030-32813-9_2
14. Johnson, Johnson, A.E., et al.: Mimic-iii, a freely accessible critical care database. Sci. Data **3**(1), 1–9 (2016)
15. Kairouz, P., et al.: Advances and open problems in federated learning. arXiv preprint arXiv:1912.04977 (2019)
16. Konen, J., Mcmahan, H.B., Ramage, D., Richtrik, P.: Federated optimization: Distributed machine learning for on-device intelligence (2016)
17. Li, Q., Wen, Z., He, B.: Practical federated gradient boosting decision trees (2020)
18. Li, T., Sahu, A.K., Talwalkar, A., Smith, V.: Federated learning: Challenges, methods, and future directions. arXiv (2019)
19. Li, T., Sanjabi, M., Beirami, A., Smith, V.: Fair resource allocation in federated learning. arXiv preprint arXiv:1905.10497 (2019)
20. Li, X., Huang, K., Yang, W., Wang, S., Zhang, Z.: On the convergence of fedavg on non-iid data. arXiv preprint arXiv:1907.02189 (2019)
21. Luo, C., et al.: AIoT Bench: towards comprehensive benchmarking mobile and embedded device intelligence. In: Zheng, C., Zhan, J. (eds.) Bench 2018. LNCS, vol. 11459, pp. 31–35. Springer, Cham (2019). https://doi.org/10.1007/978-3-030-32813-9_4
22. Luo, J., et al.: Real-world image datasets for federated learning. arXiv preprint arXiv:1910.11089 (2019)
23. Mattson, P., et al.: Mlperf training benchmark (10 2019)
24. McMahan, H., Moore, E., Ramage, D., Agera y Arcas, B.: Federated learning of deep networks using model averaging (2016)
25. Mohri, M.: Gary Sivek. Agnostic federated learning, A.T.S. (2019)
26. Mohri, M., Sivek, G., Suresh, A.T.: Agnostic federated learning. arXiv preprint arXiv:1902.00146 (2019)
27. Peng, X., Huang, Z., Zhu, Y., Saenko, K.: Federated adversarial domain adaptation. arXiv preprint arXiv:1911.02054 (2019)
28. Petersen, R.C., et al.: Alzheimer's disease neuroimaging initiative (ADNI): clinical characterization. Neurology **74**(3), 201–209 (2010)
29. Rothchild, D., et al.: Fetchsgd: Communication-efficient federated learning with sketching. In: International Conference on Machine Learning, pp. 8253–8265. PMLR (2020)
30. Song, T., Tong, Y., Wei, S.: Profit allocation for federated learning, pp. 2577–2586 (12 2019). https://doi.org/10.1109/BigData47090.2019.9006327
31. Tang, F., et al.: AIBench training: balanced industry-standard AI training benchmarking. In: 2021 IEEE International Symposium on Performance Analysis of Systems and Software (ISPASS). IEEE Computer Society (2021)
32. Thapa, C., Chamikara, M.A.P., Camtepe, S.: Splitfed: When federated learning meets split learning. arXiv preprint arXiv:2004.12088 (2020)

33. Toyoda, K., Zhang, A.N.: Mechanism design for an incentive-aware blockchain-enabled federated learning platform. In: International Conference on Dig Data (2019)
34. Triastcyn, A., Faltings, B.: Federated learning with Bayesian differential privacy. arXiv: Learning (2019)
35. Wang, G., Dang, C.X., Zhou, Z.: Measure contribution of participants in federated learning. In: International Conference on Dig Data (2019)
36. Wang, H., Yurochkin, M., Sun, Y., Papailiopoulos, D., Khazaeni, Y.: Federated learning with matched averaging. arXiv preprint arXiv:2002.06440 (2020)
37. Wang, Y., Tong, Y., Shi, D.: Federated latent Dirichlet allocation: a local differential privacy based framework. In: The 34th AAAI Conference on Artificial Intelligence, 7 February - 12 February 2020. AAAI 2020, New York City, NY, USA (2020)
38. Xie, C., Huang, K., Chen, P.Y., Li, B.: Dba: Distributed backdoor attacks against federated learning. In: International Conference on Learning Representations (2020). https://openreview.net/forum?id=rkgyS0VFvr
39. Yang, Q., Liu, Y., Chen, T., Tong, Y.: Federated machine learning: concept and applications. ACM TIST 10(2), 12:1–12:19 (2019)
40. Yurochkin, M., Agarwal, M., Ghosh, S., Greenewald, K., Hoang, T.N., Khazaeni, Y.: Bayesian nonparametric federated learning of neural networks. In: International Conference on Machine Learning (2019)

# Fake News Detection Using Knowledge Vector

Hansen He[1,2], Guozi Sun[1,2(✉)], Qiumei Yu[1,2], and Huakang Li[1,2]

[1] School of Computer Science, Nanjing University of Posts and Telecommunications, Nanjing, China
[2] Key Laboratory of Urban Natural Resources Monitoring and Simulation Ministry of Natural Resources, Shenzhen, China
hehansn@foxmail.com, sun@njupt.edu.cn, y0zzzhy@qq.com, huakanglee@163.com

**Abstract.** In recent years, social media takes the advantages of fast spreading speed, wide range and low cost to become the main channel for people to obtain news, which also makes it to be a hotbed for the proliferation of fake news, exposing users and society to huge risks. Due to the fact that there is some true information in fake news, traditional text feature detection algorithms are more difficult to detect the fake news. Therefore, it is necessary to use knowledge as auxiliary information to help detection. We propose a fake news detection framework using knowledge vectors, which can adopt existing and reliable news as knowledge sources and reduce the dependence on expert verification. The framework consists of three parts: event triple extraction based on reliable content, fusion knowledge vector and fake news detector. The experimental results on the data set show that the framework can fuse part of the knowledge information and optimize the detection performance.

**Keywords:** Knowledge representation · Vector fusion · Fake news detection

## 1 Introduction

With the continuous development of social media in the Internet age, fake news has covered all aspects of people's daily lives at an unprecedented speed. During the 2016 U.S. presidential election, a small southern European small city named Veles with a population of only 55,000 actually had at least 100 websites supporting Trump, many of which were full of sensational fake news, making money by attracting network traffic. The ultimate goal is to capture the advertising fee [1]. For fake news producers, they have obtained huge profits at a low cost by creating fake news, inducing users to click to read and earning network traffic. The fake news produced by them inevitably has a negative impact on society. Websites create fake news, while major well-known internet platforms reproduce it without review. The research results of the literature [2] show that half of the total network traffic of fake news websites comes from Facebook's

© Springer Nature Singapore Pte Ltd. 2021
W. Gao et al. (Eds.): FICC 2020, CCIS 1385, pp. 177–189, 2021.
https://doi.org/10.1007/978-981-16-1160-5_15

homepage recommendation, while Facebook has little effect on attracting websites with good reputation, which only accounts for one-fifth of the total traffic. On the one hand, the proliferation of false news originates from the non-selective dissemination of news information by media platforms. On the other hand, it is affected by the limitations of the crowd's perception of news. Due to the limited knowledge acquisition, people's knowledge structure is not enough to fully realize the identification of fake news. According to a survey conducted by YouGov, a public opinion survey agency, nearly half of people (49%) think they can distinguish fake news, but the test results show that only 4% of people can identify fake news by headline. Meanwhile, people will deliberately spread fake news for personal purposes such as mocking others or gaining economic benefits.

Fake news attracts readers' attention with curious headlines and fictitious events, which not only affects individuals' accurate perceptions of social events, hinders people's access to real news, but also has a negative impact on social and economic stability. However, manual methods are often difficult to ensure the quality and authenticity of mass news content, let alone screen and prevent the spread of fake news. The method of manually identifying fake news also has problems such as low efficiency and time lag. Therefore, the introduction of artificial intelligence technology helps to quickly and effectively reduce the spread of fake news on Internet platforms, and also provides the possibility for the improvement of the automatic detection technology of fake news on social media platforms.

## 2    Related Work

In this chapter, we briefly reviewed the content-based fake news detection model, word embedding and knowledge representation learning.

Fake news content detection models can generally be divided into two categories: knowledge-based and style-based [3].

Based on knowledge verification, namely fact checking, Etzioni et al. [4] identify consistency by matching the content extracted from the network with the statements of related documents, but this method is subject to challenges such as the credibility and quality of network data. Maria [5] proposed that the rGALA system obtains incremental knowledge from the Internet, which is simple and effective but requires specific input for different fields. Rashkin et al. [6] analyze the language of news media in the context of political fact checking and fake news detection. They compare real news with irony, pranks, and propaganda, and prove that fact checking is indeed challenging task. Jeff Z. Pan et al. [7] propose some novel methods including the B-transE model, using knowledge graphs to detect fake news content. Studies have shown that incomplete and inaccurate knowledge graphs are helpful in detecting fake news. But to a large extent, it depends on the comprehensiveness of the knowledge graph and the effectiveness of the corresponding fusion transE model. These knowledge graphs also ignore some semantic features of the word itself. The fake news detection framework proposed by Marina et al. [8] uses a combination of source and fact verification and NLP analysis, but this automatic knowledge verification relies heavily on the establishment of semantic similarity.

On the one hand, the detection based on content style is to explore the possible features of fake news from the content of the news itself, mainly including language features, vocabulary features, psycholinguistic features, etc. This includes the number of words in the sentence, syllables, part of speech, reference, punctuation, topic, etc. For example, Castillo et al. [9] extract basic semantic and emotional features from the content, and extract feature classification models of statistical features from users. Majed et al. [10] evaluated the reliability of Twitter content by extracting the length, symbols and emotional word features of the Twitter content text; Liuyang et al. [11] aggregated text features and user features through multi-size windows, and proposed a novel deep neural network and early detection of fake news. On the other hand, it starts with the objectivity-oriented extreme partisan style detection based on language features or the detection of distinctive titles (such as pornography, violent, excessive exaggeration, etc.), so misleading and deceptive clickbait titles can be used as a good indicator to identify fake news [12–14].

For word embedding, Mikolov et al. [15,16] propose word2vec, which includes skip-gram and cbow models and two approximate training methods: hierarchical softmax and negative sampling. Among them, the skip-gram model predicts the words around it through the central word. The cbow model uses surrounding words to predict the central word. Word2vec makes good use of contextual information, but when detecting fake news, the lack of factual auxiliary information often makes the model detection results less satisfactory.

Inspired by word2vec, TransE [17] is a distributed vector representation based on entities and relationships. Each triple (head, relation, tail) is trained in a simple and extensible way by treating the relation as a translation from head to tail. Compared with the previous knowledge representation models, the model is simpler and easier to be understood and works well. However, transE does not perform well in one-to-many, many-to-many relations, so other transX series algorithms that improve on this have appeared later, such as mapping different entities to hyperplanes based on different relations for vector representation. The transH [18] algorithm maps entities and relationships into different spaces, and the entities in the entity space are transferred to the relational space through the transition matrix for the vector representation of the transR [19] algorithm. The entity type is considered in transD [20] algorithm of mapping matrix to project different types of entities and relationships.

On the one side, the simple use of word vector representation without factual auxiliary information tends to have a general model effect. On the other side, the single usage of knowledge representation will make the semantic information of the word itself missing. Therefore, in order to overcome the shortcomings, we propose a fake news detection model based on the fusion of word2vec and transE. The word2vec and transE models are chosen because they are the baseline models in word vector representation learning and knowledge representation learning, respectively, which have strong representation and scalability.

## 3   Methodology

In this chapter, we will introduce our inspection process and model.

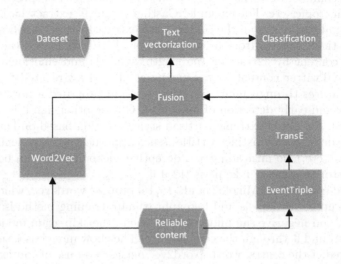

**Fig. 1.** The architecture of proposed framework.

### 3.1   Overview

The problem setting is as follows. Each sample contains the title of the news article and its corresponding true or false news tags. Our goal is to predict the tags of unlabeled news. The specific process is shown in Fig. 1. We use the method of event triple extraction to extract the reliable content, and then perform fusion training with word2vec to obtain a word vector with certain knowledge and ability, and use the word vector to vectorize the news text. Finally, a classifier is used to detect the correctness of the news.

### 3.2   Extract Event Triple

We obtain real news articles from reliable news organizations and websites that specialize in fake news verification respectively. Based on these two kinds of highly reliable news articles, we use event triple extraction dependency-based syntax analysis [21] and semantic role annotation [22]. The specific process is shown in Fig. 2.

Semantic role labeling mainly centers on the predicate of the sentence, analyzes the relationship between each component in the sentence and the predicate. Then, it finds the corresponding agent and recipient based on the predicate, while

**Fig. 2.** The architecture of event triples extracting.

analyzes the main components, namely "subject-predicate-object", of the sentence So far, the "agent-predicate-receive" event triple extracted by semantic role has been realized.

In complex sentences, the predicate and the object are often not in a one-to-one relationship, and there may be multiple predicates and multiple objects. Therefore, based on the semantic role labeling, the dependence syntax analysis can be used to analyze the dependence between the components in the language unit relationship and reveal its syntactic structure. According to the dependency syntax analysis, we can get the subject-verb relationship, verb-object relationship, complement relationship, preposition-object relationship, and attribute relationship. Through these relations, we can extend the predicate, and then get the event triple whose main component is "subject-predicate-object".

---

**Algorithm 1.** Event triple extraction

---

**Input:** news article $A$

**Output:** event triple $T$

1: Get sentence collections from news articles by segmenting sentences, $S_i \in A_i$, $\{s_1, s_2, \ldots, s_n\} \in S_i$, where $n$ is the total number of sentences
2: Get the word set by cutting the sentence $S_i$, $W_{S_i}$, $\{w_1, w_2, \ldots, w_n\} \in W_{S_i}$, where $n$ is the total number of words
3: Tag each word in the word set $W_{S_i}$), get the result set $P_{W_{S_i}}$
4: According to $W_{S_i}$ and $P_{W_{S_i}}$, perform semantic role labeling and dependency syntax analysis respectively, get the result set $R_{S_i}$ and $A_{S_i}$, and get the predicate set $P_n \in R_{S_i}$
5: For simple sentences, look for the event structure of 'agent-predicate-receiver' according to each predicate in $P_n$ and form a triplet, $(h, r, t)_{j \in N} \in T_R$
6: For complex sentences, according to the relationship between each predicate in $P_n$ and $\{sv, vo, co, po, at\} \in A_{Si}$, search the structure of the subject-predicate-object for and constitutes a triplet, $(h, r, t)_{k \in N} \in T_A$
7: **return** Event triple $T = T_R \cup T_A$

---

Algorithm 1 introduces the process algorithm of event triple extraction. First, segment the news article by segmenting words, and then tag the cut-out words by part of speech. Next, based on the words and corresponding parts of speech, we can perform semantic role tagging and dependency syntax analysis. Finally, we extract simple sentences based on the predicate as the center. And the complex sentence is a triple subject with "subject, predicate and object" as the core.

### 3.3    Fuse Word2vec and TransE

We use dependency-based syntax and semantic role labeling to extract triples, and obtain triples information. For example, "There is no evidence that confirms flies are spreading the COVID-19.", and we can get the eventtriple like this (there is no evidence, confirm, flies are spreading the COVID-19). In the process of executing the word2vec model to train the word vector, these triples of auxiliary information are added. In this way, the word vector contains certain triples of fact information, that is, prior information. According to literature [15–17], we know that the objective functions of word2vec's Cbow model and transE model based on negative sampling are formula (1) and formula (2)

$$L = \sum_{w \in C} \{\log[\sigma(\boldsymbol{x}_w^\top \boldsymbol{\theta}^w)] + \sum_{u \in NEG(w)} \log[\sigma(-\boldsymbol{x}_w^\top \boldsymbol{\theta}^w)]\} \qquad (1)$$

In the formula (1) $C$ is the corpus, $\boldsymbol{w}$ is the predicted central word, $NEG(w)$ is the negative sample of word $\boldsymbol{w}$, $\sigma(\boldsymbol{x}_w^\top \boldsymbol{\theta}^w)$ is the probability of predicting central word $\boldsymbol{w}$ when the context is context($\boldsymbol{w}$). $\sigma(\boldsymbol{x}_w^\top \boldsymbol{\theta}^w)$ && $u \in NEG(\boldsymbol{w})$ is the probability of predicting the central word u when the context is context($\boldsymbol{w}$);

$$L = \sum_{(h,r,t) \in S} \sum_{(h',r,t') \in S'_{(h,r,t)}} [\gamma + d(h+r,t) - d(h'+r,t')]_+ \qquad (2)$$

In the formula (2), $S$ is consist of triples $(h,r,t)$, $E$ is consist of two entities $h$ and $t$. $[\boldsymbol{x}]_+$ denotes the positive part of $\boldsymbol{x}$. is a margin hyperparameter. $d(\boldsymbol{x})$ is the dissimilarity measure, which we take to be the norm $\ell_1$ or $\ell_2$ and $S'_{(h,r,t)} = \{(h',r,t)|h' \in E\} \cup \{(h,r,t')|t' \in E\}$.

Therefore, our idea is to add triple information to the Cbow model. We fuse the above two objective functions. For the word w, the formula (3) is as follows:

$$L = \sum_{w \in C} \{\log[\sigma(\boldsymbol{x}_w^\top \boldsymbol{\theta}^w)] + \sum_{u \in NEG(w)} \log[\sigma(-\boldsymbol{x}_w^\top \boldsymbol{\theta}^w)]\}$$

$$+ \alpha \sum_T \sum_{T'} [\gamma + d(\boldsymbol{T}_{head} + \boldsymbol{T}_{relation} + \boldsymbol{T}_{tail}) - d(\boldsymbol{T}_{head'} + \boldsymbol{T}_{relation}, \boldsymbol{T}_{tail'})]_+$$

$$(3)$$

Among them, is used to adjust the contribution of the triad of auxiliary information, $\boldsymbol{T} = \{(w,r,t)|w \in E\} \cup \{(h,r,w|w \in E)\}$, $\boldsymbol{T'} = \{(w',r,t)|w' \in E\} \cup \{(h,r,w'|w' \in E)\}$, $\boldsymbol{T}$ corresponds to the original triple, the word w can be the head entity or the tail entity, $\boldsymbol{T'}$ is a triple of randomly replacing the head entity or tail entity corresponding to $\boldsymbol{T}$.

---

**Algorithm 2.** Training word2vec with transE

---

**Input:** corpus $C$, embedding dimension $k$, triples $T$
**Output:** wordvector $V$

1: ***Initialize*** $vocab\_size \leftarrow build\_dataset(C \cup T)$
$\qquad\qquad V \leftarrow init\_vector(vocab\_size, k)$
$\qquad\qquad \theta \leftarrow init\_vector(vocab\_size, k)$
2: **for all** $w_i \in C$ **do**
3: $\quad$ e=0
4: $\quad$ $X_w = \sum_{u \in context(w_i)} V(u)$
5: $\quad$ **for all** $u = \{w_i\} \cup NEG(w_i)$ **do**
6: $\quad\quad$ $q \leftarrow \sigma(X_{w_i}^\top \theta^u)$
7: $\quad\quad$ $g \leftarrow \eta(L^{w_i}(u) - q)$
8: $\quad\quad$ $e \leftarrow e + g\theta^u$
9: $\quad\quad$ $\theta^u \leftarrow \theta^u + gX_w$
10: $\quad$ **end for**
11: $\quad$ **for all** $u \in context(w_i)$ **do**
12: $\quad\quad$ $V(u) \leftarrow V(u) + e$
13: $\quad$ **end for**
14: $\quad$ $T_{w_i} \leftarrow get\_triples\_h\_t(T, w_i)$ //get a triplet with $w_i$ as the head entity or tail entity
15: $\quad$ $T'_{w_i} \leftarrow sample(T_{w_i})$ //sample a corrupted triplet
16: $\quad$ $(h, r, t) = T_{w_i} \cup T'_{w_i}$
17: $\quad$ **for all** $u = \{t\} \cup NEG(h, r)$ **do**
18: $\quad\quad$ $X_{h+r} \leftarrow h + r$
19: $\quad\quad$ $q \leftarrow \sigma(X_{h+r}^\top \theta^u)$
20: $\quad\quad$ $g \leftarrow \eta(L^{w_i}(u) - q)$
21: $\quad\quad$ $e \leftarrow e + g\theta^u$
22: $\quad\quad$ $\theta^u \leftarrow \theta^u + gX_{h+r}$
23: $\quad$ **end for**
24: $\quad$ **for all** $u = \{h\} \cup \{t\}$ **do**
25: $\quad\quad$ $V(u) \leftarrow V(u) + e$
26: $\quad$ **end for**
27: **end for**
28: **return** $V$

---

Algorithm 2 introduces the main optimization process of the fusion knowledge word vector algorithm. Among them, for $T'$, the amount of data set is limited, if $w$ does not appear in the entity set, a separate structure of self-directed triple information is used to meet the requirements of batch training.

## 3.4  Detect Fake News

Here we choose TextCNN [23] and Bi-LSTM [24] as the classifiers for fake news detection. We use news headlines as the input of the model, use the word vector trained by our model to vectorize the text, and then use TextCNN or Bi-LSTM to detect true and false news.

## 4  Experiment

### 4.1  Dateset

The sources of the data sets are mainly the news data of Tencent's Fact Platform and People's Daily Online. As shown in Fig. 3 and Table 1, the data set mainly

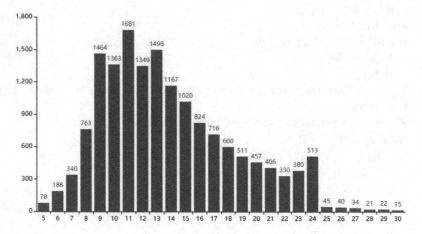

**Fig. 3.** The title words count ranging from 5 to 30.

**Table 1.** Table about samples of dataset.

| Title | Content | Label |
|---|---|---|
| Junior high school students died suddenly because of wearing a mask while running | According to the known information, there may be a correlation between the student wearing a mask and sudden death, but the causal relationship is not clear. According to media reports, he used... | Fake |
| Eat too much seafood, easy to get hepatitis B | This is an unfounded fallacy. The way of hepatitis B infection is blood transmission, and only after hepatitis B virus enters the blood circulation can it reproduce and replicate. Such as | Fake |
| . . . | . . . | . . . |

contains 3 fields, news headlines (the headline length does not exceed 30 words), and news tags (true or fake news) and news content (the news content on the authentic platform is the content of the authentic check), totaling 15,825 pieces of data. The news on Tencent's Fact Platform has been labeled "true/false", while the news on People's Daily Online is not labeled. However, considering the official nature of its official media, we believe that the news on People's Daily Online is true. Therefore, we will compare the news on the authentic platform. The authentic content of the inspection and the news of People's Daily Online are used as reliable content to extract triples.

## 4.2  Experimental Setup

The model is mainly built to test the effect of vectors incorporating knowledge information on detection performance. Three sets of controlled experiments are used. The first set is the case of using random initialization weights without using any pre-training word vectors. We name the model corresponding to this case *rand*. The second group is the case of using word2vec (Cbow) training method to train the word vector, we name the corresponding model in this case *word2vec*. The last group is the case of using the word vector of fusion knowledge, we name the corresponding model in this case *word2vec_ke*.

The evaluation method selects the 10-fold cross-validation of stratified sampling, and the performance metrics select accuracy, precision, recall and F1-score.

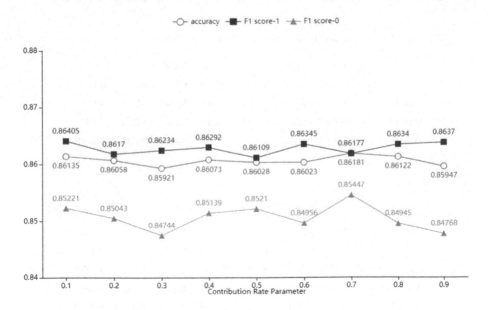

**Fig. 4.** Performance comparison of different contribution rate parameters.

**Parameter**

For the word2vec_ke, the word vector dimension in the model is 128, the window size is 2, the learning rate is set to 0.01, the margin is set to 1.0, and the adjustment model contribution rate is set to 0.7. The different values of contribution rate have an effect on the model performance, which is shown in Fig. 4. The model comprehensively considers the characteristics of convolutional neural networks (CNN) and recurrent neural network (RNN), and adopts a structure in which the features of single-layer convolution pooling and the final state features of bidirectional lstm are connected in parallel (spliced) and then sent to the fully connected layer for classification. In this model, the x-axis represents the change range of from 0.1 to 0.9, where the change interval is 0.1. There are three performance indicators, namely the accuracy, the f1 score of the true sample and the f1 score of the fake sample. Here we set the parameter to 0.7, when the both the values of the f1 score of the true sample and the f1 score of the fake sample are higher. The distance loss function of the head and tail entities and the relationship adopts L1 regularization, while the number of iterations is set to 150,000.

For the classifier model, all word vectors or randomly initialized word embedding layer dimensions are 128. All vectors are fine-tuned during training. We set the initial value of the learning rate to 0.001 and select Adam as the optimization algorithm. The output dimension of the 3-layer convolutional layer of the TextCNN model is set to 256, the size of the convolution kernel is set to [3,4,5]. In the flat layer, the dropout value is set to 0.5. Finally, the Bi-LSTM hidden layer dimension is set to 256, while the dropout value is also set to 0.5.

**Table 2.** The result of TextCNN.

| | Acc | True | | | Fake | | |
|---|---|---|---|---|---|---|---|
| | | P | R | F1 | P | R | F1 |
| Rand | 82.58 | 83.98 | 83.13 | 83.11 | 81.02 | 81.98 | 80.9 |
| Word2vec | 86.25 | 88.36 | 85.43 | 86.45 | 84.1 | 87.31 | 85.15 |
| Word2vec_ke | **86.96** | **88.9** | **86.36** | **87.27** | **84.93** | **87.63** | **85.83** |

**Table 3.** The result of Bi-LSTM.

| | Acc | True | | | Fake | | |
|---|---|---|---|---|---|---|---|
| | | P | R | F1 | P | R | F1 |
| Rand | 85.19 | 87.89 | **83.89** | 85.46 | 82.52 | 86.58 | 84.03 |
| Word2vec | **86.68** | 90.52 | 83.75 | **86.61** | **82.91** | 90.09 | 85.92 |
| Word2vec_ke | 86.62 | **90.63** | 83.5 | 86.54 | 82.77 | **90.24** | **85.94** |

**Result Analysis**

Table 2 and Table 3 respectively show the experimental results of using three different word vectors for TextCNN and Bi-LSTM models. For each type of news, calculations of accuracy, precision, recall, and F1-score are performed. We can get several observations:

Word2vec_ke achieves the best results in TextCNN, but does not perform well in Bi-LSTM. Compared with word2vec, word2vec_ke has added a part of the knowledge information. The global timing characteristics of Bi-LSTM may introduce noise to the word2vec_ke model, resulting in the difference of the prior knowledge part degrades the performance of the model.

In the experimental results of TextCNN, the results of rand and the other two models are quite different. The reason is that the weights of random initialization do not reflect the relationship between words. Therefore, for the model, such random initialization weight features lack strong discrimination.

Compared with Bi-LSTM, the experimental results of TextCNN are better in the case of rand. It can be seen that for random weights that do not contain any pre-information, the convolution features of CNN are not as important as the contextual information, which is easy to understand. For news, the contextual information is often more critical, and the information captured by the TextCNN feature is often limited because of the window.

For Bi-LSTM experimental results, the performance evaluation indicators of rand, word2vec and word2vec_ke are relatively close. The index of word2vec_ke in the fake tag is slightly better than word2vec. On the contrary, the index of word2vec in the true tag is slightly better than word2vec_ke.

In general, the result of rand is the worst, because it does not consider the relationship between words at all, while word2vec and word2vec_ke establish the relationship between words through pre-training, where word2vec_ke adds triples. Knowledge is used as auxiliary information, so the performance index of word2vec_ke is better than word2vec.

## 5  Conclusion

In this article, we study the problem of detecting fake news by using reliable information content to construct knowledge triples as auxiliary information. Due to the authenticity, timeliness, and versatility of news, it is particularly difficult to detect fake news. Therefore, our idea is to use some available and reliable content to construct simple triple knowledge information to help the model improve detection performance. Our proposed model includes triple extraction, fusion training of word vector with knowledge representation, and fake news detection. The triad extraction constructs the corresponding triad information by extracting events from reliable content. Then, these triads will be fusion trained through the cbow model and the trans model to obtain word vectors with a certain knowledge representation ability. Finally, the representation of word vectors is used to detect fake news. Comparative experiments on fake news datasets show that our method can use triple knowledge to improve detection performance to a certain extent.

**Acknowledgement.** The authors would like to thank the anonymous reviewers for their elaborate reviews and feedback. This work is supported by the National Natural Science Foundation of China (No. 61906099), the Open Fund of Key Laboratory of Urban Land Resources Monitoring and SimulationMinistry of Natural Resources (No. KF-2019-04-065).

# References

1. Visit the Fake News Factory in Macedonia. https://epaper.oeeee.com/epaper/C/html/2017-02/26/content_9585.htm. Accessed 26 Feb 2017
2. Verónica, P., Bennett, K., Alexandra, L., Rada, M.: Automatic detection of fake news. In: Proceedings of the 27th International Conference on Computational Linguistics, pp. 3391–3401. Association for Computational Linguistics, Santa Fe (2018)
3. Kai, S., Amy, S., Suhang, W., Jiliang, T., Huan, L.: Fake news detection on social media: a data mining perspective. SIGKDD Explor. **19**(1), 22–36 (2017)
4. Oren, E., Michele, B., Jin, S., Daniel, S.: Open information extraction from the web. Commun. ACM **51**(12), 68–74 (2008)
5. Maria, M.H.K.: Incremental knowledge acquisition approach for information extraction on both semi-structured and unstructured text from the open domain web. In: Jojo, S.W. (ed.) Gholamreza, H., pp. 88–96. ALTA, Proceedings of the Australasian Language Technology Association Workshop (2017)
6. Hannah, R., Eunsol, C., Jin, Y.J., Svitlana, V., Yejin, C.: Truth of Varying Shades: Analyzing Language in Fake News and Political Fact-Checking. In: Martha, P., Rebecca, H. (eds.) Proceedings of the 2017 Conference on Empirical Methods in Natural Language Processing, EMNLP 2017, pp. 2931–2937. Association for Computational Linguistics, Copenhagen (2017) . https://doi.org/10.18653/v1/d17-1317
7. Je, Z.P., Siyana, P., Chenxi, L., Ningxi, L., Yangmei, L., Jinshuo, L.: Content Based Fake News Detection Using Knowledge Graphs. In: Denny, V., Kalina, B.(eds.) The 17th International Semantic Web Conference, ISWC 2018, pp. 669–683. Springer, Monterey (2018). https://doi.org/10.1007/978-3-030-00671-6
8. Marina, D.I., Kin, F.L.: A Machine learning approach to fake news detec-tion using knowledge verification and natural language processing. In: Leonard,B., Hiroaki, N. (eds.) The 11th International Conference on Intelligent Network-ing and Collaborative Systems, INCoS 2019, pp. 223–234. Springer, Oita(2019).https://doi.org/10.1007/978-3-030-29035-1
9. Carlos, C., Marcelo, M., Barbara, P.: Information credibility on twitter. In: Sadagopan, S., Krithi, R. (eds.) Proceedings of the 20th International Conference on World Wide Web, WWW 2011, pp. 675–684. ACM, Hyderabad (2011). https://doi.org/10.1145/1963405.1963500
10. Majed, A.A., Muhammad, A., Muhammad, M.H., Atif, A.: A credibility analysis system for assessing information on twitter. IEEE Trans. Dependable Secur. Comput. **15**(4), 661–674 (2018)
11. Yang, L., Yifang, B.W.: FNED: a deep network for fake news early detection on social media. ACM Trans. Inf. Syst. **38**(3), 1–33 (2020)
12. Daoud, M.D., Samir, A.E.: An effective approach for clickbait detection based on supervised machine learning technique. Int. J. Online Biomed. Eng. **15**(3), 21–32 (2019)

13. Yimin, J.C., Niall, J.C., Victoria, L.R.: Misleading online content: recognizing click-bait as false news. In: Mohamed, A., Mihai, B. (eds.) Proceedings of the 2015 ACM Workshop on Multimodal Deception Detection, WMDD@ICMI 2015, pp. 15–19. ACM, Seattle (2015). https://doi.org/10.1145/2823465

14. Kai, S., Suhang, W., Thai, L., Dongwon, L., Huan, L.: Deep headline generation for clickbait etection. In: IEEE International Conference on Data Mining, pp. 467–476. IEEE Computer Society, Singapore (2018)

15. Tomas, M., Kai, C., Greg, C., Jeffrey, D.: Efficient estimation of word representations in vector space. In: Yoshua, B., Yann, L., (eds.) 1st International Conference on Learning Representations, ICLR 2013. ICLR, Scottsdale (2013)

16. Tomas, M., Ilya, S., Kai, C., Gregory, S.C., Jeffrey, D.: Distributed representations of words and phrases and their compositionality. In: Christopher, J.C., Léon, B. (eds.) 27th Annual Conference on Neural Information Processing Systems, pp. 3111–3119 (2013)

17. Antoine, B., Nicolas, U., Alberto, G., Jason, W., Oksana, Y.: Translating embeddings for modeling multi-relational data. In: Christopher, J.C., Léon, B. (eds.) 27th Annual Conference on Neural Information Processing Systems, pp. 2787–2795 (2013)

18. Zhen, W., Jianwen, Z., Jianlin, F., Zheng, C.: Knowledge graph embedding by translating on hyperplanes. In: Proceedings of the Twenty-Eighth AAAI Conference on Artificial Intelligence, pp. 1112–1119. AAAI Press, Québec City (2014)

19. Yankai, L., Zhiyuan, L., Maosong, S., Yang, L., Xuan, Z.: Learning entity and relation embeddings for knowledge graph completion. In: Proceedings of the Twenty-Ninth AAAI Conference on Artificial Intelligence, pp. 2181–2187. AAAI Press, Austin (2015)

20. Guoliang, J., Shizhu, H., Liheng, X., Kang, L., Jun, Z.: Knowledge graph embedding via dynamic mapping matrix. In: Proceedings of the 53rd Annual Meeting of the Association for Computational Linguistics and the 7th International Joint Conference on Natural Language Processing of the Asian Federation of Natural Language Processing, pp. 687–696. The Association for Computer Linguistics, Beijing (2015)

21. Weifa, Z.: A summary of chinese syntactic analysis research. Inf. Technol. **000**(007), 72–74, 78 (2012)

22. Jihong, L., Ruibo, W., Weilin, W., Guochen, L.: Automatic labeling of semantic roles in Chinese frames. J. Softw. **21**(004), 597–611 (2010)

23. Yoon, K.: Convolutional Neural Networks for Sentence Classification. In: Alessandro, M., Bo, P. (eds.) Proceedings of the 2014 Conference on Empirical Methods in Natural Language Processing, EMNLP 2014, pp. 1746–1751. ACL, Doha (2014). https://doi.org/10.3115/v1/d14-1181

24. Mike, S., Kuldip, K.P.: Bidirectional recurrent neural networks. IEEE Trans. Sign. Process. **45**(11), 2673–2681 (1997)

# A Reconfigurable Electrical Circuit Auto-Processing Method for Direct Electromagnetic Inversion

Jun Lu(✉)📛

Institute of Physics, Chinese Academy of Sciences, Beijing National Laboratory
for Condensed Matter Physics, Beijing, China
lujun@iphy.ac.cn
http://www.iop.cas.cn/rcjy/yjdwfgj/?id=1784

**Abstract.** Extracting information as much and precise as possible from nondestructive measurements remains a challenge, especially when advanced test applications are emerging in electromagnetic encephalography and high throughput physical property characterization of materials genome chips. To solve the inversion problem, various soft algorithms have been developed such as finite element method, machine learning and artificial neural network, whose performance is limited by indirectly processing of intermediate layers in digital computers. This paper proposes a novel direct method of analog network entity with reconfigurable electrical circuit auto-processing (RECAP), which mainly consists of voltage controlled elements, measurement unit, and automatic feedback unit. During each inversion process, after the test results are input, the circuit network performs initialization, choosing topology, automatically tuning the property for each network element, and finally approaching a convergent solution to user request after some cycles of self-adjustment. Principles and advantages are introduced with several instances, showing high accuracy and stable convergence ability, as well as helping judge whether the topology is suitable for optimization. This method can not only invert purely loss components, but also invert circuits containing reactive components. Based on the verification from cases, it is also found that the inversion efficiency of RECAP is linearly dependent on the number of elements N, which is better than the usual mature inversion algorithms. Therefore, it is then concluded as a promising tool for high performance inversion.

**Keywords:** Nondestructive testing · Topological electrical circuit · Electromagnetic inversion

## 1   Introduction

The inversion problem was described by Langer as determination of equation coefficients followed by finding solution of the problem in 1930s [15]. Calderon

This work is under supports from National Natural Science Foundation of China (No. 51327806), Fujian Institute of Innovation and Youth Innovation Promotion Association of Chinese Academy of Sciences (No. 2018009).

clarified it as inverse boundary value problem [4]. Later, thanks to the introduction of regularization, complex practical inversion problems can also be simplified and solved [10]. Global uniqueness of inversion results is also possible if boundary has Lipshitz continuity for a 2D problem [19], 3D object [3] or system with higher dimension [28]. Inverse processing has been widely used in various engineering problems, especially electromagnetic related inversion, such as process control [9], nondestructive detection or exploration [5,14,21,26], atomic resolution structural investigation [5,7,8], and biomedical characterization [12,22,31]. Many kinds of methods has been developed to solve electromagnetic inversion problems, including direct inversion [27] or asymptotic treatment [1], principle component analysis [29], statistical simulation [6], approximate finite elements analysis [11], and machine learning [17] even deep learning [16] in recent years. It is true that these inversion methods are pure digital algorithms where preprocessing and post-processing of data is necessary in computer. Moreover, calculation complexity of these inversion methods is polynomial dependent on number of basic blocks. Benner and Mach have proposed hierarchical process to reduce the time and storage cost, where the inverse complexity has been optimized to $O(N*logN)$ [2]. In the case of large number of basic blocks, it is desirable to further reduce the processing complexity, since computation resources are limited. This paper is going to describe a new method for electromagnetic inversion, which can reduce the complexity further and hence possess high efficiency. The new method is based on reconfiguable electric circuits with precision measurement and automatic feedback. The concept of this reconfigurable electrical circuit auto-processing (RECAP) inversion process is merging test devices and the circuit under test, so as to directly invert the original problem by self-adjusting each component in the network. Equivalent electrical circuits are usually used in simulation [13] and fitting software [24,32], or even for decryption [30]. Murai and Kagawa demonstrated 2D electric circuit network for a finite element solver [18], and Selleiri developed algorithms for topological optimization of circuit networks [25]. However, as far as the current author can concern, reconfigurable electrical circuits for automatical inversion have not been reported. Since quantum algorithms are crucial for the next generation computation and still under developing [20], the current paper is expected to provide a possible way for realizing the electromagnetic inversion of quantum optimization [23].

## 2   Principle Design

It has been an interesting but challenging problem to invert all components of a electrical network without breaking any connection. The difficulty is not only from limited information on network topology, also from inevitable intercoupling, since single component can influence all impedance measurements and every two-node impedance result comes from the whole circuit network. Correspondingly, the proposed method compose of two levels of nested loops including main part level in the internal loop and topology selection level in the outer loop. The main inversion circuit contains measurement comparing part and auto feedback adjustment part, where reconfigurable electrical network is composed of voltage tunable electronic components. As shown in Fig. 1, the outer loop performs topology selection, and the inner loop performs self-adjustment of component parameters for the set topology. In each set of topological structures, the parameters of the components can be adjusted by voltage, and the in-situ electrical properties at both ends of these variable electronic components are measured one by one. The measurement result is simultaneously compared with the measurement result of the inverse object circuit. The feedback voltage of the comparison result then controls the variable element to automatically adjust, and finally makes the in-situ measured electrical properties at both ends of the variable element is equal to that of the corresponding two end points of the inversion object circuit. The inversion operation of other components is subsequentially performed by the same procedures. When the comparison and feedback of all components in the set topology are finished, a round of RECAP inversion iteration is carried out. After that, it is judged whether to perform a new round of RECAP inversion iteration according to the relative error between the in-position measurement result and the inverse object, and the cycle repeats until the inversion result meets the stop error or reaches the maximum allowable number of iterations. When making optimization among multiple topological structures, it is necessary to compare the final relative error of the inversion iteration results of each structure, and select the topological structure with the lowest relative error. Finally, when the topological structure and all circuit components show acceptable consistent with the inversion object based on measurement of circuit properties, the whole inversion operation is completed.

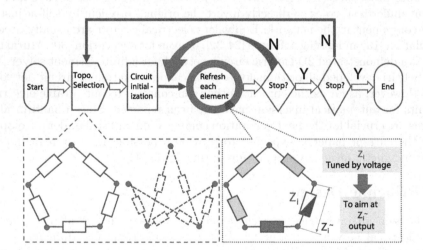

**Fig. 1.** Principle diagram of reconfigurable electrical circuit network for direct electromagnetic inversion.

# 3   Method Demonstration

In order to make the RECAP inversion method easier to be understood and applied, this section demonstrates the RECAP in a simple resistor network as an example for analysis. As shown in Fig. 2, a six-node, seven-element circuit network, each element is set with a resistance value Ri, and the resistance Ri' is directly measured in situ. The simulation results are shown in the table below the figure, where the first two data lines indicate that the difference between Ri and Ri' is very large. However, through the RECAP method, the set resistance value Ri of each element itself can be recovered after 3 iterations of inversion, and the average relative error $\sum_{i=1}^{N}[(Ri' - Ri)/Ri]/N$ is less than 1%, where N is the total number of elements and i is each natural number between 1 and N (N = 7 in the current demonstration). More cycles of inversion could be performed later, so as to obtain a lower relative error. The RECAP starts with initialization via selecting a measuring component. Under the constraint that all components are set to the same resistance, that is, the control terminals of all voltage variable resistors are connected to the same control terminal, and the voltage is adjusted to change the resistance of all components so that the apparent resistance of the two nodes of the measuring element is equal to the resistance of the corresponding nodes of the inversion object. Here, as the starting point, the resistance measured between nodes I and VI is $0.9\,\Omega$. By simulation tools such as LTSPICE one can find that when all resistances in the network are adjusted to $1.227\,\Omega$, the resistance measurement between nodes I and VI is $0.9\,\Omega$s. Then RECAP continue moving the measurement probes of the resistance meter to the next element. After such initialization, the control ends of reconfigurable circuit components for all subsequent iterations are separated independently, that is, the RECAP component parameters are controlled and adjusted one by one. For example, the second comparison feedback element is connected to the nodes I and II, where the measured result of resistance between the nodes I and II of the inversion object circuit is $1.6\,\Omega$. To address the target resistance, the RECAP component between the nodes I and II is varied while all other resistances remain unchanged at $1.227\,\Omega$. Circuit simulation results show that the current iteration of the element is finished when the resistance increases to about $3\,\Omega$. Then the measurement probes is moved to nodes V and VI. As in the RECAP inversion procedures for the previous component connected between nodes I and II, the inversion result of the current iteration of the element is $4.75\,\Omega$. Sequentially, after performing such parameter reconstruction operation for all components of the target inversion circuit, the first round of inversion iteration is completed. At this time, it is judged whether the average relative error between the nodes of RECAP circuit and the corresponding nodes of the inversion object is acceptable. When the average relative error is larger than the set limit, such as 1% , the next round of inversion iteration is then proceeded. It can be seen from the inversion result data of this example that after the third round of inversion iteration, the average relative error of the inversion object has dropped below 1%. If more precision convergence is required, more inversion iterations will help to improve the inversion results.

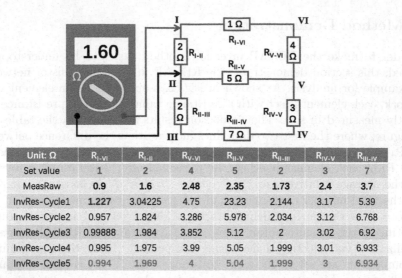

| Unit: Ω | $R_{I-VI}$ | $R_{I-II}$ | $R_{V-VI}$ | $R_{II-V}$ | $R_{II-III}$ | $R_{IV-V}$ | $R_{III-IV}$ |
|---|---|---|---|---|---|---|---|
| Set value | 1 | 2 | 4 | 5 | 2 | 3 | 7 |
| MeasRaw | 0.9 | 1.6 | 2.48 | 2.35 | 1.73 | 2.4 | 3.7 |
| InvRes-Cycle1 | 1.227 | 3.04225 | 4.75 | 23.23 | 2.144 | 3.17 | 5.39 |
| InvRes-Cycle2 | 0.957 | 1.824 | 3.286 | 5.978 | 2.034 | 3.12 | 6.768 |
| InvRes-Cycle3 | 0.99888 | 1.984 | 3.852 | 5.12 | 2 | 3.02 | 6.92 |
| InvRes-Cycle4 | 0.995 | 1.975 | 3.99 | 5.05 | 1.999 | 3.01 | 6.933 |
| InvRes-Cycle5 | 0.994 | 1.969 | 4 | 5.04 | 1.999 | 3 | 6.934 |

**Fig. 2.** A typical circuit for inverse demonstration, where each resistor component is in situ measured, and the entire measurement result is input to the reconfigurable circuit network. Under the RECAP inverse strategy, after 3 runs of auto adjustment, the inverse circuit inverts the set parameters within 1% uncertainty, and more iteration cycles afterwards can further improve the precision.

## 4    Evaluation and Discussion

### 4.1    Topological Determination

The determination of the topological structure is an interesting problem in electromagnetic inversion. It is found through the simulation results of cross-topology inversion that the RECAP method can not only select the best one from multiple topological structures based on the average relative error information, and it can also automatically detect whether the selected topology is correct, through the parameter evolution in the iterative process of topological structures with higher symmetry. This paper uses a four-node circuit network example to analyze, where the three topological networks a, b, and c of the four-node circuit shown in Fig. 3 have 4 components, 5 components, and 6 components, respectively. In order to better analyze the topological structures, each component parameter is set to equal resistance (here all resistances are 1 Ω). The resistance between all two combination of the four nodes of the three topological structures has been listed in the table at the bottom of Fig. 3. As a result, it can be seen that the symmetry of the three topological structures with the same component parameters gradually increases from a to b to c.

When we use three different topological structures for the pairwise combination between RECAP and the three target circuits to be inverted, we obtain a 3 × 3 matrix, as shown in Fig. 4. Except for the diagonal combination of the topological cross matrix, where the inversion result is completely consistent with the set circuit, it is found that if a topology with a lower symmetry is used to invert a target circuit with a higher symmetry, systemic errors will obviously occur. The error cannot be automatically judged without adding measurement information. Conversely, if a topology with higher symmetry is used to invert a circuit with lower symmetry, the results of the inverted components are all correct, and the "excess" components will exhibit exponential divergence as the number of iterations increases. It can be seen that when a topology with a higher symmetry is used to invert a circuit with a lower symmetry, RECAP can automatically determine that it uses an over-high symmetry. Obviously, for the topology inversion of RECAP, the recommended way is to start with high symmetry and reduce the symmetry gradually through clue from divergent components, so that the inversion results just do not appear divergent components.

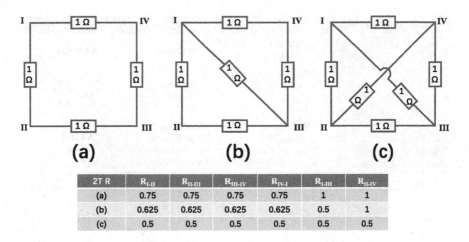

| 2T R | $R_{I-II}$ | $R_{II-III}$ | $R_{III-IV}$ | $R_{IV-I}$ | $R_{I-III}$ | $R_{II-IV}$ |
|---|---|---|---|---|---|---|
| (a) | 0.75 | 0.75 | 0.75 | 0.75 | 1 | 1 |
| (b) | 0.625 | 0.625 | 0.625 | 0.625 | 0.5 | 1 |
| (c) | 0.5 | 0.5 | 0.5 | 0.5 | 0.5 | 0.5 |

**Fig. 3.** Four-node electrical circuit network with three topologies (a) a close loop with four resistors, (b) co-edge double triangles with five resistors, (c) a tetrahedron with six resistors, and the bottom table shows the raw resistance between nodes for three configurations.

## 4.2   Convergence Performance

When the topological structure is fixed, the convergence behavior during the electromagnetic inversion iteration process is a very interesting topic. It not only shows the evolution of the inversion accuracy, but also reflects whether the initialization process is reasonable. Through different set resistance parameters of the same topology, especially the change of the ratio of the maximum resistance to the minimum resistance, the simulation results are studied. As shown in Fig. 5, ratios of the maximum to minimum resistance values of the inversion object are set 1, 10, and 1000 in Fig. 5 a, b, and c, respectively. For the resistor network with the same resistance, after the initialization operation of RECAP, the inversion result can be obtained from the beginning, and as the maximum to minimum resistance ratio increases, that is, the asymmetry of the inversion component parameters increases, and the circuit inversion iteration becomes slower. By changing the position of the initialization, the RECAP result also found that for the inversion object with low parameter symmetry, if the initial resistance is exactly selected at element where the smallest resistance locates, the result may not be converged. Nevertheless, when the initial position is selected at the highest resistance, as shown in Fig. 5 d and e, it is rather useful to obtain a stable iteration result. It is easier to perform RECAP initialization at a larger resistance to obtain a stable convergence result, mainly because the two-node resistance in the low resistance network has an upper limit of measured resistance, and it is impossible to obtain a higher resistance by adjusting the resistance of the component connected to these two nodes.

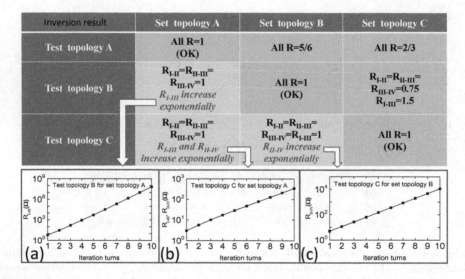

**Fig. 4.** Matrix show of inverse result for mixed topologies. When test topology matches the set topology by coincidence, shown in the diagonal cell of the table, inversion process can be performed normally, but error could happen as the test topology is oversimplified. If test topology is more complicated than set topology, excess inverted components go divergent, as show in (a) divergence of excess resistance between node I and III of test topology B for set topology A, (b) divergence of excess resistance between node I(II) and III(IV) of test topology C for set topology A, and (c) divergence of excess resistance between node II and IV of test topology C for set topology B

## 4.3   Inversion with Admittance

The potential application of RECAP is actually not only in pure loss networks, so the current section discusses electromagnetic inversion with energy storage elements. In order to focus on the inversion of reactive components, the number of nodes used for the inversion of reactive components is set to three, and only typical circuit units are discussed, that is, capacitors with resistors in parallel and inductors with resistors in series as inversion units. As shown in Fig. 6, the circuit to be inverted and the RECAP inversion circuit are shown in Fig. 6a and Fig. 6b, respectively. The RECAP tool is composed of three nodes. The variable unit between each two nodes is composed of four components, saying variable capacitors with variable resistance in parallel, and variable inductance with resistance in series. The RECAP inversion process and results are presented in the table in the lower left corner of Fig. 6. In order to highlight the inversion of the reactive element, the inversion process of the resistance element is not displayed. It can be noticed that there are a total of 12 components in three nodes to be inverted. It seems that the amount of information available for inversion is too small. However, there is no problem in principle by making full use of frequency dependent complex impedance measurement in the inversion process of reactance components. In order to focus on the feasibility of the RECAP inversion of reactance, this article does not involve the problem of mining as many circuit components in unit as possible from the complex impedance spectrum, but only the lower frequency 100 Hz and the higher frequency of 1 MHz is chosen for inversion tests. The complex impedance of the

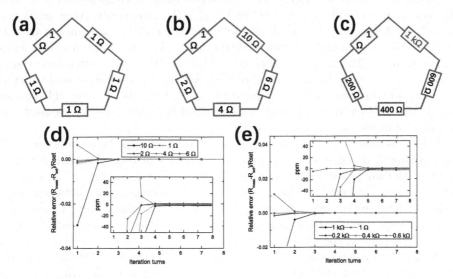

**Fig. 5.** Convergence process dependence of resistance differences for the same topology, where maximum to minimum ratio is equal to (a) 1 (b) 10 and (c) 1000. The convergence of the (a) is obviously one-step convergence, while the convergence process of (b) and (c) is shown in (d) and (e), respectively.

dual frequency point is implemented here by LCR measurement. At each frequency point one can invert two component parameters by complex impedance measurement, so dual frequency measurement results of the three nodes can achieve exactly 12 component parameters in total. It is not difficult to understand that the inductance element has obvious contribution to impedance at high frequencies but less contribution to impedance at low frequencies, which is equivalent to being short-circuited. Capacitive elements have a greater contribution to impedance at low frequencies, but are similar to being short-circuited at high frequencies. Therefore, this work uses in-parallel RC measurement 100 Hz for the separate inversion of parallel resistance and parallel capacitance, and the series RL measurement result at 1 MHz corresponds to the separate inversion of series inductance and series resistance. Like the simulation example discussed in the second section of this paper, the first row of data in the inversion process table in Fig. 6 is the set component parameters, the second row of data is the capacitance or inductance corresponding to the two-node measurement results, and the third row is initialization result. Nodes pair with the largest impedance modulus among the three-node combination was chosen for initialization of the high and low frequency components. That means, initialization starts from the parallel capacitors at the III-I node with the smallest capacitance 100 Hz (the parallel resistance is not introduced when the parallel capacitor is initialized, and all the series inductance and series resistance are short-circuited), and then the parallel resistance is initialized similarly to the method. The series inductance is initialized at the II-III node with the largest inductance at 1 MHz. Then similar rules were used to initialize the series resistance. After initialization, the parallel capacitance measured 100 Hz and the series inductance measured at 1 MHz between each two nodes are 667 nF and 2.57 mH, respectively. After initialization, the first iteration of RECAP also follows the parallel capacitance of the I-II node 100 Hz, the parallel resistance 100 Hz, the series inductance and the series resistance at 1 MHz; then the II-III node and the III-I node. After each round of inversion, whether the next round of RECAP inversion iteration is also determined by whether the average relative error is less than the set limit. This paper found that the inversion of a three-node parallel RC plus series RL element can be lower than the average relative error limit of 1.

When checking the frequency spectra of complex impedance between node III and node I, one may find the stable convergence picture as shown in Fig. 7. Because the initialization process begins from the low-frequency capacitance between node III and I, initial parameters of the components between other nodes deviate greatly from the set value. Therefore, initially only the complex impedance spectrum at the low-frequency point is consistent with that of the set parameters, while it completely deviates from the multiple turning characters in the middle frequency range due to LC resonances. As the RECAP iterative process progresses, from the first round, the second round to the fourth round, the turning curves in the middle steadily approaches the complex impedance spectrum of set parameters. After the fourth round of inversion iteration, there is no obvious difference between the RECAP result and the frequency spectrum of set parameters from the spectrum comparison. When we check the convergence process of multiple iterations, as shown in Fig. 8, the two curves of solid circles and open squares respectively represent the average relative error between the component parameter and the set parameter and the LCR testing result of corresponding node pairs. The relative error of the component parameters is obviously greater than the relative error of the AC properties of the corresponding node pair, because the adjustment of the component parameter is based on the measurement and comparison results of the AC performance between the every node pair, and the primary measurement comparison error is amplified. However, from the average relative error attenuation curve of the component parameters. It can also be seen that at the fourth round of RECAP, the coarse component self-adjustment is transformed into the fine adjustment of components, thereby gradually approaching the set value. It is not difficult to see that after the 12th cycles of iteration, the average relative error afterwards is close to one thousandth.

### 4.4   Complexity Analysis

Inversion efficiency and process complexity are very important indicators for application. From the instances demonstrated in previous sections, it is found that the RECAP method directly compares the measurement results of the learning circuits with the inverted circuit, and the entire process is truly observable. Moreover, it is easy to be understood in real physical scenes. That is not like

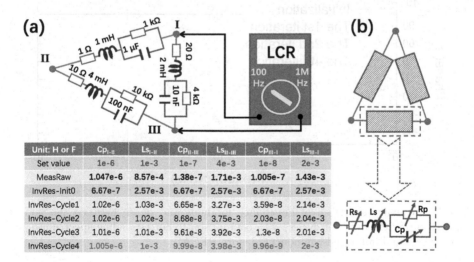

| Unit: H or F | $Cp_{I-II}$ | $Ls_{I-II}$ | $Cp_{II-III}$ | $Ls_{II-III}$ | $Cp_{III-I}$ | $Ls_{III-I}$ |
|---|---|---|---|---|---|---|
| Set value | 1e-6 | 1e-3 | 1e-7 | 4e-3 | 1e-8 | 2e-3 |
| MeasRaw | 1.047e-6 | 8.57e-4 | 1.38e-7 | 1.71e-3 | 1.005e-7 | 1.43e-3 |
| InvRes-Init0 | 6.67e-7 | 2.57e-3 | 6.67e-7 | 2.57e-3 | 6.67e-7 | 2.57e-3 |
| InvRes-Cycle1 | 1.02e-6 | 1.03e-3 | 6.65e-8 | 3.27e-3 | 3.59e-8 | 2.14e-3 |
| InvRes-Cycle2 | 1.02e-6 | 1.02e-3 | 8.68e-8 | 3.75e-3 | 2.03e-8 | 2.04e-3 |
| InvRes-Cycle3 | 1.01e-6 | 1.01e-3 | 9.61e-8 | 3.92e-3 | 1.3e-8 | 2.01e-3 |
| InvRes-Cycle4 | 1.005e-6 | 1e-3 | 9.99e-8 | 3.98e-3 | 9.96e-9 | 2e-3 |

**Fig. 6.** A typical circuit including inductors and capacitors for inverse demonstration, where capacitance with resistance in parallel and inductance with resistance in series is measured 100 Hz and 1 MHz, respectively, for node pair from I to II, from II to III and from III to I. Then the entire measurement result is input to the reconfigurable circuit network. Under the RECAP inverse strategy, after 4 runs of auto adjustment, the inverse circuit inverts the set parameters within 1% uncertainty.

normal numerical methods which convert the measured results in the real physical world into a digital mode, and at the end of the inversion, it is converted back into observable and understandable data for the real physical world. The intermediate processed data is difficult to be directly accessible by the real world. In addition, from previous examples, it can be seen that RECAP's dependence on the time or space resources required for inversion increases linearly with the increase of number of elements N, because the number of iterations is limited (about 3–5 rounds based on examples in this article), a total of N operations are performed on the variable elements one by one in each round. Furthermore, if block or hierarchical designs are introduced in future, there is still potential for improvement, which is more attractive than the usual inversion algorithm with the super-linear dependence of N. Therefore, from the perspective of better comprehensibility and resource consumption dependency, RECAP does provide a new, simpler and more direct way for inversion.

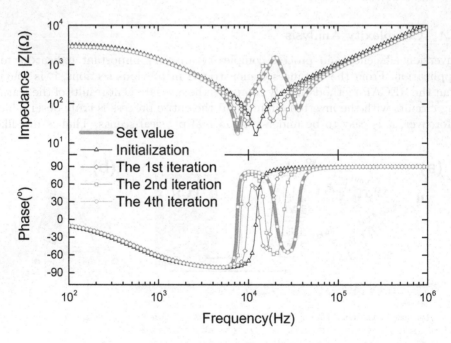

**Fig. 7.** Frequency spectrum of impedance modulus and phase between the node I and III during the different cycles, with comparison to corresponding spectrum of set components and initialized components before RECAP.

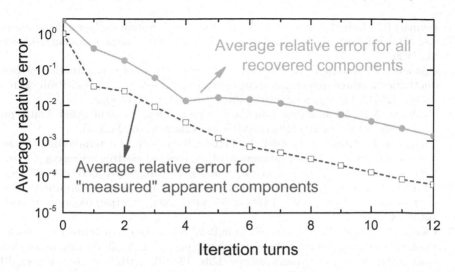

**Fig. 8.** Convergence behavior of average relative error for recovered components and measured apparent components during the RECAP inversion process.

## 5    Conclusion

This paper proposes a reconfigurable circuit auto-processing method for electromagnetic inversion. The method uses the measured properties of the inverted circuit as a criterion, automatically adjusts the component parameters, and gradually approximates the inverted circuit component parameters after several iterations. Through several simulation demonstrations of typical circuit inversion, this method shows high accuracy and stable convergence ability, and can help judge and optimize whether the topology is suitable. This method can not only invert purely loss components, but also invert circuits containing reactive components. Based on the verification from cases, it is also found that the inversion efficiency of RECAP is linearly dependent on the number of elements N, which is better than the usual mature inversion algorithms. RECAP combines precision measurement and circuit flexibility with switchable topology, avoiding the complicated transformation between the real world and mathematical forms of pure numerical or digital methods, and provides a new type of efficient and fast hardware for solving simulation and inversion for real world problems. This method is expected to be used to guide the design and verification of direct inversion with integrated processing chips.

## References

1. Bao, G., Ammari, H., Fleming, J.: An inverse source problem for maxwell's equations in magnetoencephalography. SIAM J. Appl. Math. **62**(4), 1369–1382 (2002). https://doi.org/10.1137/S0036139900373927

2. Benner, P., Mach, T.: The preconditioned inverse iteration for hierarchical matrices. Numer. Linear Algebra Appl. **20**(1), 150–166 (2013). https://doi.org/10.1002/nla.1830

3. Bikowski, J., Knudsen, K., Mueller, J.: Direct numerical reconstruction of conductivities in three dimensions using scattering transforms. Inverse Prob. **27**(1), 015002 (2011). https://doi.org/10.1088/0266-5611/27/1/015002

4. Calderon, A.P.: On an inverse boundary value problem. Comput. Appl. Math. pp. 2–3 (1980). https://www.scielo.br/pdf/cam/v25n2-3/a02v2523.pdf

5. Chapman, J., Batson, P., Waddell, E., Ferrier, R.: The direct determination of magnetic domain wall profiles by differential phase contrast electron microscopy. Ultramicroscopy **3**, 203–214 (1978). https://doi.org/10.1016/S0304-3991(78)80027-8

6. Christen, J., Fox, C.: Markov chain monte carlo using an approximation. J. Comput. Graph. Stat. **14**(4), 795–810 (2005). https://doi.org/10.1198/106186005X76983

7. Close, R., Chen, Z., Shibata, N., Findlay, S.: Towards quantitative, atomic-resolution reconstruction of the electrostatic potential via differential phase contrast using electrons. Ultramicroscopy **159**, 124–137 (2015). https://doi.org/10.1016/j.ultramic.2015.09.002

8. Cultrera, A., Callegaro, L.: Electrical resistance tomography of conductive thin films. IEEE Trans. Instrum. Measur. **65**(9), 2101–2107 (2016). https://doi.org/10.1109/TIM.2016.2570127

9. Dickin, F., Wang, M.: Electrical resistance tomography for process applications. Measur. Sci. Technol. **7**(3), 247–260 (1996). https://doi.org/10.1088/2F0957-0233/2F7/2F3/2F005

10. Engl, H.W., Hanke, M., Neubauer, A.: Regularization of Inverse Problem. Kluwer Academic Publishers, Boston (1996)

11. Gehre, M., Jin, B., Lu, X.: An analysis of finite element approximation in electrical impedance tomography. Inverse Prob. **30**(4), 045013 (2014). https://doi.org/10.1088/0266-5611/30/4/045013

12. Hähnlein, C., Schilcher, K., Sebu, C., Spiesberger, H.: Conductivity imaging with interior potential measurements. Inverse Prob. Sci. Eng. **19**(5), 729–750 (2011). https://doi.org/10.1080/17415977.2011.598522

13. Jifu, H., Ke, W.: A unified tlm model for wave propagation of electrical and optical structures considering permittivity and permeability tensors. IEEE Trans. Microw. Theory Tech. **43**(10), 2472–2477 (1995). https://doi.org/10.1109/22.466182

14. Lambot, S., Slob, E., van, d., Stockbroeckx, B., Vanclooster, M.: Modeling of ground-penetrating radar for accurate characterization of subsurface electric properties. IEEE Trans. Geosci. Remote Sens. **42**(11), 2555–2568 (2004). https://doi.org/10.1109/TGRS.2004.834800

15. Langer, R.: An inverse problem in differential equations. Bull. Amer. Math. Soc. **39**(10), 814–820 (1933), https://projecteuclid.org/euclid.bams/1183496974

16. Li, X., et al.: A novel deep neural network method for electrical impedance tomography. Trans. Inst. Meas. Control **41**(14), 4035–4049 (2019). https://doi.org/10.1177/0142331219845037

17. Liu, S., Jia, J., Zhang, Y., Yang, Y.: Image reconstruction in electrical impedance tomography based on structure-aware sparse bayesian learning. IEEE Trans. Med. Imaging **37**(9), 2090–2102 (2018). https://doi.org/10.1109/TMI.2018.2816739

18. Murai, T., Kagawa, Y.: Electrical impedance computed tomography based on a finite element model. IEEE Trans. Biomed. Eng. Bme **32**(3), 177–184 (1985). https://doi.org/10.1109/TBME.1985.325526

19. Nachman, A.: Global uniqueness for a two-dimensional inverse boundary value problem. Ann. Math. **143**(1), 71 (1996). https://doi.org/10.2307/2118653
20. Nielsen, M., Chuang, I.: Quantum Computation and Quantum Information. Cambridge University Press, Cambridge (2000)
21. Norton, S., Bowler, J.: Theory of eddy current inversion. J. Appl. Phys. **73**(2), 501–512 (1993). https://doi.org/10.1063/1.353359
22. Pascual-Marqui, R., et al.: Assessing interactions in the brain with exact low-resolution electromagnetic tomography. Philos. Trans. Royal Soc. A: Math. Phys. Eng. Sci. **369**(1952), 3768–3784 (2011). https://doi.org/10.1098/rsta.2011.0081
23. Rehman, O., Rehman, S., Tu, S., Khan, S., Waqas, M., Yang, S.: A quantum particle swarm optimization method with fitness selection methodology for electromagnetic inverse problems. IEEE Access **6**, 63155–63163 (2018). https://doi.org/10.1109/ACCESS.2018.2873670
24. Ren, H., Zhao, Y., Chen, S., Yang, L.: A comparative study of lumped equivalent circuit models of a lithium battery for state of charge prediction. Int. J. Energy Res. **43**(13), 7306–7315 (2019). https://doi.org/10.1002/er.4759, https://onlinelibrary.wiley.com/doi/abs/10.1002/er.4759
25. Selleri, S.: Genetic algorithms for the automatic synthesis of equivalent lumped-element circuits. Microw. Opt. Technol. Lett. **29**(5), 356–359 (2001), https://onlinelibrary.wiley.com/doi/abs/10.1002/mop.1177
26. Soleimani, M., Lionheart, W.: Absolute conductivity reconstruction in magnetic induction tomography using a nonlinear method. IEEE Trans. Med. Imaging **25**(12), 1521–1530 (2006). https://doi.org/10.1109/TMI.2006.884196, http://ieeexplore.ieee.org/document/4016173/
27. Somersalo, E., Cheney, M., Isaacson, D., Isaacson, E.: Layer stripping: a direct numerical method for impedance imaging. Inverse Prob. 7(6), 899–926 (1991), https://doi.org/10.1088/2F0266-5611/2F7/2F6/2F011
28. Sylvester, J., Uhlmann, G.: A global uniqueness theorem for an inverse boundary value problem. Ann. Math. **125**(1), 153–169 (1987). https://doi.org/10.2307/1971291, https://www.jstor.org/stable/1971291?origin=crossref
29. Vauhkonen, M., Kaipio, J., Somersalo, E., Karjalainen, P.: Electrical impedance tomography with basis constraints. Inverse Prob. **13**(2), 523–530 (1997). https://doi.org/10.1088/0266-5611/13/2/020, https://iopscience.iop.org/article/10.1088/0266-5611/13/2/020
30. Wang, K.: An encrypt and decrypt algorithm implementation on FPGAs. In: 2009 Fifth International Conference on Semantics, Knowledge and Grid, pp. 298–301 (2009). https://doi.org/10.1109/SKG.2009.74, http://ieeexplore.ieee.org/document/5370112/
31. Xu, Y., He, B.: Magnetoacoustic tomography with magnetic induction (mat-mi). Phys. Med. Biol. **50**(21), 5175–5187 (2005). https://doi.org/10.1088/0031-9155/50/21/015, https://iopscience.iop.org/article/10.1088/0031-9155/50/21/015
32. Zhu, L., Wu, K.: Numerical de-embedding procedure and unified circuit model for planar integrated circuits. In: Hwu, R.J., Wu, K. (eds.) Terahertz and Gigahertz Photonics. vol. 3795, pp. 400–411. International Society for Optics and Photonics, SPIE (1999). https://doi.org/10.1117/12.370186, https://doi.org/10.1117/12.370186

# Implementing Natural Language Processes to Natural Language Programming

Yi Zhang[1], Xu Zhu[1(✉)], and Weiping Li[1,2]

[1] Nanjing Audit University, Nanjing 211815, Jiangsu, China
1033805063@qq.com, w.li@okstate.edu
[2] Oklahoma State University, Stillwater, OK 74078, USA

**Abstract.** We exhibit a detailed implementation on the natural language processing for natural language programming in python. We break the natural language text into active verbs and plural nouns to realize the sentence breaker and loop finder. Various examples of the implementation are presented. The realization of the natural language text into a computer programming does benefit in understanding the structure of the natural language processing and also the construction of the natural language programming.

**Keywords:** Natural language processing · Natural language programming · Noun phrase · Verb phrase · Direct object · Loop variable

## 1 Introduction

Challenges in natural language processing mostly involve speech recognition, natural language understanding, natural language generation, and human-computer interaction. Pargman et al. (2019) [9] build a dialogue on the Future of Computing and Wisdom to encompass workshop participants' preparatory work with writing "fictional abstracts" (abstracts of yet-to-be-written research papers that will be published in 2068) and to include the voices of the future researchers of 2068 who wrote the abstracts in question as well as the voices of the organisms, individuals, intelligent agents and communities who are the subjects, victims, beneficiaries and bystanders of (un)wise future computing systems. Mihalcea et al. (2006) [8] try to understand the interaction through languages between human and computer, and demonstrate that natural language can be mapped onto program structures through steps and loops. In this paper, we realize the aspect on turning the natural language text (process) into a natural language programming by converting an English text into a computer program automatically with various natural language identifications.

The idea of English or any natural language as a programming language was initially proposed by Samment (1966) [11] to tie a program language with the

© Springer Nature Singapore Pte Ltd. 2021
W. Gao et al. (Eds.): FICC 2020, CCIS 1385, pp. 204–215, 2021.
https://doi.org/10.1007/978-981-16-1160-5_17

natural language. At that time, the goal of natural language programming was to produce complete computer programs that could be interpreted or compiled. Dijkstra (1978) [3] first attacked the study of natural language programming and derived much consolation: quoted "I suspect that machines to be programmed in our native tongues – be it Dutch, English, American, French, German, or Swahili – are as damned difficult to make as they would be to use". Led to the suspension of natural language programming research. Researchers start to take a look at natural language programming differently and come up with some ideas that might work. Price et al. (2000) [10] created NaturalJava, a prototype for an intelligent, natural language based user interface that allows programmers to create, modify, and examine Java programs. Liu and Wu (2019) [7] describe three case–studies that demonstrate the functionalities of this program synthesis framework and show how natural language alleviates challenges for novice programmers to conduct software development, scripting, and verification. Mihalcea et al.(2006) [8] transformed the natural language text into a program framework through the design of step finder, loop finder and annotation identification. We follow their ideas to realize the natural language process through the *python* programming language to implement sentence breaker, text preprocessing, sentence decomposition, realization of loop finde and results displaying by C programming language syntax rules.

**Table 1.** The result of part-of-speech tagging.

[[ ('Write' , 'VB') , ('a' , 'DT') , ('program' , 'NN') , ('to' , 'TO') , ('read' , 'VB') , ('the' , 'DT') , ('text' , 'NN'),('10' , 'CD') , ('lines' , 'NNS') , ('of' , 'IN') , ('text' , 'NN') , ('and' , 'CC') , ('then' , 'RB') , ('write' , 'VBZ') , ('the' , 'DT') , ('number' , 'NN') , ('of' , 'IN') , ('words' , 'NNS') , ('contained' , 'VBN') , ('in' ,'IN') ,('those' , 'DT') , ('lines' , 'NNS') , ('.' , '.') ]]

The purpose of the step finder is to identify the step performed by the program and the function name corresponding to that step. Mihalcea et al. (2006) [8] proposed to implement part-of-speech tagging with Brill mark. Then, using active verbs as a boundary for this step, they use a shallow parser to find the noun phrase that plays a role of a direct object and identify the head of this noun phrase as the object of the corresponding action. Finally, Mihalcea et al. (2006) [8] extract the active verb and its direct object to form the function name of this step. From the implementing perspective, we take the verb under the active voice as the beginning of the programming step, where the NLTK tool is used for tagging part-of-speech. In generating function name to help people understand the function in programming, it is difficult to extract the direct object of the verb. It is key that the direct object is correctly determined or the direct object does not matter with normal operation in the programming. If the function name is designed to conform to the rules of the programming language, then the program can run normally. The structure of English sentences is VSO and VOS (subject S, verb V, object O). The object of the action is after the verb. All nouns after the verb are the objects of this action. Hence, we extract the verb and the first noun after the verb to generate the function name. We realize sentence breaker first through the step finder and text preprocessing, and carry the loop finder in natural language programming.

**Table 2.** Part-of-speech annotation for NLTK.

| Tab | Part-of-speech | Tab | Part-of-speech |
|---|---|---|---|
| CC | Coordinating conjunction | PRP\$ | Possessive pronoun |
| CD | Cardinal number | RB | Adverb |
| DT | Determiner | RBR | Adverb comparative |
| EX | Existential there | RBS | Adverb superlative |
| FW | Foreign word | RP | Particle |
| IN | Preposition or subordinating Conjunction | SYM | Symbol |
| JJ | Adjective | TO | to |
| JJR | Adjective comparative | UH | Interjection |
| JJS | Adjective superlative | VB | Verb base form |
| LS | List item marker | VBD | Verb past tense |
| MD | Modal | VBG | Verb gerund or present |
| NN | Noun singular or mass | VBN | Verb past participle |
| NNS | Noun plural | VBP | Verb non-3rd person singular |
| NNP | Proper noun singular | VBZ | Verb 3rd person singular present |
| NNPS | Proper noun plural | WDT | Wh-determiner |
| PDT | Predeterminer | WP | Wh-pronoun |
| POS | Possessive ending | WP\$ | Possessive wh-pronoun |
| PRP | Personal pronoun | WRB | Wh-adverb |

In terms of loop finder, its function is to find out what needs to be repeated after the natural language text is converted into a program. "Needs to be repeated" is called a loop which Mihalcea et al. (2006) [8] proposed to use plural nouns the basis for the existence of loops. The plural nouns must be a head of the noun phrase with a number as an indicator of the number of repetitions. Mihalcea et al. (2006) [8] challenges the assumption that the necessity of a formal programming language for communicating with a computer is always for granted, and tackle what are perceived hardest par on steps and loops, and develop some techniques mapping linguistic constructs onto programmatic semantics. We break the natural language text into the active verbs and plural nouns to express repetition and modification, and exclude plural nouns that have the function of modification. The rest of plural nouns is used to determine if there is a loop. When a structure like "NN+ NN" and "the+NN+of+NN" appears, there is a modifier noun (NN is for noun). For example, apple cider, the number of integers. For our implementation, modified nouns are not able to determine. This reduces the accuracy of loop identification, but ensures the correctness of the final program.

## 2    Brief Natural Language Process

Language (spoken or written, human or computer) is essential to communicate, and natural language processing (NLP) is crucial in communication among human beings and machines for both past, current and future. NLP is the analysis of mathematical and computational modeling of various aspects of language and the developments. Chomsky (1959) [2] initiated the formal grammars in the mathematical and computational modeling of grammars, and introduced a hierarchy of grammars (finite state grammars, context–free grammars, context–sensitive grammars, and unrestricted rewriting systems) and studied their linguistic adequacy.

**Table 3.** Programming implementation of sentence segmentation.

Since natural language text may not be a sentence, the outermost loop is added to enhance the program's robustness.

```
funlist=[]
    for tag in tags:
        i =0
        while i < len(tag):
            if tag[i][1] == "VB" or tag[i][1] == "VBZ"
            or tag[i][1] == "VBG" or tag[i][1] == "VBP" :
                a=[ ]
                a.append(tag[i])
                i=i+1
                while i < len(tag) and tag[i][1] !="VB" and tag[i][1] !="VBZ"
                    and tag[i][1] != "VBG" and tag[i][1] != "VBP" :
                    a.append(tag[i])
                    i=i+1
                funlist.append(a)
            else:
                i=i+1
```

Many NLP systems are based on context–free grammars (CFG). A CFG consists of a finite set of nonterminals (S: sentence, NP noun phrase, VP verb phrase, V verb, ADV adverb), a finite set of terminals and a finite set of rewrite rules of the form $A \rightarrow W$ with A nonterminal and W a string of zero or more nonterminals and terminals. In CFG, the dependency between a verb and its two arguments (subject (NP) and object (NP)) is specified by means of two rules of the grammar. It is not possible to identify this dependency in a single rule without giving up the VP (verb phrase) in the structure. We use a rule that $S \rightarrow NPVNP$ (subject verb object) to specify the dependency. Joshi (1991) [5] discussed grammars and parsing (an active theoretical area in NLP), statistical approaches to NLP (entails the use of very large quantities of data in the development of the theories) and multilingual processing (a rich domain for testing current and new formalisms in all aspects of NLP). Hardeniya et al. (2016) [4] present how to break natural language text down into its components for spelling correction, feature extraction, and phrase transformation as well as NLP concepts with simple and easy-to-follow programming into some research topics of NLP.

**Table 4.** The result of part-of-speech tagging.

[('read' , 'VB'), ('the' , 'DT'), ('text' , 'NN'), ('10' , 'CD'), ('lines' , 'NNS'), ('of' , 'IN'), ('text' , 'NN'), ('and' , 'CC'), ('then' , 'RB')]

[('writes','VBZ'), ('the','DT'), ('number','NN'), ('of','IN'), ('words','NNS'), ('contained','VBN'), ('in','IN'), ('those','DT'), ('lines','NNS'), ('.','.')]

Liddy (2001) [6] summarized natural language processing approaches into four categories: symbolic, statistical, connectionist, and hybrid. Symbolic approaches perform deep analysis of linguistic phenomena and are based on explicit representation of facts about language through well-understood knowledge representation schemes and associated algorithms. We would like to mention a frequently used statistical model – the Hidden Markov Model (HMM) inherited from the speech community. HMM is a finite state automaton that has a set of states with transition probabilities attached between states. Connectionist models combine statistical learning with various theories of representation to allow transformation, inference and manipulation of logic formulae. Each approach has advantage or disadvantage depending on the task. Researchers develope hybrid techniques that utilize the strengths of each approach in an attempt to address NLP problems more effectively and in a more flexible manner.

## 3   Implementing Natural Language Programming

The implementation of Natural Language Programming is to carry a sequences of computer action statements to operate various data structures. Mihalcea et al. (2006) [8] propose an outline to execute steps and loops to verify those natural language programmings, and divide the procedure of natural language programming into three steps: step finder, loop finder and comment recognition components. We realize sentence breaker first through the step finder and text preprocessing, and carry the loop finder in natural language programming in this section.

**Table 5.** The code that determines the scope of the loop

```
z=[]
i =0
while i < len(a1):
    z1=[]
    z1.append(i)
    for k in a1[i]:
        j=i-1
        while j >=0:
            if k in a1[j]:
                z1.append(j)
            j=j-1
    z1.sort()
    z.append(z1)
    i=i+1
```

## 3.1  Sentence Breaker

The function of sentence breakers is to decompose natural language text into phrases and identify phrases that need to be translated into a programming language. Each phrase corresponds to a function in the programming language, "Sentence Breaker" needs to generate names of these functions (methods) in natural language programming. In programming, functions (methods) consists of names and actual tasks. The aim and purpose of the function task is realized by the actual function body. If the natural language text contains only one noun, then one can convert the noun into a programming statement as an instruction. Therefore, there is no need to convert a programming language to a function (method) for a phrase that only has one noun. To be more specific, we pre-process the text by using NLTK's tagger to tokenize and tag part-of-speech as described in Brill (1995); Hardeniya et al. (2016) [1,4]. We identify verbs (active voice) in each sentence and use these verbs to cut the sentence into phrases. Third, we need to find phrases that can be translated into a programming language. Meanwhile, we need to find verbs (active voice) in these phrases. We take the first object after the direct verb, and generate a function name by using both verbs and direct objects.

This is first to input the natural language text with nouns and active verb and then to break this text into a series of natural language statements that correspond to computer programming statements such that each statement can be turned into a program function with a function parameter associated with direct objects.

**Text Preprocessing.** First, we use the sent_tokenize tool in NLTK to clause the language text. Second, the word_tokenize tool in NLTK is used to separate words. Finally, we use the pos_tag tool in NLTK to mark the words in the sentence as part-of-speech. We illustrate an example in this article to realize the steps and natural language programming. Let the natural language text to be "write a program to read the text 10 lines of text, and then write the number of words contained in those lines." for our example. For this language text, we use NLTK to tagging the part-of-speech annotation (stored in the collection) are shown in Table 1. Table 1 breaks the sentence into 22 words and 1 punctuation mark with tags. The tag associated to each word is based on the NLTK given in Table 2. The interpretation of part-of-speech tagging is shown in Table 2 (data from https://wenku.baidu.com/view/c63bec3b366baf1ffc4ffe4733687e21af45ffab.html or see Hardeniya et al. (2016) [4] for NLTK)

**Table 6.** Turn a plural noun into a singular noun

```
lemmatizer = ns.WordNetLemmatizer()
n_lemma = lemmatizer.lemmatize(word, pos='n')
```

**Sentence Decomposition and Function Name (Method).** We work with the explicit implementation to realize the natural language programming. The active of verb is used to syncopate with the original natural language text. We only save phrases that have verbs (active voice). We look for the number of nouns in each phrase. If the number of nouns in a sentence is greater than 1, then we save them into the new list based on nouns. Finally, we extract the first word (verb) and the first noun in the phrase to map into a function name. Using this described method, we complete the decomposition of the sentence and associate the function name for each component.

The implementing code is essential to the realization of the natural language programming. Since the implementation of the first function is an important step of this section, we describe our programming ideas and code in detail. Our aim is to use verbs (active voice) to divide sentences into phrases. We define a list called *funlist* to be used to store these phrases. First, we find verbs in active voice (VB, VBZ, VBG, and VBP all represent verbs in the active voice), where the verb (active voice) is used as the beginning of a phrase. To do this, we need to write a loop in our code. We define a list called $\mathcal{A}$ to store the verb (active voice). Second, we need to look up where the phrase ends. If we come across the next verb (active voice) or the end of a sentence, then we consider that the phrase is over. In this process, we record each word in the list $\mathcal{A}$ (except for the last word). Finally, we add the list $\mathcal{A}$ to the list *funlist*. Now we look at how the program works.

Define an empty list as *funlist*. Go over the outer loop to screen through data in Table 1. When the screening traverses a word with tagging "VB", "VBZ", "VBG", or "VBP", we stop and perform the following:

1. Define a new empty list which is named $\mathcal{A}$, add this word to list $\mathcal{A}$ and increment the value of counters by plus one.
2. We continue to go over the data in Table 1 by using the counters (inner loop) and add each word to the list $\mathcal{A}$. As long as a word is reached with marks "VB", "VBZ", "VBG", or "VBP", we stop.
3. Add $\mathcal{A}$ to the list named *funlist*. The specific code is shown in Table 3.

Table 3 illustrates the programming implementation for a natural language text.

```
Identify the statements to be converted to the programming language (c):
[('read', 'VB'), ('the', 'DT'), ('text', 'NN'), ('10', 'CD'), ('lines', 'NNS')]
[('of', 'IN'), ('text', 'NN'), (',', ','), ('and', 'CC'), ('then', 'RB')]

[('write', 'VB'), ('the', 'DT'), ('number', 'NN'), ('of', 'IN'), ('words', 'NNS')]
[('contained', 'VBN'), ('in', 'IN'), ('those', 'DT'), ('lines', 'NNS'), ('.', '.')]

The function name : ['read_text()', 'write_number()']
for(i=0;i< 10 ;i++){
    \\read the text 10 lines of text , and then
    read_text() ;
}
    \\write the number of words contained in those lines .
    write_number() ;
```

**Fig. 1.** "Write a program to read the text 10 lines of text, and then write the number of words contained in those lines" for the Natural Language text to generated Natural Language Programming.

## 3.2   Realization of Loop Finder

The purpose of this component is to determine if there is a loop after a natural language text is converted into a natural language programming. If there is indeed a loop, then we need to determine the loop's scope in the natural language text. We further determine the number of repetitions (loop).

First, we use plural nouns (plural nouns are not modifiers) to determine if there is a loop. Specifically, we need to determine if there are plural nouns in the phrase. If there are plural nouns in the phrase, then we need to exclude cases where the noun might be a modifier. We set a rule that the nouns in two constructions "noun + noun" and "the + noun + of + noun" may be modifiers. By doing so, we run the risk that phrases that should be repeated (loop) are not repeated (loop). The reason we chose to do this is to ensure that the resulting programming language is correct. It should be noted that we need to define a list named $\mathcal{A}$ to store the sequence number of repeated steps.

Second, we need to settle the scope of the loop. If the steps have the same noun, then the steps are in the same loop. The implementation of this feature is a little bit complicated and we show the ideas of the natural language programming and some codes. To start, all the nouns are extracted from steps which are required to repeat. Next to this, we translate all nouns into the corresponding singular form. To convert a plural noun to a singular noun, we need to use the NLTK tool. How to use this tool is shown in Table 6. We make a list of nouns for each step and collect them to a new list $\mathcal{A}_1$. Thirdly, we define a list called Z to store the sequence number of each step that are required to repeat. We compare each noun in the list $\mathcal{A}_1$ with the words in the list $\mathcal{A}_1$. If the same word exists, then the sequence number of the step is added to the group. Our code in detail is given in Table 5. Finally, we extract the first number between the beginning of the phrase (step) and the plural noun as the number of repetitions.

### 3.3   Results Display

In Subsect. 3.1 and 3.2, we obtained the function name for each step, determined whether there was a loop in the step, and the scope of the loop. But displaying the information according to the syntax rules of the programming language (in this case, the C programming language syntax rules) is still not an easy task. The final result on displaying is shown in Fig. 1.

First, in a list $z1$, we determine the scope of the loop at each step. For example, 0 and 1 are in the same loop, $z1$ is $[[0, 1], [1, 0]]$. The same loop is shown twice and we must to solve this problem. We solved this problem by sorting and then de-duplication.

Second, from the computer programming perspective, comments do not play an executive role in a language programming. But they are important in the computer program to add detailed information on programming statements. Since functions and loops in the natural language programming are obtained from Table 4, we can directly convert those statements from Table 4 into comments of those functions.

Thirdly, we define a list named *list_number* to be used to store the sequence number of all steps.

```
Identify the statements to be converted to the programming language (c):
[('generate', 'VB'), ('10000', 'CD'), ('random', 'JJ'), ('numbers', 'NNS'), ('between', 'IN')]
[('0', 'CD'), ('and', 'CC'), ('99', 'CD'), ('inclusive', 'NN'), ('.', '.')]

[('count', 'VB'), ('how', 'WRB'), ('many', 'JJ'), ('of', 'IN'), ('times', 'NNS')]
[('each', 'DT'), ('number', 'NN')]

[('write', 'VB'), ('these', 'DT'), ('counts', 'NNS'), ('out', 'RP'), ('to', 'TO')]
[('the', 'DT'), ('screen', 'NN')]

The function name : ['generate_numbers()', 'count_times()', 'write_counts()']
for(i=0;i< 10000 ;i++){
    \\generate 10000 random numbers between 0 and 99 inclusive .
    generate_numbers() ;
    \\count how many of times each number
    count_times() ;
}
    \\write these counts out to the screen
    write_counts() ;
```

**Fig. 2.** "Write a program to generate 10000 random numbers between 0 and 99 inclusive. You should count how many of times each number is generated and write these counts out to the screen" for the Natural Language text to generated Natural Language Programming.

Finally, we first print a step in list z1 to the screen and remove its sequence number from the list *list_number*. The list z1 contains that all steps are printed out. Then prints the steps in the list *list_number*.

### 3.4 Testing

In Mihalcea et al. (2006)'s [8] paper, they used a case to illustrate the effect of their method. We use this example to verify and carry out our program. The example is "Write a program to generate 10000 random numbers between 0 and 99 inclusive. You should count how many of times each number is generated and write these counts out to the screen". In order to compare our results with Mihalcea et al. (2006)'s [8] results, we put Mihalcea et al. (2006)'s [8] results in Fig. 3.

In the first place, we validated the function of "Sentence Breaker". Sentence breaker not only automatically divides sentences into phrases but also automatically recognizes phrases that need to be translated into a programming language. In this example the "Sentence Breaker" should tell us that there are only three phrases needs to be translated into the programming language. They are "generate 10,000 random Numbers between 0 and 99", "count the number of times each number is generated" and "write these counts to the screen". The sentence breaker not only does this but also prints the function name corresponding to the phrase. We show the end result of our program execution in Fig. 2.

Next, we test the function of the "loop finder" by entering these three phrases into the loop finder. Loop finders can help us identify phrases that need to be

```
#===============================================================
# Write a program to generate 10000 random numbers between 0 and
# 99 inclusive. You should count how many of times each number
# is generated and write these counts out to the screen.
#===============================================================

for($i = 0; $i < 10000; $i++) {

    # to generate 10000 random numbers between 0 and 99 inclusive
    &generateNumber(number)

    # You should count how many of times each number is generated
    &count()
}

foreach $count (@counts) {

    # write these counts out to the screen
    &writeCount(count)

}
```

**Fig. 3.** Mihalcea et al. (2006)'s sample output produced by the natural language programming system

executed repeatedly. Reading these phrases, we can know that all phrases have plural nouns and these plural nouns have no function of modifying. So that all three phrases need to be looped (executed repeatedly). The first and second phrases have the same object, so they should be in the same loop. Since the object in the third phrase is different from the object in the other phrases, the third phrase should be in a separate looper. Finally, these three phrases should be repeated in two different loops. We only found the number of repetitions in

```
Identify the statements to be converted to the programming language (c):
[('reads', 'VBZ'), ('a', 'DT'), ('string', 'NN'), ('of', 'IN'), ('keyboard', 'NN')]
[('character', 'NN'), ('and', 'CC')]

[('writes', 'VBZ'), ('the', 'DT'), ('characters', 'NNS'), ('in', 'IN'), ('reverse', 'JJ')]
[('order', 'NN'), ('.', '.')]

The function name : ['reads_string()', 'writes_characters()']
        \\reads a string of keyboard character and
        reads_string() ;
        \\writes the characters in reverse order .
        writes_characters() ;
```

**Fig. 4.** "Write a program that reads a string of keyboard characters and writes the characters in reverse order." for the Natural Language text to generated Natural Language Programming.

the first phrase. So, only the first phrase exists in a loop that can be translated into a programming language's for loop. We show the end result of our program execution in Fig. 2. Our results were similar to those of Mihalcea et al. (2006) [8], but we wanted the program's output to run in C instead of pseudocode. In addition, we have also used other cases to test our program, with good results. Such as "write a program that reads a string of keyboard characters and writes the characters in reverse order." We use the Fig. 4 to show the result.

## 4  Conclusion

The study of natural language processing for natural language programming and vice versa is certainly an important step to connect the community of human and the computer machine from the language perspectives. The interactive or interpreted relation between natural language processing and natural language programming would be beneficial for both fields. In this paper, we present how to use the structure of natural language processing to implement a computer language with results displayed by automatically generated natural language program skeletons. The finer structure natural language text exhibits, the better constructive natural language programming codes. Advances in one NLP help the better performance or understanding in another NLP.

## References

1. Brill, E.: Transformation-based error driven learning and natural processing. Comput. Linguist. **21**(4), 543–566 (1995)
2. Chomsky, N.: On certain formal properties of grammars. Inf. Control **2**, 137–167 (1959)
3. Dijkstra, E.W.: On the foolishness of "natural language programming". In: Bauer, F.L., et al. (eds.) Program Construction. LNCS, vol. 69, pp. 51–53. Springer, Heidelberg (1979). https://doi.org/10.1007/BFb0014656
4. Hardeniya, N., Perkins, J., Chopra, D., Joshi, N., Mathur, I.: Natural Language Processing: Python and NLTK. Packt Publishing Ltd. (2016)
5. Joshi, A.: Natural language processing. Science **253**, 1242–1249 (1991)
6. Liddy, E.D.: Natural language processing. In: Encyclopedia of Library and Information Science, 2nd Ed. NY. Marcel Decker Inc (2001)
7. Liu, X., Wu, D.: From natural language to programming language. In: Innovative Methods, User-friendly Tools, Coding, and Design Approaches in People-oriented Programming, pp. 110–130 (2019)
8. Mihalcea, R., Liu, H., Lieberman, H.: NLP (natural language processing) for NLP (natural language programming). In: Gelbukh, A. (ed.) CICLing 2006. LNCS, vol. 3878, pp. 319–330. Springer, Heidelberg (2006). https://doi.org/10.1007/11671299_34
9. Pargmana, D., et al.: The future of computing and wisdom: insights from human-computer interaction. Futures **113**, 102434 (2019)
10. Price, D., Riloff, E., Zachary, J., Harvey, B.: NaturalJava: a natural language interface for programming in Java. In: Proceedings of the 2000 ACM Intelligent User Interfaces Conference, pp. 207–211 (2000)
11. Samment, J.E.: The use of English as a programming language. Commun. ACM **9**(3), 365230–365274 (1966)

# AI and Block Chain

# LSO: A Dynamic and Scalable Blockchain Structuring Framework

Wei-Tek Tsai[1,2,3,4,5]($\boxtimes$), Weijing Xiang[1], Rong Wang[1], and Enyan Deng[3]

[1] Digital Society & Blockchain Laboratory, Beihang University, Beijing, China
tsai7@yahoo.com
[2] Arizona State University, Tempe, AZ 85287, USA
[3] Beijing Tiande Technologies, Beijing, China
[4] Andrew International Sandbox Institute, Qingdao, China
[5] IOB Laboratory, National Big Data Comprehensive Experimental Area, Guizhou, China

**Abstract.** This paper proposes a dynamic and scalable blockchain system framework for structuring a blockchain systems (BC). Traditionally a BC maintain multiple nodes with a smart-contract engine (SC) running on nodes, possibly with one or more Oracles Machines (OMs). However, many sophisticated applications require a much flexible yet still secure and scalable system architecture. The new framework LSO (Ledgers, Smart contracts, Oracles) with an inter-unit collaboration protocol that can be used for system registration, identification, communication, privacy protocols, and scalability management. The LSO framework is a dynamic and scalable framework because BCs, SCs, and OMs can be added without affecting the overall structure and without performance degradation. This framework support this by running a collection of cooperative Collaboration Layers (CLs) that acts like a DNS (Domain Name System) in Internet but this time to interconnect various BCs, SCs, and OMs. This LSO framework is a part of ChainNet initiative where BCs are used as building blocks in information and communication systems.

**Keywords:** Blockchains · Smart contracts · Oracles · ChainNet · System architecture · LSO Framework

## 1 Introduction

### 1.1 A Subsection Sample

This paper proposes a new framework for structuring a blockchain (BC) system. Traditionally a BC maintains multiple nodes with a smart-contract engine (SC) running on some nodes, possibly with one or more oracles machines (OMs). This is used in cryptocurrency systems such as Bitcoins and Ethereum, as well as stablecoins and CBDC Central Bank Digital Currency) systems.

However, this architecture has been shown to be inadequate for many applications.

A BC stores and protects data once they are entered into the system, however, it does not guarantee that the data are true or real. The data are as true as the data supplied to

W. Gao et al. (Eds.): FICC 2020, CCIS 1385, pp. 219–238, 2021.
https://doi.org/10.1007/978-981-16-1160-5_18

the BC, thus many propose to use OMs to address this problem and an OM can receive data from external world, at the same time it can send data from the BC to the external world.

A BC does not guarantee that any computation done are correct, e.g., those computations done by SCs running on top of the BC. Two ways to improve this situation are 1) those SCs can use data from the BC only as the BC provides some assurance that data have not been altered once they entered the system; 2) any computation done should be done by multiple SCs, and the results need to go through the consensus process. These two mechanisms are far from adequate in assuring that the computation done by SCs are correct, but they provide minimum support.

Regulators such as central banks are concerned with lack of transaction visibility and potential risk of money laundry and other financial risks. For example, in a report published by Bank of Canada in 2017 suggested many BC systems including those specifically designed for financial applications do not support regulators to supervise transactions.

Institutions are concerned with the lack of scalability and performance needed for modern financial applications. While many designs have been proposed to address this issue, but few solutions are available. This problem is a challenging task because not only the system needs to perform well with scalability, but also the system needs to inform regulators in a timely manner to monitor these transactions.

Clients, both individuals and corporations, are concerned with the lack of privacy for their identification and transaction data.

System scalability needs to addressed from the overall system architecture point of view, rather than from individual component systems, whether the component system is a BC, SC, or OM. For example, many have advocated a BC system should connect to multiple SCs, and multiple OMs; while at the same time an SC should be able to connect to multiple BCs; and an OM should be able to connect to multiple SCs and BCs. These new requirements come from the fact that a data source (OM) may serve multiple BCs, and a modern financial system require sophisticated SCs as suggested by CFTC (Commodity Futures Trading Commission).

For these reasons, one needs a new system architecture framework with three major components, OM for interfacing external systems or entities, SC for computation, and BC for storing data. The new framework is called LSO model, as L represents a ledger system (provide by BCs), S represents SC computation, and O for OMs.

This paper makes the following contributions: 1) Propose a new structuring framework LSO to cover the whole spectrum of computation, data storage, and communication, and the framework is dynamic as it allows BCs, SCs, and OMs to join the ChainNet, and it can scale to millions of systems by a hierarchical structure; 2) Propose an Event-Driven Architecture (EDA) to support SC computation in an asynchronous manner; 3) Propose a privacy mechanism to work within OMs so that privacy can be preserved in the LSO framework; 4) Illustrate the framework with an application system, a regulation compliance system with a bigdata platform interconnecting BCs.

This paper is organized in as follows: Sect. 2 presents related work; Sect. 3 presents the LSO framework including overall architecture, protocols, event architecture; Sect. 4 presents the privacy mechanisms; Sect. 5 concludes this paper.

## 2   Related Work

### 2.1   ChainNet

ChainNet is a project started in early 2020 where BCs and related technology are used in information and communication systems. In other words, these systems will be blockchainized. For example, OS (operating systems), databases, network protocols will be modified to support BC operations directly. Currently, these computing and communication infrastructures have been developed for conventional applications. They can be used to support BC operations, but they are not efficient in supporting BC applications. For example, in high-speed trade clearing experiments done in 2017, the entire operations can be done using 36 servers running as a 4-node system, 90% of system computation and communication resources were used to support BC operations, i.e., encryption, decryption, and consensus protocols, leaving only 10% of resources for application processes. This means that the current infrastructures are poorly suited to support BC-related functions (Fig. 1).

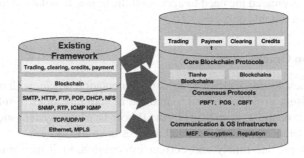

**Fig. 1.** ChainNet restructuring current technology stack

The ChainNet Initiative updates current network protocols, operating systems, databases, and application structures, so that BCs and related systems can be running efficiently. We have proposed several new designs in OS, databases, and network protocols, and application architecture in ChainNet [Tsai 2020].

### 2.2   ABC/TBC Architecture for Scalability and Privacy

This is a protocol designed for scalability and privacy protection for financial application. This approach allows BCs to scale in two dimensions, one way is horizontal scaling by adding resources into account BCs (ABCs), the other way is by adding new trading BCs (TBCs). This approach is used because scaling an account system and a trading system require two different scalability mechanisms. By providing two different scaling mechanism, a BC system can be scaled. This protocol has been used to process trade clearing in a commodity trading firm in 2017 with a BC system with bigdata capabilities [4] (Fig. 2).

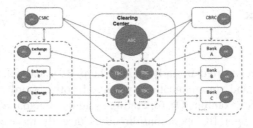

**Fig. 2.** ABC/TBC architecture for scalability and privacy

As each financial institution maintains her own BC systems, thus privacy of their clients can be ensured, while a regulatory agent can have nodes in each financial BCs, thus they can have access to all the data. If a financial institution needs to make an inter-institution trade, it will send a request to a TBC that will be responsible for trading, and the results will be sent back to their respective ABCs. An intra-intuition can be made locally without working with external TBCs. Many central banks including Bank of Canada have expressed the need to access all the transaction data in all the BCs, and this approach is a suitable one.

### 2.3 Blockchain Oracles

An OM can be a software, hardware, human, computation oracle, with or without consensus mechanisms, provide inbound, outbound or both directions of information, and related to specific contracts in an SC [1]. An OM can be a large system such as a banking system or an insurance platform, or small devices in an IoT system.

An OM can provide timely and correct information, or it may provide incorrect or even misleading or malicious information. Thus, the LSO system needs to have a dynamic evaluation mechanism to track the reliability of participating OMs.

Also, several design patterns are associated with an OM: Immediate-Read, Publish-Subscribe, and Request-Response. All these are all related to Observer design pattern, event-driven architecture (EDA). Related patterns include Saga, Domain events, CQRS, Event Sourcing, and Audit Logging.

### 2.4 Event-Driven Architecture

EDA has been practiced extensively before. For example, IBM's EDA is an integration model built around the publication, capture, processing, and storage (or persistence) of events. Specifically, when an application or service performs an action or undergoes a change that another application or service might want to know about, it publishes an event—a record of that action or change—that another application or service can consume and process to perform one or more actions in turn. In a simple example, a banking service might transmit a 'deposit' event, which another service at the bank would consume and respond to by writing a deposit to the customer's statement. But event-driven integrations can also trigger real-time responses based on complex analysis of huge volumes of data, such as when the 'event' of a customer clicking a product on an e-commerce site generates instant product recommendations based on other customers' purchases.

Similarly, many microservice systems also has Event Source model. This model persists the state of a business entity such an Order or a Customer as a sequence of state-changing events. Whenever the state of a business entity changes, a new event is appended to the list of events. Since saving an event is a single operation, it is inherently atomic. The application reconstructs an entity's current state by replaying the events. The event list in the event source is stored in a centralized database, and it can also be stored in a distributed database (Fig. 3).

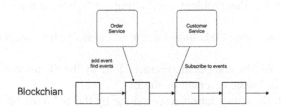

**Fig. 3.** ChainNet restructuring current technology stack

An important point in microservices is each service has its own database. Some business transactions, however, span multiple service so you need a mechanism to implement transactions that span services. From this, they proposed a solution called Saga, which Implements each business transaction that spans multiple services. A saga is a sequence of local transactions. Each local transaction updates the database and publishes a message or event to trigger the next local transaction in the saga. If a local transaction fails because it violates a business rule then the saga executes a series of compensating transactions that undo the changes that were made by the preceding local transactions (Fig. 4). There are two ways of coordination sagas that each local transaction publishes domain events that trigger local transactions in other services, and an orchestrator (object) tells the participants what local transactions to execute.

**Fig. 4.** Coordination sagas

These two models show the interaction between events and the database, the coordination between events. We will design a protocol to realize the information exchange between systems and design a model to complete the interaction between events and the BC storage system that can support large-scale applications.

## 3   LSO System Structuring Framework

Traditionally a BC has an associated SCs and connect to a collection of OMs. The BC can receive data from the SC and OMs; the SC can receive data from the BC and OM, and can send data to the BC and OMs; the OM can send data to the SC and the BC. However, this structuring rule is not rigorously enough as the SC can receive data from OMs directly for execution, but the data may be faulty or lost. First different data may send to different nodes within the SC and this will surely lead to different computation results, and stop the system from working correctly. Furthermore, data sent from OMs have not been validated before applications.

One way to address this problem is enforcing the following rules:

- Input rule: All data from OMs must be sent to the BC before sending to the SCs for execution.
- Consensus rule: All the computation results done by the SC must go through the BC consensus protocol to be accepted by the BC.
- Output rule: All the computation results done by the SC must be stored in the BC before sending the results to clients or OMs.

This is three SC principles and they address the problems mentioned earlier. This set of principles work fine when there is only one BC and only one SC. When a system has multiple BCs, SCs, and OMs, more sophisticated communication must be available.

### 3.1   LSO System Framework

The new BC framework also have BCs, SCs and OMs, but now BCs can work with multiple SCs, and at the same time, one SC system can work with multiple BCs as illustrated in Fig. 5 with three different configurations.

Ledger system (L): This is the conventional BC system where a set of transaction data with timestamps is grouped as a block, the hash of the block is used as a part of data in the next block. In this way, blocks of data are chained together to make sure data cannot be modified without being alerted. Furthermore, for a BC system, it has a collection of minimally four nodes, each contains the same information, these nodes use a Byzantine General consensus protocols or similar protocols to make sure that all the honest nodes take in the same blocks of data. This definition allows for both permissioned BCs and public BCs. But for modern digital financial applications, often only permissioned BCs can be used as evidence by Facebook's Libra 2.0 where the public BC route is abandoned (Fig. 6).

A major distinction is that a BC can work with multiple SCs, each SC is responsible for one set of related functions, e.g., one SC responsible for KYC, other set responsible for AML as illustrated in Fig. 7.

One advantage of this structure is that one set of SCs can work with multiple BCs, each take care of its own transactions. With this approach, the output from the SC computation need to record 1) the data source information (OMs, BCs, timestamps); 2) the associated SC information (set of rules used, timestamps); 3) the IDs of the BCs that will store the results.

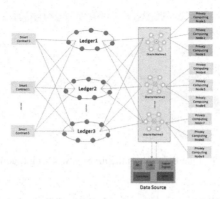

**Fig. 5.** LSO system architecture

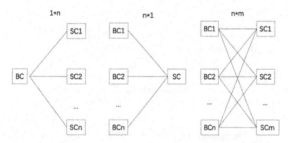

**Fig. 6.** 1-n, n-1, n-m relationship between BCs with SCs

Similarly, each SC can work with multiple BCs, each will record all the data produced to and from the SC. Like before, all the data from OMs are sent to the resident BC.

With the new framework, the working rules are updated:

- Input rule: Only OM, associated SCs, or authenticated clients can send data to a BC system.
- Consensus rule: All the computation results done by the SC must go through the consensus protocol of participating BC consensus protocol.
- Output rule: Only associated SCs and OMs, or authenticated clients can receive data from the BC system.

SC system (S): This is like the traditional SC system except each SC system can work with multiple BCs.

OM systems (O): This is the traditional oracle system, and it can send/receive data from/to multiple BCs. Each OM must register with a BC before it can send or receive data to or from the BC.

### 3.2  Collaboration Layer to Support Registration

The LSO framework allows multiple BCs, SCs, and OMs to collaborate with each other, even their relationship can be dynamically created via a Collaboration Layer (CL), and

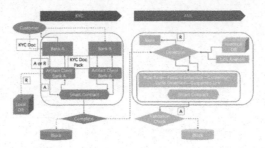

**Fig. 7.** One BC with multiple SCs working together

the CL sits among The CL sits among multiple BCs, SCs and OMs. The CL becomes a component of the framework infrastructure. Figure 8 illustrates the diagram.

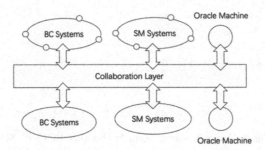

**Fig. 8.** Schematic diagram of collaboration layer

The design of the CL assumes that there are N (N = 0, 2, …) BCs, M (M = 0, 2, …) SCs, and W (W = 0, 1, …) OMs (M + N + W > 1), in particular, each BC may have some SCs and OMs. Traditionally a BC has minimally four nodes, but an OM can be any system such as a BC or an IoT system. For example, a BC network may connect to multiple SCs and multiple OMs, but the same SCs and OMs may be connected to another BC network. All BCs, SCs, and OMs that wish to collaborate need to register at the CL.

The positioning and affiliation of nodes and systems of the CL can be determined by the following three-dimensional coordinates:

The CL provides unique type identification for the three types of systems: BC, SC, and OM systems;
The CL provides unique system identification for different systems. The CL provides a unique identifier for each registered system; For example, two BCs cooperating, BC A has A-1, A-2, A-3, A-4 nodes, and BC B has B-1 B-2, B-3, B-4 nodes.

For example, there are two BCs, two SCs, and two OMs. Each BC and SC contains 4 nodes, and the OM contains 1 to 4 nodes. Each system and node first register with the CL, and after registration, the CL returns the corresponding three-dimensional identification

(Table 1). The three types of systems marked by the CL are BC, SC, and OM. There are 6 registered systems in the CL, which are BC1, BC2, SC1, SC2, OM1, and OM2.

**Table 1.** Three-dimensional identification of each node

| System type | Node identification |
|---|---|
| BC | <BC, BC1, BC1-1> <BC, BC1, BC1-2> <BC, BC1, BC1-3> <BC, BC1, BC1-4> |
| | <BC, BC2, BC2-1> <BC, BC2, BC2-2> <BC, BC2, BC2-3> <BC, BC2, BC2-4> |
| SC | <SC, SC1, SC1-1> <SC, SC1, SC1-2> <SC, SC1, SC1-3> <SC, SC1, SC1-4> |
| | <SC, SC2, SC2-1> <SC, SC2, SC2-2> <SC, SC2, SC2-3> <SC, SC2, SC2-4> |
| OM | <OM, OM1, OM1-1> <OM, OM2, OM2-1> |

The tasks recorded by the CL include account, transaction, accountValidation, transfer, accountInfo, and credit. Each system has corresponding functions and stored data (Table 2).

**Table 2.** Functions and tasks of each system

| System type | Function | Task |
|---|---|---|
| BC1 | Record account data | {BC1:[account], BC2:[transaction]} |
| BC2 | Record transaction data | |
| SC1 | Account verification | {BC1:[account], BC2:[transaction], SC1:[accountValidation], SC2:[transfer]} |
| SC2 | Transaction processing | |
| OM1 | Collect account personal information | {BC1:[account], BC2:[transaction], SC1:[accountValidation], SC2:[transfer], OM1=[accountInfo], OM2=[credit]} |
| OM2 | Collect user credit information | |

Transaction data are processed by the CL in the following three steps:

1. The data collected by the oracle machine needs to be transmitted to the smart contract for account verification. OM1 and OM2 respectively initiate a collaboration request for the accountValidation task to the CL. The CL returns the node information of the SC1 system and records the collaboration relationship: <OM1, SC1, accountValidation>, <OM2, SC1, accountValidation>.

2.  When SC1 performs account verification, it needs to query account information and initiate an account communication request to the CL. The CL returns the BC1 system and node information, and SC1 establishes communication with BC1. The CL records the collaboration relationship <SC1, BC1, account>. After the SC1 S completes the account verification, it passes the result to the transaction chain, initiates a transaction collaboration request to the CL, and the CL returns the BC2 information and records <SC1, BC2, transaction>.To process transaction data in BC2, transfer processing needs to be performed through a S, and the processed result is stored in BC2, so the collaboration <BC2, SC2, transfer>, <SC2, BC2, transaction> is created.
3.  The CL provides the registration service so that BCs, SCs, and OMs to work together. It provides standardized interfaces for collaborative management, including node and system registration, change or cancellation, system task registration or task assignment change, query corresponding system and node information, and broadcast changes.

The CL acts as an Interface, and it has its own BCs with SCs, and the BC keeps track of requests, and SC acts to verify and validate requests, and collaboration information to form an overall collaboration graph (Fig. 9).

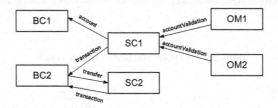

**Fig. 9.** Sample collaboration graph

### 3.3 Multi-level CL Network

If the BC network is large, say hundreds of thousands to million participating BCs, SCs and OMs, one CL will not be able to support all the operation. In this case, multiple CLs can be created, and they form a CL network. Figure 10 shows a 2-level CL network, where the BC network is divided into regions, each CL operates in one region, and they communicate with each other on a regular basis. This 2-level network can be extended to K level once the participating BCs, SCs, and OMs keep on increasing.

### 3.4 Dynamic Trust Evaluation

OMs are third-party services that provide smart contract with external information. They serve as bridges between BCs and the outside world. Oracles are a tool that can provide interoperability between different BCs and communicate with external data sources.

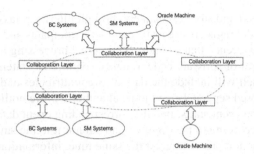

**Fig. 10.** Two-layer collaboration architecture

This section proposes a scalable OM supporting privacy computing, that not only can ensure that the data collection process is reasonable, but also can carry out privacy computing and maintain the scalability of system services. In addition, we also design a reputation evaluation strategy for each node to determine whether each node is a malicious node. After each feed, the reputation value of each node is updated. In addition, a OM uses a BC to preserve data to protect content providers. As a data feed service of SCs, it can provide offline data faithfully so as to ensure the authenticity of data on ledger and realize the interaction between SCs and the outside world.

### 3.5 Event-Driven Architecture (EDA)

CFTC has recommended that SC code should be standardized, and International Swaps and Derivatives Association (ISDA) has proposed many SC standards. Furthermore, CFTC recommends that each SC perform only limited functionality only, for example, one of many steps necessary to complete a financial transaction.

The standardization process splits a transaction into multiple steps, and each step is implemented by a fragmented and standard SC service. This will subvert the previous SC development process, from the customization of the entire SC development to the integration of standardized SCs, and realizes the preprocessing of transaction data. In the event registration phase, EDA has three-step process as shown in Fig. 11.

1) Enable SC participants to register for events;
2) Standardize the events in the natural language contract and preprocess the contract;
3) Pack and store the data including event attributes and contract attributes separately.

Note that the EDA proposed here are different from traditional EDA systems. Traditional EDA system employ various design patterns such as Observers, Subcribe, Request/Response, and Event Streaming. Even machine learning technology can be used in EDA systems. They are also flexible and scalable, yet they do not use BCs or SCs. The EDA system proposed in this paper employ BCs and SCs, and provide CL-based registration services. CLs also use BCs and SCs as a part of technology base.

For contracts submitted by participants, EDA can extract relevant attributes through machine learning technology. Analyze the text structure of standardized contracts, and then extract features of different attribute types. Finally, the contract attributes and event attributes are packaged and stored separately through filtering, classification, and clustering. The event model registers events by contract participants and standardize natural language contracts as events, that is, under the internal processing logic of the contract (expressed in the form of software code, roughly conditional statements), the contract is preprocessed, which will include the data of event attributes and contract attributes are packaged and stored separately, and judicial authorities are notified to intervene for notarization. After the event starts, find the corresponding event data package and contract data package according to the event ID and submit it to the smart contract system for automated transaction processing. At the same time, information related to account assets is submitted to the core ledger system for asset verification or asset certification in advance, and the corresponding account's the credit records is searched If it involves asset fraud or the participant is on the list of untrustworthy parties, the contract participant or regulatory agency can be notified to terminate the contract and initiate a termination event (Fig. 12).

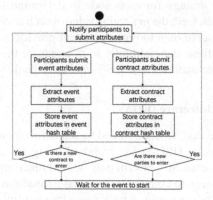

**Fig. 11.** Event registration process

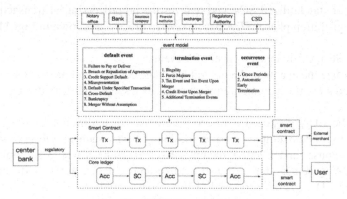

**Fig. 12.** EDA model

In the event of a default, the party whose fault or defect occurred between the parties is called the "defaulting party" and the other party is called the "non-defaulting party". The event of a default needs to be defined in the SC in advance. Once a default event occurs, the parties can choose to terminate the transaction. However, the termination event is different from the event of the default. The termination event is intended to occur when neither party is at fault. Therefore, the parties affected by the incident are called "affected parties." In the event that events prevent or hinder both parties from fulfilling their respective obligations, both parties may become affected parties. Otherwise, in the case of only one affected party, the other party is called the "unaffected party". The termination event may also affect only certain transactions. These are called "affected transactions." For example, in an embodiment, it may be illegal for a party to continue to pay based on a certain type of transaction. In this case, the parties can determine that only these types of transactions will be terminated, and other transactions will proceed normally.

The EDA model should consider the overall financial transaction infrastructure, interacting with many institutions, financial or non-financial institutions. On this infrastructure, for example, a new event occurs in a financial institution, and the event must be transmitted to other units through this infrastructure, including the BCs and SCs running at these institutions. Other institutions can be regulatory authorities, and the same event represents the same information in these BCs or SCs and cannot be changed. Participating institutions of this model can be banks, insurance companies, exchanges, financial institutions, notary offices, regulatory authority, CSD (Central Securities Depository System), Bureau of Industry and Commerce, and National Taxation Bureau. These financial institutions participate in the formulation of contract terms and involve different trading activities in reality. When a financial institution makes the first event occur through the event model, the contract execution is triggered. Due to the contract process, according to different execution conditions, other related events may also occur. Related events may also trigger other events, triggering the execution of multiple SCs until the end of the process. Therefore, the occurrence of an event will trigger the occurrence of multiple related events. At this time, the node needs to be notified again to start the related event.

As the initiation of an event in the EDA model may cause the domino effect triggered by the event, and this results in triggering multiple SCs to be activated. Therefore, such processing needs cannot be met in traditional BCs with SCs running on the BC platforms, without any regulatory authorities involved. To handle a large number of transaction requests efficiently and quickly, the relationship between the SC system (S) and the BC system (or the ledger system L) can be considered in the following three ways:

- S and L can be processed in parallel, that is, the SC platform and the BC platform process transactions at the same time, e.g., S does settlement and L does transaction;
- S is processed before L, that is, after the SC system is executed, it is handed over to the BC system for processing, e.g., S processes customer information first to ensure that the customer information is correct before trading in the L system;
- S is processed after L completes its tasks, e.g., transactions are carried out in the L system, and after the transaction, the S system is cleared outside of the L system.

After the event model receives the corresponding event start signal, it separates the event attributes, contract attributes and other data and packs them for storage. The storage method adopts the form of establishing a hash table, that is, establishing an event attribute hash table and a contract attribute hash table. The event ID and contract ID are directly mapped to the corresponding positions of the hash table. The contract participant initiates the event, and the event model maps the event data package according to the event ID and classifies the event. Event attributes include event name, event ID, and event participant information. The information of the event participants is divided into personal name, age, certificate type, certificate number, phone number, social account number, address, etc. (or company name, company location, legal representative, registered capital, business scope, business qualification, number of employees, company URL, contact information, etc. Type of incident, content of incident, date of incident, contract ID, arbitration tribunal). Contract attributes include contract ID, contract duration, contract asset types, asset quantities, asset certificates, contract content, contract location, contract date, etc. As soon as the event occurs, the corresponding data packet is retrieved from the database and submitted to the S system for processing (Fig. 13).

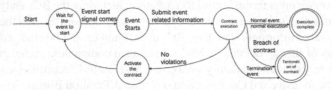

**Fig. 13.** Event-initiated automatic state machine

In the event model, next we can give an example of account A deposits in bank B. First, Bank B enters the LSO system through registration events. A submits a deposit request to bank B, and B initiates a corresponding deposit event.

**Contract:** Account A deposits a sum of money N in bank B for n days.
**Event:** A account exists
**Event:** Calculate interest rate
**Event:** Update account A balance
**Event:** Bank failure
**Event:** The contract has not expired
**Event:** Deposit transactions are insured
**Event:** Insurance company pays

The processing logic is expressed in the form of software code (pseudo-code), including data standards or logic standards, and the event and contract data transmitted by the event model in parentheses.

---

**Contract logic**

---

    **If** correct  (correct) {

        **If**  (A account exists) will be executed

        { {Confirm deposit term}

        **Then** (Calculate interest rate)

        **Then** ( Update account A balance)

        **If**  ( Bank failure) || ( The contract has not expired) will be executed

            {{Confirm whether the deposit is secured}

            **If** (Deposit transactions are insured) will be. executed (Insurance company pays)

            }

        }

    Otherwise { Notify account A to open an account at bank B

  }

  }

---

### 3.6   Oracle Machine Operation

OMs collect off-chain data, and the collection process can be managed by their own SCs. OMs, together with their associated SCs, PMs, and multiple data-providing modules complete this job. The OM computing model as shown in Fig. 14. To ensure the accuracy of data, the OM uses a dynamic credit value for each data provider to measure its credibility. The OM's credibility for the data provider is calculated by the accuracy of the data provided by its history, The OM weights the data provided by multiple data providers according to their trust degree and delays the calculation of reputation value. If and only when the total trust value of data results reaches the trust threshold required by the data request, the out of chain result data will be sent to the user's SC through the OM's SC.

    The BC node transmits the encrypted data on the chain to the off-chain PM node through the OM. The blockchain node uses the off-chain PM call component pre-deployed at the client to transfer the encrypted data to the off-chain privacy computing node. The data on the chain is transmitted to the PM node under the chain. By creating an off-chain TEE, off-chain privacy computing nodes can implement deployment operations on off-chain contracts and call execution operations after deployment and ensure data security and privacy protection during the operation.

    The OM provides accurate data for the OM's SC, and the OM performs trust management on the data provider. Off-chain PM nodes can prove the authenticity of the data they obtain on the chain based on verifiable computation technology. The OM uses a dynamic credit value for each data provider to measure its credibility. The OM's credibility for the data provider is calculated and generated by the accuracy of the data provided by the oracle engine. The data provided by the data provider is weighted and postponed according to its trust level, and only when the total trust value of the data result reaches the trust threshold required by the data request, the out-of-chain result data will be passed through the OM's SC sent to the user's smart contract.

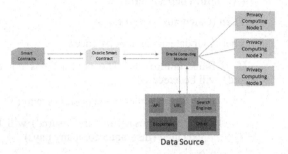

**Fig. 14.** Oracle computing architecture

The OM is responsible for reasonably assigning the received private computing tasks to the off-chain PM clusters to achieve load balancing of each node. The oracle engine encrypts the data and transmits it to the off-chain private computing node, and the off-chain trusted computing node deploys an off-chain trusted execution environment for performing private computing on the on-chain data. Its trusted execution environment is a trusted execution environment based on CPU hardware that is completely isolated from the outside. The PM node performs calculations in an off-chain trusted execution environment, and encrypts the calculation result and feeds it back to the oracle engine. The OM gathers the calculation results of each private computing node and sends it to the OM's SC.

# 4   Applications

## 4.1   BDL System in LSO

A BDL (Blockchain Data Lake) shown in Fig. 16 is a bigdata platform interconnecting with BCs for regulation compliance and monitoring. This section applies the LSO framework model to a BDL system. This system uses big data technology for regulatory analysis and supports the access of multiple BCs. The BCs realize data interconnection through a CL, and supports complex query, data mining and data analysis functions.

The BDL is proposed to address these problems: 1) transaction data are stored in BCs and they cannot be easily modified without being noticed; 2) regulation compliance need to collect data from these BCs so to find potential frauds and money laundering, and this requires centralized processing. However, current BC solutions tend to make transaction data in isolation, but data stored in a centralized database can be altered.

The solution proposed is to have a centralized bigdata platform that interconnect various BCs, but at the same time, data and their hashes collected and stored in the platform are written back into originating BCs, these originating BCs can verify that data stored are correct. If the platform altered the data, the hash of these data will be different, and the originating BC can identify this data alternation. This is the double interlocking mechanism, i.e., BC data including original data and their hash values are stored in the platform, and at the same time, the platform data and their hash values are stored back into originating BCs. In this way, the platform cannot modify the data without being noticed (Fig. 15).

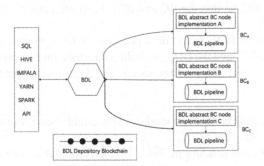

**Fig. 15.** Oracle computing architecture

This BDL system can be supported by the LSO framework. First, BDL can identify a BC registered at CLs, and its trust levels as well as other information, such as other BCs, SCs, and OMs that work with the BC.

The system includes: (1) Mirror BC data module, including data acquisition and transmission modules; (2) BC data pipeline that provides functions such as BC data sending and receiving, BC data conversion and processing; (3) BDL core, including BC database, data analysis component, data security and access component, and BDL BC.

The abstract BC data module collects data from different BCs and sends the data to the BC data pipeline. The BC data pipeline formats and encrypts the original data and sends to CL. After receiving the request, the CL queries the BDL system and node information corresponding to the task and establishes a communication connection. After receiving the data, the BDL core needs to verify the validity of the data source to verify whether it has obtained the access authorization and whether it has the legal IP address, so as to ensure the authenticity of the data, and store the data in the database and conduct relevant data analysis.

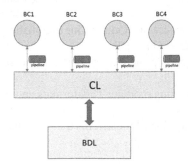

**Fig. 16.** BDL system

A component of BDL is the abstract BC data module. The node can be an actual node or an agent node that is only responsible for contacting the BDL. This node includes the following Module:

- New data collection module in BCs: get the data of the new block regularly;
- New data sending module in BCs: regularly send data to the BC data pipeline;
- Data on-chain module from BDL: When the BDL system sends information, it is registered on the BC, so that the BC and BDL keep each other's data.

Another component of BDL is the data pipeline of BCs, this is responsible for the connection between BCs and the BDL, including data transmission and reception of BCs and BDL, and data conversion and processing. The data pipeline includes the following modules:

- BC node and BDL link connection module: This establishes a communication channel between BCs and the BDL;
- BC data receiving and sending module: This is responsible for receiving data from BCs and BDL;
- BC data conversion processing module: These formats and encrypts data sent from the BC node;
- BC Data Security Transmission Module: This transmits formatted and encrypted data to the BDL;

The BDL core, including a BC database, data analysis component, data security and access component, and BDL BC.

- BC database: This supports the storage and fast retrieval of massive BC data;
- Data analysis component: This supports a variety of data analysis tools, including but not limited to SQL, Hive, Impala, Spark;
- Data security and access components: This is responsible for block link access authorization and BDL access control;
- BDL BC: This is used to chain key BDL data and provide BDL own data verification function;
- BDL data sending component: This is responsible for sending relevant BDL data back to the relevant BC and maintaining the BDL-BC to maintain the same counterparty data. As one BDL can interconnect to numerous BCs, related authority management is important.
- BC node and BDL link connection module: This establishes a communication channel between participating BCs and the BDL;
- BC data receiving and sending module: This is responsible for receiving data from BC nodes and the BDL;
- BC data conversion processing module: This formats and encrypts data sent from BC nodes;
- BC Data Security Transmission Module: This transmits formatted and encrypted data to BDL.

The resulting BDL system can be visualized as in Fig. 16 where a centralized bigdata platform is used by regulators to monitor transactions that is interconnected to many BCs. The platform can be scaled by adding more servers and communication links. As regulators can see all the transactions done in participating BCs, they can perform KYC, AML and other analysis easily as all the data are gathered and stored at the platform. The data stored in the BDL platform are also written back to participating BCs so that these BCs can verify that the data stored have not been altered by regulators. In this manner, regulators can monitor all the related transactions, and all participating BCs can be assured that data reported are stored at the BDL without any modifications.

## 5 Conclusion

This paper proposes a new BC system structuring framework LSO where a collection of BCs, SCs, and OMs work together via CLs. This framework is a dynamic framework because BCs, SCs, and OMs can be added to it without affecting the overall structure. This framework supports multiple BCs work with multiple SCs, as well as multiple OMs. In this way, modern financial application can be performed.

The framework is scalable as it can handle arbitrary number of BCs, SCs, or OMs. The LSO framework supports scalability via a collection of cooperative CLs that acts like a DNS system to interconnect various BCs, SCs, and OMs.

This paper illustrates this LSO framework using a BDL system where a bigdata platform work together with BCs, potentially hundreds to millions of BCs to support transaction supervision and regulation compliance.

The LSO framework performs all these without changing the functionality and features of traditional BCs such Merkel tree, Byzantine General Consensus protocols, encryption, and digital signatures.

**Acknowledgment.** This work is supported by National Key Laboratory of Software Environment at Beihang University, National 973 Program (Grant No. 2013CB329601) and National Natural Science Foundation of China (Grant No. 61672075) and (Grant No. 61690202). This work was also supported by Chinese Ministry of Science and Technology (Grant No. 2018YFB1402700). This is also supported by LaoShan government. This is also supported by Major Science and Technology Innovation Projects in Shandong Province (Grant No. 2018CXGC0703).

## References

1. Beniiche, A.: A Study of Blockchain Oracles, 14 July 2020
2. Youtradefx: Blockchain Oracles: The Key to Scalability and Interoperability, 9 August 2020. https://www.youtradefx.com/blockchain-oracles-the-key-to-scalability-and-interoper ability/#more-73
3. Tsai, W.T., Blower, R., Zhu, Y., Yu, L.: A system view of financial blockchains. In: Proceedings of IEEE Service-Oriented System Engineering (2016)
4. Tsai, W.T., Deng, E., Ding, X., Li, J.: Application of blockchains to trade clearing. In: Proceedings of IEEE Sympoisum on Software Quality, Reliability, and Security Companion, 1 July 2018

5. Tsai, W.T., et al.: Blockchain application development techniques. J. Softw. **28**(6), 1474–1487 (2017)
6. Tsai, W.T., Yu, L.: Lessons learned from developing permissioned blockchains. In: 2018 IEEE International Conference on Software Quality, Reliability and Security Companion (QRS-C). IEEE (2018)
7. Lian, Y., Tsai, W.T.: State synchronization in process-oriented chaincode. Front. Comput. Sci. **13**(6), 1166–1181 (2019)
8. Tsai, W.T., et al.: COMPSACC: a data-driven blockahin evaluation. In: Proceedings of IEEE SOSE (2020)
9. Tsai, W.T.: ChainNet: a new approach for structuring system architecture and applications. Keynote Presentation on August 6, 2020 with Video Recording
10. He, J., Wang, R., Tsai, W.T., Deng, E.: SDFS: a scalable data feed service for smart contracts. In 2019 IEEE 10th International Conference on Software Engineering and Service Science (ICSESS), pp. 581–585. IEEE, October 2019
11. Guarnizo, J., Szalachowski, P.: PDFS: practical data feed service for smart contracts. In: Sako, K., Schneider, S., Ryan, P.Y.A. (eds.) ESORICS 2019. LNCS, vol. 11735, pp. 767–789. Springer, Cham (2019). https://doi.org/10.1007/978-3-030-29959-0_37
12. Wang, S., et al.: Blockchain-enabled smart contracts: architecture, applications, and future trends. IEEE Trans. Syst. Man Cybern. Syst. **49**(11), 2266–2277 (2019)
13. Daniel, F., Guida, L.: A service-oriented perspective on blockchain smart contracts. IEEE Internet Comput. **23**(1), 46–53 (2019)
14. Ali, S., et al.: Secure data provenance in cloud-centric internet of things via blockchain smart contracts. In: 2018 IEEE SmartWorld, Ubiquitous Intelligence & Computing, Advanced & Trusted Computing, Scalable Computing & Communications, Cloud & Big Data Computing, Internet of People and Smart City Innovation (SmartWorld/SCALCOM/UIC/ATC/CBDCom/IOP/SCI). IEEE (2018)
15. Nelaturu, K., et al.: Verified development and deployment of multiple interacting smart contracts with VeriSolid. In: 2020 IEEE International Conference on Blockchain and Cryptocurrency (ICBC). IEEE (2020)
16. Tsai, W.-T., et al.: Requirement engineering in service-oriented system engineering. In: IEEE International Conference on e-Business Engineering (ICEBE 2007). IEEE (2007)
17. Bai, X., Wang, Y., Dai, G., Tsai, W.-T., Chen, Y.: A framework for contract-based collaborative verification and validation of web services. In: Schmidt, H.W., Crnkovic, I., Heineman, G.T., Stafford, J.A. (eds.) CBSE 2007. LNCS, vol. 4608, pp. 258–273. Springer, Heidelberg (2007). https://doi.org/10.1007/978-3-540-73551-9_18
18. Ramamoorthy, C.V., Tsai, W.-T.: Advances in software engineering. Computer **29**(10), 47–58 (1996)
19. Ramamoorthy, C.V., Tsai, W.T.: Adaptive hierarchical routing algorithm. In: Proceedings of IEEE Computer Society's International Computer Software & Applications Conference. IEEE (1983)

# CISV: A Cross-Blockchain Information Synchronization and Verification Mode

Yu Gu[1,2], Guozi Sun[1,2](✉), Jitao Wang[1,2], Kun Liu[1,2], Changsong Zhou[1,2], and Xuan You[1,2]

[1] School of Computer Science, Nanjing University of Posts and Telecommunications, Nanjing, China

[2] Key Laboratory of Urban Natural Resources Monitoring and Simulation Ministry of Natural Resources, Shenzhen, China

gy_blockchain@163.com, sun@njupt.edu.cn, joten_wang@qq.com, liukun_it@163.com, eric_zhou97@163.com, you3xuan@163.com

**Abstract.** Although the advanced technology stack of blockchain leads to the prosperity of its applications, the underlying mechanism of blockchain interoperability remains stagnant. Targeting at the efficiency and verification problems of existing cross-blockchain mechanisms, we design a Cross-Blockchain Information Synchronization and Verification (CISV) mode. In CISV, Cross-chain Information Synchronization (CIS) uses the design of touch block to provide high synchronization rate within low network delay, while Cross-chain Information Verification (CIV) applies ECC cryptography algorithm and Secure Multi-party Computation (SMC) to guarantee the accuracy and privacy of cross-chain information. Experiments executed on three mainstream blockchains (Ethereum, Hyperledger Fabric, and EOS) show that CISV has a reliable performance in blockchain interoperability with strong expansibility.

**Keywords:** Cross-blockchain · Blockchain interoperability · Synchronization · Secure multi-party computation

## 1 Introduction

The development of blockchain technology has formed a relatively complete technology stack, which has been widely concerned and studied mainly because of its unique characteristics: decentralization, anonymity, transparency and traceability [1]. On one hand, from the perspective of data, blockchain is a distributed database that is almost impossible to be changed. The concept of *distribution* is not only reflected in the distributed storage of data, but also in the distributed record of data [2], which is jointly maintained by system participants. On the other hand, from the perspective of technology, blockchain is not a single unit, but the result of the integration of multiple technologies. These technologies are combined with new structures to form a new way of data recording, storage and expression.

Although the characteristics of blockchain trigger the prosperity of engineering-oriented projects (DApps) [3], the blockchain's underlying mechanism is somehow

© Springer Nature Singapore Pte Ltd. 2021
W. Gao et al. (Eds.): FICC 2020, CCIS 1385, pp. 239–252, 2021.
https://doi.org/10.1007/978-981-16-1160-5_19

underestimated. At present, the research process of the underlying technology of blockchain has come to a bottleneck period [4], while most blockchain practitioners have great expectations for its popularization. Due to the variety of blockchain platforms, the fundamental design of each blockchain platform has its own unique features. However, if a multi-chain system wants to record several types of relevant information to different chains at the same time, different consensus mechanisms and varied methods of block generation will lead to weakness of synchronization [5], which will produce many inevitable problems in the multi blockchain system [6]. Therefore, it is a serious problem to keep relevant data synchronized between different blockchains [7].

The technical problems to be solved in this paper are to overcome several common limitations of the existing blockchain interoperability technology such as low synchronization rate, limited efficiency, lack of information verification, along with security issues like data privacy and cyberattack threats. Many researchers have put their strength unilaterally on limited aspects, causing the other factors to be more serious.

To tackle the shortcomings above, we design a Cross-blockchain Information Synchronization and Verification (CISV) mode, including the synchronization mode and the verification mode. The simplified process of the CISV mode is shown in the following steps:

**Step 1.** The Cross-chain Information Synchronization (CIS) module obtains the data information to be synchronized, then uses the consensus mechanism of the blockchain system to package the data into a new block, where the blockchain system marks the new block as a candidate block of **touch block**;

**Step 2.** After the candidate of touch block is generated, the signal sent by the Cross-chain Information Verification (CIV) module outside the blockchain system is continuously detected. When the end signal is detected, the nearest block in the current main chain is obtained;

**Step 3.** Point the nearest block in the current main chain to the candidate of touch block, then the candidate rises to the real touch block;

**Step 4.** Query synchronous signal in touch block and broadcast the information onto each blockchain system to finish the synchronization and verification.

Our contribution can be included as follow:

(1) We design and implement a Cross-chain Information Synchronization (CIS) mode, which provides high synchronization rate within an acceptable network delay.
(2) We effectively guarantee the privacy and accuracy of cross-chain information during the synchronization by applying the Cross-chain Information Verification (CIV) mode. Cryptography technology and secure multi-party computation method are used in to ensure the security of interaction.
(3) We provide strong expansibility atop this mode. By making experiments on several mainstream blockchains (Ethereum, Hyperledger Fabric, and EOS), we prove the applicability and efficiency of CISV mode, which is well grounded to be expanded among other blockchain systems.

## 2 Related Work

### 2.1 Blockchain Underlying Storage Mechanism

Since the underlying mechanism fundamentally influences the performance of blockchain storage and the effectiveness of cross-chain synchronization and verification, we firstly study the researches on storage mechanisms of several mainstream blockchains, including Ethereum, Hyperledger Fabric, and EOS.

A performance prediction method [8] is proposed targeting at offering a more accurate estimation of prospective storage performance of Ethereum. By executing contract atop transaction volume, the researchers calculate the effectiveness of its transmission speed, which is not optimistic. Similarly, mLSM [9] points out that although Ethereum guarantees authenticated storage, the communication efficiency tends to be unacceptable in that every single read requires a returned value with a proof, resulting in high resource consumption. Merkelized LSM (mLSM), therefore, is proposed to reduce the read and write amplification as well as providing the client verification. From the perspective of storage and data transmission, Ethereum is reliable on its accuracy and security, while cannot avoid the weakness of low transmission efficiency, which may lead to a comparatively negative impact on cross-blockchain information synchronization.

Compared with Ethereum, Hyperledger Fabric, a highly scalable authenticated platform, has a better performance on data transmission efficiency. Fabric executes data operations by key-value format, which applies several peers to manage the blockchain instead of a certain administrator. Generalized Stochastic Petri Nets (GSPN) [10] is adopted to evaluate the performance of a Fabric-based system to estimate the computational power of Hyperledger. Similarly, a hierarchical model approach [11] is applied to analyze the performance of Fabric, which is advanced in focusing on the underestimated transaction endorsement failure as well as ignored block timeout. We note that the Fabric-based systems tend to perform better than the Ethereum-based systems [12], which emphasizes that the blockchain's basic storage efficiency should be considered as a critical factor of cross-blockchain information synchronization and verification. Also, the transaction endorsement failure should also be calculated to determine the synchronization rate.

EOS blockchain overcomes the low throughput limitation of PoW-based systems by adopting the Delegated Proof of Stake (DPoS) consensus mechanism. A large-scale dataset of EOS [13] is gathered to characterize the activities upon this blockchain, pointing out that EOS has comparatively high throughput efficiency with several remained vulnerabilities. Similarly, Xblock-EOS [14] is a well-processed EOSIO dataset, which is analyzed from the perspective of internal and external transfer action. The analysis of data transmission in Xblock-EOS provides us with a wider horizon on cross-blockchain information synchronization, while existing blockchain interoperability methods encounter various problems, which are demonstrated in Section B.

### 2.2 Blockchain Interoperability

Originally, the basic way to obtain the interoperability among different blockchain systems is to trade data (json flow or tokens) on centralized storage mechanisms, which provides the real applications with functionalities. To deal with third-party trust crisis, decentralized exchanges like 0x [15] take place. Furthermore, a Republic protocol [16] including a decentralized dark pool data transmission mechanism, which increases the security of cross-blockchain information exchange. However, the mentioned methods or protocols tend to face the problems of low transmission efficiency and low synchronization rate.

Another unique mode to handle cross-blockchain synchronization is PolkaDot [17]. PolkaDot aims to provide a series of techniques to share information between different blockchain systems and apply an *inter-chain communication protocol* (ICMP). PolkaDot designs a set of separated blockchains named parachains and a single trusted blockchain to meet the requirement of blockchain interoperability. However, the generation of the self-designed blockchain increases the complicity of its application, while the trusted single chain still meets the problem of trust crisis on centralized data transmission.

Furthermore, Metronome [18] and Dextt [19] define the cross-blockchain tokens like Metronome tokens (MET) and Deterministic Cross-Blockchain token (DeXT), which provide the mechanism of cross-blockchain transmission. However, these methods relying on tokens still cannot resolve the limitation of cross-chain data size, which can be considered as another format of low efficiency.

Take the researches above into consideration, it is not difficult to note that there are several common limitations on the existing blockchain interoperability methods, including low synchronization rate, limited efficiency, lack of information verification, along with security issues like data privacy and smart contract vulnerability. The mentioned limitations of the existing cross-blockchain motivate us to take a deeper investigation on blockchain interoperability mode.

## 3   Cross-Blockchain Information Synchronization and Verification

To handle the limitations on blockchain interoperability, we design a Cross-Blockchain Information Synchronization and Verification (CISV) mode, which can be divided into two main parts: Cross-chain Information Synchronization (CIS) and Cross-chain Information Verification (CIV).

Section 3.1 mainly introduces the definitions of several critical terms. Section 3.2 contains the implementation of CIS mode, as well as the security and efficiency analysis. Section 3.3 illustrates the CIV mode including its implementation and the algorithm details.

### 3.1   Definitions

**Earliest Block:** The block that is generated the earliest and exists on the main chain among all the blocks in the blockchain (the earliest generated block may not be connected to the main chain).

**Nearest Block:** The block that is generated most recently and exists block on the main chain among all the blocks in the blockchain (the most recently generated block may not be connected to the main chain).

**Blocks to be Generated:** Blocks that will point to the nearest block in all blocks of the blockchain.

**Touch Block:** The block used to record cross-blockchain synchronous data information in all blocks of the blockchain, which is usually pointed to by a certain block in the main chain.

## 3.2 Cross-Chain Information Synchronization (CIS)

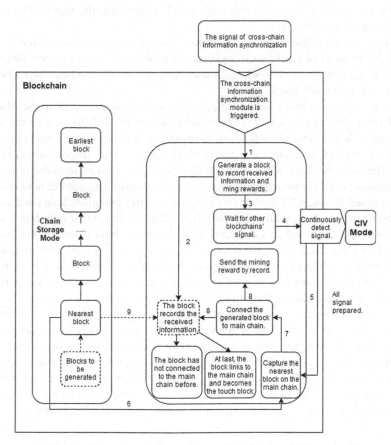

**Fig. 1.** Cross-chain information synchronization (CIS) mode

1) When the cross-chain information synchronization processing module is triggered, it will get the information to be synchronized, and then the obtained data will be packaged in a new block through the consensus mechanism of the blockchain system, which is consistent with the way the main chain generates blocks. However, the block generated by the module is not connected to the main chain as soon as it is generated. At this time, it exists in the blockchain system as an *isolated block*, and the *isolated block* is specially

marked to identify it as a candidate block of **touch block**. The generation of this block, like the main chain block, requires miners and mining rewards. These contents are not disclosed in the blockchain system at this time, but these information are temporarily stored, which is called the *delayed reward*. The whole process of generating touch block candidate does not affect the generation of blocks on the main chain, while the information of main chain has been recorded in the block generated by mining. The generated block of the main chain and the generated block by CIS module are isolated and do not affect each other (Fig. 1).

2) After the candidate block of touch block is generated, the signal sent by the Cross-chain Information Verification (CIV) module outside the blockchain system is continuously detected. Once the prepared signal is detected, the nearest block in the current main chain is obtained.

3) After obtaining the current nearest block, the module finds the previously generated candidate block of touch block, and makes the obtained current nearest block point to the hash value of the touch block candidate. At this time, the candidate block of touch block rises to real touch block.

4) After the touch block is born, the miners and mining rewards in the block are queried from the "touch block", and the delayed reward will be sent to the miners immediately. At this time, the whole information synchronization process is finished.

The touch block is connected to the main chain as a small branch, which not only ensures the security of the whole blockchain system, but also realizes the information synchronization of the cross-chain system within a certain and acceptable network delay.

It is proved that the CIS module has a higher security guarantee than the traditional blockchain system. Technically, the traditional blockchain system only has a smooth main chain, and the standard for determining the main chain is simply the length of branch. By adding touch block, the whole blockchain system is more sophisticated, and the number of touch blocks is also applied as the standard to determine the main chain. Then, for attackers, more factors need to be considered, which leads to the increase of attack difficulty, thus making the whole blockchain system more secure.

From the perspective of network delay in cross-chain information synchronization, the consensus mechanism of multiple blockchain systems may lead to varied amounts. Even if the same consensus mechanism is adopted, the mining speed of each blockchain system is also different. By mining and preparing blocks in advance, when each blockchain system has prepared such blocks, it is almost synchronized at the user level by making the current nearest block point to the candidate block of touch block, and the maximum time difference will not be greater than the network delay.

### 3.3   Cross-Chain Information Verification (CIV)

After the cross-chain information synchronization process starts, each blockchain system will continuously send a signal to judge whether the other blockchain systems are ready or not. Each blockchain system only needs to care whether every blockchain is ready for the final signal, while does not need to analyze the signal sent by other blockchain systems in the process. In order to achieve such effect, the CIV mode adopts a secure multi-party computation method for signal processing. Secure multiparty computation is that there is a group of participants who do not trust each other, but they hope to

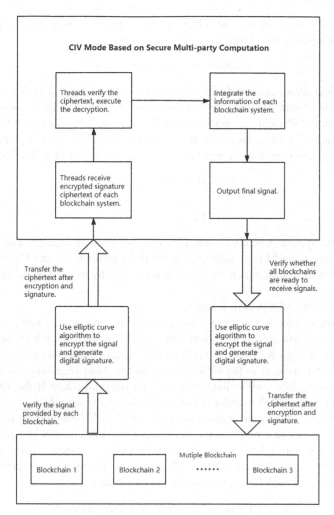

**Fig. 2.** Cross-chain information verification (CIV) mode

get the correct result when calculating a contract function while the output data of each participant is confidential. Therefore, secure multi-party computation is suitable for the verification module (Fig. 2).

Before sending the prepared signal, each blockchain system needs to encrypt and sign the signal content with cryptography algorithm. The purpose is to ensure the privacy of each blockchain system signal, and the verification module can verify whether the received signal is sent by the corresponding blockchain system through digital signature.

The cryptography algorithm is elliptic curve algorithm (ECC). Elliptic curve cryptography is a public key encryption algorithm based on elliptic curve mathematics. The signal encryption process is as follows:

Before the blockchain system A sends the signal content, the verification module selects an elliptic curve EP (a, b) in advance, and selects a point on the elliptic curve as the base point G; then, the system selects a private key, which is called a private key, and generates a public key through the private key, that is,

$$publicKey = privateKey\_G \tag{1}$$

The verification module sends the elliptic curve EP (a, b), public key and base point G to blockchain system A; after receiving the information, blockchain system A encodes the signal content to be sent to a point M on EP (a, b), and generates a random integer r. Blockchain system A calculates C1 and C2 through M and r. C1 and C2 are the points on two rectangular coordinate systems calculated according to M and r, that is,

$$C1 = M + r * publicKey \tag{2}$$

$$C2 = r * G \tag{3}$$

Blockchain system A sends C1 and C2 to the verification module. The verification module decrypts point M (the point can be obtained through C1 and C2) by receiving C1, C2 and private key, and then the signal content sent by blockchain system A can be obtained.

The process of ECDSA is as follows:

Blockchain system A generates a group of public and private key pairs, the base point G = g (x, y) and selects a random number r. Then, through the message M, hash value H and private key K, S = (H + Kx/R) is calculated. Finally, the message M and the signature {rG, S} are sent to the verification module. After receiving the information, the verification module verifies the signature with the public key. If the verification is successful, it proves that the sender is indeed blockchain system A.

In this module, the signal transmission of each blockchain system is continuous, thus we apply multiple threads in the verification module process the encrypted and signed signal content ciphertext sent by each blockchain system. After verifying the digital signature of the ciphertext, the verification module decrypts the ciphertext. And the signal content of each blockchain system is calculated, that is,

Signal 1 && Signal 2 && $\cdots$ && Signal N.

The result is encrypted by elliptic curve and digitally signature, then sent to each blockchain system. The purpose of encryption and digital signature here is to prevent the attacker from tampering with the information where the final signal received by each blockchain system will be different, resulting in abnormal synchronous transmission of data and information.

The CIV mode uses secure multi-party computation to ensure the privacy of each blockchain system signal under the multi blockchain system, and uses elliptic curve cryptography algorithm to encrypt and sign the signal content, so as to ensure the security of the signal content in the transmission process.

The combination of the cross-chain information synchronization and the external cross-chain verification mode of the blockchain can achieve the effect of multiple blockchains information synchronization. What is more, under the guarantee of cryptography technology, it can be applied to the cross-blockchain network system composed of different blockchain systems.

# 4 Experiments and Analysis

In order to determine the correctness and efficiency of the cross-blockchain information synchronization and verification mode, we carry on the experiments including four steps: on-chain data processing, blockchain storage performance test, cross-chain information synchronization, and cross-chain information verification. The experiments are executed on three mainstream blockchains: Ethereum, Hyperledger Fabric, and EOS. These blockchains apply different underlying consensus mechanism with varied throughput, providing a reference value to calculate the expansibility.

## 4.1 On-Chain Data Processing

Before the information is transferred onto the blockchain, raw data should be pro-processed to ensure the correctness of on-chain data format. ECC algorithm and Secure Multi-party Computation (SMC) parameters are deployed to ensure the security of data transmission. The process steps of the raw data can be concluded as follows:

1) Dynamically modify the data volume and calculate the origin JSON Size of the raw data.
2) Add the secure multi-party computation parameters onto the raw data, which is prepared for CIV mode.
3) Use ECC algorithm to encrypt the data, protecting the security of private data.
4) Convert ciphertext to Base64 format for more unified data transmission.
5) Add the secure multi-party computation parameters onto the raw data, which is prepared for CIV mode.
6) Encapsulate the data into the final request body and calculate the size.

From Fig. 3, it is obvious that the size of the original JSON streams change greatly after adding the SMC parameters. Specifically, if the SMC parameters are added to 10000 pieces of data, the volume will increase by about 288 kb. The reason is that after adding the SMC parameters, the original business parameter JSON is taken as a sub JSON field, which paraphrases the original JSON, thus a large number of data entries will be added.

At the same time, it can be noted that the volume of the encrypted data is almost unchanged. However, after converting the binary ciphertext into Base64 format, the volume increases by 30%. Furthermore, With the increase of data volume, the percent of increment tend to decrease. Finally, the increment stabilizes at about 50%.

## 4.2 Blockchain Storage Performance Test

After the data is processed, we transfer the processed information onto Ethereum, Fabric and EOS to estimate the fundamental performance and response time of different blockchain systems, which are critical factors as a reference of relative synchronization efficiency. We make the test on Ubuntu 16.04 with Geth 1.8.27, Fabric 1.4.6 and EOS 1.0.9. The test process can be concluded as follows:

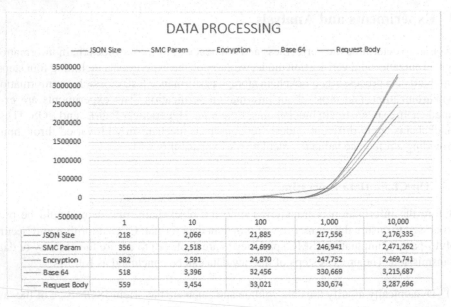

**Fig. 3.** Data processing

1) Deploy these three blockchains.
2) Determine the transaction number to be 100, which is suitable to record the response time.
3) Calculate the origin JSON Size of the raw data, make sure that every blockchain has the same data volume.
4) Calculate the Request Body size, make sure that every blockchain has the same data volume.
5) Execute the transactions, record the On-chain Time and Query time of each blockchain.
6) Calculate the failure rate of every blockchain data storage as a reference.

As illustrated in Fig. 4, 100 transactions are delivered onto three blockchains, while Ethereum is proven to have the lowest efficiency in data on-chain time and query time and the highest correctness. Fabric and EOS perform better in data transmission efficiency according to on-chain time and query time, which confirms the research results of their storage mechanism. These factors of blockchains themselves are an important reference on the synchronization response time and success rate.

## 4.3  Cross-Chain Information Synchronization

After testing the blockchain storage performance, we make the experiments on cross-chain information synchronization. The CIS mode and CIV mode works together to ensure the correctness of the whole operation process, the touch blocks are chosen according to CIS standard to synchronize different blockchain platforms. We execute

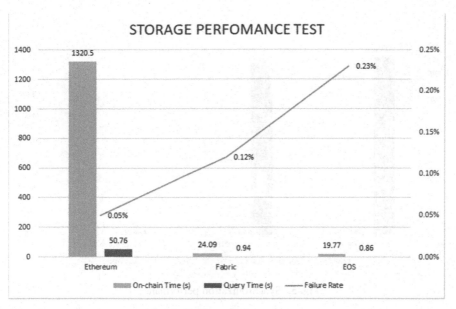

**Fig. 4.** Storage performance test

the synchronizations between Ethereum, Fabric and EOS, the synchronization process can be concluded as follows:

1) Deploy the CIS and CIV mode onto three experimental blockchains.
2) Determine the transaction number to be 100, which is suitable to record the response time.
3) Calculate the origin JSON Size of the raw data, make sure that every synchronization process has the same data volume.
4) Calculate the Request Body size, make sure that every synchronization process has the same data volume.
5) Execute the CIS and CIV modules to ensure the correctness of synchronization, calculate the synchronization time by recording the total on-chain time and query time.
6) Calculate the failure rate and synchronization rate of every synchronization process.

The results from Fig. 5 prove the hypothesis that the blockchains' own storage performance makes a difference on the synchronization efficiency. According to the results of Ethereum synchronization, it is obvious that the data on-chain time and query are mainly determined by the blockchain with low efficiency, as the experimental data here is quite close to that in Table II (on-chain time and query time). Take Eth & Fabric synchronization as an example, the on-chain time is only 2.15 s more than the one of Ethereum, which also proves that the synchronization delay is at an acceptable level. Furthermore, the synchronization rate are proven to be influence by blockchain's own transaction endorsement failure.

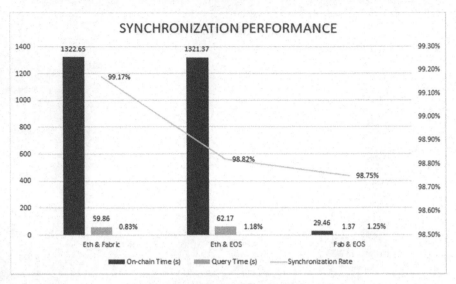

**Fig. 5.** Synchronization performance

## 4.4 Cross-Chain Information Verification

The CIV mode works with the CIS mode during synchronization to guarantee the security and correctness of information synchronization, while it is also important to evaluate the response performance of verification results. After the information is transferred between different blockchains, several steps are taken to get the targeted information. We estimate the response time of ECC Decryption, JSON Serialization Data Verification and Deserialization to evaluate the verification performance, steps are listed as follows:

1) Extract the data from three blockchains during the period of synchronization.
2) Apply the ECC algorithm for decryption, and record the response time of different data volumes.
3) Execute the JSON Serialization, and record the response time of different data volumes.
4) Check the data verification operation flow, and record the response time of different data volumes.
5) Execute the deserialization module, and record the response time of different data volumes (Fig. 6).

As shown in Table IV , the response time of are limited at the level of microsecond and millisecond, which proves the reliability of verification efficiency. The verification time are mainly influenced by the data volume, which includes the data itself and the signal for verification. Experiment results show that the performance of CIS and CIV modes are comparatively optimistic.

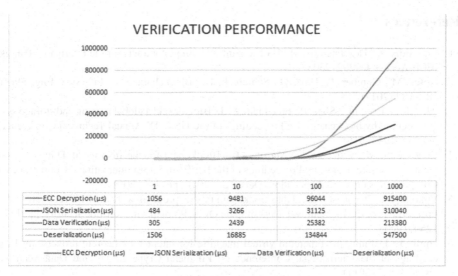

**Fig. 6.** Verification performance

# 5 Conclusion

In conclusion, we have presented CISV, a Cross-Blockchain Information Synchronization and Verification mode. In CISV, Cross-chain Information Synchronization (CIS) uses the design of touch block to provide high synchronization rate within low network delay, while Cross-chain Information Verification (CIV) applies ECC cryptography algorithm and Secure Multi-party Computation (SMC) to guarantee the accuracy and privacy of cross-chain information.

Our experiments on three mainstream blockchains (Ethereum, Hyperledger Fabric, and EOS) show that the response time of synchronization mainly depend on the blockchain with the lowest storage efficiency, where CIS provides a comparatively high synchronization rate and low network delay. From the perspective of cross-blockchain verification, experimental results show that CIV guarantees the security and correctness of information transmission in a relatively effective way.

In future work, we will investigate the methods to further improve the synchronization rate, while reduce the network latency. More encryption algorithms will be compared and then applied onto CISV so as to increase the execution efficiency and security index of the whole system. Furthermore, more blockchain platforms will be chosen to expand the scope of application. Finally, we will expand the concept of touch block to other cross-blockchain issues.

**Acknowledgement.** The authors would like to thank the anonymous reviewers for their elaborate reviews and feedback. This work is supported by the National Natural Science Foundation of China (No. 61906099), the Open Fund of Key Laboratory of Urban Land Resources Monitoring and Simulation, Ministry of Natural Resources (No. KF-2019-04-065).

# References

1. Christidis, K., Devetsikiotis, M.: Blockchains and smart contracts for the Internet of Things. IEEE Access **4**, 2292–2303 (2016)
2. Nofer, M., Gomber, P., Hinz, O., Schiereck, D.: Blockchain. Bus. Inf. Syst. Eng. **59**(3), 183–187 (2017)
3. Ali, M., Nelson, J.C., Shea, R., Freedman, M.J.: Blockstack: a global naming and storage system secured by blockchains. In: Proceedings of the USENIX Annual Technical Conference, pp. 181–194 (2016)
4. Conoscenti, M., Vetró, A., De Martin, J.C.: Blockchain for the Internet of Things: a systematic literature review. In: Proceedings of the IEEE/ACS 13th International Conference on Computer Systems and Application, December 2016, pp. 1–6 (2016)
5. Vukolić, M.: The quest for scalable blockchain fabric: proof-of-work vs. BFT replication. In: Proceedings of the International Workshop on Open Problems in Network Security, pp. 112–125 (2015)
6. Hirai, Y.: Defining the Ethereum virtual machine for interactive theorem provers. In: Proceedings of the International Conference on Financial Cryptography and Data Security, pp. 520–535 (2017)
7. Herlihy, M.: Atomic cross-chain swaps. In: Proceedings of the ACM Symposium on Principles of Distributed Computing, pp. 245–254 (2018)
8. Zhang, H., Jin, C., Cui, H.: A method to predict the performance and storage of executing contract for Ethereum consortium-blockchain. In: Chen, S., Wang, H., Zhang, L.-J. (eds.) ICBC 2018. LNCS, vol. 10974, pp. 63–74. Springer, Cham (2018). https://doi.org/10.1007/978-3-319-94478-4_5
9. Raju, P., et al.: mLSM: making authenticated storage faster in Ethereum. In: HotStorage 2018 (2018)
10. Yuan, P., Zheng, K., Xiong, X., Zhang, K., Lei, L.: Performance modeling and analysis of a Hyperledger-based system using GSPN. Comput. Commun. **153**, 117–124 (2020)
11. Jiang, L., Chang, X., Liu, Y., Mišić, J., Mišić, V.B.: Performance analysis of hyperledger fabric platform: a hierarchical model approach. Peer-to-Peer Netw. Appl. **13**(3), 1014–1025 (2020). https://doi.org/10.1007/s12083-019-00850-z
12. Kakei, S., Shiraishi, Y., Mohri, M., Nakamura, T., Hashimoto, M., Saito, S.: Cross-certification towards distributed authentication infrastructure: a case of hyperledger fabric. IEEE Access **8**, 135742–135757 (2020)
13. Huang, Y., et al.: Characterizing EOSIO Blockchain. CoRR abs/2002.05369 (2020)
14. Zheng, W., Zheng, Z., Dai, H.N., Chen, X., Zheng, P.: XBlock-EOS: extracting and exploring blockchain data from EOSIO. CoRR abs/2003.11967 (2020)
15. Warren, W., Bandeali, A.: 0x: an open protocol for decentralized exchange on the Ethereum blockchain. https://0xproject.com/pdfs/0x_white_paper.pdf. Accessed 31 May 2019
16. Zhang, T., Wang, L.: Republic protocol: a decentralized dark pool exchange providing atomic swaps for Ethereum-based assets and Bitcoin. https://releases.republicprotocol.com/whitepaper/1.0.0/whitepaper_1.0.0.pdf. Accessed 31 May 2019
17. Wood, G.: PolkaDot: vision for a heterogeneous multi-chain framework. https://polkadot.network/PolkaDotPaper.pdf. Accessed 31 May 2019
18. Metronome: Owner's manual. https://www.metronome.io/pdf/owners_manual.pdf. Accessed 31 May 2019
19. Borkowski, M., Sigwart, M., Frauenthaler, P., Hukkinen, T., Schulte, S.: Dextt: deterministic cross-blockchain token transfers. IEEE Access **7**, 111030–111042 (2019)

# A Formal Process Virtual Machine for EOS-Based Smart Contract Security Verification

Zheng Yang$^{(\boxtimes)}$ and Hang Lei

School of Information and Software Engineering, University of Electronic
Science and Technology of China, Chengdu, China
zyang.uestc@gmail.com

**Abstract.** With the rapid development of blockchain technology, the reliability
and security of blockchain smart contracts is one of the most emerging issues of
greatest interest for researchers. In this paper, the framework of formal symbolic
process virtual machine, called FSPVM-EOS, is presented to certify the reliability
and security of EOS-based smart contracts in Coq proof assistant. The fundamental
theoretical concepts of FSPVM-EOS are abstract formal symbolic process virtual
machine and execution-verification isomorphism. The current version of FSPVM-
EOS is constructed on a formal virtual memory model with multiple-level table
structure, a formal intermediate specification language, which is a large subset of
the C ++ programming language based on generalized algebraic datatypes, and
the corresponding formal definitional interpreter. This framework can automatically execute the smart contract programs of EOS blockchain and simultaneously
symbolically verify their reliability and security properties using Hoare logic in
Coq.

**Keywords:** Formal verification; EOS · Smart contracts · Higher-order logic
theorem proving · Coq

## 1 Introduction

With the rapid development of blockchain technology [1], a number of smart contracts
have been deployed in many critical domains. The smart contract is a kind of script
program that represents digital contract where the code is the law. However, this characteristic of smart contracts makes them susceptible to attack and result in economic loss.
Therefore, the reliability and security analysis of such programs is an urgent problem
should be resolved. The formal method is one of the most rigorous technologies for
verifying the reliability and security of programs to build trustworthy software systems.

Currently, most of researches about the reliability and security analysis of smart
contracts have focused on Ethereum platform. However, there is few work concentrate
on EOS [2] which is also a popular blockchain platform. In order to fill this gap, based
on our previous work [3], in this paper, we present a formal verification framework in

© Springer Nature Singapore Pte Ltd. 2021
W. Gao et al. (Eds.): FICC 2020, CCIS 1385, pp. 253–263, 2021.
https://doi.org/10.1007/978-981-16-1160-5_20

Coq [4] for the security and reliability of EOS-based smart contract, and we have made following contributions.

First of all, we present a brief introduction of an application extension of Curry-Howard isomorphism (CHI), denoted as execution-verification isomorphism (EVI) and it is the basic theory for combining higher-order logic systems, supporting CHI, and symbolic execution technology to construct a formal symbolic process virtual machine for solving the problems of automation, consistency and reusability in higher-order logic theorem proving. Second, we take EVI as the fundamental theoretical framework to develop an FSPVM for EOS (FSPVM-EOS) including a g formal memory framework with multiple-level table structure called MLTGM, a formal intermediate formal specification language for C++ formalization based on generalized algebraic datatypes denoted as Lolisa-EOS, and the corresponding formally verified interpreter of Lolisa-EOS denoted as FEos. The FSPVM-EOS is entirely developed in Coq. The FSPVM-EOS is adopted to semi-automatically verify the security and reliability properties of EOS-based smart contracts. Finally, to demonstrate the power of our purposed framework, we have presented simple case studies to illustrate the application process of FSPVM-EOS.

The remainder of this paper is organized as follows. Section 2 introduces related work on security analysis for smart contracts. Section 3 briefly illustrates the basic concepts of EVI. Section 4 describes the overall implementation of FSPVM-EOS. Section 5 demonstrates the power of the purposed framework using simple cases of its application. Finally, Sect. 6 presents preliminary conclusions and directions for future work.

## 2   Related Work

In [1], the authors provide a framework using Isabelle/HOL proof assistant to verify the bytecode of existing Ethereum smart contracts and create a sound program. EtherTrust developed by Grishchenko [6] provides the static analyzer for EVM bytecode. KEVM [7] is a formal semantics for the EVM based on the K-framework, like the formalization conducted in Lem [8]. It is an executable framwork, so it can run the validation test suite provided by the Ethereum foundation. Osiris [9] and Mythril [10] are dynamic analysis tools. However, they have not yet been proven to be effective at increasing the security and reliability of smart contracts. VeriSol (Verifier for Solidity) [11] is a formal certification model based on model checking technology. It provides a formal verification and analysis system for smart contracts developed using the Solidity programming language.

However, as introduced above, these works represent currently available programming verification tools for Ethereum smart contracts, but they are not well suited to other blockchain platforms, such as EOS. In this paper, based on our previous work, we presented a formal verification framework for EOS-based smart contracts verification at source code level.

## 3   Foundational Concepts

The theoretical framework of FSPVM-EOS is a FSPVM which consists of two core components: 1) EVI, which is an application extension of CHI, proving the isomorphic relationship between symbolic execution and higher-order theorem proving; 2) the theoretical basis of constructing an FSPVM in a formal logic system that supports CHI. The motivations of them are explained as follows.

For the first component, briefly, P. Wadler's [12] summarizes that CHI proposes that a deep correspondence exists between the worlds of logic and computation which is listed as Property 1.

**Property 1** (Isomorphism): This correspondence can be expressed according to three general principles as follows.

(1) *types correspond to propositions*;
(2) *proofs correspond to $\mathcal{P}$rograms*;
(3) *proofs correspond to the evaluation of $\mathcal{P}$rograms*;

These deep correspondences make CHI very useful for unifying formal proofs and program computation. As a result, $\mathcal{P}$rograms implemented using functional specification languages (FSLs) in proof assistants supporting CHI, such as Coq, can be directly evaluated and defined as lemmas and theorems for property certifications in higher-order theorem proving systems that support CHI. This is summarized below as *property 1*.

**Property 2** (Self-certification): All $\mathcal{P}$rogram s with specifications defined in a higher-order theorem proving context $\Gamma$ that supports CHI can be proven directly. This self-certification property is represented as $\Gamma$, $\mathcal{F}(\mathcal{P}rogram) \vdash \mathcal{V}(\mathcal{P}rogram)$.

However, CHI is limited in that most mainstream $\mathcal{L}$, such as C++ and Java, employed in the real world are not designed based on higher-order typed lambda calculus, and are therefore far different from FSLs. Moreover, nearly all FSLs of higher-order logic theorem-proving assistants, such as Gallina, which is implemented in Coq, do not directly support complex values, such as arrays and mapping values, owing to the strict typing system of the trusted core and the adoption of different paradigms. Hence, $program_{rw}$ cannot be directly expressed by FSLs in proof systems, and CHI cannot be directly applied to unify property verification and $program_{rw}$ execution.

Next, Dijkstra-style Weakest Precondition Calculus (DSWPC) [13], which is widely used in program verification based on symbolic execution technique. In general, if we assume there is a symbolic execution engine $\mathcal{EI}$ developed by operational semantics, any hypothesis H defined as the form of Hoare triple {Pre}Code{Post} can be equivalently represented by the reachability chain H: $\{Pre\} \wedge \mathcal{EI}(Code) \xrightarrow{\text{execution}} \mathcal{EI}(\emptyset) \wedge \{Post\}$, where $\emptyset$ is a pattern representing the empty program. In other words, a Hoare logic proof derivation is equivalent to the evaluation of formal operational semantics under a logic system, and thereby provides a mechanical translation from Hoare logic proofs to formal operational semantics proofs [14]. This method can efficiently analyze and verify the programs developed by $\mathcal{L}$. Nevertheless, the correctness certification of $\mathcal{EI}$ is a very difficult mission for most symbolic execution tools adopted DSWPC, because most $\mathcal{EI}$ are developed by general-purpose programming languages that cannot be evaluated by proof assistants directly.

Fortunately, we note that, according to the Turing-Church thesis [15], a $\mathcal{L}$ based on Turing machine and a FSL based on higher-order typed lambda calculus have equivalent computing ability. The $\mathcal{L}$ can be equivalently formalized as $\mathcal{FL}$ using the FSL. This theory helps us to combine CHI and DSWPC to alleviate their limitations.

To be specific, here we assume $\Gamma_{HOL}$ is the context of a higher-order logic proof system which supports CHI and provides a formal specification language $FSL_{HOL}$ based on the higher-order typed lambda calculus.

**Definition** (Abstract EVI model). An abstract EVI model contains three elements $\langle \mathcal{FL}, \mathcal{FM}, \mathcal{FI} \rangle$ where.

1. $\mathcal{FL} : Stt$ is the formal version of $\mathcal{L}$ mechanized in $\Gamma_{HOL}$ using $FSL_{HOL}$ with type$Stt$;
2. $\mathcal{FM} : M$ is a formal memory model developed in $\Gamma_{HOL}$ using $FSL_{HOL}$ to store logic state of variables with type$M$;
3. $\mathcal{FI} : Stt \rightarrow M \rightarrow M$ is a formal symbolic execution engine developed in $\Gamma_{HOL}$ using $FSL_{HOL}$ with function type$Stt \rightarrow M \rightarrow M$.

First of all, an $\mathcal{FL}$ that lies intermediate between real-world languages and the formal language system must be defined to directly represent real-world values with an equivalent syntax as that of the formal language system, and translate real-world values into the native values of $FSL_{HOL}$ that can be evaluated in $\Gamma_{HOL}$.

Secondly, the $\mathcal{FI}$ should be constructed by the $S_{exe}$ of $\mathcal{FL}$ based on $\mathcal{FM}$. As defined above, $\mathcal{FI}$ is developed in $\Gamma_{HOL}$ using $FSL_{HOL}$, so it is also a special $\mathcal{P}rogram$ that takes program$_{formal}$ written by $\mathcal{FL}$ and precondintion $\sigma_{init}$ based on $\mathcal{FM}$ as parameters. According *to* Property 1*Isomorphism*, $\mathcal{FI}$ can be unified into verification hypotheses, and the evaluation process corresponds to the proof construction process of verification. Hence, a hypothesis in $FSL_{HOL}$ can be defined with the DSWPC reachability chain form.

$$H' : \{Pre\} \wedge \mathcal{FI}( \text{program}_{formal}) \xrightarrow{\text{execution}} \mathcal{FI}(\emptyset) \wedge \{Post\}$$

In this manner, driven by the proof kernel, $\mathcal{FI}$ can apply $S_{exe}$ to symbolically execute a program$_{formal}$ written by $\mathcal{FL}$ with a precondintion $\sigma_{init}$ in $\Gamma_{HOL}$, and generate $\sigma_{final}$ for the postcondition verification.

Thus, program$_{rw}$ execution and property verifications can be unified in a $\Gamma_{HOL}$ in the form of program$_{formal}$ by evaluating $\mathcal{FI}$, and this method alleviates the limitation of CHI. This yields the following rules from *Property 1*.

(1) the evaluation of $\mathcal{FI}$ corresponds to the execution of program formal;
(2) execution of *program$_{formal}$* corresponds to the verification of properties;
(3) proofs correspond to the evaluation of $\mathcal{FI}$ corresponds to the verification of properties.

Moreover, because $\mathcal{FI}$ is a $\mathcal{P}rogram$, according to *Property 2 self-certification*, we can get the expression $\Gamma_{HOL}, \mathcal{F}(\mathcal{FI}) \vdash \mathcal{V}(\mathcal{FI})$ that the correctness of $\mathcal{FI}$ can be verified in $\Gamma_{HOL}$ directly. Therefore, this method also alleviates the limitation of DSWPC.

The Coq proof assistant is one of the proof assitants that support CHI and the development of DSWPC. In this manner, FSPVM-EOS accurately simulates the execution behavior of program$_{rw}$ in the real world.

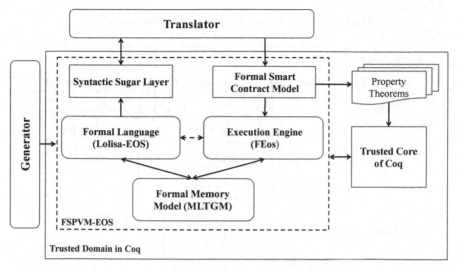

**Fig. 1.** Overall architecure of FSPVM-EOS.

# 4    Overview of FSPVM-EOS

## 4.1    Architecture

Taking EVI as the fundamental theoretical framework, all components of the FSPVM-EOS framework are developed in Coq proof assistant. As shown in Fig. 1, the architecture of FSPVM-EOS is constructed by three components. Specifically, the basic component of FSPVM-EOS is a general formal memory model with multiple-level table structure, denoted as MLTGM. This memory model simulates physical memory hardware structure, including a low-level formal memory space, and provides a set of application programming interfaces to support different formal verification specifications. The second part is a formal intermediate specification language, denoted as Lolisa-EOS, which is a large subset of the C++ programming language mechanized in Coq. The semantics of Lolisa-EOS are based on MLTGM framework. The third part is a formal verified interpreter for Lolisa-EOS, called FEos, which connect MLTGM, Lolisa-EOS and trusted core of Coq (TCOC) together to automatically and symbolically execute and verify the smart contracts of EOS.

## 4.2    Formal Memory Model with Multi-level Table Structure

The basic component of FSPVM-EOS is an extensible formal memory model developed in Coq, denoted as MLTGM. Specifically, to avoid virtual memory scalability and execution efficiency problem in Coq, we define the abstract specifications of formal memory model using multi-level table structure and corresponding memory operations in Coq, through combing the polymorphic instantiation and iterative nesting of modules technology.

Different from conventional formal memory models defined with inductive datatype, such as list and tree structure, MLTGM is a record datatype based formal memory

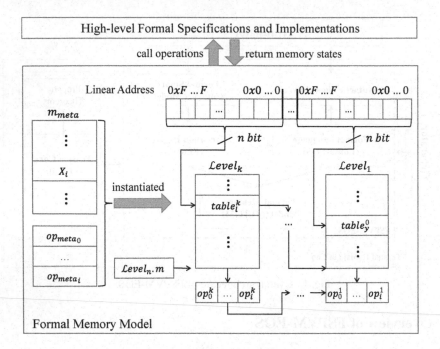

**Fig. 2.** Architecture of MLTGM.

model. Because all fields of a record datatype in Coq are accessor functions with type $\tau_{record} \rightarrow \tau_{field}$, which allows an instantiated record term $r : \tau_{record}$ to directly access a specific field using the respective field identifier $id_{field}$ with the form $id_{field}\,r$. In this manner, the memory operations of MLTGM can modify a logic memory unit directly through its unique field identifier instead of searching nodes one by one.

As shown in Fig. 2, the MLTGM framework includes two main components: a formal meta memory model using polymorphic technology, and instantiated formal memory model based on iterated modular definitions. The first component is the generic template of record type based meta memory space $m_{meta}$ and corresponding meta memory operations $op_{meta}$ using polymorphic definitions. The second component is the instantiated memory table of different levels. This component is constructed by modular system recursively. This model structure can avoid the naming conflict and ill-formed recursive problems and simulate the recursive nestification of multiple level table structure defined as rule 1 by iterative instantiation of polymorphism in different modules.

$$m_{level_n} := m_{meta}(m_{meta}(\ldots(m_{meta}(value)))) \tag{1}$$

Particularly, although these assistant tools are implemented in the general domain using general-purpose programming languages, according to EVI introduced in Section III, the relationship between the assistant tools and the respective generated results satisfies the non-aftereffect property [2]. As such, the verified results are not influenced by the assistant tools implementation.

The workflow of the MLTGM framework can be defined as follows. First, users set initial memory size, and the assistant tools generate the respective specifications. Next, according to *Property 2*, the generated specifications of MLTGM are certified according to the correctness properties employed in Coq. If the formal memory model satisfies all required properties, then a MLTGM framework with specific requirements has been constructed successfully. The users can then formalize high-level formal specifications based on the generated MLTGM formal memory model.

### 4.3  Formal Intermediate Specification Language

The formal intermediate specification language of the FSPVM-EOS framework is an extended version of Lolisa [16], denoted as Lolisa-EOS. In our previous work, Lolisa is the specification language of our previous work FSPVM-E [2]. Based on our previous work, we have extended Lolisa as a subset of the C++ programming language, denoted as Lolisa-EOS. It includes the most important characteristic components of C++, such as pointer arithmetic, struct, field access, lambda definition, basic template and class mechanism, but it also contains general-purpose programming language features. Similar to Lolisa original version, explicit unit of EOS, and non-structured forms of switch such as goto statement and Duff's device [17] are omitted in Lolisa-EOS.

Particularly, one of the most important features of Lolisa-EOS is its formal syntax is formalized with generalized algebraic datatypes (GADTs) [18] theory, which gives imparts static type annotation to all the values, expressions and statements of Lolisa-EOS. Taking the expression formalization as an example, formal syntax of expressions are defined with GADTs which are annotated by two type signatures according to the rule 2 below.

$$expr : \tau_0 \rightarrow \tau_1 \rightarrow Type \tag{2}$$

Here, $\tau_0$ refers to the current expression type and $\tau_1$ refers to the final type of the expression after evaluation. In this manner, the formal syntax of expressions becomes more clear and abstract, and allows the type safety of Lolisa-EOS expressions to be maintained strictly. Besides, combining the two static type limitation can facilitate the formalization of a very large number of different types of expressions using identical inductive constructors. As such, Lolisa-EOS has a stronger static typing judgements mechanism compared with C++ for checking the construction of programs. And it is impossible to construct ill-typed terms in Lolisa-EOS.

The formal semantics of Lolisa-EOS is formalized using small-step operational semantics. Because Lolisa is employed as the equivalent intermediate language for C++, it should be able to be parsed, executed and verified in Coq. Therefore, the semantics of Lolisa-EOS are deterministic and based on the MLTGM memory model, and it has completely mechanized into Coq.

### 4.4  Formal Executable Definitional Interpreter for EOS Verification

The final component of FSPVM-EOS is the respective formal verified interpreter of Lolisa-EOS, denoted as FEos. It is the core component for constructing the FSPVM-EOS, which is the updated version of FEther [19] built on the specifications of MLTGM

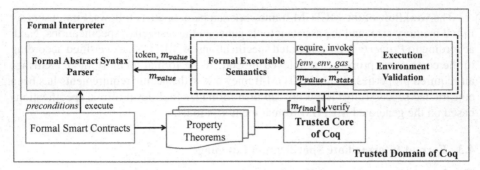

**Fig. 3.** Architecture of the FEos framework.

memory model and Lolisa-EOS. This component is used to automatically verify and symbolically execute EOS-based smart contracts.

To be specific, as the architecture of FEos shown in Fig. 3, the FEos is the bridge that connects TCOC, MLTGM and Lolisa-EOS, and it is constructed by two parts. The first part is the parser which analyzes the syntax of formal smart contracts written in Lolisa-EOS, and extracts tokens to invoke corresponding executable semantics. The second part is the in-built formal instruction set architecture constructed by executable semantics $\mathcal{ES}$ of Lolisa-EOS. The workflow of FEos is defined as follows. First of all, the parser will analyze and operate the formal abstract syntax of formal smart contract models written in Lolisa-EOS and generate the tokens of the current invoking statement. Second, the formal executable semantics $\mathcal{ES}$ using big-step operational semantics, which is strictly equivalent with the inductive semantics $\mathcal{S}$ of Lolisa-EOS, will be invoked by corresponding tokens. And the instruction set $\mathcal{I} \backslash f$ of Lolisa-EOS is defined as rule 3. A new intermediate memory state $m_{state}$ will be generated by TCOC and returned to the formal interpreter as a new initial memory state for the next execution iteration. And the FEos will repeat these process until it is satisfied certain conditions, such as the programs stopped normally or breaking off, and output the final optional memory state $m_{final}$.

$$\mathcal{Ins} \stackrel{\text{def}}{=} \forall i \in \mathbb{N}. (\cup_{i=0}^{n} \mathcal{S}_i) \leftrightarrow (\cup_{i=0}^{n} \mathcal{ES}_i) \tag{3}$$

Similar to our previous work, FEos also combines the gas mechanism of EOS platform and Bounded Model Checking (BMC). To be specific, first of all, the combination of symbolic execution and higher-order theorem proving facilitates our use of BMC to verify formal smart contracts. We employ BMC notion to set a limitation into the implementation of FEos that it only can execute K times for avoiding the infinite execution situation defined in rule 4, where the context of the formal memory space is denoted as $M$, $\sigma$ represents the current memory state, the context of the execution environment is represented as $\Omega$ and $\mathcal{F}$ represents the formal system of verification. In addition, the initial environment $env$, which are initialized by the helper function $init_{env}$, and the initial gas value of $env$ is set by $set_{gas}$. These rules represent two conditions of Lolisa-EOS programs $P(stt)$ execution, denoted as $\Downarrow_{P(stt)}$. Here, $opars$ represents an optional arguments list. The workload of the updating for FEos is about 3000 Coq lines.

$$\frac{\Omega \vdash env \quad M \vdash \sigma, b_{infor} \quad \mathcal{F} \vdash opars \quad \Omega, M, \mathcal{F} \vdash P(stt)}{env = set_{gas}\big(init_{env}(P(stt))\big) \quad fenv = init_{env}(P(stt))}{\sigma = init_{mem}(P(stt))}$$

$$\Omega, M, \mathcal{F} \vdash \langle \sigma, env, opars, \Downarrow_{P(stt)} \rangle \xRightarrow{\text{execute}, \infty} \langle \sigma', env' \rangle \wedge env'.(gas) \leq gasLimit \xRightarrow{T} \langle \sigma', env' \rangle \tag{4}$$

## 5  Case Study

To illustrate the power of our purposed framework, FSPVM-EOS is adopted to specify and verify some simple EOS-based smart contracts written in C++ in the Coq proof assistant. As shown in Fig. 6, it is code segment of EOS smart contracts written in Lolisa-EOS which includes basic statements, such as lambda and function definitions. As shown in Fig. 4 and 5, the properties of the code segment can be verified with Hoare style and verified in Coq directly using FSPVM-EOS. Besides, as shown in Fig. 7, the intermediate state during verification can be checked in the proof context. In short, FSPVM-EOS makes the verification process of EOS smart contracts become much more efficiently. The current version of FSPVM-EOS has already supported basic EOS smart contracts verification, and it is able to semi-automatically verify the EOS smart contracts in Coq.

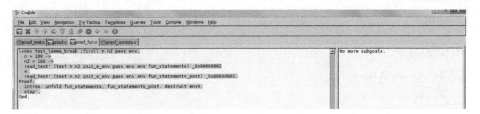

**Fig. 4.** Execution and verification of the properties written in Lolisa-EOS using FSPVM-EOS.

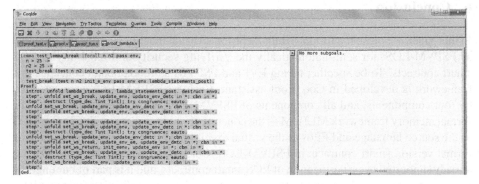

**Fig. 5.** The formal verification process of EOS-based smart contracts using FSPVM-EOS.

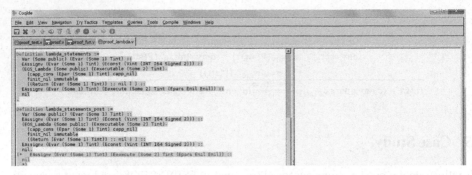

**Fig. 6.** The formal version of the case study written in Lolisa-EOS.

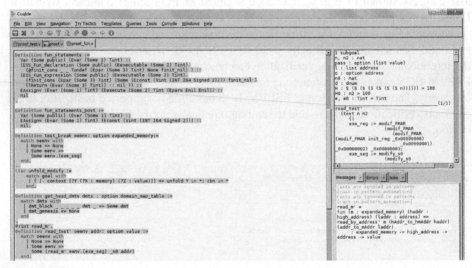

**Fig. 7.** Formal intermediate memory states during the execution and verification process in Coq.

## 6   Conclusion

In this paper, we present a formal symbolic process virtual machine framework, denoted as FSPVM-EOS, for semi-automatically the verifying security and reliability of EOS smart contracts. To be specific, taking EVI and FSPVM as fundamental concepts, this framework is developed in Coq proof assistant. Besides, FSPVM-EOS is constructed by four components, and all components of FSPVM-EOS are constructed in Coq. The formal memory framework MLTGM as the concrete formal memory model, Lolisa-EOS as the source language and FEos as the virtual execution kernel to symbolically execute formal version smart contracts in FSPVM-EOS. Current version of FSPVM-EOS has already supported basic verification of EOS smart contracts, and it is part of our ongoing project which aims at developing a general automatic formal verification framework for multiple blockchain platforms.

# References

1. Narayanan, A., Bonneau, J., Felten, E., Miller, A., Goldfede, S.: Bitcoin and Cryptocurrency Technologies: A Comprehensive Introduction, 1st ed., pp. 35–72. Princeton University Press, New Jersey (2016).
2. EOS blockchain platform. https://eos.io/, Accessed 20 Oct 2020
3. Yang, Z., Lei, H., Qian, W.Z.: A hybrid formal verification system in coq for ensuring the reliability and security of ethereum-based service smart contracts. IEEE Access **8**(1), 21411–21436 (2020)
4. The Coq Proof Assistant Reference Manual, https://coq.inria.fr/distrib/current/refman/, Accessed 20 Oct 2020
5. Amani, S., B´egel, M., Bortin, M., Staples, M.: Towards verifying ethereum smart contract bytecode in isabelle/hol. In: 7th ACM SIGPLAN International Conference on Certified Programs and Proofs, pp. 66–77 (2018)
6. Grishchenko, I., Maffei, M., Schneidewind, C.: Ethertrust: Sound static analysis of ethereum bytecode, Technische Universität Wien, Technical Report (2018)
7. Hildenbrandt, E., et al.: Kevm: a complete formal semantics of the ethereum virtual machine. In: the 31st Computer Security Foundations Symposium (CSF), pp. 204–217. IEEE (2018)
8. Hirai, Y.: Defining the ethereum virtual machine for interactive theorem provers. In: Brenner, M., et al. (eds.) FC 2017. LNCS, vol. 10323, pp. 520–535. Springer, Cham (2017). https://doi.org/10.1007/978-3-319-70278-0_33
9. Torres, C.F., Schütte, J., State, R.: Osiris: hunting for integer bugs in ethereum smart contracts, In: the 34th Annual Computer Security Applications Conference, pp. 664–676 (2018)
10. Mueller, B.: Smashing Ethereum Smart Contracts for Fun and Real Profit. HITB SECCONF Amsterdam (2018)
11. Lahiri, S.K., Chen, S., Wang, Y., Dillig, I.: Formal specification and verification of smart contracts forazure blockchain. https://arxiv.org/abs/1812.08829, Accessed20 Oct 2020
12. Wadler, P.: Propositions as types. Commun. ACM **58**, 75–84 (2015)
13. Dijkstra, E.W.: Hierarchical ordering of sequential processes. Acta Informatica **1**, 115–138 (1971)
14. Roşu, G., Ştefănescu, A.: From hoare logic to matching logic reachability. In: Giannakopoulou, D., Méry, D. (eds.) FM 2012. LNCS, vol. 7436, pp. 387–402. Springer, Heidelberg (2012). https://doi.org/10.1007/978-3-642-32759-9_32
15. Cleland, C.E.: Is the church-turing thesis true? Minds Mach. **3**(3), 283–312 (1993)
16. Yang, Z., Lei, H.: Lolisa: formal syntax and semantics for a subset of the solidity programming language. https://arxiv.org/abs/1803.09885, Access 20 Oct 2020
17. Duff, T.: Message to the Comp.lang.c Usenet Group. https://www.lysator.liu.se/c/duffs-device.html, Accessed 20 Oct 2020
18. Xi, H., Chen, C., Chen, G.: Guarded recursive datatype constructors. ACM SIGPLAN Notices **38**, 224–235 (2003)
19. Yang, Z., Lei, H.: FEther: an extensible definitional interpreter for smart-contract verifications in coq. IEEE Access **7**, 37770–37791 (2019)

# AVEI: A Scientific Data Sharing Framework Based on Blockchain

Liangming Wen[1,2] (iD), Lili Zhang[1] (iD), Yang Li[3] (iD), and Jianhui Li[1(✉)] (iD)

[1] Computer Network Information Center, Chinese Academy of Sciences, Beijing 100190, China
lijh@cnic.cn
[2] University of Chinese Academy of Sciences, Beijing 100049, China
[3] Library of Chengdu Sport University, Chengdu 610041, Sichuan, China

**Abstract.** Scientific data sharing faces some dilemmas in process of practice. Blockchain technology has the characteristics of decentralization, openness, independence, security and anonymity, it provided a new solution to solve the dilemmas of scientific data sharing. So, exploring the application of blockchain technology in scientific data sharing is of great significance to expand the applications of blockchain and improve the effectiveness of scientific data sharing. This paper introduced the basic concepts of blockchain, summarized the practical dilemmas faced by scientific data sharing, analyzed the coupling between blockchain and scientific data sharing. Finally, we proposed a scientific data sharing framework which called AVEI based on blockchain, and analyzed the performance of AVEI from three perspective. AVEI covers three processes, they are user identity role authenticate process, data verify process and data exchange process, and we also introduced incentive system to enhance the enthusiasm of participating nodes. In theory, AVEI can solve practical dilemmas such as inconsistent standards, violation of privacy and security, insufficient motivation, etc., and it has a good performance in data quality, data security, and sharing effects.

**Keywords:** Blockchain · Scientific data · Data sharing · Privacy and security · Incentive system

## 1 Introduction

Scientific data is the record of relevant facts such as original and derived data used in scientific research activities, which is considered to be the basis of facts, evidence, or reasoning that confirm scientific discoveries or scientific opinions [1]. At present, scientific data has become one of strategic resources for reshaping the world structure, once master scientific data resource, you had mastered the initiative in science and technology development [2]. In March 2018, the State Council of China issued the *"Measures for Administration of Scientific Data"* (*M4ASD*) [3]. Subsequently, the Ministry of Science and Technology of China and the Ministry of Finance of China set up 20 national scientific data centers in the fields of high-energy physics, genome, microbiology, space, astronomy, and polar [4]. On April 9, 2020, the State Council of China issued the *"Opinions on Building a More Perfect System and Mechanism for the Market-based*

© Springer Nature Singapore Pte Ltd. 2021
W. Gao et al. (Eds.): FICC 2020, CCIS 1385, pp. 264–280, 2021.
https://doi.org/10.1007/978-981-16-1160-5_21

*Allocation of Factors*", and this is data first appeared in Chinese government document as a new type of production factor [5]. Big data, especially scientific big data, has quite distinct synergistic attributes [6], so opening and sharing is an effective way to release the value of scientific data. Although many countries have formed a variety of scientific data sharing models in long-term practice, the comprehensive ecology of scientific data sharing still needs to be improved [7].

In recent years, blockchain technology has solved problems of highly dependent on central node, low efficiency of data sharing, insecure data storage and lack of supervision in centralized systems [8]. Because of these advantages, blockchain technology provides a technical basis for constructing a multi-party trusted and transparent data security sharing scheme [9]. Blockchain technology also provides a new solution to the problems of scattered data distribution, inefficient data utilization, flood of data garbage, and frequent data leakage [10]. On October 24, 2019, the Political Bureau of the Communist Party of China Central Committee conducted a leadership meeting to discuss blockchain technology, President Xi Jinping pointed out that "It is necessary to play the role of blockchain technology in promoting data sharing, and to explore the data sharing model using blockchain technology to realize the joint maintenance and utilization of data across departments and regions" [11]. In the academic field, researchers have also made some explorations, such as the application advantages of blockchain technology in scientific data management [12], data quality transaction on different distributed ledger technologies [13], blockchain-based data sharing system with Inter Planetary File System (IPFS) and Ethereum [14], scientific data sharing hierarchical blockchain architecture [15], medical imaging data sharing framework via blockchain consensus [16], efficient privacy preserving scheme blockchain-based [17], credible big data sharing model based on blockchain and smart contract [18]. Although these papers discuss related issues of blockchain and scientific data sharing in some ways, but the topics involved are relatively single or the framework lacks theoretical support. So, based on the practical dilemmas faced by scientific data sharing, this paper builds a theoretical framework for scientific data sharing, we want to expand the application scenarios of blockchain and improve the effectiveness of scientific data sharing.

This paper focuses on how to applicate blockchain into scientific data sharing. The remainder of this paper is organized as follows. Section 2 analysis the practical dilemmas, coupling between blockchain and scientific data sharing. Section 3 provides a blockchain-based scientific data sharing framework. Section 4 analysis the framework's performance. Finally, Section 5 concludes the paper.

## 2  Blockchain and Scientific Data Sharing

### 2.1  Blockchain Technology Overview

In 2008, Satoshi Nakamoto in his article "*Bitcoin: A Peer-to-Peer Electronic Cash System*" proposed an electronic payment system that does not require credit intermediaries to solve the problem of double payment [19]. Blockchain is exactly one of the underlying support technologies of Bitcoin [20], and bitcoin has become one of the most successful applications of blockchain. Currently, there are multiple definitions about the concept of blockchain. From the perspective of organizational structure view, blockchain is a decentralized sharing system that blocks are combined into a specific structure in the form of chains in chronological order, and cryptography ensures that data cannot be tampered with or forged [21]. From an application point of view, blockchain is a decentralized infrastructure and distributed computing paradigm, using distributed nodes consensus mechanism to generate and update data, using encrypted chain block structure to verify and store data, and using smart contract to edit and manipulate data [22].

As we can see from the overviews above, blockchain is not a completely innovative emerging technology, but a complex of multiple existing mature technologies such as cryptography, distributed storage, consensus mechanism, smart contract and so on. Blockchain has created a low-cost computing paradigm and collaboration model in an untrustworthy competitive environment [23]. Specifically, blockchain mainly contains three basic elements: transaction, block and chain [24]. Transaction is the behavioral operation of participate nodes, block is the record of all operations and statuses that have occurred in a period of time, chain is a log record formed by blocks in series according to sequence of operation.

## 2.2 Practical Dilemmas of Scientific Data Sharing

**Inconsistent Scientific Data Sharing Standard.** In addition to the general "4V" characteristics of big data, such as huge data volume, complex data types, rapid data process and unlimited data value, scientific data also has much unique properties, such as repeatable calculate verify, long-period storage and process, multi-source heterogeneous integrate, and promote social development [25]. Faced with such complex data characteristics, standardization is one of the guarantees to realize scientific data sharing, and sharing standard is also a necessary condition for reflect the value of scientific data. At present, the inconsistency of scientific data sharing standards is mainly reflected in the inconsistent of data structure standard, data semantic standard, data transmit standard, and data open standard. The inconsistent data sharing standards have led to various deviations in understanding of data by various groups [26], and it also has a negative impact on timeliness, accuracy, completeness, and consistency [27, 28]. Even data that has been shared, it is difficult to be effectively reuse or it is even unusable, which affects the release, cite and evaluate of scientific data.

**Invasion of Privacy Security.** In May 2018, *"General Data Protection Regulation"* *(GDPR)* came into effect, which giving data subject the rights to know, to access, to oppose, to carry and to be forgotten data [29]. In some sense, GDPR aroused people's attention to privacy and security issues, such as data transfer, data utilize, data storage and so on. First of all, private data may be leaked intentionally or unintentionally during data transferring and sharing, certain private data is often abused without the owner's knowledge [30]. Secondly, after small sample data aggregates into big data, even though the private data is treated anonymously and fuzzily, the data owner's private information and behavior can still be obtained through data mining. Finally, scientific data center or repository also faces straits such as system attack, system down, software copyright, software upgrade, hardware restriction, and network transfer restriction, lawbreaker can achieve his malicious goals by stealing or abusing scientific data in server [31].

**Insufficient Motivation for Scientific Data Sharing.** First of all, when data has become an asset [32], the confusion of its ownership makes allot of data interests difficult, and scientific data sharing is in trouble [33]. Secondly, there are interest entanglements in heart of the scientific data sharing participates. On one hand, scientific data owner believes that data sharing will lead to loss of his own data rights. On the other hand, some scientific data is regarded as core competitiveness by institutions, and data owner is unwilling to share or only share low-quality data. In addition, the intrinsic value of data is difficult to evaluate [34], which is not only manifested that unstructured data is difficult to evaluate in multiple dimensions, but mean that data set and dynamic data is also difficult to evaluate in real time. Finally, supporting incentive mechanism is missing, scientific data owner will consider the returns of labor, time and capital when sharing data [35], and lack of incentive mechanism will reduce data owner's enthusiasm.

Through the above analysis, the practical dilemmas of scientific data sharing faced can be summed up as unwilling to share, inability to share, and afraid to share. In order to solve this strait, there must develop unified scientific data standards to regulate data format so that the data content is completed with less noise data and lower data usage barriers. Data storage and transfer devices must have strong anti-risk capabilities, privacy sensitive data cannot be collected and disseminated wantonly. There must be a clear boundary to divide the rights and responsibilities of each interest community, to protect data owners' legitimate rights and interest, and to punish the illegal acts of data infringers.

## 2.3  Coupling Between Blockchain and Scientific Data Sharing

**Subject Coupling.** The essence of blockchain system is a distributed database, which contains general node, compute node, record node, verify nodes. All those nodes are connected to each other in a topology flat to share data, each node stores complete historical data for mutual backup. The update and maintain of entire data chain require coordination of most nodes to complete. There are no central node or hierarchical structure with absolute power, that is blockchain's most notable feature [36]. Scientific data sharing is a diversified, multi-level and multi-functional data sharing ecosystem [37], which involving many stakeholders, such as data producer, data user, data processor, data manager, etc., the functional positioning of different stakeholders in different sharing steps is different (as shown in Table 1). Hidden into blockchain, scientific data stakeholders can act as various nodes in blockchain, each participate node performs operations such as data exchange, data transfer, and data storage on an equal basis [38]. Thus, it can be seen that there is a subject coupling relationship between blockchain and scientific data sharing [39].

**Object Coupling.** At present, the application scenarios of blockchain are mainly concentrated in the fields of digital certificate generate, supply chain management, traceability of agriculture product, intellectual property protect, cross-border e-commerce and so on [40]. According to the differences between implement methods and apply purposes, the above application scenarios can be divided into three types: value transfer, data storage and authorize management. Value transfer refers to the transfer of digital assets between different accounts, data storage refers to record of data to blockchain, and authorize

**Table 1.** Stakeholders involved and their responsibilities in each process of scientific data sharing.

| Sharing process | Stakeholder | Stakeholder's branch | Stakeholder' responsibilities |
|---|---|---|---|
| Step 1: Propose sharing requirement | Data consumer | Metadata user, primary data product user, full data product user | Propose sharing requirement and use data |
| Step 2: Evaluate sharing requirement | Data owner | Raw data producer, data processor, funding supporter | Evaluate data requirement |
| Step 3: Review data content | Data supervisor | Administrative department, industry alliance, data owner unit | Review data sharing process |
| Step 4: Open data resource | Data saver | Domain/industry data center, institution/project repository, personal computers/hard drive | Save data and perform data sharing operation |
| Step 5: Use data resource | Data consumer | Metadata user, primary data product user, full data product user | Propose sharing requirement and use data |

management refers to use contract mechanism to control the access and use of data [41]. These scenarios have data-driven features. For scientific data sharing, the shared object is generally including basic data, materials, system-processed data products and related information [42]. It can be seen that the application scenarios of blockchain and the objects of scientific data sharing are all kinds of scientific data resources, and they are coupled in terms of object.

**Function Coupling.** The biggest contribution of blockchain is to solve the byzantine general problem through decentralized idea, block chain structure, smart contract, encrypt algorithm, and other technologies to achieve distributed nodes without relying on trusted third-party participant [43]. Consensus' trust is built on technical endorsement, and the contribution of participate nodes is measured through a consensus mechanism to encourage nodes to actively participate. The trust mechanism of blockchain is based on technical endorsement, and the contribution of participate nodes is measured through consensus mechanism to encourage nodes to actively participate. At present, the scientific data sharing models are mainly divided into large scientific device model, monitoring network model, open platform model, federal service model, data publishing model and so on [44]. Different sharing models mean different driving mechanisms, different responsibility relationship, different management methods and quality performance. Generally speaking, it is necessary to maintain the trust foundation of each participants through technical means, drive various stakeholders through corresponding incentive mechanism and supervision. Through combining and applying the blockchain's core technologies, the practical dilemmas faced by scientific data sharing can be solved (Fig. 1). Therefore, there is a function coupling relationship between blockchain and scientific data sharing.

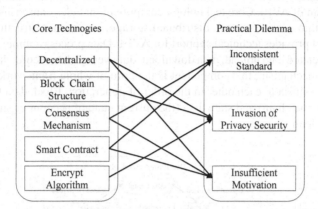

**Fig. 1.** Scientific data sharing problems that blockchain technology can solve.

**Scenario Coupling.** According to whether authorization is required to join or not, blockchain can be divided into two categories [45]: license chain and non-license chain (public chain). The join and leave of license chain need to be authorized by the privilege advantage nodes, while the non-license chain is free to enter and retreat. According to the concentration degree of authority control, license chain can be divided into alliance chain and private chain, alliance chain's control group is more disperser than private chain. Different type of blockchains have their own characteristic and scope of application. Articles 4, 20, 25 of the *M4ASD* provide detailed provisions on confidentiality level, confidentiality period, opening condition, opening object, examinate and verify procedures of scientific data [46]. Therefore, according to the requirements of different sharing scenarios for sharing system control authority and operating information disclosure, different blockchain types can be selected to limit the type of participate nodes and to clarify the sharing range, sharing degree, sharing level and so on. It can be seen that both blockchain and scientific data sharing need to limit use conditions according to use object, there is a scenario coupling relationship between blockchain and scientific data sharing.

## 3  Construction of Scientific Data Sharing Framework Based on Blockchain

### 3.1  Overall Framework Construct

According to the practical dilemmas, requirements of scientific data sharing, and combined with the characteristics of blockchain technology, this paper constructs a theoretical framework for scientific data sharing, we can call this framework AVEI for short. The specific details of AVEI are shown in Fig. 2, which includes user identity authenticate process, data verify process, data exchange process and incentive mechanism.

AVEI's body consists of four components: user node component, core technology component, data process component, and blockchain component. User node component is composed of data user, data owner, data saver, data supervisor, this component is the main participant of AVEI. Core technology component includes consensus mechanism, encrypt algorithm, smart contract, distributed storage, block chain structure and so on, this component provides technical support for AVEI. Data process component is mainly composed of related data upload, data download, data verify, data record, data exchange, and incentive mechanism, this component is the core of whole framework. Blockchain component is connect to each other in the order of generation. Block data (every operation that occurs on chain) and state data (the current state of account or smart contract) are stored on chain.

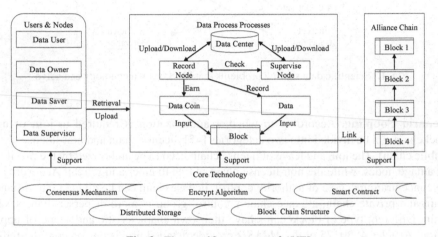

**Fig. 2.** The specific structure of AVEI.

Although decentralization means that all nodes to join or retreat freely in principle, but complete freedom is not the best option for scientific data sharing. Therefore, this paper recommends using alliance chain to restrict the scope of data sharing [47]. Nodes with certain conditions are allowed to join alliance chain after verify, data is fully open to all nodes in chain and conditionally share to nodes outside the chain.

## 3.2 User Identity Role Authenticate Process

Due to trust and authority issues between nodes, it takes a lot of costs on data verify and data confirm in a centralized data sharing system [48]. In order to better manage the identities of various participate nodes, this paper designed a user identity role authenticate process, as shown in Fig. 3, which mainly realizes the identity authentication and role authority assignment of multiple participate nodes.

The user identity role authentic process mainly includes steps such as user register, identity create, permission grant, supervise and management. The specific process is as follows:

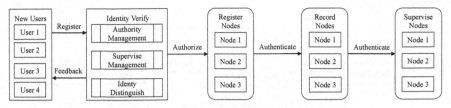

**Fig. 3.** User identity authenticate process.

*Step 1*. User register. A new user submits a register application to chain, and smart contract automatically verifies whether the application information already exist in chain or not. If application information has existed in chain, there will return message "Registered, Please Login" to user. If there is no user information in chain, the compliance of submitted information will be checked. After information audit passed, register operation will be performed and user will be notified of "Successfully Registered". The user who has successfully registered becomes registered node in chain by default.

*Step 2*. Identity creates. Establish a virtual identity mapping relationship, that is, assign a unique public key address to a virtual identity corresponding to a real identity. Virtual identity is obtained from the attribute information of the node through hash operation. A complete node information includes user ID, public key address, data coin balance and so on.

*Step 3*. Permission grant. Some registered node can become record node after consensus authenticate of most node in chain, and some record node can become supervise node after consensus authenticate. Different type of node has different operate permissions, registered node has the lowest permission, supervise node has the highest permission, node's permission matches the role position of each stakeholder.

*Step 4*. Supervise and manage. In addition to has the authority that recording node has, supervise node has the highest authority, includes monitoring whether the behavior of other nodes conforms to the rules in chain and give immediate feedback to all nodes.

### 3.3 Data Verify Process

In order to ensure data quality effectively, this paper designed a data verify process, as shown in Fig. 4. This process mainly realizes the verification of newly uploaded data and forms a data standard library. The data verify process is of great significance to data sharing: on one hand, it controls data quality from the source, and prevents the inflow of low-quality data; on the other hand, the legal rights of other nodes are protected.

The data verify process mainly includes the steps of standard library construct, data verify, and standard library update. The specific process is as follows:

*Step 1*. Standard library construct. Data user, data saver, and data supervisor propose to all nodes in chain on data structure standard, semantic standard, transfer standard, open standard, etc., and propose to form a scientific data standard library in chain after all nodes vote on consensus.

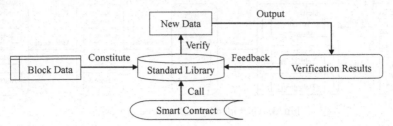

**Fig. 4.** Data verify process.

*Step 2.* Data verify. When data producer uploads new data to chain, record node and supervise node call smart contract to verify the data's volume, format, content, evaluate the data quality and feedback the verify result to the upload node. The specific verify results can be divided into several cases as shown in Table 2.

**Table 2.** Types of data verify results.

| Verify dimension | Situation classification | Situation content |
|---|---|---|
| Data volume | Oversize | The memory occupied by uploaded data exceeds the maximum limit of system |
| | Small | The memory occupied by uploaded data exceeds the minimum limit of system |
| | Suitable size | The memory occupied by uploaded data is within the range of system |
| Data format | Wrong format | The uploaded data format is outside the system limit |
| | Correct format | The uploaded data format meets the system limit |
| Data content | Repeated | Completely consistent data already exists in data standard library |
| | Analogous | Similar data already exists in data standard library |
| | Content defect | The data semantics is incomplete, and the data elements are missing |
| | Invalid content | Data is outdated or mixed with a lot of useless information, almost meaningless |
| | Content compliant | Data is not duplicated, non-similar, semantically complete, and useful |

*Step 3.* Standard library update. Non-compliance data is not allowed to be entered into alliance chain for time being. The verify node will feedback the verify results to upload node and give suggestions for modify, the modified data can be re-uploaded and

accepted for verify. Compliance data can be added into chain, and data standard library will collect and update the compliance data immediately.

## 3.4 Data Exchange Process

On one hand, the process of scientific data sharing is complicated, and on the other hand, it involves many stakeholders. In order to solve the trust problems between different stakeholders, a data exchange process is designed, as shown in Fig. 5.

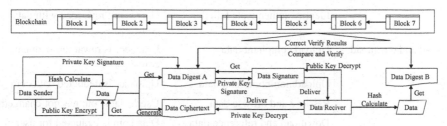

**Fig. 5.** Data exchange process.

Data exchange process mainly includes data send, data receive, data verify, operation record. The specific process is as follows:

*Step 1.* Data send. Data owner performs hash calculate, data encrypt, and digital signature he owns, then delivers the data ciphertext and digital signature to data receiver.

*Step 2.* Data receive. Data user decrypt digital signature to obtain data digest A, decrypt data ciphertext to obtain original data, and then hash calculate the original data to obtain data digest B.

*Step 3.* Data verify. Data supervisor calls smart contract to compare and verify data digest A and data digest B. If the comparison result is consistent, it means that the original data is complete and has not been tampered with. If the comparison result is inconsistent, it means that the original data has been modified.

*Step 4.* Operation record. The operation that does not pass comparison will be reject, and returns to *Step 1* to re-execute. After the operation that pass comparison and agreed by all nodes, smart contract will automatically send confirm information to on-duty record node, then record confirm information in newly generated block and embed the block in chain.

## 3.5 Incentive System

In order to quantify users' behaviors, a behavior-based incentive system is designed, that is, a specific amount of data coin is issued in chain to motivate nodes. The total issuance of data coin can be set according to actual scenarios. Assuming that a private chain is built based on a certain scientific research project (such as CASEarth), the number of data coins in chain can be equal to the project funding. All nodes form a unified consensus on the increase and decrease rules of data coin, and specific increase or decrease operations are automatically executed by smart contract. If a node makes positive contribution, it

**Table 3.** Data coin vary rule corresponding to different user behaviors.

| Classification of user behaviors | User behaviors | Data coins change rule |
|---|---|---|
| Register and authenticate | Identity register is successfully | Registered node is rewarded with data coin |
| | Identity register is unsuccessfully | Registered node is not rewarded with data coin |
| | Identify authenticate passed | Reward user with data coin |
| | Identity authenticate failed | User is not rewarded with data coin |
| Upload and download | Upload compliant data | Uploader is rewarded with data coin |
| | Upload non-compliant data | Uploader is deducted data coin |
| | Retrieve and browse data | User's data coin has not changed |
| | Download and reference data | Data coin is deducted for user, and data coin is rewarded for uploader |
| Supervise and verify | Verification between supervise node and register node is correct | Both supervise node and registered node are rewarded with data coin |
| | There is a difference in verification between supervise node and register node | Based on the verify result of most nodes, correct result node will be rewarded with data coin, wrong node will be deducted with data coin |
| | The block signed by record node is accurate | Record node is rewarded with data coin |
| | There is an error in block that recorded by record node | Data coin is deducted for register node, and reward data coin for supervise node |
| Continuously participate | Node continue to participate actively | Node is continuously rewarded with data coin |
| | Node continue to participate passively | Node is continuously deducted with data coin |
| | Node phased actively participate | Node is periodically rewarded with data coin |
| | Node phased passively participate | Node is periodically deducted with data coin |
| | Node behavior is irregular | If the contribution is positive, data coin will be rewarded. If the contribution is negative, data coin will be deducted |

will be rewarded with data coin, and if a node has bad behaviors, it will be deducted data coin. We believe that the ideal situation is that all nodes perform good behaviors and continuously participate. Integrating the specific participate behaviors of nodes and consider the degree of persistence, the rule of data coin increase or decrease for different node participate behaviors is designed, as shown in Table 3.

Table 3 only shows the rule for increase or decrease of data coin, and the specific increase or decrease will depend on application scenario. For example, according to data quality, the data uploaded by node can be divided into different levels such as excellent, good, qualified, and unqualified. On one hand, the verify results of different levels determine whether these data can be entered into chain, and on the other hand, they are also related to the number of data coin rewards that upload node can earn.

# 4 Performance Analysis of AVEI

## 4.1 Data Quality Performance Analysis

Nowadays, the consensus algorithm supported by blockchain include Proof of Work (PoW) [49], Proof of Stake (PoS) [50], Delegated Proof of Stake (DPoS) [51] and Practical Byzantine Fault Tolerance (PBFT) [52], etc. In some cases, different algorithms can be used in combination [53]. On one hand, these consensus algorithms make information stored by all nodes completely consistent, and on the other hand, they can also ensure the information published by a single node can be recorded into block by other nodes [54]. The formation of data standard library and smart contract require the consensus of all nodes. After smart contract signed, it can be embedded in any tangible or intangible assets to form a software-defined asset. Since smart contract is a set of programmed logic defined in digital form [55], once contract meets the trigger condition, it will be automatically executed until the end, and no single node can change it, which realizes the goal of "code is law" [56]. Therefore, AVEI supports the verify of data format, data volume, data content, and can automatically record and process verify results, it can also automatically assign node permission, automatically complete the increase or decrease of data coin. All operations eliminate human interference, maintain fairness and justice, and ensure the quality of scientific data sharing.

## 4.2 Data Security Performance Analysis

Data block is the basic unit of blockchain [57]. A block contains two parts: block header and block body. Block header encapsulates current block version number, merkel tree root value, timestamp, difficulty value, random number, leading block signature value and other information [58]. Block body encapsulates all operation information in current block. Current block points to both the signature value of forward block and next block, all blocks form an orderly connected data structure chain in chronological order, all behaviors information is stored in the form of merkel tree. Once the operation information changes, the root value of merkel tree will be different from the original value. As long as the version number of last block is saved, it can be verified whether the merkel tree root value is correct and whether the data has been modified [59], and the modified data block can be quickly locked according to timestamp. Therefore, AVEI can not only identify whether data has been tampered with and realize the traceability of data, but also encrypt and decrypt data with the aid of encrypt algorithms, thus ensuring the security of data transfer and sharing.

### 4.3  Sharing Effect Performance Analysis

The most important feature of blockchain is decentralization. Each peer node can complete system maintenance and self-governance through cooperation without relying on third-party trusted node [60], which embodies the technical characteristic of "equality and freedom" [61]. Based on consensus algorithms such as PoW, PoS, DPoS and PBFT to screen specific nodes to exercise account power, record node is responsible for packing the data into block and broadcasting to all nodes, and supervise node verify the received data. In addition, consensus mechanism is also an important criterion for measuring the degree of contribution of different nodes to alliance chain. Incentive system set up for different degrees of contribution can call smart contract to automatically implement rewards and punishments. Therefore, AVEI not only gives more nodes the opportunity to participate in data sharing process, but also realizes the openness and transparency of event participants, and safeguards the legitimate rights and interests of participate nodes. It also mobilizes the enthusiasm of participate node and ensures the efficiency of data sharing.

## 5  Discussion and Conclusion

### 5.1  Data Quality Performance Analysis

Because of its technical characteristics such as decentralization, non-tampering, traceability, security and controllability, blockchain can matches the quality demand, security demand and responsibility benefit demand of scientific data sharing. Blockchain and scientific data sharing have some certain degree of coupling relationship in terms of participant subject, sharing object, function and application scenario. We can rely on the technical characteristics of blockchain to solve the problems of data island, technological island and right island faced by scientific data sharing. This paper sorts out the practical dilemmas faced by scientific data sharing, and analysis the coupling relationship between blockchain and scientific data sharing. Integrate the core elements of blockchain, such as block chain structure, encrypt algorithm, smart contract and consensus mechanism, a theoretical framework for scientific data sharing based on blockchain is designed. The framework cover the processes of user identity authenticate, data quality verify, data upload and download, data exchange, data sharing, and introduces incentive system according to the behavioral characteristic of node. The framework can meet the requirements of scientific data sharing for quality performance, safety performance and effect performance, which provide solution for solving the current difficulty faced by scientific data sharing.

Compared with other research results, there are two main innovations in this article: one is to theoretically analyze the applicability of blockchain technology in scientific data sharing, and the other is that designed a theoretical framework, which not only include several important processes but also introduce incentive mechanisms. The deficiency of this paper is that it only constructs a macro-framework for whole science subject from theoretical perspective, but has not been realized by practical application scenario verify. The FAIR principle advocates that in process of opening and sharing of scientific data, efforts should be made to realize data Findable, Accessible, Interoperable and Reusable [62]. Follow-up research can be guided by FAIR principle, deeply analyze the technical characteristics of different types of blockchain, combine specific application scenarios to study and design a sharing framework system cover the whole life cycle of scientific data. In user authenticate, data verify, data record and other process, artificial intelligence technology can be used to achieve more automatically operation. In terms of user incentive mechanism, we can refer to economic theory and game theory to design detailed and specific rule for data coin vary. Future scientific data sharing activity will not be limited to "sharing" data, but will form an integrated, intelligent and FAIRness scientific data sharing ecosystem around the whole life cycle of scientific data [63]. This ecosystem integrates multiple functions such as aggregated data acquire, data storage, data distribute, data compute, data analysis, and data service applications.

**Acknowledgement.** The work described in this paper was supported by the Strategic Priority Research Program of the Chinese Academy of Sciences (No. XDA19020104) and the Informatization Program of the Chinese Academy of Sciences (No. XXH13503).

# References

1. Pilat, D., Fukasaku, Y.: OECD principles and guidelines for access to research data from public funding. Data Sci. J. **6**, 4–11 (2007)
2. Guo, H.: Scientific big data: a footstone of national strategy for big data. Bull. Chin. Acad. Sci. **33**(8), 768–773 (2018)
3. Measures for administration scientific data. https://www.gov.cn/xinwen/2018-04/02/content_5279295.htm. Accessed 23 Apr 2020
4. List of national science and technology resource sharing service platforms, https://www.gov.cn/xinwen/2019-06/11/content_5399105.htm. Accessed 23 Apr 2020
5. Opinions on building a more perfect system and mechanism for the market-based allocation of factors. https://www.gov.cn/zhengce/2020-04/09/content_5500622.htm. Accessed 15 May 2020
6. Zhang, L., Li, X., Li, A., et al.: Data sharing and transaction: the quality, value and price of data. Commun. Chin. Comput. Fed. **15**(2), 30–35 (2019)
7. National Science and Technology Infrastructure Center: National scientific data resources development report (2017). Scientific and Technical Documentation Press, Beijing (2018)
8. Azaria, A., Ekblaw, A., Vieira, T., et al.: MedRec: using blockchain for medical data access and permission management. In: 2016 2nd International Conference on Open and Big Data (OBD), pp. 25–30. IEEE, Piscataway (2016)
9. Wang, J., Gao, L., Dong, A., et al.: Blockchain based data security sharing network architecture research. J. Comput. Res. Develop. **54**(4), 742–749 (2017)

10. Zhang, J., Wang, F.: Digital asset management system architecture based on blockchain for power grid big data. Electr. Power Inf. Commun. Technol. **16**(8), 1–7 (2018)
11. President Xi Jinping presided over the 18th collective study and speech of the politburo of the Communist Party of China. https://www.gov.cn/xinwen/2019-10/25/content_5444957.htm. Accessed 17 May 2020
12. Wen, L., Zhang, L., Li, J.: Application of blockchain technology in data management: advantages and solutions. In: Li, J., Meng, X., Zhang, Y., et al. (eds.) First International Conference, Big Scientific Data Management (BigSDM 2018), LNCS, vol. 11473, pp. 239–254. Springer, Heidelberg (2019). https://doi.org/10.1007/978-3-030-28061-1_24
13. Wu, C., Zhou, L., Xie, C., et al.: Data quality transaction on different distributed ledger technologies. In: Li, J., Meng, X., Zhang, Y., et al. (eds.) First International Conference, Big Scientific Data Management (BigSDM 2018), LNCS, vol. 11473, pp. 301–308. Springer, Heidelberg (2019). https://doi.org/10.1007/978-3-030-28061-1_30
14. Wang, P., Cui, W., Li, J.: A framework of data sharing system with decentralized network. In: Li, J., Meng, X., Zhang, Y., et al. (eds.) First International Conference, Big Scientific Data Management (BigSDM 2018), LNCS, vol. 11473, pp. 255–262. Springer, Heidelberg (2019). https://doi.org/10.1007/978-3-030-28061-1_25
15. Hao, S., Xu, W., Tang, Z.: Blockchain model of scientific data sharing and its realization mechanism. Inf. Stud. Theor. Appl. **41**(11), 57–62 (2018)
16. Patel, V.: A framework for secure and decentralized sharing of medical imaging data via blockchain consensus. Health Inf. J. **25**(4), 1398–1411 (2019)
17. Fan, K., Ren, Y., Wang, Y., et al.: Blockchain-based efficient privacy preserving and data sharing scheme of content-centric network in 5G. Inst. Eng. Technol. Commun. **12**(5), 527–532 (2017)
18. Li, Y., Huang, J., Qin, S., et al.: Big data model of security sharing based on blockchain. In: 2017 3rd International Conference on Big Data Computing and Communications (BIGCOM), pp. 117–121. IEEE, Piscataway (2017)
19. Bitcoin: a peer-to-peer electronic cash system. https://bitcoin.org/bitcoin.pdf. Accessed 17 May 2020
20. Risius, M., Spohrer, K.: A blockchain research framework. Bus. Inf. Syst. Eng. **59**(6), 385–409 (2017)
21. Li, X., Liu, Z.: Study on supply chain intelligent governance mechanism based on blockchain technology. Chin. Bus. Mark. **31**(11), 34–44 (2017)
22. Zyskind, G., Nathan, O., Pentland, A., et al.: Decentralizing privacy: using blockchain to protect personal data. In: 2015 IEEE Security and Privacy Workshops, pp. 180–184. IEEE, Piscataway (2015)
23. Liu, Z.: Research on the full-process academic evaluation based on blockchain technology. J. Library Data **1**(3), 38–44 (2019)
24. Xia, Q., Zhang, F., Zuo, C.: Review for consensus mechanism of cryptocurrency system. Comput. Syst. Appl. **26**(4), 1–8 (2017)
25. Li, J., Li, Y., Wang, H., et al.: Scientific big data management technique and system. Bull. Chin. Acad. Sci. **33**(8), 796–803 (2018)
26. Stanford, N.J., Wolstencroft, K., Golebiewski, M., et al.: The evolution of standards and data management practices in systems biology. Mol. Syst. Biol. **11**(12), 851 (2015)
27. Guo, Z.: Research on data quality and data cleaning: a survey. J. Softw. **13**(11), 2076–2082 (2002)
28. Kleindienst, D.: The data quality improvement plan: deciding on choice and sequence of data quality improvements. Electron. Mark. **27**(4), 387–398 (2017). https://doi.org/10.1007/s12525-017-0245-6
29. Zhang, L., Wen, L., Shi, L., et al.: Progress in scientific data management and sharing. Bull. Chin. Acad. Sci. **33**(8), 774–782 (2018)

30. Li, X., Wang, P.: Informed consent in big data era. Med. Philos. (A) **37**(5), 9–12 (2016)
31. Ferreira, M.B., Alonso, K.C.: Identity management for the requirements of the information security. Ind. Eng. Eng. Manage. **3**, 53–57 (2013)
32. Zhu, Y., Ye, Y.: Defining data assets based on the attributes of data. Big Data Res. **4**(6), 65–76 (2018)
33. Wen, L., Zhang, L., Li, J.: Research on ethical issues of scientific data sharing in the big data era. Inf. Doc. Serv. **40**(2), 38–44 (2019)
34. Niu, C., Zheng, Z., Wu, F.: Personal data transactions: from protection to pricing. Commun. Chin. Comput. Fed. **15**(2), 39–47 (2019)
35. Huang, R., Qiu, C.: Review of research of the scientific data sharing in foreign countries. Inf. Doc. Serv. **34**(4), 24–30 (2013)
36. Yuan, Y., Wang, F.: Parallel blockchain: concept, methods and issues. Acta Automatica Sinica **43**(10), 1703–1712 (2017)
37. Zhang, L.: Scientific data sharing governance: model selection and scenario analysis. J. Library Sci. Chin. **2**, 54–65 (2017)
38. Zhang, Y.: A model of e-commerce information ecosystem based on blockchain. Res. Library Sci. **6**, 33–44 (2018)
39. Wang, C., Wan, Y., Qin, Q., et al.: A model of logistics information ecosphere of supply chain based on blockchain. Inf. Stud. Theor. Appl. **40**(7), 115–121 (2017)
40. Blockchain white paper. https://www.caict.ac.cn/kxyj/qwfb/bps/201809/P020180905517892312190.pdf. Accessed 7 Jun 2020
41. Zhu, J., Fu, Y.: Progress in blockchain application research. Sci. Tech. Rev. **35**(13), 70–76 (2017)
42. Sun, J., Huang, D., Li, X.: The new progress of science data managing and sharing service in China. World Sci-tech Res. Develop. **5**, 15–19 (2002)
43. Zhang, N., Zhong, S.: Mechanism of personal privacy protection based on blockchain. J. Comput. Appl. **37**(10), 2787–2793 (2017)
44. Zhang, L., Li, J.: Research data openness: development, models and new exploration. Big Data Res. **6**, 25–33 (2016)
45. Peters, G.W., Panayi, E.: Understanding modern banking ledgers through blockchain technologies: future of transaction processing and smart contracts on the internet of money. In: Tasca, P., Aste, T., Pelizzon, L., et al. (eds.) Banking Beyond Banks and Money: A Guide to Banking Services in the Twenty-First Century, pp. 239–278. Springer, Heidelberg (2016). https://doi.org/10.1007/978-3-319-42448-4_13
46. Xing, W., Hong, F., Li, X.: Interpretation of scientific data management rule from the two-dimensional perspective of data life cycle and responsibility stakeholder's. Library Inf. Serv. **63**(23), 30–37 (2019)
47. Gu, J., Xu, X.: Design and implementation of a humanities and social sciences data sharing model: a case study of consortium blockchain. J. Chin. Soc. Sci. Tech. Inf. **38**(4), 354–367 (2019)
48. Tian, Y.: Semantic sharing mechanism of heterogeneous government data based on blockchain. J. Library Inf. Sci. Agric. **32**(1), 12–22 (2020)
49. Proof-of-Work Proves Not to Work. https://www.cl.cam.ac.uk/~rnc1/proofwork.pdf. Accessed 13 Jun 2020
50. PPCoin: Peer-to-Peer Crypto-Currency with Proof-of-Stake. https://peercoin.net/assets/paper/peercoin-paper.pdf. Accessed 15 Jun 2020
51. Delegated Proof-of-Stake (DPoS) Explained. https://www.mycryptopedia.com/delegated-proof-stake-dpos-explained/.htm. Accessed 15 Jun 2020
52. Practical Byzantine Fault Tolerance. https://dts-web1.it.vanderbilt.edu/~dowdylw//courses/cs381/castro.pdf. Accessed 20 Jun 2020

53. Han, X., Liu, Y.: Research on the consensus mechanisms of blockchain technology. Net Inf. Secur. **9**, 147–152 (2017)
54. Garay, J., Kiayias, A., Leonardos, N.: The Bitcoin backbone protocol: analysis and applications. In: Oswald, E., Fischlin, M. (eds.) EUROCRYPT 2015. LNCS, vol. 9057, pp. 281–310. Springer, Heidelberg (2015). https://doi.org/10.1007/978-3-662-46803-6_10
55. Gatteschi, V., Lamberti, F., Demartini, C.G., et al.: Blockchain and smart contracts for insurance: is the technology mature enough? Fut. Internet **10**(2), 20 (2018)
56. Swan, M.: Blockchain: Blueprint for a New Economy. O'Reilly Media, Sebastopol (2015)
57. Shao, Q., Jin, C., Zhang, Z., et al.: Blockchain: architecture and research progress. Chin. J. Comput. **41**(5), 969–988 (2018)
58. Zheng, Z., Xie, S., Dai, H., et al.: An overview of blockchain technology: architecture, consensus, and future trends. In: 6th International Congress on Big Data, pp. 557–564. IEEE, Piscataway (2017)
59. Narayanan, A., Bonneau, J., Felten, E.: Bitcoin and Cryptocurrency Technologies. Princeton University Press, Princeton (2016)
60. Zhu, L., Gao, F., Shen, M., et al.: Survey on privacy preserving techniques for blockchain technology. J. Comput. Res. Develop. **54**(10), 2170–2186 (2017)
61. Yuan, Y., Zhou, T., Zhou, A., et al.: Blockchain technology: from data intelligence to knowledge automation. Acta Automatica Sinica **54**(10), 2170–2186 (2017)
62. Wilkinson, M.D., Dumontier, M., Aalbersberg, I.J., et al.: The FAIR guiding principles for scientific data management and stewardship. Sci. Data **3**(1), 160018 (2016)
63. Li, Y., Wen, L., Zhang, L., et al.: The status and trends of scientific data sharing systems. J. Agric. Big Data **1**(4), 86–97 (2019)

# SCT-CC: A Supply Chain Traceability System Based on Cross-chain Technology of Blockchain

Yong Wang[1,2], Tong Cheng[1,2(✉)], and Jinsong Xi[1,2]

[1] School of Computer Science and Information Security,
Guilin University of Electronic Technology, Guilin 541004, China
1132482532@qq.com
[2] Guangxi Key Laboratory of Cryptography and Information Security,
Guilin University of Electronic Technology, Guilin 541004, China

**Abstract.** In the traditional supply chain traceability model, a centralized management method is prevalent in supply chain traceability system. Problems such as falsification of production data, difficulty in information sharing, and opaque data have become more and more prominent. Blockchain technology has the characteristics of decentralization, non-tampering, openness, transparency and traceability, and can overcome many shortcomings of centralized traceability systems. Building supply chain traceability system based on blockchain technology provides an effective way solving the problems arousing in the supply chain safety domain. We propose SCT-CC, a supply chain traceability system based on cross-chain technology of blockchain. First, we use cross-chain technology to design a multi-chain architecture. Then, we design smart contracts for different agencies in the supply chain. Finally, we use Hyperledger fabric as a development framework and built an experimental environment using Docker technology. We also test the query interface and the write interface. Through the analysis of the test results, the system can meet the needs of actual applications.

**Keywords:** Supply chain · Blockchain · Cross-chain · Hyperledger fabric

## 1 Introduction

Supply chain traceability refers to the tracing of information at the stages of processing, production, transportation, and sales of commodities [1]. Over the years, the global supply chain has become more and more complex, the competition between countries for control of the supply chain has become increasingly fierce, the continuous adjustment of the supply chain system and the accumulation of risks in the global supply chain have become the main cause of trade frictions among countries. The problem of counterfeit and shoddy products has always been a pain point to be solved urgently in all walks of life. The emergence of these

© Springer Nature Singapore Pte Ltd. 2021
W. Gao et al. (Eds.): FICC 2020, CCIS 1385, pp. 281–293, 2021.
https://doi.org/10.1007/978-981-16-1160-5_22

counterfeit and shoddy products is closely related to the product supply chain management of the transformation of primary raw materials into end products. Therefore, managing product quality from the source is the most important link in the later product quality assurance. At present, the supply chain traceability system is a centralized system, and all data is stored in a central database, which makes the traceability information may be tampered with.

The emergence of blockchain technology provides a solution to the problem of supply chain traceability. Blockchain technology has the characteristics of decentralization, non-tampering, openness and transparency, and traceability [2]. However, the blockchain technology also has bottlenecks, and the biggest constraint comes from its poor scalability [3]. This paper proposes a supply chain traceability system based on blockchain cross-chain technology. The system adopts a multi-chain architecture design. The system architecture consists of a main chain and multiple parallel side chains. The system is developed on the basis of the Hyperledger Fabric framework, and the system performance is tested [4].

## 2    Related Work

### 2.1    Blockchain Technology

In 2008, Satoshi Nakamoto published the article "Bitcoin: A Peer-to-Peer Electronic Cash System", thus proposing the concept of blockchain [5]. The core advantage of blockchain technology is decentralization, which can realize decentralized transactions in distributed systems through the use of data encryption, time stamping, distributed consensus, and economic incentives. Figure 1 represents the structure of the block in the chain. In the blockchain, each data block contains two parts: a block header and a block body. The block header encapsulates information such as the previous block hash, nonce, timestamp, and merkle root. The block body records the transaction data, each transaction in the set is converted into a hash value and the hash values are further combined to obtain the merkle root [6, 7].

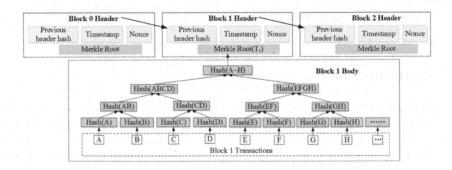

**Fig. 1.** Structure of blockchain.

## 2.2   Blockchain in Supply Chain

In recent years, supply chain traceability issues have received increasing public attention because more and more products have quality and safety issues. The emergence of blockchain technology provides new ideas for supply chain research [8]. At present, there have been some researches on supply chain traceability based on blockchain technology. For example, Tian F. [9] proposed an agri-food supply chain traceability system Based on RFID & Blockchain Technology. This system realizes the monitoring, tracing and traceability management for the quality and safety of the agri-food "from farm to fork". Chaodong et al. [10] designed and implemented the supply chain traceability system based on sidechain technology. The system realizes cargo management, information sharing and product traceability in the supply chain through the Ethereum smart contract. Jamil et al. [11] propose a novel drug supply chain management using Hyperledger Fabric based on blockchain technology to handle secure drug supply chain records.

## 2.3   Cross-chain Technology

The cross-chain technology is an important technical means of blockchain, which has the characteristics of interoperability and scalability. Buterin V. [12] summarized three cross-chain technologies: notary schemes, side chains/relays, and hash-locking. In recent years, with the continuous enrichment of blockchain application scenarios, more and more blockchain projects have proposed cross-chain requirements and solutions, and cross-chain technology has gradually developed. For example, Ethereum designed BTCRelay [13], realized the cross-chain access of Ethereum to the Bitcoin blockchain. Kwon proposed a network architecture that supports multiple block linking and interoperability Cosmos [14]. Its goal is to create a blockchain Internet that allows a large number of blockchains to extend and interact with each other. Spoke et al. [15] propose a design plan for Aion network, the Aion Network is designed to support custom blockchain architectures, while providing a trustless mechanism for cross-chain interoperability.

To the best of our knowledge, many cross-chain projects developed by blockchain companies are mainly used in the financial industry, and cross-chain technology is rarely used in supply chain traceability projects. Therefore, we design a supply chain traceability system based on cross-chain technology of blockchain. This system has the following innovations:

(1) Use blockchain technology to solve the problems of information fraud and data falsification in traditional supply chain traceability systems.
(2) Design corresponding smart contracts according to different participants, which not only ensures the openness and transparency of information, but also greatly reduces the workload of corresponding departments.
(3) Design a blockchain network with a multi-chain architecture. The product information in the supply chain is stored in parallel side chains, and the hash value generated by the product information is synchronized to the main chain to reduce the storage pressure of the main chain.

## 3    Design and Implementation of SCT-CC

### 3.1    System Architecture

The system architecture of SCT-CC is shown in Fig. 2. The system consists of five layers: Acquisition Layer, Application Layer, Contract Layer, Network Layer and Data Layer.

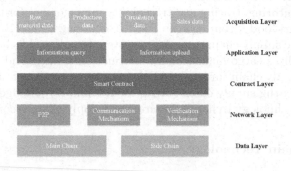

**Fig. 2.** System architecture.

- Acquisition Layer: The supply chain consists of supplier, manufacturer, transporter and retailer that collect information about the raw materials, production, distribution and sales of goods through IoT devices, and then upload the information about the products.
- Application layer: The application layer provides the interface for query and upload. The query interface is used to query the traceability information of product and display the product information to users through website, app and other forms. The upload interface is responsible for uploading the information collected by the collection layer to the blockchain.
- Contract layer: Smart contracts are deployed on blockchain nodes. That provide rules for institutions in the supply chain that can access the blockchain through smart contracts.
- Network layer: The blockchain nodes communicate with each other through P2P network protocol, and use the communication mechanism and verification mechanism to spread and verify the block data.
- Data layer: In order to reduce the storage pressure of SCT-CC, multi-chain architecture is used to build the data layer, which consists of a main chain and multiple side chains.

As shown in Fig. 3, the network architecture of SCT-CC includes main chain network, side chain network and external network.

- Main Chain Network: It is the heart of the system and is run by regulators in the supply chain. Each side chain network needs to be connected to the main chain network.

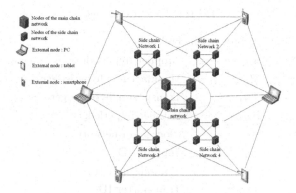

**Fig. 3.** Network topology diagram.

- Side Chain Network: SCT-CC has multiple side chain networks, and each side chain network is jointly operated by suppliers, manufacturers, logistics companies, and retailers. These companies upload the collected product information to the side chain network.
- External Node: It is an open network, and any organization or individual can access the external network according to certain rules. However, the external network only has access rights, and visitors can query product information of the supply chain through smart phones, computers and other tools.

### 3.2 Smart Contract

Smart contract is a program that runs in blockchain node and can be automatically executed according to preset conditions [16,17]. SCT-CC has four types of participants, including suppliers, manufacturers, transporters, retailers and ordinary users. The smart contract algorithm is mainly divided into the following five parts:

**Raw Material Input Smart Contract:** Suppliers can log in to SCT-CC by $S_{ID}$, if there is no identity information of this supplier in the blockchain network, it is not allowed to enter the system. After the login is successful, the supplier can create, update and query the raw material information. The automatically execution of supplier in smart contracts is shown in Algorithm 1. The abbreviations used in the algorithm are shown in Table 1.

**Product Processing Smart Contract:** The manufacturer processes the raw materials provided by the supplier and then produces the product. During the production process, production data will be generated, and relevant personnel need to upload the production data of the product to the chain. The smart contract algorithm is shown in Algorithm 2.

**Table 1.** Abbreviations.

| Abbreviation | Full name |
|---|---|
| $S_{HL}$ | Supplier hyperledger |
| $M_{HL}$ | Manufacturer hyperledger |
| $T_{HL}$ | Transporter hyperledger |
| $R_{HL}$ | Retailer hyperledger |
| $B_N$ | Blockchain network |
| $S_{ID}$ | Supplier ID |
| $M_{ID}$ | Manufacturer ID |
| $T_{ID}$ | Transporter ID |
| $R_{ID}$ | Retailer ID |
| $R_{info}$ | Raw material information |
| $P_{data}$ | Production data |
| $T_{info}$ | Transporter information |
| $P_{ID}$ | Product ID |

---

**Algorithm 1.** Algorithm on Supplier Working.

---

**Input:** Raw Material Info($R_{info}$)
**Output:** Get access to $S_{HL}$ transactions
1: **if** $S_{ID} \in B_N$ **then**
2:    **if** $R_{ID} \in S_{ID}$ **then**
3:       $Update(S_{ID},R_{ID},R_{info})$
4:       $Query(S_{ID},R_{ID},R_{info})$
5:    **else**
6:       $Create(S_{ID},R_{ID},R_{info})$
7:    **end if**
8: **else**
9:    $NotExist(S_{ID})$
10: **end if**

---

**Algorithm 2.** Algorithm on manufacturer Working.

---

**Input:** Production Data($P_{data}$)
**Output:** Get access to $S_{HL}$ transactions
1: **if** $M_{ID} \in B_N$ **then**
2:    **if** $Exist(P_{data})$ **then**
3:       $R_{ID} \leftarrow Query(M_{ID})$
4:       $Update(M_{ID},R_{ID},P_{data})$
5:    **else**
6:       $R_{ID} \leftarrow Query(M_{ID})$
7:       $Create(M_{ID},R_{ID},P_{data})$
8:    **end if**
9: **else**
10:    $NotExist(M_{ID})$
11: **end if**

---

**Product Transportation Smart Contract:** The transporter records the logistics information of the product. When the product is shipped, transport personnel will upload logistics information. As shown in Algorithm 3.

---
**Algorithm 3.** Algorithm on transporter Working.

---
**Input:** Transport information($T_{info}$)
**Output:** Get access to $S_{HL}$ transactions
1: **if** $T_{ID} \in B_N$ **then**
2:    **if** $Exist(T_{info})$ **then**
3:       $UpdateTransportInfo(T_{ID}, T_{info})$
4:    **else**
5:       $CreateTransportInfo(T_{ID}, R_{ID}, T_{info})$
6:    **end if**
7: **else**
8:
9: **end if**

---

**Product Sales Smart Contract:** The retailer is responsible for product sales and upload product sales information to the blockchain. Since the entry of product sales information is similar to the entry of transportation information, retailer's smart contract algorithm reference Algorithm 3.

**Product Query Smart Contract:** Ordinary users can query product information without identity authentication, but ordinary users only have the authority to query. The smart contract algorithm is shown in Algorithm 4.

---
**Algorithm 4.** Algorithm on ordinary user Working.

---
**Input:** Product ID($P_{ID}$)
**Output:** Get access to product information
1: **if** $Exist(P_{ID})$ **then**
2:    $Query(P_{ID})$
3: **else**
4:    $NotExist(P_{ID})$
5: **end if**

---

### 3.3 Cross-chain Mechanism

SCT-CC has multiple side chain networks. After suppliers, manufacturers, transporters, and retailers upload product information to the side chain, the side chain needs to synchronize the product information to the main chain. In order

to reduce the storage pressure of the main chain, the product raw material information, production data, transportation information, retail information, etc. are
used as the leaf nodes of the merkle tree, and the hash value (merkle root) is
calculated, and then the merkle root, product ID and side chain address information will be synchronized to the main chain. When users need to query product
information, SCT-CC verifies the merkle root of the main chain and the product information of the side chain. After the verification is passed, the product
information in the main chain and the side chain are consistent and have not
been tampered with. In order to ensure the consistency and non-tampering of
product information, we have designed a cross-chain mechanism. In our proposed
mechanism, we adopted two stages: data synchronization and data verification.
The process of mechanism is described briefly below:

**Data Synchronization:** After the user initiates a write request, the product
information is processed on the side chain, and then the side chain initiates a
synchronization request. The steps to synchronize product information from the
side chain to the main chain are described as follows:

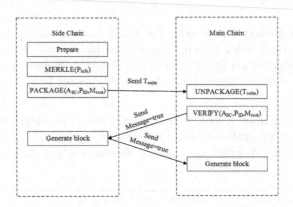

**Fig. 4.** Data synchronization mechanism.

(1) The side chain receives the Write request, and then gets into prepare stage,
and will use the merkle algorithm [18] to hash the product information,and
finally obtains the hash value of the merkle root.
(2) The hash value of merkle root ($M_{root}$), the address number of the side chain
($A_{SC}$), and the product ID ($P_{ID}$) will be packaged into $T_{write}$ by the side
chain, and then $T_{write}$ will be sent to the main chain.
(3) After receiving the $T_{write}$ ,the main chain will unpack $T_{write}$ into $A_{SC}$, $P_{ID}$
and $M_{root}$, verify that the transaction is reasonable. However, the product
information will not be written into block before the result is confirmed by
side chain. Then the main chain send the verification result to the side chain.

(4) After the side chain receives the verification result, the product information will agree to be written into the block. Finally, the side chain sends the confirmation result to the main chain.

(5) After the confirmation result of the side chain received by the main chain, $A_{SC}$, $P_{ID}$ and $M_{root}$ are written into the block, and finally completes the synchronization of product information to the main chain.

**Data Verification:** After the user initiates a query request, the side chain obtains product information from the main chain, and then verifies the product information. The steps to verify product information are described as follows:

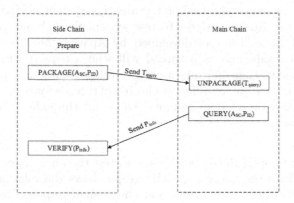

**Fig. 5.** Data verification mechanism.

(1) The side chain receives the query request, and then gets into prepare stage. $A_{SC}$ and $P_{ID}$ will be packaged into $T_{query}$ by the side chain, and then the $T_{query}$ will be sent to the main chain.

(2) After receiving $T_{query}$, the main chain will parse $T_{query}$ into $A_{SC}$ and $P_{ID}$, and query the corresponding product information. The main chain returns the product information $P_{info}$ to the side chain.

(3) After receiving the $P_{info}$, the side chain needs to compare and verify the $P_{info}$ with the local product information. After the verification is passed, the product information is displayed to the user.

## 4   Experiment Analysis

### 4.1   Experimental Environment

In the experiment, Hyperledger Fabric [19] is used as the underlying framework for the SCT-CC. At the same time, We set up the side chain network and the main chain network, and put all the nodes installing the smart contract into the Docker container to run. The experimental environment configuration is shown in Table 2.

**Table 2.** Experimental environment configure.

| Environment | Configuration |
| --- | --- |
| CPU | Intel(R) Xeon(R) Gold 6128 CPU @ 3.40 GHz |
| Memory | 16 G |
| System | CentOS 7 |
| Test tool | JMeter |

## 4.2    Performance Analysis

Throughput is an important basis for evaluating system performance. In this paper, we will use Apache JMeter to test system throughput. Apache JMeter is a java-based stress test tool developed by Apache [20]. We use JMeter to open multiple threads, and each thread will send a request within a specified time, thereby simulating multiple users initiating transaction requests to the blockchain. In the test, we set different number of threads for query test and wirte test, and recorded the median response time and throughput under different number of threads.

### A.Query Test

We use JMeter to open 10 to 200 threads to test the query operation. Figure 6 shows the result of the query test. The graph shows the relationship between the throughput of query operations and the median response time. When the number of threads is between 0 and 120, the response time increases slowly, and the TPS gradually increases. When the number of threads reaches 120, TPS gradually stabilizes and fluctuates around 110, while the response time increases rapidly, indicating that the system throughput has reached its peak at this time. Through the analysis of the experimental results, we get that the system throughput of query operation is 110, which can basically meet the needs of practical applications.

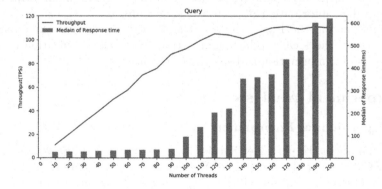

**Fig. 6.** SCT-CC query test.

## B. Write Test

Write operation is much more complicated than query operation. Write operation needs to generate blocks. The block generation process is affected by parameters such as Consensus Protocol, Logging Level, Message Count, Block Size, etc. Therefore, before testing, we configure the experimental parameters, as shown in the Table 3.

**Table 3.** Experimental parameter configuration.

| Consensus protocol | Logging level | Message count | Block size(KB) |
|---|---|---|---|
| Solo | INFO | 10 | 512 |

To test the write operation, we use Jmeter to start 10 to 200 threads to store data in the blockchain. Figure 7 shows the test results of write. We can see the change in throughput and median response time under different number of threads. When the number of threads is between 0 and 100, the response time increases slowly and the throughput gradually increases. When the number of threads is between 100 and 200, the response time increases rapidly and the throughput increases slowly. When the number of threads reaches 120, the throughput gradually stabilizes and fluctuates around 102. Through the analysis of the write test results, we get that the system throughput of the write interface is 102TPS, which can basically meet the needs of practical applications.

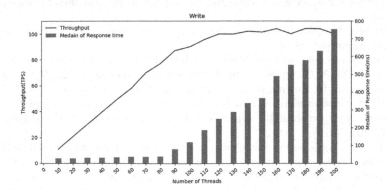

**Fig. 7.** SCT-CC write test.

## 5   Conclusion

In this paper, we propose a supply chain traceability system based on blockchain cross-chain technology called SCT-CC. The system can provide services for governments, enterprises, and consumers. We design smart contracts for suppliers, manufacturers, transporters, retailers and ordinary users in the supply chain. Institutions and users in the SCT-CC access the blockchain through smart contracts, thus ensuring the security and stability of the system. On this basis, we build a multi-chain architecture and design a cross-chain mechanism, which not only ensures the consistency and immutability of product information, but also reduces the storage pressure on the main chain. Finally, through experimental test and analysis, the feasibility of the scheme is verified, and the scheme can meet the needs of practical applications.

**Acknowledgment.** This research was supported by Guangxi Ministry of Education, Guangxi Colleges and Universities Key Laboratory of Cloud Computing and Complex System(YF17105), Guangxi Key Laboratory of Cryptography and Information Security and Guangxi Key Laboratory of Trusted Software.

# References

1. Wei, Y.: Research on information sharing and security in supply chain management. Value Eng. **36**(9), 86–88 (2017)
2. Pilkington, M.: Blockchain technology: principles and applications. In Research Handbook on Digital Transformations. Edward Elgar Publishing (2016)
3. Yu, H., Nikolić, I., Hou, R., Saxena, P.: Ohie: blockchain scaling made simple. In: 2020 IEEE Symposium on Security and Privacy (SP), pp. 90–105. IEEE (2020)
4. Carvalho, H., Azevedo, S., et al.: Trade-offs among lean, agile, resilient and green paradigms in supply chain management: a case study approach. In: Xu, J., Fry, J., Lev, B., Hajiyev, A. (eds.) Proceedings of the Seventh International Conference on Management Science and Engineering Management, pp. 953–968. Springer, Berlin (2014)
5. Nakamoto, Satoshi: Bitcoin: A peer-to-peer electronic cash system. Technical report, Manubot (2019)
6. Bitcoin sourcecode. https://github.com/bitcoin/bitcoin/ Accessed 18 July 2020
7. Yuan, Y., Ni, X., Zeng, S., Wang, F.: Blockchain consensus algorithms: the state of the art and future trends. Acta Automatica Sinica **44**(11), 2011–2022 (2008)
8. Saberi, S., Kouhizadeh, M., Sarkis, J., Shen, L.: Blockchain technology and its relationships to sustainable supply chain management. Int. J. Prod. Res. **57**(7), 2117–2135 (2019)
9. Tian, F.: An Agri-food supply chain traceability system for china based on RFID & blockchain technology. In: 2016 13th International Conference on Service Systems and Service Management (ICSSSM), pp. 1–6. IEEE (2016)
10. Chaodong, Z., Baosheng, W., Wenping, D.: Sidechain scheme based supply chain traceability system design and implementation. Comput. Eng. **45**(11), 1–8 (2019)
11. Jamil, F., Hang, L., Kim, K.H., Kim, D.H.: A novel medical blockchain model for drug supply chain integrity management in a smart hospital. Electronics **8**(5), 505 (2019)
12. Buterin, V.: Chain interoperability. R3 Research Paper (2016)
13. Chow, J.: Btc relay. btc-relay (2016)
14. Kwon, J., Buchman, E.: Cosmos: a network of distributed ledgers. https://cosmos. network/whitepaper (2016)
15. Spoke, M., NE Team, et al.: Aion: Enabling the decentralized internet. AION, White Paper, July 2017

16. Wang, S., Ouyang, L., Yuan, Y., Ni, X., Han, X., Wang, F.-Y.: Blockchain-enabled smart contracts: architecture, applications, and future trends. IEEE Trans. Syst. Man Cybern. Syst. **49**(11), 2266–2277 (2019)
17. Buterin, V., et al. A next-generation smart contract and decentralized application platform. White paper 3(37), (2014)
18. Szydlo, M.: Merkle tree traversal in log space and time. In: Cachin, C., Camenisch, J.L. (eds.) EUROCRYPT 2004. LNCS, vol. 3027, pp. 541–554. Springer, Heidelberg (2004). https://doi.org/10.1007/978-3-540-24676-3_32
19. Androulaki, E. et al.: Hyperledger fabric: a distributed operating system for permissioned blockchains. In: Proceedings of the thirteenth EuroSys conference, pp. 1–15 (2018)
20. Halili, E.H.: Apache JMeter: A Practical Beginner's Guide to Automated Testing and Performance Measurement for Your Websites. Packt Publishing Ltd, Birmingham (2008)

# Game-Theoretic Analysis
# on CBDC Adoption

Chenqi Mou[1,3](✉), Wei-Tek Tsai[2,4], Xiaofang Jiang[2], and Dong Yang[2]

[1] LMIB – School of Mathematics and Systems Science, Beihang University,
Beijing 100191, China
chenqi.mou@buaa.edu.cn
[2] SKLSDE – School of Computer Science and Engineering, Beihang University,
Beijing 100191, China
tsai7@yahoo.com, {jxf120,yangdong2019}@buaa.edu.cn
[3] Beijing Advanced Innovation Center for Big Data and Brain Computing,
Beihang University, Beijing 100191, China
[4] School of Computing, Informatics, and Decision Systems Engineering,
Arizona State University, Tempe, USA

**Abstract.** As an important blockchain application, CBDC (Central
Bank Digital Currency) has received significant worldwide attention as
it can restructure financial market, affect national currency policies, and
introduce new regulation policies and mechanisms. It is widely predicted
that CBDC will introduce numerous digital currency competitions in
various aspects of the global financial market, and winners will lead the
next wave of digital currency market. This paper applies the game theory to study the competitions between different countries, in particular
to analyze whether they should adopt the CBDC program. We propose
two game-theoretic models for CBDC adoption, both analyzing whether
to adopt the CBDC program via the Nash equilibrium. Both game-
theoretic models draw the same conclusion that each country should
adopt the CBDC program regardless of the choices of other counties.
In other words, current currency leaders should adopt CBDC because
it may lose the premier status, and other countries should adopt CBDC
otherwise they risk of getting even further behind in the digital economy.
According to our game-theoretic models, the current market leader who
has 90% of market shares may lose about 19.2% shares if it is not the
first mover.

**Keywords:** Central Bank Digital Currency · Game theory ·
Blockchain · Currency competition

This work was partially supported by National Basic Research Program of China
(Grant No. 2013CB329601), National Natural Science Foundation of China (Grant
No. 61672075 and 61690202), Chinese Ministry of Science and Technology (Grant No.
2018YFB1402700), Key Research and Development Program of ShanDong Province
(Grant No. 2018CXGC0703), and LaoShan government of QingDao Municipality.

# 1    Introduction

CBDC (Central Bank Digital Currency) has received significant attention recently, and it is an important blockchain application [4,9,17–19]. Many central banks including US Federal Reserve, ECB (European Central Bank), PBOC (People's Bank of China), Bank of England have all announced their CBDC projects. Numerous financial institutions such as World Bank, IMF (International Monetary Fund), BIS (Bank for International Settlements) have initiated their own CBDC projects. In 2017, IMF recommended that each country develop her own CBDC program to compete with the cryptocurrencies.

Bank of England (BoE) is the first central bank to propose her own CBDC program, and since 2014 has published numerous research reports such as RTGS (Real-Time Gross Settlement) blueprints and conducted many experiments. One of earliest conclusions of their studies is that there may be a competition between commercial banks and central banks for deposits because now central banks offer CBDC, and customers may prefer to save their money as CBDC as it is without any credit risks, while money in commercial banks still carries risks. This is one of many competitions that may happen when CBDC is introduced.

Since 2015, BoE has suggested that CBDC will have a profound impact to the financial world, affect national currency policies, and change the way financial transactions are made. Indeed, the importance of CBDC cannot be underestimated. In 2017, Bordo and Levin predicted that if a country does not develop its own CBDC program, the country will incur financial risks in future [5][1].

In November 2019, Rogoff of Harvard University said currently there is a new currency war, but this is not a conventional currency war, but a digital version [15]. He also mentioned that the new war involves of both regulated financial market and underground market, where cryptocurrencies such as Bitcoins are often used. He also mentioned that technology is the key driver for these changes.

In addition to the central bank-commercial bank competition for deposits, IMF have suggested stablecoins issued in the private sector can have a profound impact to the financial world. Stablecoins are those digital currencies backed by fiat currencies or other assets such as bonds, but they have stable price, and can be used do cross-border transactions. The IMF July 2019 report [1] claims if a stablecoin is supported by a central bank, it behaves like a CBDC. This new digital currency is called synthetic CBDC (short as sCBDC hereafter), rather than CBDC, but like CBDC the reserve money is stored in a central bank so that sCBDC does not have any credit risks. But in this manner, the IMF report says that there will be competitions between those sCBDC with fiat currency as people may prefer using those stablecoins rather than their fiat currency. This

---

[1] "Given the rapid pace of payment technology innovation and the proliferation of virtual currencies such as bitcoin and Ethereum, it may not be wise for the central bank to adopt a negative attitude in dealing with CBDC. If the central bank does not produce any form of digital currency, there is a risk of losing monetary control, and the possibility of a serious economic recession is greater. Because of this, central banks act quickly when considering adopting CBDC.".

will also create another competition between commercial banks and stablecoin entities, both seeking for deposits.

In August 2019, former BoE governor Mark Carney said that a synthetic hegemony digital currency may replace US dollar as the world's reserve currency. This is another competition where an sCBDC may compete with US dollars or any other fiat currency to be the world's reserve currency.

In June 2020, US Federal Reserve published a research paper [7] supporting the claim of deposit competition between central banks and commercial banks. If this is the case, the financial market will indeed be fundamentally changed due to introduction of CBDC or sCBDC.

Furthermore, stablecoins or sCBDC may compete with each other in the market place, for example, a large stablecoin will have strong competitive edge over other stablecoins. For example, in May 2020, ECB issued a report predicting that Facebook's Libra may have over 3 trillion Euros in deposit and it may become the largest fund in Europe. The ECB made this prediction based on the fact that Facebook is widely used in the world, many Facebook users will become Libra users.

Thus, many competitions are created due to CBDC or sCBDC:

- Between central banks and commercial banks for deposits;
- Between stablecoin entities and commercial banks for deposits;
- Between stablecoins and fiat currencies;
- Between stablecoins and stalecoins; and
- Between different CBDCs or sCBDCs.

This paper is our first attempt in a series to theoretically study various competitions introduced by CBDC or sCBDC from the viewpoint of game theory [8]. As done in the game-theoretic analysis on software crowdsourcing [10,11,13], this paper applies the game theory to investigate whether a specific country will attempt to develop its own CBDC program. Developing a CBDC program is a challenge task because this will affect national monetary policies, regulator policies, technology solutions, security issues as well as consumer protection issues. Some countries may not have the technology know-how to develop her own technologies, and for this reason Facebook Libra 2.0 offers to assist any country to establish her own CBDC programs.

This paper makes the following contributions: game-theoretic models are applied to analyze whether a country should adopt the CBDC program with respect to beneifts and cost; and according to these models, each country will eventually commit to her own CBDC program regardless, confirming the IMF recommendation made in 2017. In particular, the detailed game-theoretic model provides explicitly, for a simplified market of two countries of 90% and 10% market shares respectively, the changes of market shares for all the four possible cases of whether they adopt CBDC or not (see Table 3 for the details). It is interesting to see that in such settings, if the current market leader does not adopt CBDC, it risks 19.2% of the economic market share. This confirms the conjecture made by Bordo and Levin in 2017 that the CBDC has a serious impact on national economy system and global market leadership.

This paper is organized as follows: Sect. 2 reviews the existing work related to CBDC; Sect. 3 and Sect. 4 present our simple and detailed game-theoretic models for CBDC adoption respectively; Sect. 5 reports our implementation and preliminary experimental results with the detailed game-theoretic model; Sect. 6 concludes this paper with remarks and future work.

## 2    Related Work

### IMF Report

IMF issued a report "The Rise of Digital Money" in July 2019 [1], and one of the main theses is that people will choose different forms of money due to convenience. The report proposes two key concepts: 1) sCBDC where reserve money will be placed in central banks so that stablecoins will be risk free; 2) proposes three stages of financial market restructuring: coexistence, complementarity, and substitution. The June 2020 Federal Reserve Report [7] further confirms that the 3rd stage, i.e., substitution, is a likely event as central banks will have monopoly of deposits. In this case, financial market structure is fundamentally changed.

### Digital Currency Areas

Brunnermeier, James and Landau of Princeton University proposed this Digital Currency Areas (DCA) theory in 2019 [6]. Some of their key findings include: 1) Platforms become the center of financial market: traditionally the centers are banks, but once Internet-based platforms become the financial center, financial markets are significantly restructured. Furthermore, those who manage those platforms will have significant economic advantages; 2) Digital dollarization, i.e., digital currencies will compete with fiat currencies; 3) Digital fragmentation: different parts of the world will run different digital currencies due to stiff competitions among different digital currencies; 4) Role of digital money: digital currencies and fiat currencies have overlapping but different emphasis as digital currencies are mainly used in transactions including cross-border transactions. This theory has received significant attention as BoE, ECB, and Federal Reserve have quoted and discussed this theory publicly.

### Federal Reserve Report

In June 2020, the Philadelphia Federal Reserve released a research report "Central Bank Digital Currency: Central Banking for all?" [7]. This report uses a game theory model, confirming the theory by BoE that there will be competition between central banks and commercial banks for deposits. It also points out that the current two-tiered system with a central bank and commercial banks came after the World War II, before, central banks can play the role of lending (currently done by commercial banks). Thus, it is not inconceivable that commercial banks provide different services. In other words, the current banking structure can be changed if necessary due to CBDC.

The models used include consumers, banks and central banks according to various scenarios. Yet, over time, people will choose to deposit their money into their central bank as the money will be risk free as money in commercial banks carries some risks. In this case, central banks have monopoly of deposit, and commercial banks need to provide different services.

**Libra Stablecoin**
On June 18th 2019, Libra Association released the whitepaper [2], and created a great of discussions in the world, especially among central banks and commercial banks. On April 16th, 2020, the Association released Libra 2.0 whitepaper [3] (hereinafter referred to as "Libra 2.0"). According to the whitepaper of Libra 2.0, if a country or region is worried that its own fiat currency can be replaced by Libra, Libra will cooperate with the central bank to establish the CBDC for the country. Libra 2.0 will no longer pursue public blockchain route, instead it will follow FATF regulations, such as Travel Rules. Libra 2.0 will also incorporate embedded supervision mechanisms to monitor transactions in real.

# 3   Simple Game-Theoretic Model

For a specific country, it has two choices to adopt CBDC, or not. To start it simple, we assume there are only two countries–$C_1$ and $C_2$, and they compete in the financial market.

Suppose that $E_1$ is the benefit country $C_1$ gained for a successful domestic CBDC project, and $H_1$ is the damage to country $C_2$ incurred of a successful foreign CBDC project in country $C_1$. They are of the similar meanings for $E_2$ and $H_2$.

The benefits gained for a successful domestic CBDC project include improving its international financial position, gaining the reserve currency status, providing the same services and sharing platform to countries that cannot develop their own CBDC as well as managing the digital currency platform. According to the DCA theory, the entity or the country that owns the platform will have significant advantage over others who do not. There are other benefits as well.

The damage incurred of a successful foreign CBDC project include deterioration of international monetary position, and possibility of a major financial crisis, lose of ability to manage national currency policies.

In digital economy, the first mover will have significant advantages over followers according to the Davidow law. This law says that the first product of a class to reach a market automatically gets a 50% market share. If CBDC economic model follows the Davidow law, this means those entities that move first will dominate the market, and this will apply to CBDC issued by central banks, or CBDC by private parties. In this case, if country (or entity) $C_1$ develops her CBDC program, while country (or entity) $C_2$ does not, $C_1$ will gain significant advantages over $C_2$ by owing more than 50% of market share. According the DCA theory, $C_1$ will have significant economic benefit over $C_2$.

Assume that $\alpha_i$ is the success rate for country $C_i$ to develop CBDC program for $i = 1, 2$. For example, if $C_1$ has superior technology, then $\alpha_1$ will be close to 1. The game model between these two countries are expressed in the following Table 1.

**Table 1.** Simple game-theoretic model for CBDC adoption of two countries $C_1$ and $C_2$

| $C_1$ | $C_2$ | |
|---|---|---|
| | Develop CBDC | Do not develop CBDC |
| Develop CBDC | $(E_1\alpha_1 - H_2\alpha_2, E_2\alpha_2 - H_1\alpha_1)$ | $(E_1\alpha_1, -H_1\alpha_1)$ |
| Do not develop CBDC | $(-H_2\alpha_2, E_2\alpha_1)$ | $(0,0)$ |

In the case that country $C_2$ decides to develop CBDC program: if $C_1$ decides to develop, it will get the payoff of $E_1\alpha_1 - H_2\alpha_2$; if $C_1$ decides not, it will get $-H_2\alpha_2$. Obviously $E_1\alpha_1 - H_2\alpha_2$ is larger than $-H_2\alpha_2$, and thus $C_1$ should develop her own CBDC program.

If $C_2$ decides not to develop her CBDC program, then $C_1$ should still choose to develop CBDC, because the benefit of choosing is $E_1\alpha_1$, while the benefit of not developing CBDC is 0, obviously $E_1\alpha_1$ is larger. Thus, no matter how $C_2$ chooses, $C_1$ should develop her own CBDC program.

Similarly, regardless whatever $C_1$ chooses, ultimately country $C_2$ needs to develop her own CBDC program. In this way both countries will develop CBDC programs as a Nash equilibrium for this model.

However, $C_1$ and $C_2$ will have different benefit $E_1$, $E_2$ and damage $H_2$, $H_1$. If $C_1$ is powerful economically, but does not develop her CBDC program, the damage that may be incurred will be significant because potentially it loses her premier status. However, if $C_1$ develops CBDC, but country $C_2$ does not, $C_2$ will be lagging further behind.

The success probability $\alpha_i$ is a function of time and increases with time as more research results will be available over time, and each country eventually will be able to develop her own CBDC program. An advanced country may reach 1 before other countries, and achieve Davidow advantages. But eventually, both $\alpha_1$ and $\alpha_2$ will be close to 1, and this means that every country will be able to develop her own CBDC program. Yet the first mover will own market shares.

# 4    Detailed Game-Theoretic Model

Next we aim at establishing a detailed game-theoretic model to analyze the benefits or losses of a country or an entity in the choice of adopting CBDC. The benefits or losses are represented merely by the change of the market shares, and this is because the cost for the economy's transformation to the existence of CBDC is negligible compared with the change of market shares.

## 4.1    Game-Theoretic Settings of the Model

As in Sect. 3, let us focus on the simplest game of 2 players $C_1$ and $C_2$ with current market shares of $M_1$ and $M_2$, where $M_1 + M_2 = 1$. Let $S_1 = S_2 = \{Y, N\}$ be the set of pure strategies for $C_1$ and $C_2$, where $Y$ and $N$ stand for adopting and not adopting CBDC respectively. Then for $i = 1$ and 2, the payoff function $p_i$ for $C_i$ is a mapping from $S_1 \times S_2$ to $[-1,1]$, and $p_i(s)$ represents $C_i$'s payoff given a profile $s = (s_1, s_2)$ of pure strategies $s_1 \in S_1$ and $s_2 \in S_2$.

For $i = 1$ and 2, a *mixed strategy* $\sigma_i$ of $C_i$ is a probability distribution over the pure strategies in $S_i$, where $\sigma_i(s)$ is the probability for $C_i$ to choose a pure strategy $s \in S_i$. Denote the space of mixed strategies of $C_i$ by $\Sigma_i$. We can define the payoff $\tilde{p}_i(\sigma)$ of $C_i$ for a mixed strategy profile $\sigma = (\sigma_1, \sigma_2) \in \Sigma_1 \times \Sigma_2$ as

$$\tilde{p}_i(\sigma) = \sum_{(s_1,s_2) \in S_1 \times S_2} \sigma_1(s_1)\sigma_2(s_2)p_i(s_1,s_2),$$

which is essentially the expected payoff of $C_i$ for the probability distribution $\sigma$.

A mixed strategy profile $\sigma^* = (\sigma_1^*, \sigma_2^*) \in \Sigma_1 \times \Sigma_2$ is called a *Nash equilibrium* of the game if $\tilde{p}_1(\sigma_1^*, \sigma_2^*) \geq \tilde{p}_1(s_1, \sigma_2^*)$ and $\tilde{p}_2(\sigma_1^*, \sigma_2^*) \geq \tilde{p}_2(\sigma_1^*, s_2)$ for any pure strategy $s_1 \in S_1$ and $s_2 \in S_2$. A Nash equilibrium is a state such that each player in the game maximizes his expected payoff under the condition that the mixed strategies of the other players are fixed, and thus anyone attempting to change his mixed strategy from the Nash equilibrium will face a reduced payoff. For a finite non-cooperative game, at least one Nash equilibrium exists [14].

In this model we will compute the Nash equilibrium of the game to reveal the probability of a certain player to adopt CBDC. To do this, we need explicit expressions for the payoff functions.

## 4.2  Construction of Payoff Functions

We consider a dynamic game in the time $t$. The players of the game $C_1$ and $C_2$, together with their pure strategies $\{Y, N\}$, keep unchanged regardless of the time $t$. But their payoff functions $p_1$ and $p_2$ indeed change with $t$, as explained as follows. To simplify our model, the time is discretized to take only non-negative integers. It may help to assume a unit interval $[t_0, t_0 + 1]$ of time to be 3 months in the real world.

For each player $C_i$ in our game, there is a probability of success $Prob_i(t)$ if he chooses to adopt CBDC. This probability has an initial value $Prob_i(0) = P_i$ and it increases with time (to mimic the increasing storage of underlying related knowledge and technologies to adopt CBDC). We assume that the increase rate is a constant $c$ in time for both $C_1$ and $C_2$, and thus we know that the probability of success for $C_i$ is $Prob_i(t) = \min\{ct + P_i, 1\}$ for $i = 1, 2$. Once a player $C_i$ chooses to adopt CBDC at a certain time $\bar{t}$, there is a preparatory duration of $d$ units of time, and this means that $C_i$ will succeed at the time $\bar{t} + d$ with a probability of $Prob_i(\bar{t})$.

Next we discuss the payoff of each player with respect to the four possible combinations of pure strategies. Basically the payoff of $C_1$ in this game is to gain the market share of $C_2$ if $C_1$ chooses to adopt CBDC and succeeds while $C_2$ fails or chooses not to adopt it, and vice versa. Suppose that at a certain time $\bar{t}$, $C_1$ finishes its preparatory duration and succeeds in adopting CBDC, then $C_1$ will take a percentage of $C_2$'s market share at $\bar{t}$ in the next time interval $(\bar{t}, \bar{t} + 1)$.

Denote by $M_1(t)$ and $M_2(t)$ the market shares of $C_1$ and $C_2$ with respect to $t$, and let $r$ $(0 < r < 1)$ be the percentage of gaining (or losing) the market share in one unit time interval. Then in the above scenario,

$$M_1(\bar{t}+1) = M_1(\bar{t}) + rM_2(\bar{t}), \qquad M_2(\bar{t}+1) = (1-r)M_2(\bar{t}).$$

And the payoff of $C_1$ in the interval $(\bar{t}, \bar{t}+1)$ is $rM_2(\bar{t})$, while that for $C_2$ is $-rM_2(\bar{t})$.

**Combination I**, $(N, N)$ at $\bar{t}$: in this case the payoffs are $p_1(N, N)(\bar{t}) = p_2(N, N)(\bar{t}) = 0$.

**Combination II**, $(Y, Y)$ at $\bar{t}$: in this case both $C_1$ and $C_2$ choose to adopt CBDC, and then their payoffs depend on which succeeds first. Since the function of success probability is $Prob_i(t) = \min\{ct + P_i, 1\}$ which linearly grows to 1, we know that the maximum possible length of time in which there exists one and only one player who succeeds in adopting CBDC is $[\bar{t}+d, T]$, where

$$T := \bar{t} + d + \max\left\{ \lceil \frac{1 - Prob_1(\bar{t})}{c} \rceil, \lceil \frac{1 - Prob_2(\bar{t})}{c} \rceil \right\}.$$

For each integer $t \in [\bar{t}+d, T]$, we can explicitly compute the probability $\tilde{Prob}_i(t)$ for $C_i$ to succeed in adopting CBDC at $t$ for the first time (which implies that $C_i$ fails in the interval $[\bar{t}+d, t-1]$). See Table 2 for an illustrative example of the computed success probabilities.

With $\tilde{Prob}_1(t)$ and $\tilde{Prob}_2(t)$ for $t \in [\bar{t}+d, T]$, we can enumerate each possible combination of the two first success time $T_1, T_2 \in [\bar{t}+d, T]$ of $C_1$ and $C_2$ respectively, together with the possibility $P_{T_1, T_2}$ for this combination to happen. In this way, we can compute the expected payoff of $C_1$, for example (that for $C_2$ is just the opposite):

$$p_1(Y, Y)(\bar{t}) = \sum_{T_1, T_2 \in [\bar{t}+d, T]} P_{T_1, T_2} \cdot \left( \begin{array}{ll} M_2(\bar{t}) r \cdot \sum_{j \in [0, T_2 - T_1]} (1-r)^j & \text{when } T_2 > T_1 \\ -M_1(\bar{t}) r \cdot \sum_{j \in [0, T_1 - T_2]} (1-r)^j & \text{otherwise.} \end{array} \right).$$

**Combination III**, $(Y, N)$ at $\bar{t}$: in this case we assume that $C_2$ does not realize the importance of adopting CBDC until he has lost a considerable percentage $m$ $(0 < m < 1)$ of his market share. In other words, $C_2$ starts to adopt CBDC only after he loses $m$ of his market share. Let $T_1$ $(\bar{t}+d \le T_1 \le \bar{t}+d+\lceil \frac{1-Prob_1(\bar{t})}{c} \rceil)$ be the time when $C_1$ succeeds in adopting CBDC and $T_2 = T_1 + \min\{t \in \mathbb{Z}_{\ge 0} : (1-r)^t < 1-m\}$ be the time when $C_2$ first loses at least $m$ of his market share. Then the earliest time for $C_2$ to succeed in adopting CBDC falls in the range $\left[ T_2 + d, T_2 + d + \lceil \frac{1-Prob_2(\bar{t})}{c} \rceil \right]$. For each $t \in \left[ T_2 + d, T_2 + d + \lceil \frac{1-Prob_2(\bar{t})}{c} \rceil \right]$, similarly to the case above we can compute the probability and $\tilde{Prob}_2(t)$ for $C_2$ to succeed in adopting CBDC at $t$ for the first time. Then the payoff of $C_1$ is

$$p_1(Y, N)(\bar{t}) = \sum_{t \in \left[ T_2+d, T_2+d+\lceil \frac{1-Prob_2(\bar{t})}{c} \rceil \right]} \tilde{Prob}_2(t) \cdot M_2(\bar{t}) r \cdot \sum_{j \in [0, t-T_1]} (1-r)^j.$$

**Combination IV**, $(N, Y)$: the dual case as III above.

With the payoff functions of $C_1$ and $C_2$ known, we can compute the Nash equilibrium of the game. Let $\boldsymbol{\sigma}^* = (\sigma_1^*, \sigma_2^*) \in \Sigma_1 \times \Sigma_2$ be the Nash equilibrium of the game, then $\sigma_1^*(Y)$ (a probability) is the tendency for $C_1$ to choose to adopt CBDC, and $\sigma_2^*(Y)$ is that for $C_2$.

# 5    Implementation and Experiments

In this section, we first fix the values for our parameters listed in the above section to have an illustrative model, with the interval $[t, t+1]$ representing 3 months in the real world in mind.

- Market shares $M_1 = 0.9$, $M_2 = 0.1$: country $C_1$ takes the overwhelmingly dominant part of the market.
- Preparation duration $d = 6$: one country needs 18 months to prepare the adoption of CBDC.
- Initial success probabilities $Prob_1(0) = 0.5$ and $Prob_2(0) = 0.3$, and the increase rate $r$ of the success rate is set to 0.05.
- The loss rate of market share (in 3 months) $r = 0.1$: this means that one country loses 10% of its market share in 3 months if the other player succeeds but it does not.
- Loss of market shares for awareness $m = 0.3$.

With the above setup, let us work out the payoff of $C_1$ in combination II in Sect. 4.2. In this case, both $C_1$ and $C_2$ choose to adopt CBDC at $t = 0$. Then at $t = 6$ (after 18 months, the preparation duration), the probability for $C_1$ to succeed is 0.5 while that for $C_2$ is 0.3. Furthermore, let us calculate the probability $\tilde{Prob}_i(t)$ for $C_i$ to first succeed at time $t$, as in the following table.

**Table 2.** An example illustrating the probabilities of first success of two countries at different time

| $t$ | $\tilde{Prob}_1(t)$ | $\tilde{Prob}_2(t)$ |
|---|---|---|
| 6 | 0.5 | 0.3 |
| 7 | $0.55 \times 0.5$ | $0.35 \times 0.7$ |
| $i$ $(8 \leq i \leq 15)$ | $(0.5 + 0.05(i-6)) \times \prod_{j=0..i-7}(0.5 - 0.05j)$ | $(0.3 + 0.05(i-6)) \times \prod_{j=0..i-7}(0.7 - 0.05j)$ |
| $i$ $(16 \leq i \leq 20)$ | $\prod_{j=0..9}(0.5 - 0.05j)$ | Same as above |
| $\geq 21$ | Same as above | $\prod_{j=0..13}(0.7 - 0.05j)$ |

For example, from this table, we will be able to read that the probability for $C_1$ to succeed at $t = 6$ AND $C_2$ to succeed at $t = 7$ for the first time is $0.5 \times 0.245 = 0.1225$. In such combination the whole payoff of $C_1$ is $(7-6) \times 3 = 3$ months of eating $C_2$'s market share, that is $r \times M_2 = 0.01$. Similarly, the payoff of $C_2$ is $-0.01$. Traversing all the possible combinations in the above table, we will be able to compute the expected payoffs of $C_1$ and $C_2$ for the pure strategy profile $(Y, Y)$.

We have implemented an algorithm to compute the explicit payoffs for $C_1$ and $C_2$ with respect to all the 4 combinations of pure strategies. For the parameter values listed above, the computed payoffs are as shown in Table 3.

**Table 3.** Payoffs of $C_1$ and $C_2$ for the listed parameter values above

| $C_1$ | $C_2$ | |
|---|---|---|
| | Develop CBDC | Do not develop CBDC |
| Develop CBDC | $(-0.020990, 0.020990)$ | $(0.031578, -0.031578)$ |
| Do not develop CBDC | $(-0.192320, 0.192320)$ | $(0, 0)$ |

It is interesting that Table 3 shows, if country $C_1$ with the current dominant market shares (90%) does not adopt the CBDC program while the other country $C_2$ with 10% market shares adopts it, then $C_1$ will lose around 19.2% of the total market shares, which are too significant to lose for $C_1$. This result indeed partially confirms that the CBDC economic model follows the Davidow law, that is the first country adopting the CBCD program will take a large amount of the market shares from those who do not adopt.

Next we input the payoff functions in Table 3 into the software Gambit [12] for computing the Nash equilibria, and the Nash equilibria is $(1, 0)$ for $C_1$ and $(1, 0)$ for $C_2$, which means that both $C_1$ and $C_2$ will choose to adopt CBDC with probability 1. This accords with the analytical result we derive in Sect. 3.

# 6    Concluding Remarks and Future Work

CBDC and sCBDC have received significant attention by major central banks as well as bigTech such as Facebook have embraced this new technology. These will not only provide technology breakthrough but also restructure financial markets and change national currency policies. This paper has analyzed whether a given country will commit to developing her own CBDC program given the benefits for developing the CBDC and potential damages due to inactivity for not developing the program. The main conclusion is that regardless if current technology status of the country, eventually the country will commit to developing the CBDC due to competitions with other countries.

The detailed game-theoretic model indicates that, in a simplified market of two countries of 90% and 10% market shares respectively, if the current market leader does not adopt CBDC, it risks 19.2% of the economic market share. This confirms the conjecture made by Bordo and Levin in 2017 that the CBDC has a serious impact on national economy system and global market leadership. Furthermore, if one country is not the current market leader, then it has all the incentives to become the first mover of CBDC as it can gain 19.2% market shares from the current leader. According to the New Lanchester Strategy [20], if an entity achieves 41% of the market shares, it may become the new market leader.

This paper assumes that benefits associated with CBDC is real, and damages is also real, and success rates improve over time. The first two assumptions are realistic as evidenced by the impact of Facebook's Libra on central banks and commercial banks in 2019. Furthermore, blockchain or related technologies have improved significantly during the last twelve months with new theories, new architecture, new frameworks, and new regulation technologies having emerged. For example, even Libra has improved her regulation mechanism to include embedded supervision. Thus, these three assumptions are indeed realistic.

As we mentioned in the introduction, this paper is our first attempt to theoretically study various competitions introduced by CBDC from the viewpoint of game theory. The impact of different values of the parameters in our detailed game-theoretic model on the expected payoffs of the players based on our further experiments will be reported in a forthcoming paper. These further experimental results will tell, for example, how the initial success rates influence the payoffs of the players in the game, which essentially reflects the importance of the underlying technological competence of the countries with respect to CBDC adoption. The simple game-theoretic model of two players can also be extended to a multiple-player game for more sophisticated analysis, which is part of our future work. We would like to mention that some computational difficulties in such extension to a multiple-player game lie in the rapidly growing numbers of possible combinations of first success time and the increasing hardness of computing Nash equilibria when the number of players is large [16].

# References

1. Adrian, T., Mancini-Griffoli, T.: The Rise of Digital Money. No. 19/001 in FinTech Notes, International Monetary Fund (2019)
2. Libra Association: Libra white paper (2019). https://libra.org/en-US/whitepaper/ Accessed Jun 2019
3. Libra Association: Libra white paper (2020). https://libra.org/en-US/whitepaper/ Accessed Apr 2020
4. Bai, X., Tsai, W.T., Jiang, X.: Blockchain design-A PFMI viewpoint. In: 2019 IEEE International Conference on Service-Oriented System Engineering (SOSE), pp. 146–14609. IEEE (2019)
5. Bordo, M., Levin, A.: Central Bank Digital Currency and the Future of Monetary Policy. Technical report No. w23711, National Bureau of Economic Research (2017)
6. Brunnermeier, M., James, H., Landau, J.P.: Digital currency areas. Vox CEPR Policy Portal (2019)
7. Fernández-Villaverde, J., Sanches, D., Schilling, L., Uhlig, H.: Central Bank Digital Currency: Central Banking For All? Technical Report No. w26753, National Bureau of Economic Research (2020)
8. Fudenberg, D., Tirole, J.: Game Theory. MIT Press, Cambridge (1991)

9. He, J., Wang, R., Tsai, W.T., Deng, E.: SDFS: a scalable data feed service for smart contracts. In: 2019 IEEE 10th International Conference on Software Engineering and Service Science (ICSESS), pp. 581–585. IEEE (2019)
10. Hu, Z., Wu, W.: A game theoretic model of software crowdsourcing. In: IEEE 8th International Symposium on Service Oriented System Engineering, pp. 446–453. IEEE (2014)
11. Li, W., Huhns, M.N., Tsai, W.T., Wu, W. (eds.): Crowdsourcing. PI. Springer, Heidelberg (2015). https://doi.org/10.1007/978-3-662-47011-4
12. McKelvey, R., McLennan, A., Turocy, T.: Gambit: software tools for game theory (2006)
13. Moshfeghi, Y., Rosero, A.F.H., Jose, J.: A game-theory approach for effective crowdsource-based relevance assessment. ACM Trans. Intell. Syst. Technol. 7(4), 1–25 (2016)
14. Nash, J.: Non-cooperative games. Annals of Mathematics pp. 286–295 (1951)
15. Rogoff, K.: The high stakes of the coming digital currency war. Project Syndicate (2019)
16. Roughgarden, T.: Algorithmic game theory. Commun. ACM **53**(7), 78–86 (2010)
17. Tsai, W.T., Ge, N., Jiang, J., Feng, K., He, J.: Beagle: a new framework for smart contracts taking account of law. In: 2019 IEEE International Conference on Service-Oriented System Engineering (SOSE), pp. 134–13411. IEEE (2019)
18. Tsai, W.T., Wang, R., Liu, S., Deng, E., Yang, D.: COMPASS: a data-driven blockchain evaluation framwework. In: 2020 IEEE International Conference on Service Oriented Systems Engineering (SOSE), pp. 17–30. IEEE (2020)
19. Tsai, W.T., Zhao, Z., Zhang, C., Yu, L., Deng, E.: A multi-chain model for CBDC. In: 2018 5th International Conference on Dependable Systems and Their Applications (DSA), pp. 25–34. IEEE (2018)
20. Yano, S.: New Lanchester Strategy. Lanchester Press Inc. Pennsylvania (1995)

# Design of Experiment Management System for Stability Control System Based on Blockchain Technology

Xiaodan Cui[1,2]([✉]), Ming Lei[1], Tianshu Yang[1], Jialong Wu[1], Jiaqi Feng[1], and Xinlei Yang[1]

[1] NARI Group Corporation (State Grid Electric Power Research Institute), Nanjing 211106, China
cuixiaodan@sgepri.sgcc.com.cn

[2] State Key Laboratory of Smart Grid Protection and Control, Nanjing 211106, China

**Abstract.** Stability control system (SCS) plays a very important role in the operation of power systems, so it is correspondingly essential to ensure the reliability of the SCS during its experimental and verification process. However, there are numerous problems in the management of the paper form based experiment mode of SCS today, such as long circulation cycle caused by low work efficiency, insecure and diseconomy on storage. More importantly, it is difficult to query and review historical records quickly, and it may even lead to the loss of records for various reasons. In this paper, taking advantage of blockchain technology, all the information of each experimental stage of SCS, including the approval information, experimental environment information, experimental equipment information of each manufacturer, experimental test reports and conclusions, are packaged and distributed managed. And then, a blockchain based experimental management system framework with the unified data form and module interface is proposed, in which the whole process of process record can be traced and cannot be tampered with. The experimental management system can be applied in the whole process of the SCS, including the experiment scheme making, experiment environment constructing, experiment processing, result evaluating, and results storage, etc. The blockchain based mechanism and system framework is expect to take many advantage to experiment management of SCS.

**Keywords:** Stability control system(SCS) · Blockchain · Experiment and verification · Management system design

## 1 Introduction

### 1.1 A Subsection Sample

Security and stability control system (referred to as "stability control system" or "SCS") plays an extremely important role in power system security and stability in china [1]. An incorrect action of SCS under serious failure of power system may directly cause an

W. Gao et al. (Eds.): FICC 2020, CCIS 1385, pp. 306–317, 2021.
https://doi.org/10.1007/978-981-16-1160-5_24

instability event or even a blackout accident of power system [2]. In order to ensure the security and stability of power system, and to prevent the occurrence of blackout accident, the SCS has become fundamental configuration of major power transmission and power supply construction engineering. Within the rapid development of China's power grid, especially the development of UHV AC and DC transmission construction, the security and stability characteristics of power system are becoming more and more complicated, and the risk of losing systemic stability under large disturbances of power grid is really high [3–5]. In order to adapt the evolution of stability problems like involving larger area or more stability issues, the architecture and control strategy of the SCS are getting more and more complicated, which may bring potential hidden dangers to the SCS itself [6–10]. Therefore, it is necessary to carry out experimental verification of SCS meticulously to ensure the reliability.

At present, there are some relevant standards to guide the management of SCS experiment [11]. However, the current management methods are still dominated by manual and administrative commands, which are relatively extensive, not automated enough, and there are problems such as low safety and reliability, poor traceability. In view of the defects in the current management of SCS verification, it is necessary to propose a new advanced and applicable technology of experimental verification management of SCS by integrating the advanced information technology. So, blockchain technology is applied to the experimental verification management of SCS in this paper to improve the reliability and automation level of SCS experiment. On the basis of necessity and feasibility analysis, an experimental management system is designed, by which all information needs to be stored in a secure and tamper-resistant manner and the history data and operations can be traced conveniently.

## 2 Necessity and Feasibility of Blockchain Technology Application on SCS Experiment Management

### 2.1 Brief Introduction of Blockchain Technology

The concept of blockchain comes from the article "Bitcoin: A Peer-to-Peer Electronic Cash System" published by a self-proclaimed Satoshi Nakamoto in 2008 [12], which was originally invented as the underlying technology and infrastructure of Bitcoin. The blockchain is a distributed ledger, and collectively maintains a reliable database through decentralization and trust [13–16], without relying on a third party, the information is verified and uploaded to distributed nodes through an encryption algorithm to form a chain structure storage [17]. Blockchain technology combines multidisciplinary achievements such as cryptography principles, probability theory, and computers [18, 19], which has the advantages like, data cannot be tampered, no trust is required, and it has a mechanism of decentralization [13, 14], which has been applied in the fields of finance, entities, industry services, etc. [20].

### 2.2 Necessity Analysis

The experimental verification management of SCS mainly includes four stages, namely, experimental system construction, experimental scheme, experimental results evaluation. The major element of each stage are as follows.

1) In the stage of determination of experimental boundary conditions, the SCS, the power grid the SCS defensing for, the content to be verified and analyzed, and the requirements for the construction of the experimental environment according to the experimental needs, are determined. All of these information is the prerequisite and basis for experimental scheme and experimental system construction.

2) In the stage of experimental scheme formulation, the equivalent scheme of SCS, the power grid, and verification scenarios are formulated according to the experimental boundary conditions.

3) In the stage of experimental environment construction, a software and hardware environment including the primary power grid model and functionally equivalent SCS is built, which passed the joint debugging and structural integrity assessment.

4) In the stage of experimental results evaluation, the content and result of all tests are confirmed, recorded and preserved in some form.

Since the experimental verification of SCS involves many physical originations, like dispatch management departments, dispatch departments (users of SCS), equipment manufacturers, third-party verification organization, the current mode of experimental management of SCS is extremely difficult.

1) Trust issues. For example, equipment manufacturers are reluctant to disclose the actual parameters and internal test details of their own equipment under unnecessary circumstances to prevent competitors from imitation or infringement.

2) Efficiency issues. Under the paper documents relied mode, the management of various work needs to be carried out through the way of approval layer by layer. The workload of each link is large, the efficiency is low, and the circulation period is extremely long for the reasons of human or material resources.

3) Security and cost issues. The storage of paper application forms is unsafe and takes up a lot of storage and management costs. It is also difficult to meet the needs of quick query and retrieval of historical records, and the records may lost due to various reasons [21, 22].

It can be seen that above issues are directly related to the effectiveness and efficiency of the experimental verification of SCS, and it is urgent to change the current management model and improve the management technology significantly.

## 2.3 Feasibility Analysis

The experiment process involves many interested or responsible institution or departments. The data security, operational standardization, and reliability of the experimental results in the whole process of experimental verification are the focus of attention of every institution or departments.

For example, when the test system environment is constructed, the device model or software program version of SCS directly affects the test verification results of its function or performance. Due to inadequate management or lack of effective means, some equipment manufacturers may tamper with samples or data during the experiment process for the purpose of speeding up the project progress or enhancing product reputation, and even there is a situation where the test sample system is inconsistent with site operation system, which seriously violates the conventions for fairness that many relevant institutions or departments need to comply with. For another example, whether the experimental boundary conditions are clear and the experimental operation is standardized are all required to be transparently visible and arrived at a consensus by the regulatory agency, SCS users, and equipment manufacturers.

It can be seen that the experiment of SCS requires multiple efforts and due diligence to achieve fairness and justice, and the significance and role of the experiment can be fully exerted. In view of the characteristics of the blockchain's own structure, support for data encryption, and the trust and sharing mechanism [23–25], it is an effective way to solve many of the focus issues mentioned above.

## 3    Design of Stability Control Experiment Management System Based on Blockchain

Based on the above analysis, a management system for SCS experiment management is designed based on blockchain technology in this section, including the definition of related concepts, system architecture and specific business processes.

### 3.1    Definition of Related Concepts

(1)    Alliance members

The management system for SCS experiment management will be designed based on the alliance chain technology application, which mainly involves the members of the alliance chain such as the dispatch management department, equipment manufacturers, dispatch department, and third-party verification institution. The responsibilities of each member are as follows.

1)    The dispatch management department is the approval and supervision agency in the alliance chain of the experimental verification process, responsible for the approval and supervision of the block information entered into the chain in the four stages of the experimental process.

2)    The dispatching department is the initiator in the alliance chain of the experiment verification process, approving the experiment plan and experiment result evaluation, and supervising the blockchain information in other stages.

3)    The third-party verification institution is the program formulation unit in the alliance chain of the experimental verification process, responsible for the construction of experimental test environment and experimental testing and forming experimental results, supervising the experimental boundary conditions and software version of the SCS.

4)    The equipment manufacturers are participating units in the alliance chain of the stability verification system experimental verification process, providing the software version of the SCS, confirming and supervising the experimental scheme system and experimental result information in the alliance chain.

(2)   Consensus mechanism

Different from the consensus of the decentralized longest chain mechanism achieved by the original Bitcoin blockchain through "mining" calculation [12, 26], the reliability of the nodes involved in the SCS experimental management system is relatively high, and the main precautions are to avoid situations like carrying out the experiment without confirming the experimental boundary, modifying the experimental scheme without permission after being determined, starting the experiment before the experimental version and test environment being well established, etc. So the consensus mechanism designed mainly includes the following aspects.

1)   After receiving the experimental boundary application from the dispatching operation department, including the data of power grid and the SCS, the content need to be verified and analyzed, and the requirements of experimental environment construction, the dispatch management department is responsible for checking if the applicant is reasonable and the data is intact, and whether the applicant have the application authority according to the pre-authorization setting. If all information is correct, the dispatch management department then verify and approve the role signature from the applicant, and send confirmation information to the equipment manufacturer, dispatch operation department, and third-party verification department.

2)   The equipment manufacturer is responsible for checking if any problems are found after receiving any request or data from other members. If not, send a confirmation message, otherwise broadcast an alarm. The request or data from other members maybe the experimental boundary condition information from the dispatching operation department, the experimental plan approved by the dispatch management department and the dispatching operation department, and the test environment evaluation result sent by the third-party verification department.

3)   The dispatching and operating department will send the experimental boundary condition information to the dispatching and management department for approval, and at the same time receives the dispatching and management department's experimental boundary condition approval information, as well as the experimental plan and test environment assessment results sent by the third-party verification department.

4)   The third-party verification department will receive the experimental boundary condition information, formulate the experimental scheme, the equipment manufacturer's SCS software version, and send the experimental plan to the dispatching and operating department, the test environment construction information to other members. The third-party verification department will send the experimental results to other members, after confirmation form to the dispatch management department and dispatch operation department, the new block will be issued on the chain notification and new blocks will be added.

(3)   Smart contract

Smart contracts record the rights and obligations of participants in the form of contracts defined by digital logic. In the process of the experimental verification management of SCS, the smart contract is constructed by the consensus algorithm of the blockchain and written into the blockchain in a digital form.

(4)  Data block structure

The data block structure contains the content involved in the experimental verification process of SCS, as shown in Fig. 1.

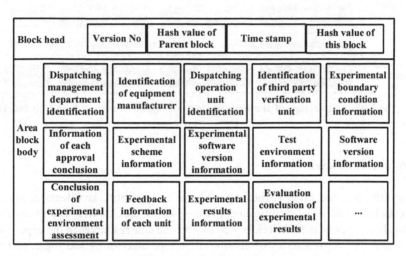

| Block head | Version No | Hash value of Parent block | Time stamp | Hash value of this block |
|---|---|---|---|---|

| Area block body | Dispatching management department identification | Identification of equipment manufacturer | Dispatching operation unit identification | Identification of third party verification unit | Experimental boundary condition information |
|---|---|---|---|---|---|
| | Information of each approval conclusion | Experimental scheme information | Experimental software version information | Test environment information | Software version information |
| | Conclusion of experimental environment assessment | Feedback information of each unit | Experimental results information | Evaluation conclusion of experimental results | ... |

**Fig. 1.** Block data structure

## 3.2  System Architecture Design

The basic architecture of the experimental management system is built based on the blockchain technology, as shown in Fig. 2, which is composed of the infrastructure layer, the data network service layer, the consensus layer, and the application layer connected in sequence from the bottom up. The data network service layer is the foundation of the experimental process data management system. Each blockchain node connects the infrastructure layer, consensus layer and application layer. Blockchain technology is applied to encapsulate the information involved in the experimental management process, therefore, the data management of experimental verification process could be distributed stored and managed safely and reliably.

Specifically, mature blockchain solutions can be learned in the actual construction of SCS experimental management system, avoiding the development of the underlying blockchain technology from zero. Currently, BaaS (Blockchain as a Service) is a feasible and mature new cloud service [27–29], which can quickly build the development environment required by the system construction described in this article, while providing block-based Chain search, query, analysis and other series of operation services.

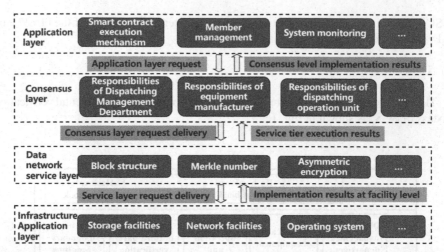

**Fig. 2.** The basic architecture of the experimental management system based on blockchain technology

(1)  Infrastructure layer

The infrastructure layer is the bottom layer of the entire architecture, including storage facilities, network facilities, operating systems, etc., as well as a decentralized network. Each node of the network is used to store the data resources in the experimental management. Through a series of distributed storage algorithms and logic, combined with the upper layer data network service layer, consensus layer, and application layer, the reliability, consistency, and integrity of the data in the infrastructure layer are realized.

(2)  Data network service layer

The data network service layer provides blockchain-related services. The core purpose is to convert the experiment management data into the corresponding block and recorded data structure, and to achieve the digital signature function of the blockchain.

Specifically, it mainly provides blockchain ledger maintenance and network transmission functions. The experimental management information is stored in the infrastructure layer, and the key information is also chained through the Merkle tree, in the way of which the unique identification of the experimental management information cannot be tampered, and any attempt behavior to tamper with the experimental management information will be detected.

(3)  Consensus layer

The consensus layer implements the responsibilities of each member. Under the premise of decentralization, each node can reach consensus on the validity of the block data. The consensus mechanism is explained in 2.3 above.

(4)  Application layer

The application layer includes the actual application scenarios of experimental management, in which smart contracts ensure the normal performance of all members' responsibilities.

## 3.3  Analysis of System Operation Process

The overall operation process of SCS experiment management based on blockchain is shown in Fig. 3 below.

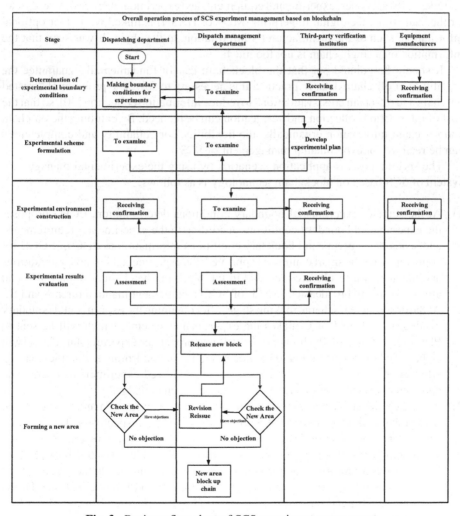

**Fig. 3.**  Business flow chart of SCS experiment management

## 4  Analysis of System Operation Flow Based on Actual Application

The following is an example for illustrating the application process of the proposed management system for actual SCS experiment management. The alliance members include the power dispatching center of SGCC(PDC of SGCC) as the dispatch management department, equipment manufacturers (A and B), East China Power Grid Corporation(ECPG) as the dispatch department, and System Protection Laboratory of SGCC(SPL of SGCC) as a third-party verification department. Blockchain nodes are built for each member to form an alliance chain. And business cooperation relationships are established between all members, and customized services can be carried out on the developed management system. Information that needs to be uploaded to the chain includes software codes, experimental equipment, and experimental process records, etc. At the same time, the digital fingerprint (MD5 or HASH) is linked with other optional information, such as the manufacturer's information, to the chain. It was seen that the information on a single chain is not too much.

It should be pointed out that the blockchain cannot fundamentally guarantee the credibility of off-chain data. In practice, it is necessary to solve the problem of "trusted on-chain and off-chain cross-validation" to reduce information collection [30], so that the cost of information collection and verification can be reduced. By combing the on-chain and off-chain information organically, the flexibility, controllability and completeness can be realized for experimental management of SCS.

The specific process application scenarios by using the experimental management system of SCS based on blockchain technology is as follows.

1) At the stage of experimental boundary conditions determination, ECPG proposes the experimental boundary conditions according to the experimental requirements, and sends the experimental boundary conditions information to PDC of SGCC for approval. After the smart contract is approved, the experimental boundary conditions and the approval results are recorded in the block on the chain account book, and afterwards being broadcasted to SPL of SGCC, equipment manufacturer A and B.

2) At the experimental scheme making stage, after obtaining the experimental boundary conditions, SPL of SGCC makes the experimental scheme, which will be sent to PDC of SGCC and ECPG for approval. After approval, the experimental scheme will be provided to equipment manufacturer A and B. for confirming and implementing. After the execution of the smart contract, the approved experimental scheme and approval conclusion will be recorded on the blockchain ledger.

3) At the experimental environment construction stage, equipment manufacturer A and B send their software versions of SCS to SPL of SGCC respectively. The software version and confirmation information are also recorded in the blockchain ledger, which will be broadcasted for supervision to all stakeholders. On this basis, SPL of SGCC builds a test system environment, encapsulates and sends the results of the rationality assessment of the environment construction to PDC of SGCC and ECPG for confirmation, and the process information will be recorded in the blockchain ledger after approval and confirmation.

4) At the stage of the experimental results evaluation, SPL of SGCC sends the experimental results to PDC of SGCC and ECPG, which will be recorded in the blockchain

ledger. Eventually, all the relevant information mentioned above is added to the new blockchain and goes on-chain.

The whole process of experiment management is shown in Fig. 4.

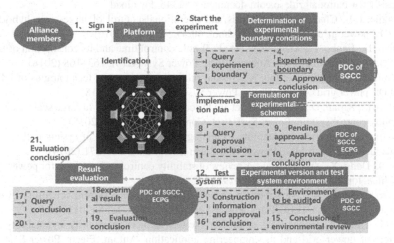

**Fig. 4.** Flow chart of system operation based on actual application

## 5 Conclusion

In order to make up the deficiencies on the current SCS experimental and verification management, the blockchain technology is applied on the design of experimental and verification management system of SCS to create a new model management for SCS. The distributed storage, asymmetric encryption, consensus mechanism, smart contract and other characteristics of the blockchain technology may effectively resolve the related problems of low security trust, low work efficiency, poor automation, etc. By constructing an experimental and verification management system with unified basic data, unified system interface, traceable whole process recording process and non-tampering, the whole cycle management of the experimental management of SCS is realized.

The proposed design method of the experimental system of SCS based on the blockchain technology also has a universal reference significance for the experimental management of other similar large-scale control systems and equipment. The current application of blockchain technology in the experimental management is still in the exploratory stage, and there are few lessons to be learned. It is necessary to continuously innovate or improve its architecture and function design based on the application effect.

**Acknowledgments.** This work is supported by State Grid Corporation of China (Research on System Protection Test and Verification Technology Based on Multi-mode Real-time Co-simulation and Flexible Reconfiguration of Control Strategy, No. 5100-201940008A-0-0-00).

# References

1. Jinhai, S., Xueming, L., Changan, J., et al.: Current situation and development prospect of security and stability control equipment. Autom. Electr. Power Syst. **29**(23), 91–96 (2005)
2. National Grid ESO. Technical report on the events of 9 August 2019 [EB/OL]. [2019-09-06]. https://www.nationalgrideso.com/document/152346/download
3. Mingjie, L.I.: Characteristic analysis and operational control of large-scale hybrid UHV AC/DC power grids. Power Syst. Technol. **40**(4), 985–991 (2016)
4. Xijian, D., Jianbo, L., Xiaodan, C., et al.: Whole control time and its constitution of security and stability control system. Autom. Electr. Power Syst. **42**(5), 163–168 (2018)
5. Guoping, C., Mingjie, L., Tao, X.: System protection and its key technologies of UHV AC and DC power grid. Autom. Electr. Power Syst. **42**(22), 2–10 (2018)
6. Jianbo, L., Xijian, D., Xiaodan, C., et al.: Discussion on reliability of large scale security and stability control system. Power Syst. Prot. Control **46**(8), 65–72 (2018)
7. Yongjie, F.: Reflections on stability technology for reducing risk of system collapse due to cascading outages. J. Mod. Power Syst. Clean Energ. **2**(3), 264–271 (2014)
8. Luo, J., Dong, X., Xue, F., et al.: A review of stability control technologies for power systems in China. Iop Conf. **227**, 032020 (2019)
9. Jiang, K., Singh, C.: New models and concepts for power system reliability evaluation including protection. IEEE Trans. Power Syst. **26**(4), 1845–1855 (2011)
10. Qi, G., Yihua, Z., Dongxu, C., et al.: Remote test method for security and stability control system of power grid and its engineering application. Autom. Electr. Power Syst. **44**(1), 152–159 (2020)
11. Technical guide for electric power system security and stability control: GB/T 26399 -2011. Standards Press of China, Beijing (2011)
12. Nakamoto, S.: Bitcoin: a peer-to-peer electronic cash system (2009). https://bitcoin.org/bitcoin.pdf
13. Bitcoin: Bitcoin core integration [EB/OL] (2017). https://github.com/bitcoin/bitcoin
14. Swan, M.: Blockchain: Blueprint for a New Economy. O'Reilly, Sebastopol (2015)
15. Tapscott, D., Tapscott, A.: Blockchain Revolution: How the Technology Behind Bitcoin and Other Cryptocurrencies is Changing the World. Portfolio, New York (2016)
16. Wattenhofer, R.: The Science of the Blockchain. CreateSpace Independent Publishing Platform, USA (2016)
17. Tschorsch, F., Scheuermann, B.: Bitcoin and beyond: a technical survey on decentralized digital currencies. IEEE Commun. Surv. Tutorials **18**(3), 2084–2123 (2016)
18. Buterin, V.: A next-generation smart contract and decentralized application platform [EB/OL] (2017). https://github.com/ethereum/wiki/wiki/White-paper
19. Tao, W., Yanyun, Z.: Theory and architecture design of statistical blockchain. Stat. Decis. Making **534**(18), 5–9 (2019)
20. Liang, C., Qilei, L., Xiubo, L.: Blockchain technology advancement and practice. People's post and Telecommunications Press, Beijing (2018)
21. Xingpei, J., Yizhong, W.: Reaserch and implementation of the system of multi-disciplinary flow integration and design of experment. Comput. Sci. **40**(11A), 369–373 (2013)
22. Jindong, C., Shengwen, W., Yechun, X.: Research on technical framework of smart grid data management from consortium blockchain perspective. Proc. CSEE **40**(3), 836–847 (2020)
23. Swan, M.: Blockchain: Blueprint for a New Economy. Reilly Media Inc, USA (2015)
24. Markus, D., Vera, D.: Financial Technology: The Promise of Blockchain. IW-Kurzberichte, no. 1 (2017)
25. Pilkington, M.: Blockchain Technology: Principles and Applications. Edward Elgar Publishing, Northampton (2016)

26. Yong, Y., Xiao-Chun, N., Shuai, Z., et al.: Blockchain consensus algorithms: the state of the art and future trends. Acta Automatica Sinica **44**(11), 2011–2022 (2018)
27. Hyperledger. About Hyperledger [EB/OL]. The Linux Foundation Projects, USA (2017). https://www.hyperledger.org/about
28. Dewen, W., Zhiquan, L.: Regional energy transaction model and experimental test based on smart contract. Power Syst. Technol. **043**(006), 2010–2019 (2019)
29. Bo, L., Dan, W., Pengfei, S.: Research on peer-to-peer energy transaction at user-side for microgrid based on baas platform. Electr. Power Constr. **40**(006), 13–22 (2019)
30. Chen, C.: Key technologies of alliance blockchain and regulatory challenges of blockchain. Power Equipment Management, pp. 20–22 (2019)

# A Cross-chain Gateway for Efficient Supply Chain Data Management

Chenxu Wang[1,2]([✉]), Xinxin Sang[1], and Lang Gao[3]

[1] School of Software Engineering, Xi'an Jiaotong University, Xi'an 710049, China
cxwang@mail.xjtu.edu.cn
[2] MoE Key Lab of Intelligent Network and Network Security, Xi'an, China
[3] China Zheshang Bank Co., Ltd., Hangzhou, China

**Abstract.** With the development of economic globalization, a product usually involves different organizations from the origin to retail. These organizations form a supply chain to collaborate to provide products to end-users. The arrival of the digital age offers excellent convenience for the supply chain management. However, it is difficult for existing supply chain management systems to share credit, which leads to end-users' concerns about the quality of products. The blockchain technology, characterized by tamper-proof, traceability, and distributed sharing, is expected to solve the it. Also, supply chain management also increasingly depends on the use of the Internet of things. However, the throughput limitation of current blockchain systems significantly constrains their applications to IoT data storage and management. In this paper, we propose a cross-chain solution based on IOTA and Fabric for supply chain data management. We design and implement a supply chain data management system based on a cross-chain gateway of blockchain. The system has three layers, including blockchain layer, business layer, and application layer, which provide essential logistics functions in supply chain management. Finally, we conduct experiments to validate the effectiveness of the implemented system.

**Keywords:** Blockchain · Tangle · Cross-chain gateway · Supply chain data management

## 1 Introduction

### 1.1 Background and Significance

With the emergence of global economic integration, inter-organizational cooperation is becoming more and more common. The production and manufacturing of a product may involve multiple enterprises that form a complex supply chain around the world [1]. Supply chain management systems record how the product is transported from the places of factories to the final recipients. Most current supply chain management systems have a centralized architecture characterized by monopoly, asymmetry, and opacity. This often causes trust issues as the

© Springer Nature Singapore Pte Ltd. 2021
W. Gao et al. (Eds.): FICC 2020, CCIS 1385, pp. 318–333, 2021.
https://doi.org/10.1007/978-981-16-1160-5_25

administrator can tamper with the data at no cost. Besides, a centralized system is prone to a single point of failure and vulnerable to intra-attacks.

Recently, blockchain technology provides a new paradigm for the design of the supply chain management system. A blockchain is composed of a series of blocks that are orderly linked from far to near [2]. In this way, the change of the transaction content of one block will result in the invalidation of all consequential blocks. In addition, blockchain uses standard communication protocols to form a distributed multi-node network, with each network node maintaining a blockchain ledger. The blockchain ledger has the characteristics of distributed sharing, non-tampering, and data traceability, which can solve the problems of trustiness and single point of failure in centralized supply chain management systems. A blockchain-based supply chain management system can record the products flow from the origin to end users seamlessly, eliminating the need to export data from one organization's database to another. End users can track products through the records in the blockchain ledger.

The current centralized access control system is designed to satisfy the traditional man-machine oriented IoT scenarios, and the devices are located in the same trust domain. However, IoT devices in a supply chain are more dynamic than traditional ones. IoT devices are continually moving with products and belong to different management communities in their life cycle. In addition, IoT devices can be managed by multiple parts at the same time [3]. Decentralized access control systems based on blockchain connect geographically dispersed sensor networks while eliminating single point control failures. In conclusion, in supply chain management, the use of blockchain and IoT technology ensures the credibility of the two dimensions on and off the supply chain data chain, further amplifies the data trust in supply chain management and makes product traceability easier.

## 1.2 Research Contents

However, blockchain development is still in its initial stage, and the mainstream blockchain platforms such as Bitcoin [4] and Ethereum [5] suffer low throughput limitation. To improve the performance of blockchain, a variety of extended blockchain platforms have been designed and implemented. Plasma [6] is built on a typical main-and-side chain architecture. Hyperledger Fabric [7] uses a typical sub-parent chain scheme. Although these blockchain platforms have improved the performance, the consensus algorithms used by these platforms cannot support large-scale deployment of blockchain nodes. Simultaneously, another distributed ledger technology based on the DAG (Directed Acyclic Graph) structure is designed and implemented. A typical implementation is the IOTA Tangle network, and the network performance increases with the number of participated nodes [8]. However, the IOTA Tangle network cannot obtain the global state of the ledger. Therefore, it does not support smart contracts and cannot build sophisticated applications.

Considering the low transaction volume of blockchain and the complexity of building a system on the DAG structure's distributed ledger, a feasible solution is to improve the scalability of the system through the cooperation of multiple blockchain networks. However, each blockchain platform is constituted by a closed network, and there is no interoperability between various blockchains. To enable the interaction between different blockchain networks, cross-chain technology has become a research hotspot in blockchain. However, most of existing cross-chain schemes are designed for the financial area. These schemes transfer the certificates from one chain to another. In the supply chain scenario, participants or devices in the supply chain can be deployed to different blockchain platforms. Moreover, the interaction between various chain networks is more about data transfer. Unfortunately, existing cross-chain solutions cannot be applied to this scenario directly.

Based on the above analysis, this paper proposes a supply chain data management model based on a cross-chain gateway of IOTA and Fabric. Firstly, IOTA is employed to collect and store IoT data, which solves the current blockchain system's performance problem. Second, the Hyperledger Fabric is used to build different applications, aiming to solve the problems caused by the centralized IoT access control system. Third, a cross-chain gateway is designed and implemented to connect the two different blockchain platforms. Enabling the applications in the Fabric network could access the data in the IOTA network under specific policies.

The organization of this paper is as follows: Sect. 2 presents the related work. Section 3 describes our solution of the cross-chain gateway. Section 4 shows the experiments to validate our method. We conclude the work in Sect. 5.

## 2    Related Work

While blockchain is still in its early stages of development, researchers see the great potential of blockchain in improving supply chain management. Some literature focuses on the possibility of blockchain in the supply chain, and others analyzing the specific problems in the supply chain, such as product traceability and visibility, and provide solutions for the supply chain management system.

From the perspective of implementation, the underlying platforms contain two kinds of architectures, including single-chain and cross-chain. When designing a single-chain based supply chain system, the underlying blockchain platform is mainly Ethereum or Fabric. Helo et al. [9] proposed an Etherea-based architecture, which clearly separates the blockchain and application layer. The client connects to a web server responsible for caching transactions, executing queries, and maintaining the latest blockchain status. The server connects to the local Ethereum node, which acts as a bridge between the application layer and the blockchain. Toyoda et al. [10] proposed a product ownership supply chain management system based on the RFID technology for anti-fraud. The authors implemented the system based on Ethereum smart contracts. Salah et al. [11] constructed the soybean traceability system using the Ethereum blockchain and the distributed file storage system IPFS [12]. The Ethereum smart contracts were used to manage the business IPFS file storage system to store the crops' growth details. Lin et al. [13] used Ethereum smart contracts to build a food safety traceability system to detect and avoid food safety problems and track accountability effectively. Fabric is another blockchain platform with the most attention. In the industrial IOT system, the product data directly stored on the blockchain will cause the problem of efficiency and privacy of data management. To solve this problem, Qi et al. [14] proposed a blockchain based CDPs architecture and implemented the CDPs prototype on Fabric. CDPs realizes efficient private data sharing through point transaction and data transaction supporting data access on-chain or out-chain, tree-based data compression mechanism and hybrid access control mechanism. In order to solve the problem of price discrimination in ride-hailing service, Lu et al. [15] propose a smart price auditing system named Spas. The system uses smart contract technology to automatically audit and compensates the order price and a built-in accountability system. A trustworthy and transparent distributed price audit system is implemented on Fabric. Malik et al. [16] established a reputation and trust model of the supply chain based on the Fabric platform. Gao et al. [17] proposed a food trade and traceability system based on Fabric. This system gathers all suppliers, including grain storage enterprises, food processing enterprises, and food retailers, to form a complete food supply chain and provide reliable food tracking. To ensure drug safety and prevent fake drugs, Jamil et al. [18] developed a drug supply chain management system based on the Fabric platform. Wal-mart has built a food supply chain based on Fabric, enabling pork from China to upload certificates to the blockchain, which improves the transparency and traceability of food [19]. Biswas et al. [20] constructed a wine supply chain traceability system using a private blockchain to solve the problems of counterfeiting, fraud, and excessive use of chemicals and preservatives in wine supply [21].

However, sometimes the single-chain blockchain architecture fails to meet business requirements. Hence, some literature have proposed a hybrid cross-chain system architecture to improve scalability and enhance privacy protection. Wu et al. [22,23] proposed a blockchain model consisting of a group of private chains, a central public chain, and a central server, which provides directory services for the interaction between the private chain and central open chain nodes. This model technically aims to solve the interoperability of blockchains. In this cross-chain architecture, the directory server faces a centralization problem. In this paper, the cross-chain connected by cross-chain gateway consensus group to solve the problem. To build the product traceability system, Ding et al. [24] proposed a two-layer blockchain architecture based on licensed blockchains to solve the performance problem. The leading blockchain used an alliance chain, while the sub-blockchains use private chains. Leng et al. [25] designed a double-chain model for the agricultural supply chain management system to solve the privacy protection problem. The double-chain model is composed of user information chain and business chain. Malik et al. [26] proposed an extensible product blockchain framework to solve the traceability problem of the supply chain. The model realized the cross-chain mode of the main side chain using sharding technology. Although the above cross-chain scheme considers the business requirements of the supply chain, it does not focus on the acquisition of data under the chain. In this paper, we design and implement a cross-chain scheme based on IOTA and Fabric. The IOTA network is a distributed ledger technology for IoT. IoT devices use iota distributed ledger to carry out large-scale deployment and data storage. Fabric network can obtain and perform access control on IoT data. Finally, the data is traceable throughout the supply chain, thus ensuring its authenticity and correctness.

With the development of cloud platform technology at home and abroad, companies such as Microsoft, IBM, Tencent, and Huawei have built blockchain cloud platforms. These cloud platforms offer multiple blockchain technologies. For example, the Microsoft Azure cloud platform provides Hyperledger Fabric, Ethereum, Corda, and other blockchains. Therefore, developers only need to focus on the specific logic of the supply chain. Figorilli et al. [27] designed an electronic tracking system for a wood chain based on the blockchain using Microsoft azure cloud platform. This system, based on RFID sensors and open source technology, simulated the entire forest wood supply chain and the process from stumps to final products. Supported by IBM cloud and IBM blockchain, TradeLens is a blockchain-based platform for tracking shipping containers and related documents [28]. Although the cloud-based blockchain platform improves blockchain usability, it is not adapted to the requirements of multi-scene blockchain.

## 3   Our Solution

### 3.1   The Cross-chain Framework of IOTA and Fabric

This paper presents a cross-chain supply chain data management solution based on IOTA and Fabric. Figure 1 shows the architecture diagram of the cross-chain scheme. The model includes five layers, including perception layer, data layer, extension layer, control layer, and application layer.

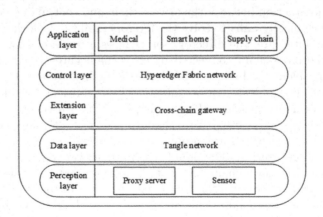

**Fig. 1.** The cross-chain solution diagram of IOTA and Fabric

The Perception Layer The perception layer consists of IoT devices and IoT proxy servers. IoT devices are connected to IoT proxy servers. IoT devices monitor the temperature, location and other data of the products in the supply chain. The data is uploaded to the data layer by IoT proxy servers.

The Data Layer The data layer uses IOTA Tangle network to store IoT data (i.e., product information is collected by the perception layer). The Tangle network provides a MAM channel tool to encrypt and store data in the Tangle network, which provides the privacy protection of IoT data.

The Extension Layer The extension layer is a cross-chain gateway middleware, which is responsible for connecting different blockchain platforms and reliably forwarding cross-chain transactions between different blockchains. That is, the extension layer not only can route the requests from Fabric network to the Tangle network, but also can route the requests in the Tangle network to the Fabric network. Thus, the IOTA Tangle network and Fabric network are connected.

The Control Layer The control layer uses the Fabric network, which provides services for the application layer, and performs access control and intelligent computing through smart contracts. For data query requests, product delivery and receipt request and logistics information from the data layer will be sent to end users through the cross-chain gateway. Other requests will interact directly on the Fabric blockchain network, and the calculation results will finally be returned to the application layer. Therefore, our model has the characteristics of privacy protection, permission control, smart contract and so on, which meets the needs of enterprise business.

The Application Layer The application layer is composed of the application ecology of cross-chain architecture, which can be used to realize different business requirements, such as intelligent health, smart home, supply chain and so on.

Figure 2 shows the physical architecture diagram of our cross-chain scheme. The Tangle network and the Fabric network are connected through the cross-chain gateway. A cross-chain gateway connects a whole-node of both the Tangle network and the Fabric network, which ensures that the cross-chain gateway node has two ledger states at the same time. When the Fabric network ledge status changes affect the IOTA network ledger status, the cross-chain gateway can make the corresponding route. After changing the ledger status on the tangle end, the gateway routes the result back to the Fabric end. The nodes in the Fabric network are divided into three types of nodes, including submission nodes, endorsement nodes, and sorting nodes. The sorting nodes sort the transactions; the submission nodes accept the transaction blocks generated by the sorting nodes. Only the endorsement nodes have the right to write the account book. In the process, the cross-chain gateway needs to have write permission for two ledgers. Therefore, the Fabric nodes connected by the cross-chain portal are endorsement nodes and the whole-nodes connected to the Tangle network.

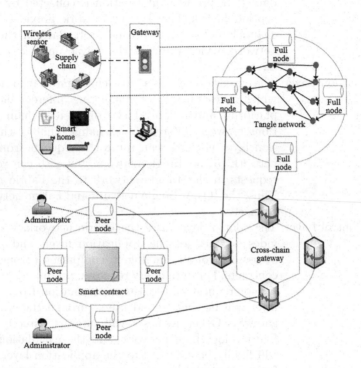

**Fig. 2.** Physical architecture diagram of the cross-chain scheme

## 3.2   The Cross-chain Gateway

**The Cross-chain Gateway Transaction Process.** The cross-chain gateway is the core for connecting different blockchains. The cross-chain gateway's design improves system performance and provides a feasible scheme for the future system to connect more blockchains and increase the ecosystem. As shown in Fig. 3, the cross-chain process includes the following steps:

Step S1: the source chain submits a cross-chain transaction and sets the target chain for routing in the transaction. After the transaction is finished, the blockchain platform will generate cross-chain events with a cross-chain identity and routing direction.

Step S2: in the extended layer, the cross-chain gateway nodes listen to the cross-chain event thrown by the source chain. When they receive a cross-chain event, they put the event in the subscription message queue.

Step S3: in the extension layer, the cross-chain gateway nodes analyze the cross-chain event. They obtain the transaction ID that generates the cross-chain event and use the ID for consensus.

Step S4: The system randomly selects a cross-chain gateway node to fetch the cross-chain transaction content from the source chain. Then it analyzes the cross-chain transaction content and converts it into the transaction format of the target chain. Finally, it identifies the converted transaction as cross-chain operations, sets the target chain for routing, and submits it to the target chain.

After the target chain confirms the transaction, a cross-chain transaction event is generated, triggering the cross-chain process from the target chain to the source chain. The cross-chain process is executed in the sequence of steps S1, S2, S3, and S4. A complete cross-chain transaction is generated when the cross-chain transaction flow from the target chain to the source chain is completed. In addition, the target chain and source chain are abstractions to express convenience. The above source chain and the target chain are regarded as the Hyperledger Fabric network of the control layer and the Tangle network of the data layer.

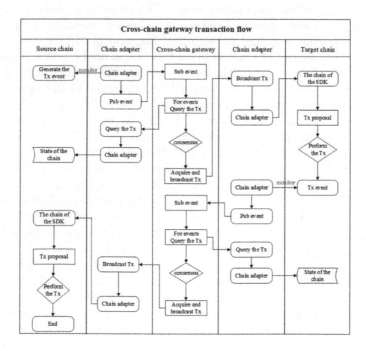

**Fig. 3.** Cross-chain gateway transaction flow

326 C. Wang et al.

**The Cross-chain Gateway Implementation.** The realization of Cross-chain gateway mainly includes three components: adapter, message queue and concurrent lock mechanism.

Adapter In this paper, the Cross-chain is oriented towards IOTA and Fabric blockchain platform. The adapter's implementation is first to encapsulate the IOTA and Fabric clients and then to implement the two clients as a unified interface for the adapter. For the Fabric client, interaction with the smart contract is the core of the client. The Fabric chain code calls the ChaincodeInvoke function to create the chain code to invoke the entity. The entity provides functionality for the Fabric client. For the IOTA client, interaction with the channel is the client's core, so this paper gives the logic of adding data and querying data in the channel. Through the MAMTransmit function adds MAM data in the channel, through the Fetch function queries MAM data in the channel. For the adapter's creation, the factory function is built to create the adapter and save it.

Message queue For a distributed system, the message queue can be used as the middleware of communication between nodes. Generally, a message queue contains two roles: producer and consumer. This paper uses NATS message middleware to implement these two roles. The following describes the production function and subscription function of NATS based message queue. The subscription function is used to consume messages in the queue. Through the connect2Nats function connects to the Nats server. Subscribe to messages in the message queue according to the topic. The producer function adds a message to the message queue. Again, connect to the NATS server via the function connect2Nats, and then send message data to the message queue via the Publish function.

Concurrent lock In a distributed system, system nodes usually need to share some data. The synchronous access of shared data by the system's nodes should be mutually exclusive. That is, one node obtains the access rights of the data, and other data needs to wait for the data to release the access rights. In this paper, etcd middleware is used to implement lock, unlock, and data update functions. In the locking function, the system first gets the mutex through the NewMutex function, then gets the channel credential iotaPayload based on the IOTA address, and then changes the locked value to true. The Update function adds channel credential data to the etcd based on the IOTA address. When the prompt for a cross-chain transaction is successful, the Unlock function through the mutex. Unlock function releases the lock.

## 3.3    The Cross-chain Data Management Scheme

This section describes how to build supply chain data management based on the cross-chain scheme. We mainly focus on the multi-party logistics scenario in supply chain management. The supply chain abstracts business roles, such as dealers, retailers, logistics, and end-users, as sellers, logistics, and buyers, respectively. We use the Hyperledger Fabric blockchain platform to form the license type of supply chain network. Tangle network provides the MAM channel tool. The data obtained by temperature sensors is encrypted and stored in the Tangle network through the MAM channel. When MAM channel is created, it is necessary to provide secret encryption key and generate a channel address. When accessing the MAM channel, the user needs to give the secret key and the channel address. When the MAM channel is operated, the channel credentials are stored in the Fabric network. The system uses smart contracts to control the operations on the channel credentials in the Fabric network. The multi-party logistics scenario includes the following steps:

Step S1: in the supply chain management application, sellers need to transport a product to the buyer. Fabric smart contracts at the control layer provide an interface; the sellers use the interface to generate a logistics order. Meanwhile, the status of the logistics order is changed to requested. The logistics company is entrusted with transporting the product.

Step S2: the logistics company needs to pack the product after receiving the seller's authorization. In transit, the logistics company needs to ensure the product quality and allows the buyers and sellers to monitor the status of the products in real-time. The logistics company installs temperature sensors in the boxes. To upload the data to the blockchain, Fabric smart contracts at the control layer provide an interface. The logistics company uses the interface to set the logistics order status for ready-transit. Meanwhile, a cross-chain gateway of the extension layer is triggered to send a cross-chain transaction to the data layer to create a MAM channel.

Step S3: after the data layer completes the cross-chain transaction, it triggers the cross-chain gateway to send the cross-chain transaction storing the secret key and address of the MAM channel to the control layer. Meanwhile, the logistics company sets the logistics order status for in-transit. In the sensing layer, the temperature sensors periodically monitor the surrounding environment temperature. The system uses the MAM channel to encrypt data and store it in the data layer.

Step S4: when the product is transported to the buyer place, the buyer needs to sign it. The buyer sets some conditions for the product, such as refusing to sign for the product if the temperature is above a preset threshold. The system obtains the IoT data in the MAM channel through the cross-chain. The logistics company sets the status of the logistics order for waiting-sign. To obtain MAM channel data, the system triggers the cross-chain gateway of the extended layer to send the cross-chain transaction to the data layer.

Step S5: after the data layer completes the cross-chain transaction, it triggers the cross-chain gateway of the extended layer to send the cross-chain transaction of IoT data to the control layer. After the smart contract completes the calculation of the IoT data, if the buyer's conditions are satisfied, the logistics order status is updated to sign. Otherwise it is rejected. The result is finally returned to the application layer.

## 4   Scheme Validation

In this paper, the proposed IOTA and Fabric cross-chain schemes for supply chain management are verified from the aspects of performance test and analysis.

### 4.1   Performance Testing and Analysis

The blockchain nodes and other subsystems in this paper run in the docker containers, and all docker containers run on one Linux server. Table 1 shows the system test environment where the container is located.

Table 1. System test server configuration.

| Soft hardware | Model |
|---|---|
| Mainboard | Superfine X10DRi |
| CPU | 2 Intel(R) Xeon(R) CPU e5-2650 v4 |
| Memory | 125 GB |
| Operating system | Ubuntu 16.04 LTS |
| Docker | 18.09.0 |

In this paper, an IOTA and Fabric cross-chain collaboration blockchain network is preliminarily built to test the IOTA and Fabric cross-chain solution's connectivity. The blockchain network consists of a Fabric blockchain network, an IOTA private Tangle, a cross-chain gateway node, and a NATS message queue node. The Fabric blockchain network consists of four Peer nodes, four chain code nodes, and one orderer node. The IOTA private Tangle consists of three IRI nodes and one compass coordinator node. The cross-chain gateway node is connected to the four Peer nodes of Fabric and the three IRI nodes of IOTA for forwarding cross-chain transactions. The test includes creating a MAM data channel in the IOTA network for the Fabric end and saving the data channel's relevant information to the Fabric blockchain network. In addition, the test also includes the capability test of the cross-chain gateway. In the test, a Fabric blockchain client sends 200 transactions to the Fabric network at a rate of 40 TPS (number of transactions per second), while monitoring the number of cross-chain transaction confirmation events from IOTA. We measure transaction throughput, transaction confirmation latency, memory footprint, CPU usage, and network load.

The throughput of MAM data channel cross-chain creation is 5.7 TPS, with an average transaction confirmation delay of 22.1 s, a maximum transaction confirmation delay of 30 s, and a minimum transaction confirmation delay of 1.7 s. The usage of memory, CPU, and network traffic are shown in Fig. 4. The usage of memory and CPU includes the maximum and average values, and the network load provides upload and download. The experimental results show that the Peer node memory and CPU usage of Fabric network are relatively small, and the chain code takes up less memory. Compared with the Fabric network, the memory and CPU usage of the IRI node and the compass coordinator node is high, so the devices with limited resources like IoT cannot directly deploy IRI nodes, high performance of the IoT gateway to deploy the IRI all nodes to provide access for IoT devices maintenance IOTA network security. The poor performance of the IoT gateway and IoT devices should choose the IOTA network node, or via the Internet protocols, have IRI node is connected to the Internet gateway. The cross-chain gateway occupies most of the CPU resources provided by the system in forwarding transactions, so the configuration of its computing resources should be considered when the cross-chain gateway is deployed. The blockchain network requires a large amount of communication traffic. It can be seen that the traffic of the Fabric network is relatively large, and the bandwidth resources should be considered when deploying Fabric nodes.

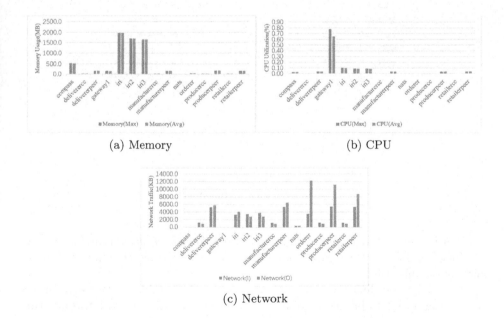

(a) Memory

(b) CPU

(c) Network

Fig. 4. Memory, CPU, and network traffic usage

## 4.2  System Implementation

Non-logistics ordinary users upload their product information to the blockchain for digital management and trade with other users. They can add products and view their own products through the product management interface. Logistics users provide logistics services to other users. To improve logistics services, logistics users use the IoT to monitor the environment of products in real-time and upload the environmental data to the blockchain. Logistics users can add containers, query container details, and query all container operations through the container management interface. Logistics users add and query data according to the MAM channel information.

Logistics users can view all the logistics information, including the tracking number, product id, product type, seller id, seller address, buyer id, buyer address, logistics carrier id, logistics carrier address, delivery time, receiving time, and status the products. The interface for viewing all logistics information is shown in Fig. 5.

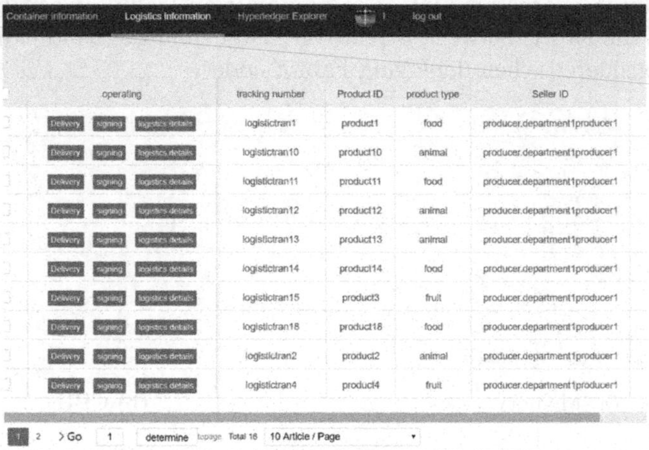

**Fig. 5.** Logistics management

Non-logistics ordinary users can view all product information, including the product ID, product type, product description, release time, commodity status, and commodity seller.

Non-logistics ordinary users need to transport the product to the buyer after the transaction. Non-logistics ordinary user forms the logistics order through the request logistics in the product management interface. All ordinary users can use the logistics management interface for logistics management. Non-logistics users can query the logistics details through the logistics management interface and query the status of the product according to the MAM channel information. Logistics users manage the delivery and receipt of goods entrusted by other users through logistics management and inquire about the status of the products. Logistics users need to choose an empty networked container to package the product and enable the monitoring function of the sensors when delivering the goods. The IoT gateway is then used to query all the containers in operation and subscribe to the data generated by the containers according to the container ID. The IoT gateway wraps the data into the right format required by the interface on the MAM channel and uses the interface to perform on-chain operations on the transformed data as the product arrives at the end-user, the logistics user signs and receives the product. When there is no product quality loss in the transportation process, the logistics agent signs and gets it on behalf of the end-user. Otherwise, the logistics agent refuses to sign into compensate for the product. The logistics users can see all the containers' information, including container key, logistics key, timestamp, position, the status of use.

## 5   Conclusions

As supply chain management has the characteristics of involving many enterprises and dispersing industries, there are some problems in supply chain management, such as the difficulty of coordination, credit, and sharing. The system implemented in this paper solves the problems and makes the product flow traceability, tamper-proof, and distributed sharing. We first introduce the IOTA and Fabric cross-chain schemes for supply chain data management. Each organizational entity in the supply chain interacts with the Fabric Alliance's blockchain, while IoT devices interact with the IOTA private Tangle network. The cross-chain gateway is responsible for forwarding cross-platform interactive content. We test the IOTA and Fabric collaboration scheme to verify its connectivity. We evaluate the performance of the implemented system by assessing the throughput, transaction validation delay, usage of CPU, memory, and network load.

**Acknowledgement.** The research presented in this paper is supported in part by National Natural Science Foundation (No. 61602370, 61672026, 61772411, U1736205), Postdoctoral Foundation (No. 201659M2806, 2018T111066), Fundamental Research Funds for the Central Universities (No. 1191320006, PY3A022), Shaanxi Postdoctoral Foundation, Project JCYJ20170816100819428 supported by SZSTI, CCF-NSFOCUS KunPeng Research Fund (No. CCF-NSFOCUS 2018006).

## References

1. Feng, T.: An agri-food supply chain traceability system for China based on RFID & blockchain technology. In: 2016 13th International Conference on Service Systems and Service Management (ICSSSM), pp. 1–6. IEEE (2016)
2. Bashir, I.: Mastering Blockchain. Packt Publishing Ltd., Birmingham (2017)
3. Novo, O.: Blockchain meets IoT: an architecture for scalable access management in IoT. IEEE Internet Things J. **5**(2), 1184–1195 (2018)
4. Nakamoto, S., A Bitcoin: A peer-to-peer electronic cash system. Bitcoin (2008). https://bitcoin.org/bitcoin.pdf

5. Vogelsteller, F., Buterin, V., et al.: Ethereum whitepaper. Ethereum Foundation (2014)
6. Poon, J., Buterin, V.: Plasma: Scalable autonomous smart contracts. White paper, pp. 1–47 (2017)
7. Androulaki, E., et al.: Hyperledger fabric: a distributed operating system for permissioned blockchains. In: Proceedings of the Thirteenth EuroSys Conference, pp. 1–15 (2018)
8. Popov, S.: The tangle. cit. on, p. 131 (2016)
9. Helo, P., Hao, Y.: Blockchains in operations and supply chains: a model and reference implementation. Comput. Ind. Eng. **136**, 242–251 (2019)
10. Toyoda, K., Mathiopoulos, P.T., Sasase, I., Ohtsuki, T.: A novel blockchain-based product ownership management system (POMS) for anti-counterfeits in the post supply chain. IEEE Access **5**, 17465–17477 (2017)
11. Salah, K., Nizamuddin, N., Jayaraman, R., Omar, M.: Blockchain-based soybean traceability in agricultural supply chain. IEEE Access **7**, 73295–73305 (2019)
12. Benet, J.: IPFS-content addressed, versioned, P2P file system. arXiv preprint arXiv:1407.3561 (2014)
13. Lin, Q., Wang, H., Pei, X., Wang, J.: Food safety traceability system based on blockchain and EPCIS. IEEE Access **7**, 20698–20707 (2019)
14. Qi, S., Lu, Y., Zheng, Y., Li, Y., Chen, X.: Cpds: enabling compressed and private data sharing for industrial IoT over blockchain. IEEE Trans. Ind. Inform. **17**, 2376–2387 (2020)
15. Lu, Y., Qi, Y., Qi, S., Li, Y., Song, H., Liu, Y.: Say no to price discrimination: decentralized and automated incentives for price auditing in ride-hailing services. IEEE Trans. Mob. Comput. (2020)
16. Malik, S., Dedeoglu, V., Kanhere, S.S., Jurdak, R.: TrustChain: trust management in blockchain and IoT supported supply chains. In: 2019 IEEE International Conference on Blockchain (Blockchain), pp. 184–193. IEEE (2019)
17. Gao, K., Liu, Y., Xu, H., Han, T.: Hyper-FTT: a food supply-chain trading and traceability system based on hyperledger fabric. In: Zheng, Z., Dai, H.-N., Tang, M., Chen, X. (eds.) BlockSys 2019. CCIS, vol. 1156, pp. 648–661. Springer, Singapore (2020). https://doi.org/10.1007/978-981-15-2777-7_53
18. Jamil, F., Hang, L., Kim, K.H., Kim, D.H.: A novel medical blockchain model for drug supply chain integrity management in a smart hospital. Electronics **8**(5), 505 (2019)
19. Hyperledger: How Walmart brought unprecedented transparency to the food supply chain with hyperledger fabric (2019)
20. Biswas, K., Muthukkumarasamy, V., Tan, W.L.: Blockchain based wine supply chain traceability system (2017)
21. Greenspan, G.: Multichain private blockchain-white paper (2015). http://www.multichain.com/download/MultiChain-White-Paper.pdf
22. Wu, H., Li, Z., King, B., Miled, Z.B., Wassick, J., Tazelaar, J.: A distributed ledger for supply chain physical distribution visibility. Information **8**(4), 137 (2017)
23. Li, Z., Wu, H., King, B., Miled, Z.B., Wassick, J., Tazelaar, J.: On the integration of event-based and transaction-based architectures for supply chains. In: 2017 IEEE 37th International Conference on Distributed Computing Systems Workshops (ICDCSW), pp. 376–382. IEEE (2017)
24. Ding, Q., Gao, S., Zhu, J., Yuan, C.: Permissioned blockchain-based double-layer framework for product traceability system. IEEE Access **8**, 6209–6225 (2019)

25. Leng, K., Bi, Y., Jing, L., Han-Chi, F., Van Nieuwenhuyse, I.: Research on agricultural supply chain system with double chain architecture based on blockchain technology. Future Gener. Comput. Syst. **86**, 641–649 (2018)

26. Malik, S., Kanhere, S.S., Jurdak, R.: ProductChain: scalable blockchain framework to support provenance in supply chains. In: 2018 IEEE 17th International Symposium on Network Computing and Applications (NCA), pp. 1–10. IEEE (2018)

27. Figorilli, S., et al.: A blockchain implementation prototype for the electronic open source traceability of wood along the whole supply chain. Sensors **18**(9), 3133 (2018)

28. Jensen, T., Hedman, J., Henningsson, S.: How tradelens delivers business value with blockchain technology. MIS Q. Executive **18**(4), 221–243 (2019)

# AI and Education Technology

# Automatic Essay Scoring Model Based on Multi-channel CNN and LSTM

Zhiyun Chen[1], Yinuo Quan[1], and Dongming Qian[2]([✉])

[1] School of Data Science and Engineering,
East China Normal University, Shanghai, China
chenzhy@cc.ecnu.edu.cn, qynjinoo@163.com
[2] National Institutes of Educational Policy Research,
East China Normal University, Shanghai, China
dmqian@admin.ecnu.edu.cn

**Abstract.** In essay marking, manual grading will waste a lot of manpower and material resources, and the subjective judgment of marking teachers is easy to cause unfair phenomenon. Therefore, this paper proposes an automatic essay grading model combining multi-channel convolution and LSTM. The model adds a dense layer after the embedding layer, obtains the weight assignment of text through softmax function, then uses the multi-channel convolutional neural network to extract the text feature information of different granularities, and the extracted feature information is fused into the LSTM to model the text. The model is experimented on the ASAP composition data set. The experimental results show that the model proposed in this paper is 6% higher than the strong baseline model, and the automatic scoring effect is improved to a certain extent.

**Keywords:** Automatic essay scoring · Multichannel convolution · Long Short-Term Memory

## 1 Introduction

In language teaching, students' writing level is often tested, such as Chinese writing, English writing and so on. The level of writing can reflect students' mastery of grammar, rhetoric and other aspects. At present, the main way of scoring is through the teachers' manual scoring, but there are inevitably some drawbacks in manual scoring. The subjectivity of manual scoring will lead to a large error in scoring, and the scoring results are easily affected by the personal preferences, and mentality of the raters, resulting in the occurrence of unfair phenomena. Therefore, the requirements for the professional quality of the rater are very high. Moreover, a large examination needs to organize a large number of marking teachers, which will consume a lot of manpower, material resources and time cost. With the development of natural language processing technology, there has been a breakthrough in the syntactic analysis and semantic analysis of

© Springer Nature Singapore Pte Ltd. 2021
W. Gao et al. (Eds.): FICC 2020, CCIS 1385, pp. 337–346, 2021.
https://doi.org/10.1007/978-981-16-1160-5_26

text, and it is also used in the field of automatic essay scoring. Automatic composition scoring is a process in which computer software can grade students' essays
without human intervention. It can overcome the above shortcomings of manual
scoring and provide accurate and unified scoring standards. The score results
will not be affected by subjective factors and will be more fair and objective.

Page implemented PEG (Project Essay Grader) [1] automatic essay scoring system, which extracted shallow features by manually made rules and used
linear regression model for scoring. With the development of natural language
technology, Pearson Knowledge Analysis developed the IEA (Intelligent Essay
Assessor) [2] based on LSA (Latent Semantic Analysis) [3]. The E-RATER [4]
system was developed by the American Educational Testing Research Center,
which scores essays through the three aspects: syntactic analysis module, discussion module and topic module. Rudner and Liang of the University of Maryland
developed the Besty [5] system, which integrated the shallow linguistic features
of PEG, LSA features and e-Rater features, and used Naive Bayesian model
for training, and achieved an accuracy rate of 80% on the test set of the system. Chen et al. [6] extracted simple features such as article length and complex
features such as syntactic features for automatic scoring. For automatic essay
grading studies, many studies take EASE [7] as the baseline model, which uses
SVR (Support Vector Regression) and BLR (Bayesian Linear Regression) to
train the model. The disadvantages of the above researches are that they only
analyze the lexical level of the essay and neglect the semantic level.

With the development of neural network technology, the neural network
model based on word embedding [8,9] is applied more and more. Some
researchers use neural network models to obtain the semantics of essays. Chen
et al. [10] proposed an automatic essay scoring model based on CNN (Convolutional Neural Networks) and Ordinal Regression (OR) [11,12], which has an
improved effect than EASE on a single prompt. Attention mechanism [13,14] has
made new breakthroughs in many tasks related to natural language processing.
Jin et al. [15] proposed a two-stage automatic essay scoring algorithm called
TDNN. Firstly, this automatic essay scoring model is trained by feature, and a
pseudo score is given to the target topic essay by using this model. Then, the
positive and negative samples of the pseudo fraction extreme are selected and
a deep neural network model is trained on this basis. Liu et al. [16] calculated
semantic score, coherence score and cue correlation score based on Long Short-
Term Memory (LSTM), and integrated feature engineering model, and achieved
good results on ASAP dataset.

The contribution of this paper is as follows: On the ASAP dataset, the model
proposed in this paper exceeded the baseline in all eight prompters, with the
MQWK 6% higher than the baseline. Compared with other models, the best
results are obtained on four prompters and the MQWK is the best. It shows the
effectiveness of automatic essay scoring model based on multi-channel convolution and LSTM.

## 2 MCNN-LSTM Model

In order to extract more effective text information, this paper proposes an automatic essay scoring model that merges multi-channel convolution with LSTM. The model first uses GloVe pre-training word vectors to map the text into low-dimensional vector representation as the input of the model. The model assigns a weight to the input text through a dense layer and Softmax function after the Embedding layer. After that the multi-channel convolutional neural network is used to extract the feature information of different granularity of the text, and the feature sequences generated by multi-channel convolution are spliced to form the fusion features as the input of LSTM layer. The LSTM layer can effectively acquire the relationships within and between sentences and model the text. The structure of the model is shown in the Fig. 1.

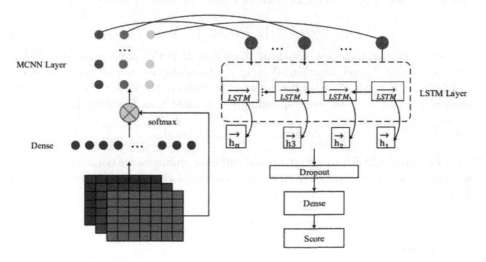

**Fig. 1.** The structure of MCNN-LSTM model.

### 2.1 Embedding-Dense Layer

A dense layer and Softmax function are added to the model. Multiplying the input of Embedding layer by the output of Softmax function is equivalent to assigning weight to the network input and capturing important semantic features of context information. As shown in formula (1) and formula (2).

$$\alpha_t = \frac{exp(e_t)}{\sum_{k=1}^{T} exp(e_k)} \tag{1}$$

$$s = \sum_{t=1}^{T} \alpha_t e_t \tag{2}$$

Among them, $e_t$ is the output of t-time Embedding layer, and $\alpha_t$ is weighted by Softmax function. Weighted the multiplication of $\alpha_t$ and the original vector $e_t$ of all nodes in a sequence layer to obtain the final text vector $S$.

## 2.2   MCNN Layer

CNN mainly learns local features through convolutional layer and pooling layer, which is mainly composed of input layer, convolutional layer, pooling layer and full connection layer. For convolution operation, choosing different sizes of convolution kernel can extract context information with different breadth, that is, sequence feature information with different granularity [17]. In the pooling layer, the feature vectors obtained by convolution are further down sampled to extract effective feature information. Suppose the feature $c_{ij}$ extracted by the convolution operation of the $i^{th}$ channel can be expressed as the formula (3).

$$c_{ij} = f(W \cdot X_{i:i+h-1} + b) \tag{3}$$

Where $h$ is the window width of the filter, $B$ is the bias term, and $f(\cdot)$ is the convolution kernel function. $X_{i:i+h-1}$ represents the local filtering window consisting of $h$ words. As the filtering window slides to the end, the characteristic sequence obtained by the corresponding $i^{th}$ channel is as formula (4).

$$c_i = [c_{i,1}, c_{i,2}, \ldots, c_{i,n-h+1}] \tag{4}$$

The convoluted feature sequences of different channels are connected to form feature sequence $c$. Assuming that the number of channels is $k$, the following formula (5) is shown.

$$c = [c_1, c_2, \ldots, c_k] \tag{5}$$

## 2.3   LSTM

The feature sequences extracted after multi-channel convolution operation are merged as the LSTM input. RNN can easily lead to gradient disappearance or gradient explosion, so it can only deal with short-term dependence problem. The LSTM is able to deal with both short-term and long-term dependencies. Three gating units are added to the LSTM, namely, the input gate, the output gate and the forgetting gate, among which the input gate and forgetting gate are the keys for the LSTM to remember long-term dependence. The input gate selectively saves the current network state information, while the forgetting gate selectively decides to discard the past information, and finally the output gate decides to output the internal state.

In LSTM, the hidden state $h_t$ and memory cell $c_t$ are functions of the hidden state $h_{t-1}$ and memory cell $c_{t-1}$ and the input vector $x_t$ of the previous cell. The specific calculation is shown in formula (6)–formula (11).

$$f_t = \sigma(W_f \cdot [h_{t-1}, x_t] + b_f) \tag{6}$$

$$i_t = \sigma(W_i \cdot [h_{t-1}, x_t] + b_i) \qquad (7)$$

$$\tilde{C}_t = tanh(W_c \cdot [h_{t-1}, x_t] + b_c) \qquad (8)$$

$$C_t = f_t \odot C_{t-1} + i_t \odot \tilde{C}_t \qquad (9)$$

$$O_t = \sigma(W_o[h_{t-1}, x_t] + b_o) \qquad (10)$$

$$h_t = O_t \odot tanh(C_t) \qquad (11)$$

Where, $x_t$ is the input vector of the model in step $t$, namely the semantic representation of the $t_{th}$ word in the sentence. $f_t$, $i_t$, $O_t$, $h_t$, $W_f$, $W_i$, $W_o$, $b_f$, $b_i$, $b_c$ represent forgetting and updating operations and parameters of hidden state and output state respectively. $\tilde{C}_t$ for step $t$ generate candidate vector, $C_{t-1}$ for the first t original vector, $h_{t-1}$ save the timing information of former $t-1$ step, is vector dot product operation. The LSTM hidden layer output matrix can be obtained as shown in the following formula (12).

$$H = \{h_1, h_2, \ldots, h_n\} \in R^{n \times d} \qquad (12)$$

Where, $n$ is the length of the sentence and $d$ is the number of LSTM hidden units.

### 2.4   Objective and Training

The essay scoring task is often viewed as a machine learning regression task. The objective of the model is to optimize the MSE (Mean Square Error) function, which is often used as the loss function in regression tasks. For $N$ essays $e_i$, MSE is used to measure the mean square error between the standard score $r$ and the prediction score $\hat{r}$. MSE is shown in formula (13).

$$\frac{1}{N} \sum_{i=1}^{N} (r_{e_i} - \hat{r}_{e_i})^2 \qquad (13)$$

Adam [18] is used as the optimizer to minimize the loss of training data.

## 3   Experiment and Analysis

### 3.1   Dataset

The dataset used in this article is Automated Student Assessment Prize (ASAP) organized by Kaggle. The dataset contains about 17,000 essays for American middle school students, divided into 8 prompts. Different essays numbers correspond to different topics, different student grades, different essay lengths and grading standards, all of which are manually graded. More details about ASAP dataset are summarized in Table 1.

**Table 1.** Details about ASAP dataset.

| Prompt | Essays | Avg length | Score range |
|---|---|---|---|
| 1 | 1783 | 350 | 2–12 |
| 2 | 1800 | 350 | 1–6 |
| 3 | 1726 | 150 | 0–3 |
| 4 | 1772 | 150 | 0–3 |
| 5 | 1805 | 150 | 0–4 |
| 6 | 1800 | 150 | 0–4 |
| 7 | 1569 | 250 | 0–30 |
| 8 | 723 | 650 | 0–60 |

## 3.2 Evaluation Metrics

In this paper, QWK (Quadratic Weight Kappa) value is used to measure the scoring effect of model. The index range is 0–1, 0 means completely inconsistent, 1 means completely consistent. After obtaining the value of QWK for each subset, the MQWK (Mean Quadratic Weight Kappa) value is calculated as the final evaluation index of the model.

**Quadratic Weight Kappa.** Assume that the essay scoring standard is 1 to $N$, the expert scoring result is $score1$, and the model scoring result is $score2$. For essay $e$, the manual scoring and model scoring results are set $(e_{score1}, e_{score2})$. An $N$ by $N$ weight matrix is calculated, in which each element $W_{ij}$ in the matrix is calculated as formula (14).

$$W_{ij} = \frac{(i-j)^2}{(N-1)^2} \tag{14}$$

The value of Quadratic Weight Kappa for the dataset is then computed as formula (15).

$$QWK = 1 - \frac{\sum_{ij} W_{ij} O_{ij}}{\sum_{ij} W_{ij} E_{ij}} \tag{15}$$

Where the $O_{i,j}$ stands for the number of essays with the manual score of $i$, the number of essays with the model score of $j$, and $E_{i,j}$ stands for the product of the probability of manual score of $i$ and the probability of the model score of $j$. $O_{i,j}$ and $E_{i,j}$ are normalized respectively, and the result of matrix addition is 1.

**Mean Quadratic Weight Kappa.** After obtaining the QWK value of each subset, the MQWK value is calculated as the final model evaluation index. Fisher transformation is applied to the QWK value to fix the Kappa value at 0–0.999, the conversion formula is as (16).

$$\check{z} = \frac{1}{2} \ln \frac{1+k}{1-k} \tag{16}$$

Due to the different scoring ranges of different prompts, it is necessary to normalize the influence of corresponding essay on the final average kappa value. The formula is as (17).

$$MQWK = \frac{e^{2\bar{z}} - 1}{e^{2\bar{z}} + 1} \tag{17}$$

### 3.3 Parameter Settings

In order to obtain rich feature information, the model adopts three-channel convolution kernel. Detailed hyperparameter settings are shown in the Table 2 below.

**Table 2.** Super parameter.

| Super parameter | Value |
| --- | --- |
| Vector dimension | 300 |
| Convolution kernel width | (2, 3, 4) |
| LSTM hidden size | 128 |
| Batch size | 64 |
| Learning rate | 0.006 |
| Dropout rate | 0.7 |

### 3.4 Experimental Results and Discussion

**Strong Baselines.** EASE is trained by extracting artificial features such as lexical features and grammatical features into the regression model, rank third in the ASAP competition. In this paper, EASE (SVR) [7], EASE (BLRR) [7] are selected as the baseline models.

**Results.** In order to evaluate the performance and effectiveness of the proposed MCNN-LSTM model, a comparative experiment is designed. The ASAP data set consists of eight prompts, in order to better evaluate the effect of the model, the experimental results in this paper give the QWK and the final MQWK for each prompt. The MCNN-LSTM model is compared with EASE(SVR), EASE(BLRR), CNN, LSTM and TDNN [15]. The comparison results are shown in the Table 3.

As you can see from the table, The MCNN-LSTM model proposed in this paper has better QWK than other models in 4 of the 8 prompts, and MQWK reached best, which is improved compared with the LSTM and CNN model alone, indicating that the model combining multi-channel convolution and LSTM improved this task. The results of each model in the $8^{th}$ prompt and the $3^{th}$ prompt datasets are relatively poor, possibly due to the imbalance in the number

**Table 3.** Experimental results.

| Models | Prompts | | | | | | | | Avg |
|---|---|---|---|---|---|---|---|---|---|
| | 1 | 2 | 3 | 4 | 5 | 6 | 7 | 8 | |
| EASE(SVR) | 0.781 | 0.621 | 0.63 | 0.749 | 0.782 | 0.771 | 0.727 | 0.534 | 0.699 |
| EASE(BLRR) | 0.761 | 0.606 | 0.621 | 0.742 | 0.784 | 0.775 | 0.73 | 0.617 | 0.705 |
| CNN | 0.797 | 0.634 | 0.646 | 0.767 | 0.746 | 0.757 | 0.746 | **0.687** | 0.722 |
| LSTM | 0.775 | **0.687** | 0.683 | 0.795 | **0.816** | 0.813 | **0.805** | 0.594 | 0.746 |
| TDNN | 0.768 | 0.685 | 0.628 | 0.757 | 0.736 | 0.675 | 0.658 | 0.574 | 0.686 |
| **MCNN-LSTM** | **0.818** | 0.683 | **0.703** | **0.805** | 0.809 | **0.820** | 0.784 | 0.652 | **0.763** |

of positive and negative essays in these prompt datasets. The method based on neural network is generally superior to the baseline method in every prompt. For the neural network method, LSTM is generally superior to CNN, which may be because the text of the data set is generally longer, and the use of LSTM can solve the long distance dependence problem.

It can be seen from Fig. 2 that, except prompt8, the QWK of the other composition subsets rapidly climbed to more than 0.6 in the first 10 epoches. With the increase of the epoch, the QWK of each composition subset fluctuated and tended to be stable. Among them, the three subsets of prompt2, prompt3, and prompt8 had large oscillations, which tended to be stable and accompanied by small shocks after more than 100 epochs. In the experiment, it was found that after 300 epochs, the MQWK on the test set was higher than that of 100 and 200 epochs, so 300 epochs were selected for training.

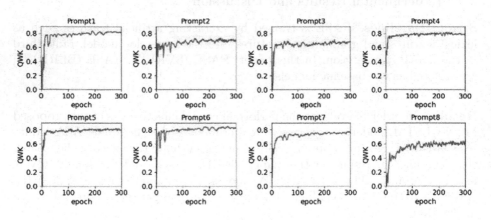

**Fig. 2.** The QWK change graph of each composition subset on the validation set.

# 4    Conclusion

In this paper, an automatic essay scoring model based on multi-channel convolution and LSTM is proposed, which is independent of feature engineering. The model first adds a dense layer to the embedding layer, obtains the weight assignment of text through softmax function, then feature information of different granularity is extracted by multi-channel convolution, and then the fused feature information is input into LSTM to code and model the text. The experimental results show that the model proposed in this paper is better than other models in some prompts of ASAP dataset, and the MQWK value is the best. The experiment also reflects some problems, such as the neural network model is difficult to make use of some spelling errors in the composition, so the follow-up work will integrate some artificial feature extraction, such as spelling errors, sentence length, etc., to improve the effect and stability of the model.

# References

1. Page, B.E.: The imminence of grading essays by computer. Phi Delta Kappan **47**(5), 238–243 (1966)
2. Landauer, T.K., Foltz, P.W.: The intelligent essay assessor. IEEE Intell. Syst. **15**(15), 27–31 (2000)
3. Foltz, P.W.: Latent semantic analysis for text-based research. Behav. Res. Methods Instrum. Comput. **28**(2), 197–202 (1996)
4. Chen, X., Ge, S.: A review of automated essay scoring. J. PLA Univ. Foreign Lang. **31**(5), 78–83 (2008)
5. Rudner, L.M., Liang, T.: Automated essay scoring using Bayes' theorem. J. Technol. Learn. Assess. **1**(2) (2002)
6. Chen, H., He, B.: Automated essay scoring by maximizing human-machine agreement. In: Proceedings of the 2013 Conference on Empirical Methods in Natural Language Processing, pp. 1741–1752 (2013)
7. Phandi, P., Chai, K.M.A., Ng, H.T.: Flexible domain adaptation for automated essay scoring using correlated linear regression. In: Proceedings of the 2015 Conference on Empirical Methods in Natural Language Processing, pp. 431–439 (2015)
8. Mikolov, T., Sutskever, I., Chen, K., et al.: Distributed representations of words and phrases and their compositionality. In: Advances in Neural Information Processing Systems, pp. 3111–3119 (2013)
9. Pennington, J., Socher, R., Manning, C.D.: GloVe: global vectors for word representation. In: Proceedings of the 2014 Conference on Empirical Methods in Natural Language Processing (EMNLP), pp. 1532–1543 (2014)
10. Chen, Z., Zhou, Y.: Research on automatic essay scoring of composition based on CNN and OR. In: 2019 2nd International Conference on Artificial Intelligence and Big Data (ICAIBD), Chengdu, China, pp. 13–18 (2019). https://doi.org/10.1109/ICAIBD.2019.8837007
11. Cheng, J., Wang, Z., Pollastri, G.: A neural network approach to ordinal regression. In: IEEE International Joint Conference on Neural Networks, pp. 1279–1284. IEEE (2008)
12. Niu, Z., Zhou, M., Wang, L., Gao, X., Hua, G.: Ordinal regression with multiple output CNN for age estimation. In: IEEE Conference on Computer Vision and Pattern Recognition, pp. 4920–4928. IEEE Computer Society (2016)

13. Bahdanau, D., Cho, K., Bengio, Y.: Neural machine translation by jointly learning to align and translate. arXiv preprint arXiv:1409.0473 (2014)
14. Vaswani, A., et al.: Attention is all you need. In: Advances in Neural Information Processing Systems, vol. 22, no. 7, pp. 139–147 (2017)
15. Jin, C., He, B., Hui, K., et al.: TDNN: a two-stage deep neural network for prompt-independent automated essay scoring. In: Proceedings of the 56th Annual Meeting of the Association for Computational Linguistics (Volume 1: Long Papers), pp. 1088–1097 (2018)
16. Liu, J., Xu, Y., Zhu, Y.: Automated essay scoring based on two-stage learning. In: arXiv preprint arXiv:1901.07744 (2019)
17. Cheng, J., Zhang, S., Li, P., et al.: DeepconvRNN for sentiment parsing of Chinese microblogging texts. In: IEEE International Conference on Computational Intelligence and Applications, pp. 265–269. IEEE (2017)
18. Dozat, T.: Incorporating Nesterov momentum into adam. In: ICLR Workshop, no. 1, pp. 2013–2016 (2016)

# Research on Knowledge Graph in Education Field from the Perspective of Knowledge Graph

Zhiyun Chen[1], Weizhong Tang[1], Lichao Ma[2], and Dongming Qian[3(✉)]

[1] School of Data Science and Engineering, East China Normal University,
Shanghai 200062, China
grandcanyon001@163.com, sxtwz@163.com
[2] Faculty of Education, East China Normal University, Shanghai 200062, China
chao_ecnu@163.com
[3] National Institutes of Educational Policy Research, East China Normal University,
Shanghai 200062, China
dmqian@admin.ecnu.edu.cn

**Abstract.** Rapid advent of the era of big data, deep integration of education and technology, and multidisciplinary nature of educational research are driving the application of knowledge graphs in educational research. We use CSSCI and CSCD journal articles as data sources and analyze the Knowledge Graph of education using the graph database Neo4j in this article. Result shows that the number of literatures in the research area show a clear upward trend and go from the stages of brewing. Highly cited literatures' research topics, subject categories, and authors are widely distributed. From the perspective of core institutions and authors, multiple normal universities have strong competitiveness. Citespace, SPSS and Bicomb are the most commonly used data processing tools for scholars. In further study of the area of knowledge graph, we should fully combine with the cutting-edge computer technology, based on big data and artificial intelligence to improve the level and quality of education research.

**Keywords:** Educational research · Knowledge graph · Network analysis · Bibliometric · Neo4j

## 1 Introduction

Knowledge Graph is a technical method that uses graph models to describe the association between knowledge. The characteristic is to build network relationships with the help of "graph structure" in order to achieve efficient and accurate queries. The "Notice of the State Council on printing and distributing the development plan for the new generation of artificial intelligence" proposes to "Establish a new generation of key common technology system of artificial intelligence, develop knowledge computing engine and knowledge service technology, and form a multi-source, multi-disciplinary and multi-data type cross-media

© Springer Nature Singapore Pte Ltd. 2021
W. Gao et al. (Eds.): FICC 2020, CCIS 1385, pp. 347–360, 2021.
https://doi.org/10.1007/978-981-16-1160-5_27

knowledge graph" [2]. For educational research, the traditional method of data collection is usually to manually read books, journals and other papers, and process them in combination with their own skill and experience. This process often produces omissions or repetitions, and is easily affected by personal subjective experience [5]. Z. Li [8] used traditional research methods to study the application of Japanese education and found that it has comprehensive applications in education assistance, information system development, commerce and other aspects.

With the improvement of computing power and the growth of data volume, it is easy to find and infer the complex relationship between data from massive and changeable literature. The calculation model is constructed by using structured and unstructured data to generate intuitive visual charts to display data, enhance understanding of data, and assist in exploring data [9], which has become the research focus of education discipline the "Sharp weapon".

With the continuous deepening of the integration of education and technology, the application of knowledge graphs in the field of education presents a trend of expanding and enriching research scope and research topics is more popular. To trace, review, summarize and reflect on the application of knowledge graphs in the field of education can not only grasp the overall context and development logic of education discipline research, but also help to further realize the deep integration of education and technology, and promote the informatization of education and scientific research to a new level. Therefore, this study analyzes the annual development trends, journal distribution, highly cited literature, publishing organizations, core author groups and their cooperation, etc. and describes the data sources, research tools, and research topics of relevant literature, and outlines the development overview and change prospect of China's educational knowledge map.

## 2    Data Source and Processing

### 2.1    Data Source

In order to ensure the comprehensiveness, scientific and accuracy of the data, we used CSSCI and CSCD journal papers in CNKI database as statistical data sources which is the most China academic authority, the language is limited to "Chinese", the "Knowledge Graph" and "education" are set as the retrieval subjects, and the journal papers related to the research topics are selected for analysis. The retrieval time is till to October 1, 2019, excluding meeting minutes, book reviews, and other non-academic literature such as journal catalogue and policy news, there are 384 literatures left as samples.

### 2.2    Data Processing

Python is used to clean data and organize structured data, and Tableau and Neo4j are used for data processing and knowledge graph visualization analysis.

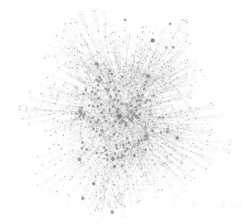

**Fig. 1.** Educational knowledge graph network (Color figure online)

Data processing process is divided into the following four steps: First, export the RefWorks format file according to the retrieval conditions, load each field by Python and convert it into a CSV file. Secondly, the author with the same name will bring some deviation to the data analysis, so the author ID in CNKI database is crawled to bind with the author entity to realize "disambiguation". Some missing data are filled manually according to the author's unit. Then, use the data preprocessing software Tableau Prep to merge and associate different CSVs, and use the visualization software Tableau to cluster and merge keywords. Finally, in view of the fact that most graph databases are not compatible with the whole import of structured data, so the solution of data conversion by python script is used to solve the data transformation. The data is decomposed into attribute graph model supported by graph database, and then the subsequent operation is carried out.

## 2.3   Graph Construction

"Graph" was first mentioned by Leonhard Euler, a Swiss mathematician, in 1736 when he solved the Seven Bridges of Königsberg [13]. It was originally used to represent the relationship between objects and solve it with mathematical algorithms. This kind of research gradually developed into "Graph Theory". In order to manage, store and retrieve the network relationship of graph more reliably, efficiently and correctly, "graph database" emerges as the times require. As a non-relational database, it optimizes tasks based on the traditional relational databases, especially in the face of complex connection query and path query. There are common used graph database solutions like Neo4j, FlockDB, AllegroGrap, GraphDB, etc. Neo4j has the characteristics of flexible structure, rich open source community resources, high performance, and meets the weight mechanism, mutual exclusion mechanism, and dynamic mechanism in the drawing process [1]. Therefore, this study uses Neo4j to construct the atlas. It is

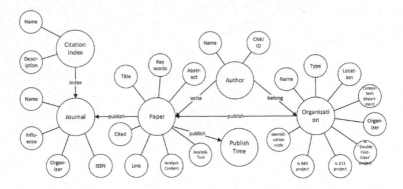

**Fig. 2.** Educational knowledge graph entity, attribute and relationship model

important to note that all relationships in Neo4j must be constructed oriented, but directionality can be ignored in query evaluation.

Figure 1 is a visual thumbnail of the knowledge graph of this study, and Fig. 2 is a model example of graph database entities, attributes and relationships of in Neo4j database. A large number of models shown in Fig. 2 are instantiated and constructed into the network of Fig. 1 based on associations, for example the blue node in Fig. 1 represents author entity, the orange node represents time, and the connected solid line represents relationship that exists between them. The author entity has two attributes: name and CNKI ID. The name is in string format. CNKI ID is composed of an array. In the graph, the same author may contain multiple CNKI ID attributes and multiple institutional entities' relationship connection due to the change of school or work unit. As an important analysis content in this study, "time" is extracted from the paper entity as a separate entity.

## 3    Research Results and Analysis

### 3.1    Annual Trends of Literature

Figure 3 outlines the changing trend of the application of knowledge graphs in educational research over time. The data shows that the earliest data analysis with the help of knowledge graph in the field of education began in 2008, and then showed an obvious upward trend. Generally speaking, it can be roughly divided into three stages:

(1) During the brewing period from 2008–2010, the number of literatures is relatively small. It mainly uses scientific measurement methods to draw a hot map of the literature of a certain journal or a certain field with a single or small number of journals as the sample source. For example, Some scholars [16] used CiteSpace software to draw the research hotspots and knowledge graphs of papers published in "Journal of Higher Education"

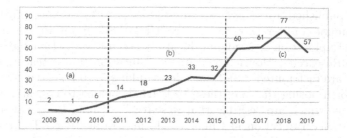

**Fig. 3.** Annual change trend chart of educational knowledge atlas

from 1998 to 2007, and use the graph to intuitively find that the research hotspots in this field mainly focus on "education reform", "teaching reform" and "colleges and universities".

(2) 2011–2015 is the development period, the number of annual publications has increased rapidly from 14 to 32. During this period, the research objects expanded rapidly, mainly focusing on higher education (29.17%), educational technology (25.83%), physical education (5.83%), teacher development (5.00%).

(3) Since 2016 has been the prime period, the number of research literature on the atlas of educational knowledge graphs has increased dramatically. Compared with 2015, the number of literature in 2016 has increased by 28, with an increase rate of 87.5%. After that, the number of papers has basically stabilized at more than 60 each year (the data in 2019 is only up to October 1st, and all the papers in the whole year have not been counted). The data sources at this stage are not limited to domestic academic journal papers, master and doctor's degree papers and SSCI journal papers, even extend to policy and regulatory texts, proposal content, school charter, President's speech, project approval and other aspects, which show that the knowledge graph has been widely concerned by education researchers, and the integration of educational research and technology applications has been continuously improved.

## 3.2   Distribution of Journals

The statistical results show that 384 articles related to the educational knowledge graph are published in 120 CSSCI and CSCD journals. Table 1 shows the top 10 journals with the highest number of published articles. A total of 107 related literatures were published in these journals, accounting for 23.13% of all the papers on the atlas of educational knowledge graph Among them, "Research on audio visual education" has the largest number of articles, up to 16, accounting for 4.17% of the total number of papers published, ranking the first. The second is "Audio Visual Education in China", with 15 articles on related topics, accounting for 3.91%. Both "China Distance Education" and "Modern Education Management" published 13 papers, accounting for 3.39%, ranking third. In addition, the number of literatures in "China's Higher Education Research", "Higher Education Exploration", "Modern Education Technology", "Heilongjiang Higher Education Research", "Journal of Distance Education" and "Teacher Education Research" are distributed in 9–11 articles, ranking 5–10 respectively.

It is worth noting that 5 of these 10 journals belong to educational technology journals, which are "e-Education Research", "China Educational Technology", "Distance Education in China", "Modern Educational Technology" and "Journal of Distance Education". There are three journals of higher education, namely "China Higher Education Research", "Higher Education Exploration" and "Heilongjiang Researches on Higher Education". This shows that there are obvious differences in the integration of different pedagogy secondary disciplines and knowledge graphs. Educational technology and higher education pay more attention to the advantages and status of knowledge graphs in academic research, while other disciplines are relatively lack.

**Table 1.** Top 10 journals in educational knowledge graph research

| No | Journal | Count | Ratio | No | Journal | Count | Ratio |
|---|---|---|---|---|---|---|---|
| 1 | e-Education Research | 16 | 4.17% | 6 | Higher Education Exploration | 10 | 2.60% |
| 2 | China Educational Technology | 15 | 3.91% | 7 | Modern Educational Technology | 10 | 2.60% |
| 3 | Distance Education in China | 13 | 3.39% | 8 | Heilongjiang Researches on Higher Education | 10 | 2.60 % |
| 4 | Modern Education Management | 13 | 3.39% | 9 | Journal of Distance Education | 10 | 2.60% |
| 5 | China Higher Education Research | 11 | 2.86% | 10 | Teacher Education Research | 9 | 2.34% |

## 3.3   Highly Cited Literature

This study counts the ten papers with the highest citations in the field of educational knowledge graph research. First of all, the paper "The Research and Application of Big Data in the Field of Online Education" comes from CSCD journals, and the remaining 9 papers are from CSSCI journals. Secondly, from the perspective of the research scope, these 10 articles not only analyze the overall research situation and hot topics of education discipline (No. 1), but also discuss specific topics in secondary disciplines such as higher education (No. 2), physical education (No. 4 and 8), special education (No. 5 and 6), and educational information technology (No. 3, 7, 9 and 10), which show the extensive application of knowledge graphs in pedagogical research; Thirdly, from the perspective of citations, there is one paper cited more than 150 times, namely " Hotspot Domains and Frontier Topics of Educational Research in the Past 10 Years——Based on the Knowledge Mapping of Key Words of Eight CSSCI Educational Journals Published in 2000–2009". The paper was published earlier (2011), which has a driving and leading role in the later application of knowledge graphs in the field of education. The number of other highly cited literatures ranged from 68 to 104. Finally, from the author's point of view, among the top 10 highly cited papers, only Guo Wenbin has two related literatures, and other scholars have one. The above data show that the research in this field shows a trend of blooming in terms of research theme, subject categories, and author distribution, which is conducive to the development of the discipline in the future.

### 3.4 Main Research Institutions and Cooperation

**Distribution of Research Institutions.** In order to intuitively describe the status of papers published by different research institutions, this study sets the "List of National Colleges and Universities in 2019" released by the Ministry of Education of the People's Republic of China as a high-weight keyword dictionary, and uses the conditional random field (CRF) algorithm to segment the entities of the authors institutions, and manually checks and corrects a small amount of errors, so as to ensure the accuracy of the obtained institutions. For example, the Department of Education of East China Normal University, the Institute of Curriculum & Instruction of East China Normal University, and the National Institutes of Education Policy Research of East China Normal University are unified and merged into "East China Normal University". On the basis of data cleaning, we use Tableau Prep to associate and match the list of institutions after formatting, and select the research institutions with a large number of original papers.

According to Price's Law [10], when the number of papers published is more than or equal to 4, it can be used as the core research institutions in this field. There are 28 institutions that meet the requirements of this study. According to the data, there are 11 institutions that have published 10 or more papers. Among them, Shaanxi Normal University has published 24 papers, ranking first. The school has five core authors, including Guo Wenbin, Yuan Liping, and Qi Zhanyong. To a certain extent, it shows that the research in the field of education knowledge graphs is relatively in-depth. Beijing Normal University, East China Normal University and Nanjing Normal University ranked 2–4 respectively, with 17–21 papers published; Henan University, Central China Normal University, Northeast Normal University, Wenzhou University, Nanjing University, Dalian University of Technology, and Wuhan University also have a relatively high number of relevant literature, all of which are at least 10.

**Institutional Cooperation.** Using Neo4j query language Cypher to generate the cooperation network of institutions based on Fig. 1's Network is shown in Fig. 4. It can be clearly found that there are three large connected areas. The largest cooperation network is composed of 41 institutions, and the other two larger cooperation networks are composed of 26 and 14 institutions respectively. In the Fig. 4, they are marked as A, B and C. Centrality is a tool to measure the importance of network nodes in the network. This paper analyzes the degree

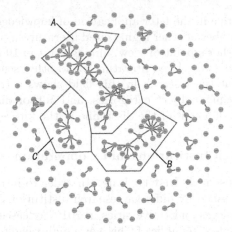

**Fig. 4.** The map of educational knowledge and institutional cooperation in the field of knowledge graph

centrality with weights and betweenness centrality. On the one hand, the degree centrality of the network nodes is calculated. The larger the value is, the more connections the node has with other nodes, and the more central the node is, the higher the reputation of the node is [11]; On the other hand, betweenness can identify the key nodes of the network. The higher the value is, the more structural holes the node occupies [14]. The node occupying this position can become a "bridge" across different networks. According to centrality calculations, Beijing Normal University, Nanjing Normal University, and East China Normal University have the highest degree centrality, while East China Normal University, Nanjing University and Nanjing Normal University have the highest betweenness centrality. These universities are at the core of the institutions cooperation in the field of educational knowledge graphs, which can promote the research and application of knowledge graphs in the field of education.

### 3.5  Analysis of Core Authors and Their Cooperation

**Core Authors Analysis.** According to Price's law, core authors refer to authors who have published more than N ($N = 0.749\sqrt{MAX}$) papers, where MAX is the number of papers published by the most authors. In this study, more than 3 articles (including 3 articles) belong to the core authors, a total of 29 people, accounting for 4.06% of all the authors.

On the one hand, in terms of the number of papers published by core authors, Lan Guoshuai has the largest number of papers on the application of knowledge graphs in the field of education research, and the number of related papers is as high as 9. His articles on the educational knowledge graph mainly focus on educational technology and international education. Cai Jiandong of Henan University has published 7 articles, ranking second. His main research fields are educational technology and ubiquitous learning. The number of articles published by Guo

Wenbin, Yu Shengquan, Yuan Liping is 5–6. The main fields of his papers on the application of knowledge graph are special education, education with big data, and the current hot issues of education which are generally concerned by the society. On the other hand, from the perspective of the institutions where the core authors work, seven of the nine core authors have worked or studied in normal universities, which fully demonstrates the "fresh force" role played by normal universities in the application of educational knowledge graph.

Of course, what cannot be ignored is that some scholars have tried to apply knowledge graphs to educational research since 2010, and selected appropriate research methods or data processing tools to assist academic research according to their own research expertise. However, there are also some scholars applied their own proficient research methods and tools to multiple research topics, forming the situation of "research topics drift from time to time", "research tools are always single", and "posting articles for the purpose of publishing". Although it is helpful to summarize and review the relevant research in different fields, it is not conducive to the development of scholars' academic career.

**Co Authors.** Network density [4] is often used to describe the density of inter-connected edges between nodes in a network. For a network with N nodes and L relations as edges, the density formula is:

$$d(G) = 2L/N(N-1) \tag{1}$$

For the research of educational knowledge graphs, the nodes N = 714, edges L = 697, network density D (g) = 0.00274 are sparse, which indicates that there is no stable cooperative author group, in this field, and the mutual cooperation between authors needs to be further strengthened. According to calculations, there are 222 groups of connected regions in this field. The largest section consists of 13 authors. The first author's units are all Beijing Normal University, 6 papers in total, namely: "The Study on Architecture and Application Model about Regional Education Big Data", "Educating People Intelligent Assistant: Key Technology and Implementation", "An Artificial Intelligence Assistant System for Educating People: The Structure and Function of 'AI Educator' ", "Design and Framework of Visualization Construction and Evolution System of Ontology", etc.

Due to the possibility of multiple collaborations between authors, this study adopts a weighted degree centrality algorithm. It is found that degree centrality is greatly affected by the number of authors in a single article. There are as many as eight authors cooperating in multiple articles, and the authors with high degree centrality generally cooperate with multiple scholars. Among them, Yu Shengquan of Beijing Normal University has the highest network centrality of 9.0, while Li Shuyu of Beijing University of Aeronautics and Astronautics and other 17 people have a network centrality of 7.0. In addition, from the perspective of betweenness centrality, Yu Shengquan (50.5), Lan Guoshuai (30.0), and Ma Ning (27.0) have high betweenness centrality, which shows that they play a bridge role in the collaboration of authors. According to the content of the papers published by the authors, these authors generally have a profound academic research background on educational technology and knowledge graphs. As a practical discipline to promote and improve the quality of education, educational technology plays an important role in solving practical educational problems by using systematic methods and tools.

### 3.6    Distribution of Data Sources

Knowledge graphs is widely used in educational research. The commonly used databases include CNKI and Web of Science(WOS). In terms of CNKI database, as shown in Table 2, the number of studies with CSSCI journal papers as a sample is the largest, up to 115, accounting for 29.95% of the total data sources, 56 core journals, accounting for 14.58%. These two kinds of sample selection methods mainly consider the contribution of sample quality to the research topic, so the relevant literatures in general journals are excluded. However, there are still 85 articles based on all relevant CNKI journals. This choice is not only to list all the relevant academic literature, but also to increase the sample size. However, to a certain extent, it will be disturbed by relatively low quality research. The interference of literature affects the reliability and scientific of research conclusions. The WOS database also has the same characteristics. There are 65 articles taking SSCI, SCI and A&HCI journal as samples, accounting for 16.93%, but there are still 14 literatures taking all WOS papers as sample sources.

Usually when querying in the database, researchers will limit the citation database, time interval, subject keywords and other elements, export them into formatted data, and clean out the content irrelevant to the research. For example, when Wang Youmei [15] explored the research status and development trend in the field of e-schoolbag in China, he limited the topic to "e-schoolbag" on CNKI, and the data source were the general network database of Chinese academic journals and excellent master and doctoral dissertations, and the time interval was 10 years before the author published the paper. Zheng Yafeng's [17] in the research on the topic structure of international flipped classroom, the data comes from the WOS core database, which includes seven sub databases: SSCI, SCI-EXPANDED, CPCI-S, CPCI-SSH, A&HCI, CCR-EXPANDED, and IC. The time interval is the first 15 years from 2000. The topic is limited to different English names in the development of flipped classroom, and 216 valid data records are obtained.

In addition, scholars have made extensive explorations on the definition of subject headings and the scientific nature, among which there are some novel researches, such as the atlas analysis of the proposal of the Teacher's Congress through natural language processing, so as to provide an objective basis for improving the internal governance of universities [6]; By using the method of word frequency statistics and knowledge graphs, this paper analyzes the inheritance and difference between Confucius and Yan Zhitui's educational thoughts in the text of "The Analects" and "Yan Family Instructions" [3]; By using MOOC platform course video subtitles and forum data to establish a knowledge graph of specific courses, and discuss the teaching reflections and model transformation in the "post-MOOC era" [12].

**Table 2.** CNKI educational literature knowledge graph author's data sources

| Source | Count | Ratio |
|---|---|---|
| CSSCI | 115 | 40.21% |
| CNKI Journals | 85 | 29.72% |
| Chinese Core Journals | 56 | 19.58% |
| Master and PhD thesis | 30 | 10.49% |

## 3.7  Distribution of Data Processing Tools

Knowledge graph is mainly through the form of vivid and intuitive pictures to dig out a large number of potential information hidden in educational resources. The commonly used graph tools include Citespace, Ucinet, NetDraw, VOSviewer and so on. In this study, the data processing tools used in the relevant research were statistically analyzed. According to the results in Table 3, there are 248 literatures using Citespace software, accounting for 70.45%, which fully shows the unique characteristics of Citespace. With the development and extension of knowledge graph research, the research methods are complex. Researchers gradually use multivariate statistical analysis, citation analysis, co-citation analysis, word frequency analysis, author cooperation analysis, social network analysis and other methods to interpret the knowledge map. Among them, SPSS (79 articles), Bicomb (62 articles), Excel (29 articles) and other software are mostly used for co-word analysis. Co-word analysis can quantify the correlation strength of keywords in the research area, so as to reveal the Knowledge structure and trend in this field. However, the authors lack a systematic analysis of other methods and tools in relevant research, which leads to the nonstandard, lax, and unsystematic use. Some scholars make subjective and artificial clustering through empirical judgments, which brings one-sided and bias to the research;

**Table 3.** Distribution table of Data Processing Tools for related research (part)

| No | Tools | Count | No | Tools | Count |
|---|---|---|---|---|---|
| 1 | Citespace | 248 | 7 | SATI | 15 |
| 2 | SPSS | 79 | 8 | Bibexcel | 13 |
| 3 | Bicomb | 62 | 9 | VOSviewer | 8 |
| 4 | Ucinet | 40 | 10 | ROSTCM | 5 |
| 5 | NetDraw | 31 | 11 | HistCite | 5 |
| 6 | Excle | 29 | | | |

some scholars directly import data by using co-word software, and do not properly merge and clean up the approximate topics, resulting in data sparse after clustering.

## 4    Conclusion and Thinking

This study reviews and summarizes the application of knowledge graphs in educational research in China, which can not only objectively present the current research status in this field, but also accumulate experience and strength, which will lay an empirical foundation for follow-up research. Based on the above analysis, the following four considerations are proposed for future research and practice.

1. Focusing on hot topics, using the "one matter, one discussion" template interpretation is more. It is common for some important problems in the field of education, even ancient themes. The research field is relatively concentrated, and there is no lack of repeated research at a lower level. At the same time, scholars lack of systematic construction of research methods, the systematization and inheritance of research results are poor, and the standardization of research needs to be enhanced. According to this data set, some researchers used the same research structure to repeat the low-quality papers in the field of subtle changes. Researchers should be guided by scientific and systematic methods and real problems, and explore the level and quality of educational research from the perspective of knowledge graphs.
2. CiteSpace has been paid close attention to by researchers in Bibliometrics, but it is not suitable for all academic researches in the field of educational. Chen Chaomei, the software designer, points out that CiteSpace is more suitable for studying the evolution of a certain topic, and it needs targeted topic retrieval. Ke Wentao [7] suggested that researchers should check whether the research field can be divided into two or three research fields before using the software, and query with keywords with high identification, fine-grained and subject matching.
3. The author's cooperation network and the organization cooperation network are relatively sparse. On the one hand, because some authors have completed the research alone, there is no connection between the author entities, which indicates that there is still a lack of interactive cooperation in this field; on the other hand, there is a large number of interdisciplinary cooperation within the research institutions, and there is a lack of Cross University cooperation, which indicates that the efforts of multi-party participation and integration in academic achievements need to be strengthened.
4. The cluster analysis of the authors of the same name needs further study. The author's identity setting will change the unit due to reasons such as further education and employment. At present, there is no better research scheme to integrate the same author. This paper crawls the link from the CNKI bibliography page to the author's personal homepage information. First, the author with the same CNKI ID is taken as an entity, and then the author

entity with the same name is merged according to the author's name, unit, historical co-author, research field and other comprehensive information on the Internet.

5. At present, there are many researches on the theoretical level of educational knowledge graphs, but the practice needs to be deepened. The research focus of education knowledge graph has gradually changed from literature metrology in the visual field to artificial intelligence using big data to help improve the quality of education. Under the support of knowledge graph and data driven to assist education: there is a better way for students to find and fill the gaps of subject knowledge points, clear learning path, and a better way of interdisciplinary knowledge modeling and fusion. The development of intelligent teaching question answering system based on knowledge map and accurate knowledge search engine can also provide great help for teachers and students. It is worth noting that at present, this field basically stays in theoretical research, and how to play its role in combination with practice needs further exploration.

# References

1. Baton: Learning Neo4j 3.x. Tsinghua University Press (2019)
2. Council, S.: Notice of the state council on printing and distributing the development plan for the new generation of artificial intelligence (2017). http://www.gov.cn/zhengce/content/2017-07/20/content_5211996.htm
3. Feng, L.: A comparative study of education philosophies of confucius and yan zhitui based on word frequency statistics. Libr. J. **37**(10), 70–77 (2018)
4. Han, Y., Fang, B., Jia, Y., Zhou, B., Han, W.: Mining characteristic clusters: a density estimation approach. J. Commun. **33**(5), 38–48 (2012)
5. Huang, W., Chen, Y.: A study on the mapping knowledge domains of china's educational economics development trends. Educ. Econ. **3**, (2010)
6. Kang, X., Liu, S., Feng, Z., Sun, Y.: University faculty interests orientation based on the proposal content analysis—a case study of the faculty congress proposals of three project 985 universities from 2007 to 2014. J. Dalian Univ. Technol. (Social Sciences) (2), 133–139 (2017)
7. Ke, W.: Disenchantment of tools: citespace's application criticism and reflection in educational research. Chongqing High. Educ. Res. **7**(5), 117–128 (2019)
8. Li, Z., Li, J., Yuan, Y., Lei, M.: Research on knowledge map educational application in Japan. In: 2019 Eighth International Conference on Educational Innovation through Technology, pp. 190–193. IEEE (2019)
9. Liu, H., Tang, W., Ren, Y.: Data visualization promotes scientization of educational decision-making: connotations, strategies and challenges. Res. Educ. Dev. **5**, 75–82 (2018)
10. Liu, K., Li, C., Bai, F.: Bibliometric study on the name authority literatures in library and information field in china. Bibliometric Study on the Name Authority Literatures in Library and Information Field in China **1**(12), 66 (2017)
11. Liu, X.: Scientometrics Methods of Science and Technology Policies Research. Science China Press (2019)
12. Lu, X., Zeng, J., Zhang, M., Guo, X., Zhang, J.: Using knowledge graphs to optimize MOOC instruction. Distance Educ. China **7**, 5–9 (2016)

13. Tsvetovat, M., K.A.: Social network analysis for startups (2013)
14. Wang, X., Liu, H.: An analysis of knowledge chain based on social network theory. An Analysis of Knowledge Chain Based on Social Network Theory **26**(2), 18–21 (2007)
15. Wang, Y., Chen, H.: Focus and direction of e-schoolbag research in china in recent ten years—based on the knowledge map analysis of co-word matrix. China Educ. Technol. **5**, 4–10 (2014)
16. Yi, G., Liu, S., Zhao, W.: Mapping of research hot-topics in journal of higher education and their intellectual base. J. Higher Educ. **10**, 74–80 (2009)
17. Zhen, Y., Huang, Z., Xu, C., Li, Y.: Topic structure and visualization analysis of international flipped classroom researches. China Educ. Technol. **10**, 53–59 (2015)

# Course Evaluation Analysis Based on Data Mining and AHP: A Case Study of Python Courses on MOOC of Chinese Universities

Hongjian Shi[✉]

Shandong University of Finance and Economics, Jinan, China
sjh17753191919@163.com

**Abstract.** Python has become a hot spot for people to learn, but beginners who lack the relevant expertise can't find the right course among the numerous learning resources of python. Therefore, based on python courses on MOOC of Chinese universities, the paper uses AHP and data mining to evaluate courses from the curriculum, student activity, students comment on several aspects and filters out python courses for beginners.

**Keywords:** Online learning · Python courses · AHP · Course evaluation

## 1 Introduction

As a product of Internet and higher education, online learning platform has become an important tool for students' daily learning, it provides rich learning resources for students. During the outbreak in 2020, the Chinese Ministry of Education organized 22 online platform to provide free online course more than 24000, countries also use online learning platform to provide services for the students to learn at home. Online learning platform played an incomparable advantages compared with offline classroom. Online learning is convenient, selectable and not subject to the site, it become an important way for the learners' learning. And more and more people will focus on the study of online learning platform.

Programming languages have always been a hot topic. Whether dealing with related applications or doing some research, programming languages are indispensable tools for us. It can help us save a lot of time and make us finish the task more efficiently. With the development of artificial intelligence techniques, python has become a hot topic for people to learn with its concise and powerful artificial intelligence algorithm library [1]. Because of the advantages of online learning, such as convenient, selectable and so on, many students choose to take Python courses on online platforms. However, because of the abundant and numerous characteristics of online learning resources, learners cannot locate the appropriate learning resources accurately, it has caused a certain amount of trouble to learners' learning efficiency. In particular, beginners may not be able to understand and absorb the course content well due to the lack of relevant

W. Gao et al. (Eds.): FICC 2020, CCIS 1385, pp. 361–368, 2021.
https://doi.org/10.1007/978-981-16-1160-5_28

professional knowledge. They may not be able to find suitable courses from a large number of "zero-basis" courses directly.

At present, China's online learning platforms mainly include Chinese university MOOC, super star learning link, rain classroom, Tencent classroom, youdao quality course, etc. Among them, Chinese University MOOC has many courses and is well-known. Therefore, this paper takes python courses on MOOC of Chinese universities[2] as an example and uses data mining, word cloud analysis and AHP methods to evaluate courses from course opening, student activity, student comments, etc. And the python courses suitable for beginners are screened out.

## 2   Related Work

At present, many researches define curriculum evaluation as evaluating the performance of students. Liu Zhijun [3], Zeng Xiaoping [4] and others have done some researches on developmental evaluation in curriculum. There are also researchers who evaluate the course itself, and the course evaluation in this paper belongs to this type.

In [5], Zhu Lingyun et al. analyzed and summarized the evaluation standards of online courses, and they summarized the research status of online course evaluation. And many scholars designed evaluation indexes of online courses based on AHP method. Wang Lizhen et al. evaluated the online course "Information Technology and Curriculum Integration" from the perspective of teaching practice based on AHP [6], and Xing Hongyu designed the online course evaluation index system by combining AHP method and Delphi method [7]. Some scholars also use fuzzy mathematics to evaluate online courses, such as [8]. Zhang Jiajian et al. believed that AHP and fuzzy evaluation were not easy to operate in course evaluation, so they proposed to use network metrology method to quantitatively evaluate online courses from the perspective of learning performance [9].

These researches mostly evaluated courses based on the content design and quality of the course, they fail to classify and analyze the learner's knowledge background, and they are not helpful for beginners to find suitable courses. Therefore, this paper considers the feature that beginners' knowledge reserve is not deep, uses data mining technology to crawl course information and comments, and combine AHP method to evaluate online courses from the curriculum, students comment on several aspects.

## 3   Evaluation Method and Process

The purpose of this paper is to crawl and analyze the python course information and course comments from MOOC of Chinese universities, and evaluate the python course to choose one suitable for beginners. This experiment mainly includes five steps of data query, data retrieval, data cleaning, data analysis and course recommended, as shown in Fig. 1.

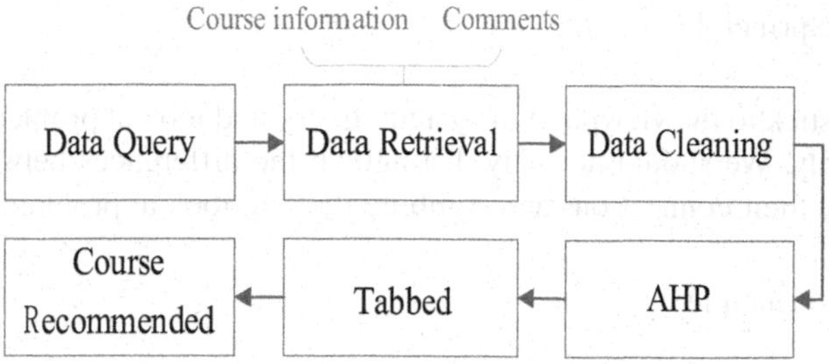

Course information     Comments

| Data Query | → | Data Retrieval | → | Data Cleaning |

| Course Recommended | ← | Tabbed | ← | AHP |

**Fig. 1.** Process

## 3.1 Data Acquisition

MOOC of Chinese universities platform applies Ajax (Asynchronous Javascript and XML), so there is a time difference between the target data transmission and other data transmission in the web page. We use the selenium which can simulate user in the browser to access web to simulate the User-Agent camouflage crawler to visit MOOC of Chinese universities. We take breadth-first strategy. We crawl the links for each course from the start page and put them in the link pool. Then program walks through each link, crawls the course information data and course evaluation and cleans the data. Specific steps are as follows:

(1) Access the starting page and use the page-turning technique.
(2) Put the links of each course into the classified link pool.
(3) Traverse each link to visit the course details page.
(4) Crawl the target data, including the teacher of the corresponding course, time, whether it is a quality course, etc.
(5) Crawl user comments, user comments adopt AJXS architecture to turn pages in the same URL.
(6) If there are still unvisited classification links in the classification link pool, repeat step 5 until all the classification links is visited.
(7) Eliminate redundant information and clean data.

## 3.2  Course Evaluation

### 3.2.1  Index Selection

The quality of the evaluation index is the key to the evaluation effect. Whether the evaluation index is scientific and reliable will affect the accuracy of evaluation, so it is necessary to follow certain principles. The evaluation index follows the following principles:

(1)  The objective(scientific) principle

The selection of indicators should be based on scientific theories. We should be carefully analyzed and combined with the characteristics of online courses to select indicators.

(2)  Systematic principle

We should stick to the viewpoint of system theory and look at problems comprehensively and deeply. We should not only distinguish the differences between indicators, but also explore their connections and combine them to look at problems.

(3)  Simplicity principle

The indicators in the evaluation model should be simple and clear, so as to avoid the difficulties caused by too many or repeated indicators. And we should make the indicators of the evaluation system easy to obtain and the operation process of evaluation more concise.

(4)  Principle of representativeness

The evaluation indicators should be representative and suitable for accurate evaluation of online course results, so that the evaluation results conform to the objective reality.

Through existing literature about course evaluation and combining with the availability of data, this paper chose the number of classes, number of participants, number of comments, and course grade four indicators to carries on the preliminary evaluation about the quality of the course.

The number of courses opened can measure the frequency and activity of course updates, and it reflects whether the course will be updated and improved. The number of participants is a direct reflection of the popularity of the course. The number of comments can also reflect the popularity of the course, but it is more likely to evaluate whether students have a tendency to communicate after participating in the course. To some extent, it can reflect the quality of the course content. Course grade can represent students' experience of the course and it is a direct indicator of the relationship between the course and students.

The correlation between the four indicators is less than 0.8, the correlation is not strong, so we can choose these four indexes.

### 3.2.2  Normalization

Due to the large data gap between the course data (the zigzag line in Fig. 2), it is necessary to normalize the data. Normalization is the scaling of data to a specific interval between the given minimum and maximum values or the conversion of the maximum absolute value of each feature to a unit size. This method is a linear transformation of the original data. The data is normalized to the middle of [0, 1]. The normalized results are shown in the flatter line in Fig. 2.

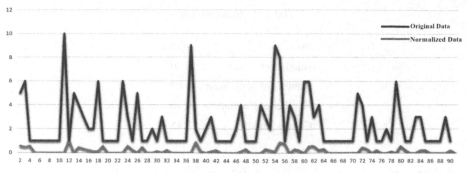

**Fig. 2.** Normalized results

### 3.2.3  Determine the Index Weight By AHP

AHP [10] analyzes the correlation between each index. It compares each element at each level in pairs and obtains the comparative scale of relative importance according to a certain scale theory. Then it establishes the judgment matrix to calculate the weight vector. The main steps are as follows:

(1)  Construct comparison matrix according to scale theory
(2)  Normalize the judgment matrix
(3)  Find the maximum eigenvalue and its eigenvector
(4)  Consistency test

The weight of each influencing factor is determined by establishing the hierarchical structure model, constructing the judgment matrix, obtaining the weight and conducting the consistency test (Table 1).

**Table 1.** Judgment matrix

|  | Number of classes | Number of participants | Number of comments | Course grade |
|---|---|---|---|---|
| Number of classes | 1 | 3 | 3 | 1/2 |
| Number of participants | 1/3 | 1 | 1/2 | 1/4 |
| Number of comments | 1/3 | 2 | 1 | 1/3 |
| Course grade | 2 | 4 | 3 | 1 |

Through consistency test, the final index weight obtained from the judgment matrix is shown in Table 2.

**Table 2.** Index weight

| Index | Weight |
|---|---|
| Number of classes | 0.30 |
| Number of participants | 0.09 |
| Number of Comments | 0.16 |
| Course grade | 0.45 |

### 3.2.4 Result

According to the weight of each index and index data, the weighted average algorithm is used to obtain the course evaluation score. We sort course according to the course evaluation score. The sorting result is shown in Fig. 3.

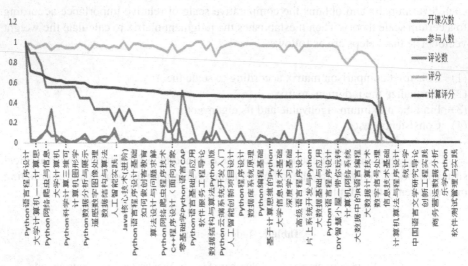

**Fig. 3.** Result

### 3.3 Comment Analysis and Results

Then we did a textual analysis of the course evaluation data (comment content, number of comment thumb up). The example of word cloud analysis results was shown in Fig. 4. Then we use TF-IDF to label the course, and the tagging example was shown in Fig. 5.

According to the results of tagging and the course evaluation score in 3.2, the python language programming course of Beijing Institute of Technology, which has a high score of 3.2 and an entry level tag, is selected as the recommended course for python beginners.

**Fig. 4.** Cloud analysis results

['老师', '课程', '非常', '讲解', '学习', 'python',
'很棒', '不错', '入门', '编程']

**Fig. 5.** Tagging example

## 4 Conclusion

Using online platform to learn python is a hot choice for many language learners. Using technology of data mining analysis processing to solve the problem is the current research trends. So this paper evaluates python courses on MOOC of Chinese universities combined with big data mining, analysis technology and AHP method. It provides an idea for online platform to evaluate the course. And according to the problem that beginners' knowledge background is not deep enough to judge the suitability of the course intuitively and quickly, this paper uses students comments for text analysis and selects python courses suitable for beginners according to labels and course quality evaluation. It solves the problem that beginners could not locate the appropriate online learning course accurately.

This paper only uses existing data of the online course (contents, introduction, comments, etc.) to evaluate the online course. To a certain extent, the result can reflect the relevant information of online course. But it is not accurate enough. In the later study, we can combine the specific behavioral data of learners to evaluate courses more detailed. It is also possible to use clustering and other methods to classify learners' behavioral preferences more specifically, so as to achieve a more complete evaluation of the course and thus achieve personalized recommendation.

## References

1. Liu, Y., Lai, X., Li, P.: Research on the Python case teaching in the background of artificial intelligence. Comput. Era **322**(04), 93–96 (2019)
2. MOOC of Chinese universities. https://www.icourse163.org/. Accessed 20 June 2020
3. Liu, Z.: On the basic concept of developmental Curriculum Evaluation. Subject Education (2003)
4. Zeng, X., Jiang, X., Wang, D., et al.: Curriculum evaluation scheme design and practice based on engineering education certification: a case study of polymer physics. Polym. Bull. **12**, 60–65 (2019)

5. Zhu, L., Luo, Y., Yu, S.: Evaluation of virtual teaching curriculum. Open Educ. Res. **000**(001), 22–28 (2002)
6. Wang, L., Ma, C., Lin, H.: The design of evaluation index system of teacher education online course: the case analysis of online course "information technology and curriculum integration." China Educ. Technol. **11**, 68–71 (2008)
7. Xing, H.: Research on the design of network course evaluation index system based on AHP method and Delphi method. China Educ. Technol. **236**, 78–81 (2006)
8. Hu, S.: Design and implementation of network course evaluation system based on fuzzy theory. E-Educ. Res. **06**, 52–55 (2006)
9. Zhang, J., Zhan, N., Li, Y.: Research on network course evaluation method based on network metrology. J. Distance Educ. **01**, 66–72 (2015)
10. Han, L., Mei, Q., Lu, Y., et al.: Analysis and research of AHP-fuzzy comprehensive evaluation method. J. Chin. Saf. Sci. **07**, 89–92+3 (2004)

# Process-Oriented Definition of Evaluation Indicators, Learning Behavior Collection and Analysis: A Case Study

Jiakuan Fan[1], Wei Wang[1(✉)], Haiming Lin[2], Yao Liu[1], and Chang Liu[2]

[1] East China Normal University, Shanghai, China
wwang@dase.ecnu.edu.cn
[2] Tongji University, Shanghai, China

**Abstract.** The burgeon of online education platforms and online training platforms represented by MOOC has brought new development opportunities for educational innovation. In so many educational and training scenarios, there is a large amount of distributed users' learning data, which leads to the non-standardization of learning data collection and the non-uniformity of storage. In this paper, we design a process-oriented model for evaluating learning indicators. Then, we take KFCoding, which is an online training platform, as a case study to gather data for the learning process. Finally, we model the collected data and analyze the relationship between learning process data and learning outcomes.

**Keywords:** Online learning platform · Process-oriented · Learning behavior · Data collection

## 1 Introduction

E-Learning learning system is a learning platform that integrates online learning resources and intelligent recommendation technology. The basic idea is to obtain students' learning behaviors through the online learning platform, use information retrieval technology to recommend learning resources that meet students' scenarios, and provide students with large-scale learning services [1]. Massive Open Online Course (MOOC), which is a typical example of an E-Learning learning system, has increasingly played an irreplaceable role. It breaks the boundaries of traditional education area and time, and realizes the relative fairness of educational resources [2]. MOOC platforms such as Coursera, Udacity, edX, and Learning Classes Online have attracted a huge number of schools, teachers, courses and students [3, 4]. The emergence and large-scale use of MOOC has changed the way content is organized while changing the way teachers and students interact. However, MOOC is still far from being able to replace traditional teaching methods. It is difficult for teachers to understand the details of the learning process of their students, which leads to trouble tracking the teaching process, evaluating the effectiveness of teaching at a consistent granular level, and personalizing teaching.

At present, these learning platforms are distributed with hundreds of millions of learners from all over the world. The learning data of their courses records the learning

© Springer Nature Singapore Pte Ltd. 2021
W. Gao et al. (Eds.): FICC 2020, CCIS 1385, pp. 369–380, 2021.
https://doi.org/10.1007/978-981-16-1160-5_29

trajectories and learning preferences of everyone. If these records can be obtained, the analysis of these data can greatly improve the efficiency of learning, and can help the learners to understand their own learning conditions more clearly. It is possible to develop a set of their own learning courses, so as to achieve diversity and personalization of education, to achieve the purpose of teaching according to their aptitude [5]. Therefore, the following problems exist: 1. Inability to measure learning against online learning data. 2. The data collection is not standardized, which is reflected in the differences in the collection methods of various platforms. 3. The data storage is not uniform, which is reflected in the large differences in storage methods of various platforms. 4. Poor reliability in data analysis, reflected in the single source of data analysis.

The main contributions of our work:

1. A quantitative learning indicator model for the online learning process is proposed to evaluate students' learning.
2. We implement a system that supports a custom approach to collecting various learning trajectories for the online learning platform, which supports uniform data format definition, standardized data storage, and multiple data source collection. It also supports real-time import of learning data and real-time monitoring of learning status.
3. We model the collected user data and analyze the relationship between learning process data and learning outcomes.

## 2   Related Work

Lee et al. [6] studied learning strategies that corresponded to learners' personality traits during online learning in different online environments; Yang and Tsai used questionnaires to collect information and explore learners' learning environment preferences and learning beliefs during online learning; Baker [7] used Web data mining techniques to model learning quality assessment in response to the fact that learning quality is poor in current online learning systems. Shuang Li [8] et al. constructed a framework for the analysis of online learning behavioral input based on the learner's behavioral input, and based on the results of the research analysis, they concluded that online learning behavioral input measurement has a greater impact on course performance. Zhuoxuan Jiang et al. [9] classified and explored online learning behaviors with respect to the characteristics of learners in the MOOC learning platform, and studied the strong relationship between online learners' learning behaviors and learning outcomes. Qiang Jiang and Wei Zhao [10] et al. comprehensively analyzed the relationship between learners' behavior and outcomes based on big data analysis in 4 dimensions: data and context (What), beneficiaries (Who), methods (How) and goals (Why). Chao-Kai He [11] used Logistic regression to predict performance by analyzing the learning behaviors that exist in large- scale online classrooms, identifying the potential learning behavioral characteristics of learners, uncovering representative learning behavioral characteristics, and thus achieving the effect of improving the quality of education and teaching. Shaodong Peng [12] identified three kinds of mining objects: server logs, platform course databases, and forum post collections, based on the intersection level perspective, and used big data analysis techniques to excavate the general framework of three kinds of learning behavior characteristics: operational behavior patterns, activity behavior patterns, and speech behavior characteristics. Yang Zong et al. [13] applied the logistic regression method to discover the relationship between learning behavior and learning outcomes and obtained key indicators to predict learning outcomes. Zhi Liu et al. [14] used discourse analysis to study forum interaction data, interaction topics, interaction emotions, etc.

## 3 System of Learning Evaluation Indicators

### 3.1 Overall Definition

Learning assessment is an important prerequisite for pedagogical decision- making, and effective learning evaluation depends on a comprehensive and reliable assessment base. Traditional teaching and learning evaluations have problems such as single subject, insufficient basis, one-sided content and subjective approach. Big Data focuses on deep mining and scientific analysis of multidimensional, large amounts of data to find the implicit relationships and values behind the data, helping to move educational evaluation from speculation based on small sample data or fragmented information to evidence-based decision making based on comprehensive, holistic data.

Traditional assessment methods, which focus on assessing students' mastery of knowledge, ignore behavior and performance in learning activities and are not conducive to diagnosis and timely intervention in learning situations.

In the design of the evaluation indicator system, it is necessary to consider not only the final effect of learning, but also the evaluation of the learning process. Therefore, on the basis of the educational goal classification theory of educators such as Bloom and Simpson, we propose to build an evaluation indicator system for the learning process from the three dimensions of learning style, learning participation and learning effect, in which learning style measures the non-intellectual factors of learners, including learning interest, learning attitude, learning methods and learning habits and other factors; learning participation measures the degree of participation of learners; learning effect measures the degree of learners' ability to contribute to the completion of learning, we proposed evaluation indicators from learners' knowledge application ability, practical ability, thinking ability, collaboration ability, self- learning ability and other aspects. The dimensions involved in the evaluation indicator system are shown in Fig. 1.

### 3.2 Detailed Definition

In this paper, Our research on learning inputs in e-learning mainly involves descriptive analysis of learning inputs in e-learning, analysis of influencing factors, and analysis of the association between learning inputs and the quality of learning. We divide the learning inputs into four dimensions of participate, focus, regularity, and interaction, each of which contains several second-level dimensions underneath. Since there are many indicators of learner engagement in learning, representative dimensions are chosen for illustration. As shown in Table 1.

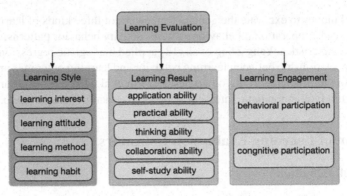

**Fig. 1.** System of process-oriented learning assessment indicators

**Table 1.** Metrics for learning input models

| Learning input dimension | Learning input indicators | Measure of learning input |
|---|---|---|
| Focus | Landing average time | Average time during each learning |
| | Quality of coursework | Quality of coursework submitted by learners |
| Participate | Login learning space | Number of login |
| | Browse courses | Number of browse |
| | Participate activities | Number of participate |
| Regularity | Learn on time | Whether to attend on time |
| | Average interval | Mean of landing time interval |
| | Landing interval | Interval between logins |
| Interaction | Post | Number of posts |
| | Upload resources | Number of resources uploaded |

## 4  Preliminary Knowledge

We use the KFCoding online training platform as a case study to implement the detailed metrics defined in Sect. 3. We adopt LRS and Experience API technologies as our implementation approach to collect the Learning data for students oriented to the learning process.

### 4.1  KFCoding

KFCoding [15] is an online training platform developed by the X-lab laboratory based on the specific landing application representative of MOOBench. The system architecture is shown in Fig. 2, and the vision is to provide one-stop training service and interactive learning experience. The platform is currently divided into six major sections: open source university, content creation, fast experience, school teaching, practical training and children's programming. In the school teaching section, the online training platform is introduced into the curriculum teaching. In terms of course content, dozens of courses including current popular deep learning, cloud computing, and big data processing systems have been produced for students to learn.

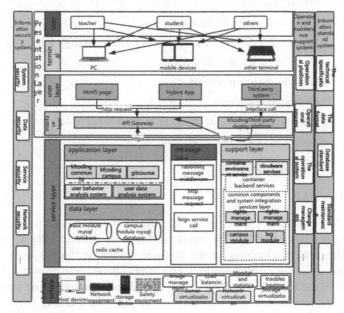

**Fig. 2.** The overall architecture of KFCoding: The system is divided into presentation layer, service layer and infrastructure layer.

The infrastructure layer mainly provides hardware resources and network control management for upper-level services and applications. Infrastructure services, including data storage and distribution at the data layer, various applications at the application layer, and cloud services and delivery of messages at the support layer. At the presentation level, it provides teachers and students with course production and course learning environment respectively.

## 4.2   LRS

Learning Record Store (LRS) is a learning database that can store activity types including mobile applications, games, simulators, and online courses. LRS is the main component of the experience API ecosystem. It is used to receive, store, and return data about learning experiences. These data can also be shared with other systems, and data analysis can be adaptively learned in the form of reports. When running an LRS system, the collected data can be simply analyzed and sorted, and the learning status of the learners can be evaluated in the form of charts to facilitate the timely development of solutions [16].

Through the unified deployment of LRS, the learning data of the MOOC platform and the training platform are injected into the LRS, which can solve the problem of non-uniform data storage.

**Table 2.** Statement property

| Property | Description |
| --- | --- |
| ID | System automatically allocates |
| Actor | Differentiate different learners to ensure the accuracy of the entire system |
| Verb | Describe the learner's action state |
| Object | Describe the learning goals |
| Result | Representing test results |
| Context | Provide specific context for declarations and refine data analysis |
| TimeStamp | Record when learning took place |

### 4.3 Experience API

Experience API (xAPI) [17] is the next generation of SCORM, a technical specification for recording learning records, with simple and flexible features [18]. It can track the recorded fields not only learning in the mobile, simulation, virtual world, and game fields, but also events in the real world, scene applications under experiential learning, learning data in the social field, offline learning and collaborative scene learning [19].

When users learn courses in the learning system, the learning events can occur anywhere, and these events can be represented by different sentence components. All these data can be recorded through xAPI. These services carry a JSON load. When users learn courses, they will generate corresponding activity records. The specific scenario description is shown in Table 2.

## 5   Implementation of Indicators

The LRS system is deployed as a storage repository, which can handle the standardized data structure formats. Generally speaking, the most basic structure that needs to satisfy the actor-verb-object, such as "Jerry is cleaning the room" is the most basic syntax structure defined. This definition of structure can be useful for understanding every event that happens. Knowing who did what at what time is the simplest way to record data for a learning process. There- fore, it is particularly important to define accurate descriptions of actions and events, which can guarantee that you will get an authentic record of the learning process. In addition, a standard set of built-in action event definitions is also available, which is also very convenient for users to call. The description of each statement is the most basic record of an event, and the records are stored uniformly in JSON format. For actors, there is an "mbox" attribute, which is a unique value describing the identity of the user; for verb, it has an"ID" attribute, which describes the specific content of the action; for object, it also has an"ID" attribute, which describes the specific content of the object. When a basic statement is defined, it can be sent to the deployed LRS system through ADL's sendStatement.

On our online training platform, we use xAPI to collect data for website users' learning data. The main structure is shown in Fig. 3. First, the LRS system is deployed on the server to store the collected data, which can solve the problem of unified storage of data. Then, pre-configured scripts for collecting data are ported to KFCoding website. In the website, extensive types of learning data are easily observable, including educational videos, course knowledge, practical sessions, and case studies, etc. When a user conducts an operation on the page, an event is triggered. Once the back-end authority control module confirms that the information can be collected, the collected information will be returned to the LRS for standardized storage through the HTTP service. For the records that have been stored in the LRS, we can use common chart analysis tools for visual analysis to understand the state of student learning. Further, we can model and analyze the stored data in a machine learning way.

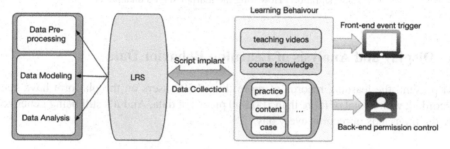

**Fig. 3.** xAPI adapts KFCoding

**Table 3.** Verb form of KFCoding LMS platform

| ID | Verb | Describe |
|---|---|---|
| 1 | Select-school | Record which school the user clicked |
| 2 | Select-course | Record which course the user clicked |
| 3 | Select-units | Record which units the user clicked |
| 4 | Select-unit | Record which unit the user clicked |
| 5 | Learn-unit | Record how long have the user learned |

It is notable that the collection of student learning events is specific to a particular scenario. In order to collect learning data across the network to comprehensively reflecting students' learning behavior, we define functions to collect data right where we initialize the course master template, which can guarantee that any courses added to the website in the future will be automatically added to the data collection system.

In order to better meet KFCoding's data collection needs, we use custom actions and event types. The action's definitions are shown in Table 3, which defines three actions including selecting a school, selecting a course, and selecting a unit. The definition of event types is shown in Table 4. There are 4 types of schools, courses, units, and chapters. In the school, there are 18 schools including open source universities and East China Normal University. The curriculum includes dozens of KFCoding platform productions.

**Table 4.** KFCoding learning behavior Activity table

| ID | Activity | Describe |
| --- | --- | --- |
| 1 | School | Include many universities |
| 2 | Course | Include a series of courses |
| 3 | Unit | Each course contains multiple units |
| 4 | Chapter | Belongs to the next level of the unit |
| 5 | Duration | How long the learner stayed on a page |

## 6   Display and Analysis of Learning Behavior Data

At present, the learning records of more than 2,500 users on the platform have been recorded, with a total of more than 100,000 pieces of data. And it's still being generated by the thousands of records every day.

**Fig. 4.**  Basic charts                          **Fig. 5.**  Highly customizable chart

Figure 4 shows the visualization functionality based on the X-lab organization. At present, the system interface provides about 10 kinds of charts that can be used, which basically includes all common types. This also implies that you can essentially demonstrate your students' behavioral data in any kind of graphical format.

It is worth mentioning that all the charts are highly configurable. As shown in Fig. 5, we can not only customize the basic information such as the horizontal and vertical coordinates and title of the chart, but also set the information and display designation in any period. More advanced, it can completely achieve more complex requirements by building custom query. In addition, what is more attractive is that all these charts are interactive in real-time, which means that the chart will refresh automatically when any of the parameters has changed.

```
"actor": {
    "name": "10194102436",
    "mbox": "mailto:10194102436@kfcoding.com",
    "objectType": "Agent"
},
"timestamp": "2020-03-12T05:50:01.689Z",
"version": "1.0.0",
"id": "632f7867-c543-4f51-8f81-b400d0d10514",
"result": {
    "score": {
        "scaled": 0.95
    },
    "success": true,
    "completion": true,
    "duration": "PT4.77S"
},
"verb": {
    "id": "http://kfcoding.com/verbs/learn-unit",
    "display": {
        "en-US": "learn-unit"
    }
},
"object": {
    "id": "http://kfcoding.com/activity/",
    "definition": {
        "name": {
            "en-US": "learn-unit:程序设计与人工智能 朱老师2班—2 数据的获取和分析"
        }
    }
```

**Fig. 6.** Data field details

As shown in Fig. 6, It is a complete record of all learning behavior events, with the raw data stored and presented in a JSON format. It mainly includes 5 fields, in which the actor filed represents the learner information, the timestamp field illustrates the time recorded, the result field has a duration attribute that represents the time spent on a page, the verb field exhibits what action did user choose, the objective field describes what did users do.

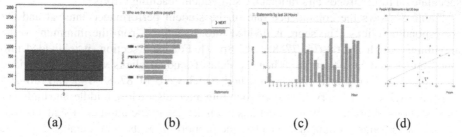

|        (a)        |        (b)        |        (c)        |        (d)        |

**Fig. 7.** (a) shows that the average learning time of the user on one page. You can see that the median is about 400 s. (b) shows that who are the most active people. (c) describes how traffic has changed over the past 24 h. (d) describes the relationship between visits and clicks in each hour over the past 30 days.

As shown in Fig. 7, due to space constraints, we extract a portion of the student's behavior data to display. (a) illustrates the percentage of time students devote to learning. As can be seen from the figure, about 2% of students spend an average of about 200 s per learning record, approximately there are 25–50% of people who spend about 400 s per study, and about 50–75% of people who spend the time is around 900 s. The last remaining quarter of the population often spent more than 1,000 s per study. Therefore, it is possible to know exactly how much time each student has spent on studying at any given time in the past. Based on this, students who usually spend little time on learning can be promptly reminded and intervened. In this way, the inappropriate learning process can be achieved in the timely detection and correction.

(b) exhibits the distribution of the number of students' click events. The horizontal coordinates represent the total number of records of students studying on the website in a week, and the vertical coordinates indicate the students' number information for all students. Only the last four digits of the student number are kept for the user's privacy and security. It is obvious that the last four digits of the student number 2126 has the highest number of learning events in a week, nearly 160. This would also mean that the total number of all the student's learning records would be recorded and updated in real time. As teachers, they'll be able to keep track of and judge how well students are doing for the week based on the amount of learning they've recorded.

(c) responds to the change in learning events over the past 24 h. The horizontal coordinates indicate the moments of the day, and the vertical coordinates denote how many learning records the students had at each moment of the day. From the data in the figure, we can see how much the students' learning records have changed over the past 24 h. Furthermore, this allows us to understand each student's learning patterns and to find out what time of day is appropriate for learning, so that we can personalize the learning process for each student.

(d) describes the relationship between visits and clicks in each hour over the past 30 days. The horizontal coordinate shows the number of people who have had a learning record, and the vertical coordinate indicates the number of learning records. From the figure, we can see how the relationship is be- tween the number of learning records and the number of records and how the overall trend is changing. Based on this, we can speculate on future trends and rationalize subsequent teaching efforts.

Table 5 shows the coincidence rate of the student performance interval and the corresponding clicks. The score is divided into 3 sections from the minimum to the maximum, which are 68–76, 77–85, and 86–94. For each section, it is divided into low, middle, and high sections. Then the data of student clicks was equally divided into 3 intervals according to the same method, which were 30–87, 88–145, and 146–205 respectively. For each interval, the corresponding intervals are low, middle, and high. For three sections of grades and three sections of clicks, count the number of students who overlap in the same section, and calculate the proportion of the total number of people who overlap in the section. From the table, it can be found that the proportion in the low and high sections is 40%, and the highest in the middle section is 54%. In addition, among the 17 people who overlaps, they account for 50% of the total 34. Therefore, we can draw a rough conclusion: students' behavior patterns and achievements on the platform should show a positive correlation. As platform data becomes more abundant, this relationship will become more apparent.

Table 5. KFCoding LMS

| Grade | Clicks | Coincidence rate |
|-------|--------|------------------|
| 68–76 | 30–87 | 40% |
| 77–85 | 88–145 | 54% |
| 86–94 | 146–205 | 40% |

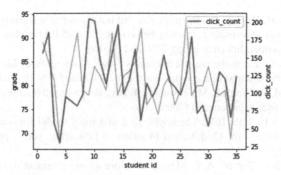

**Fig. 8.** Relationship between grades and clicks.

Figure 8 resonates the relationship between the students' academic performance and the number of learning records. We randomly select 35 students for the experiment, and the horizontal coordinates indicate the student's academic number and the vertical coordinates indicate the number of learning records of the Quantity and performance. As can be seen from the figure, the number of student records and academic performance show a rough consistency, that is, the greater the number the more likely they are to get good results.

## 7   Summary and Future Work

Firstly, we define a global perspective on the learning process-oriented evaluation indicators that include learning style, learning result, and learning engagement indicators. Secondly, we detail the design of evaluation indicators oriented to the learning process in four dimensions: focus, participation, regularity and interaction. Then, we conduct a case study on KFCoding, which is an online training platform developed by our laboratory. Based on xAPI technology, we realize the collection and unified storage of all learning data of students on the training platform. The standardized interface provided by xAPI is used to interface with the KFCoding training platform interface framework, to achieve standardized definition and collection of data. Finally, we model the collected data and analyze the relationship between data on students' learning processes and learning results.

**Acknowledgments.** Many thanks to my instructor for helping me and to my fellow X-lab students for giving me so much inspiration to complete my work. This study is supported by the National Natural Science Foundation of China (Grant No. 61672384).

## References

1. Gil-Solla, A., Blanco-Fernández, Y., Pazos-Arias, J.J.: A flexible semantic inference methodology to reason about user preferences in knowledge-based recommender systems. Knowl.-Based Syst. **21**, 305–320 (2008)

2. Huang, Y., Wu, P., Hwang, G.: The pilot study of the cooperative learning interactive model in e-classroom towards students' learning behaviors. In: 2015 IIAI 4th International Congress on Advanced Applied Informatics, pp. 279–282 (2015)
3. Zhang, M.: Micro-lesson-the prelude to china's mooc. Comput. Edu. **20**, 11–13 (2013)
4. Li, X.: Mooc: the window, or the shop? Univ. Teach. China **5**, 15–18 (2014)
5. Najjar, J., Duval, E., Wolpers, M.: Towards effective usage-based learning applications: track and learn from user experience(s) (2006)
6. Tu, L.J., Lee, W.I., Shih, B.Y.: The application of kano's model for improving web-based learning performance. In: 32nd Annual Frontiers in Education, vol. 1, pp. T3E–T3E. IEEE (2002)
7. Ryan d Baker, S.J., Corbett, A.T., Aleven, V.: More accurate student modeling through contextual estimation of slip and guess probabilities in bayesian knowledge tracing. In: Woolf, B.P., Aïmeur, E., Nkambou, R., Lajoie, S. (eds.) ITS 2008. LNCS, vol. 5091, pp. 406–415. Springer, Heidelberg (2008). https://doi.org/10.1007/978-3-540-69132-7_44
8. Li, S.: An exploration of online learning engagement models based on behavioral sequence analysis. China Electron. Educ. **3**, 88–95 (2017)
9. Yan, Z., Jiang, Z.: Learning behavior analysis and prediction based on mooc data. Comput. Res. Dev. **52**, 614 (2015)
10. Jiang, Q.: Personalized adaptive online learning analytics model and implementation based on big data. China Electron. Educ. **1**, 85–92 (2015)
11. Wu, M., He, C.: Learning behavior analysis and prediction of educational big data on the edx platform. China Dist. Educ. **6**, 54–59 (2016)
12. Peng, S.: A study of the three-cycle model of knowledge construction in blended collaborative learning. China Electron. Educ. **9**, 39–47 (2015)
13. Sun, H., Zong, Y.: Logistic regression analysis of learning behavior and learning effectiveness of moocs. China Dist. Educ. **5**, 14–22 (2016)
14. Zhang, W., Liu, Z.: A study of learner interaction discourse behavior analysis in cloud classroom forums. Research in electro-education **37**, 95–102 (2016)
15. https://kfcoding.com/
16. Wan, H., Yu, Q., Ding, J., Liu, K.: Students' behavior analysis under the sakai lms. In: 2017 IEEE 6th International Conference on Teaching, Assessment, and Learning for Engineering (TALE), pp. 250–255 (2017)
17. https://w3id.org/xapi/adl
18. Saliahhassane, H., Reuzeau, A.: Mobile open online laboratories: A way towards connectionist massive online laboratories with x-api (c-mools) (2015)
19. Zapata-Rivera, L.F., Petrie, M.M.L.: xapi-based model for tracking on-line laboratory applications (2018)

# The Reform and Construction of Computer Essential Courses in New Liberal Arts Aiming at Improving Data Literacy

Yue Bai⬤, Min Zhu⬤, and Zhiyun Chen(✉)⬤

Department School of Data Science and Engineering, East China Normal University,
Shanghai 200062, China
Grandcanyon001@163.com

**Abstract.** In the era of Big Data, all countries put innovation at the core of national development. The Higher Education Department of the Ministry of Education of the People's Republic of China calls for the comprehensive promotion of the construction of new liberal arts, with the aim of introducing the latest information technology into the traditional liberal arts teaching. As a normal university, East China Normal University is an important cradle to promote the "CS for ALL" talents. Therefore, in the computer reform of Shanghai colleges and universities, it takes the lead in the reform of computer basic teaching of New Liberal Arts. The fundamental goal of this reform is to cultivate students' Data Thinking and improve their Data Literacy. Good results are achieved in the aspects of target achievement degree and result satisfaction degree.

**Keywords:** New Liberal Arts · New Infrastructure · Data Literacy · Computer Foundation Course · CS for All · Course reform

## 1  Introduction

The development of new generation information technology, such as Big Data, Artificial Intelligence, Cloud Computing, Internet of Things and Blockchain, has had a profound impact on the world economic and social situation. China places innovation at the core of national development, attaches great importance to the development of Artificial Intelligence and promotes the implementation of the national Big Data strategy. Colleges and universities are an important base for training strategic reserve talents for the country. In the stage of general education, the level of basic computer education is closely related to the field of Big Data and artificial intelligence. It is not only related to the improvement of students' personal ability, but also an important event affecting the national development strategy and security.

The concept of New liberal arts was first proposed by Hiram College of the United States in 2017. However, China's "New liberal arts" is different from the "New liberal arts" of other countries [1], emphasizing the promotion of new technologies and the country's demand for talents. In August 2018, the Department of Higher Education of the Ministry of Education of China proposed that it is imperative to innovate and develop higher education, and comprehensively promote the New liberal arts [2], so as to provide strong support for the realization of China's education modernization in 2035.

In December 2018, China Central Economic Work Conference defined 5G, Artificial Intelligence, Industrial Internet and Internet of Things as "New Infrastructure Construction". The most important feature of the New Infrastructure is Data-Based and Intelligent [3]. It is the use of new technologies to empower the traditional infrastructure and provide a new growth point and breakthrough for the special economic situation in the post-epidemic era. Similarly, in the process of the construction of New liberal arts in colleges and universities, on one hand, it is necessary from the extension to make full use of intelligent technology to transform the teaching environment and teaching mode; on the other hand, it is necessary from the connotation to break through the barriers of traditional disciplines, and carry out "New Infrastructure", so as to empower the traditional disciplines by popularizing the new generation of information technology such as Big Data and Artificial Intelligence in the basic computer courses. By improving students' Data Literacy, as well as their ability to analyze, judge and make decisions on data, it paves the way for traditional humanities and social majors such as literature, history, philosophy, law and education, and helps traditional liberal arts majors to absorb new technologies. In the future application scenarios, students can integrate their meticulous work in the professional field with the latest development in the field of science and technology, and produce new breakthroughs and insights that cannot be obtained only by traditional research methods.

East China Normal University is a comprehensive and research-oriented university, which belongs to the first-class universities in China. Each year, more than 3000 undergraduate students are enrolled, and liberal arts students account for about one-third of the total number of students. To cultivate liberal arts students with Data Literacy is to cultivate "propaganda team"and "seeder" of "CS for ALL" (Computer Science for all people) [4].

## 2   Preparation for Teaching Reform

"*What's past is prologue*". With the rapid development of Information Technology, computer basic courses with Information Technology as the main teaching content are required to make corresponding tracking and adjustment, and the teaching content makes a "disruptive change" almost every 3–5 years. Before this round of computer teaching reform, East China Normal University has successfully completed many computer course reforms.

The last teaching reform was started in September 2014. The main work is to establish the basic computer teaching course system of liberal arts with Data Processing and Management as the core. This course system aims to guide the liberal arts students to master the basic literacy of effectively collecting, storing, processing, applying and retrieving data and information by using computer hardware and software technology through learning and practice, so as to lay a solid technical foundation for future work such as information management, E-commerce, E-government, website construction, decision support, enterprise resource planning, etc. The curriculum reform has been implemented for five years. The teaching and practice effect is good, and the expected goal has been achieved. It has trained teachers and laid the foundation for the new round of reform. However, the contents of Data Processing and Management series courses can no longer meet the requirements of liberal arts students in the Era of Big Data, so it is imperative to popularize the basic knowledge of Big Data and Artificial Intelligence.

In December 2018, according to the questionnaire of computer basic curriculum reform carried out by 47 colleges and universities in Shanghai, 68.29% of the colleges intend to choose "Data Analysis and Visualization" as an elective module on the basis of the traditional and information technology based compulsory modules. In April 2019, Shanghai Municipal Commission of education formulated *The Reference Plan for the Teaching Reform of Computer Courses in Universities and Colleges in Shanghai*, which proposed to "Actively respond to the new demand for talent training in the new era and the rapid development of the information society, with the goal of significantly improving the information literacy of college students, strengthening the Computational Thinking of college students and cultivating their ability to solve disciplinary problems by applying information technology; to promote the teaching reform of university computer course".

Under the above background, East China Normal University has become the vanguard of the computer basic teaching reform in Shanghai. Since February 2019, ECNU has started the construction and reform of new liberal arts computer basic courses aiming at improving Data Literacy.

# 3 Objectives of Teaching Reform

The fundamental goal of this round of teaching reform is to cultivate students' Data Thinking and improve their Data Literacy. The definitions of Data Thinking [5] and Data Literacy [6] are not completely unified in domestic and foreign related research, and there are different emphases in different subject area. Generally speaking, Data Thinking refers to a way of thinking in which data is used to describe, understand, and construct the world, and to explore solutions to transform the world by data; while Data Literacy refers to having data awareness and data sensitivity, being able to effectively and appropriately acquire, analyze, process, utilize and display data [7], and have the ability of critical thinking and prediction of data, and is the extension and expansion of Design Literacy and Information Literacy.

## 3.1 Improve Data Literacy

The core productivity of Artificial Intelligence era is data. All walks of life need to benefit from data collection, analysis, reasoning, prediction and insight. IDC, an international data company, once predicted that the amount of data generated by the world in 2020 will be 50 times as much as that in 2011, and the number of information sources generated will be 75 times that of 2011, and the amount of human Big Data will reach 163 ZB in 2025. These data contain great development opportunities to promote human progress. To turn the opportunity into reality, we need to train a large number of talents with Data Thinking ability and Data Literacy through basic computer education.

Jim Gray, winner of Turing Prize and founder of relational database, proposed that Big Data is not only a tool and technology, but also the fourth paradigm of scientific research and a new methodology for scientific research. Learning Big Data is the exercise and edification of an advanced way of thinking, a "liberal arts" education in the Era of Big Data, and an advanced scheme to examine and answer traditional questions from the perspective of Data Thinking and Data. Therefore, learning Big Data is also learning from Big Data, of the Data, by the Data and for the Data.

## 3.2 Enhance Social Competitiveness

Individual competitiveness is the cornerstone of national competitiveness. Daniel H. Pink, the author of *A Whole New Mind*, puts forward that "Three Senses and Three forces" [8] are the most important six factors for winning the future of individual competitiveness. The "three senses and three forces" refer to the sense of design, entertainment, sense of meaning, storytelling, synergy and empathy, which are the objects of attention and cultivation of traditional humanities. In the future, under the background of Big Data Era, it is necessary to integrate the perceptual cognition and rational Data Thinking ability, and to improve Data Insight through computer basic course training, and obtain more knowledge creation and moral cultivation, so as to enhance students' appeal, communication and persuasion in their respective professional fields.

## 3.3 Improve Comprehensive Creativity

During the construction of New Liberal Arts, we must think about the great influence of scientific and technological revolution and industrial revolution on people's mode of production and life style, how to improve the quality of human life and living state, enhance people's spiritual realm, and reveal the common ground of human beings and even the meaning of life [9]. To achieve these goals corresponding to the needs of the times, we need talents with comprehensive creativity, that is, talents with both leadership and creativity. Professor Jason Wingard of Columbia University, leadership expert, once pointed out that leadership is a kind of ability to integrate multi-disciplinary knowledge [10], while Scott page, a political science professor at the University of Michigan, believes in his book *The Diversity Bonus* that creativity is "the connection of different ideas" [11] in order to obtain the driving force of innovation in professional fields, more domain knowledge is needed, so as to have high decision-making power. The basic computer education in Colleges and universities can just "tailor-made" and at the right time to build a ladder of comprehensive creativity for liberal arts professionals in the Era of Big Data.

# 4    Problems in Teaching Reform

## 4.1    The Course is Difficult and the Students' Foundation is Weak

It is imperative to introduce Big Data into basic computer teaching in Colleges and universities, but the troika of Big Data technology is Artificial Intelligence, Statistics and Data Visualization. These three basic technologies of Big Data all have technical barriers for liberal arts students. Even students majoring in Artificial Intelligence and Big Data can basically master the above three technologies only when they are senior students or graduate students.

At the end of 2017, the Ministry of education of China officially released the Information Technology Curriculum Standards for General Senior High Schools (2017 Edition), which aims to comprehensively improve the Information Literacy of all senior high school students. However, the fact is that the time that middle school students under the pressure of college entrance examination can spend on learning information technology in high school is very limited. Compared with the threshold of learning Big Data, the vast majority of students enrolled are still "Zero basis". Therefore, on the basis of respecting the scientific law of education itself, we must design teaching content appropriately, complete teaching material construction and arrange practice link.

### 4.2 High Hardware Requirements and Difficult to Achieve the Goal

Big Data refers to large and complex data sets, including massive, time-varying, heterogeneous, distributed, etc. These characteristics make the acquisition, storage and use of big data very difficult. As a general course, the reform of basic computer teaching in Colleges and universities requires that under the condition of the original hardware equipment, every student can improve the soft power related to Big Data, and can actually operate and use big data. Therefore, it is necessary to find a new lightweight and fast tool, new approach and new form that is easy to master, can understand and complete a series of requirements such as big data storage, processing, analysis, prediction and display, etc., and design new teaching content and practice content for this purpose.

## 5  Principles of Teaching Reform

In order to achieve the goal of reform on the basis of overcoming the above problems, we set the following reform principles:

**Find the Right Position.** Limited by the foundation and class hours, the goal of computer general education course can not be ambitious. Distinguished from the training goal of professionals engaged in Data science and technology work [12], it should not pursue purely technical difficulties. In the selection of teaching and practicing content, more attention should be paid to the integrity of knowledge structure and the universality of basic principles.

**Steady and Scientific Collocation.** On the premise that the teaching content has been "revolutionary" changed, we should not pursue the large-scale demolition and construction of the teaching mode, but pay more attention to the reorganization and optimization of the traditional teaching mode when arranging the teaching schedule and teaching mode. On the one hand, we actively tries new teaching modes such as mixed teaching mode, SPOC teaching mode and online teaching mode. On the other hand, we adheres to the advantages of traditional classroom teaching, and follows the scientific and effective ways that have been proved to have universal significance [13]. According to the Theory of Desirable Difficulty [14] and H. Ebbinghaus Forgetting Curve [15], the time interval between the theoretical teaching and experimental teaching is arranged scientifically in order to achieve the best knowledge internalization and sufficient "deliberate practice" results [16], preview before class and review after class is achieved.

**Quality and Efficiency First.** In each class, we should scientifically grasp the introduction of new knowledge and the horizontal connection with the old knowledge. Researchers from the University of Arizona and Brown University have found that, as shown in Fig. 1, the error rate of 15.87% can produce the fastest learning efficiency and the most pleasant learning experience [17]. In other words, when reading and learning, when about 85% of the knowledge is already mastered by the learners, and about 15% of the knowledge is unknown, the effect is the best. This provides an important reference for us to arrange the weekly curriculum objectives, classroom teaching, after-school exercises and practical exercises.

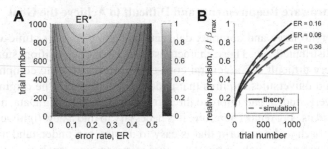

**Fig. 1.** Error rate and trial number

**Tamp the Foundation and Accumulate Steadily.** With the rapid development of Information Technology, the teaching reform must "revolutionary" adjust the teaching content and introduce the necessary new knowledge to cultivate Data Thinking. However, the contents such as Computational Thinking, computer architecture, data representation, Internet foundation and so on are still classic. In the content design of teaching reform, these underlying logic and foundation of information technology development are concerned, and still arranged enough class hours, elaborate and practice.

**Industry-University Cooperation by Openness and Inclusiveness.** Data analysis and visualization emphasize application and practice, and can't be done behind closed doors. Therefore, from the research stage of the reform, we have carried out close cooperation with Tableau, the world's leading enterprise in the field of Data Analysis and Visualization. It includes inviting senior engineers of Tableau company to participate in writing practical chapters of teaching materials, and conducting lectures on data analysis for teachers and students for many times. In addition, the president of Tableau Greater China led academic exchanges with schools with experience in Data Analysis and Visualization teaching, such as Nanyang Technological University in Singapore. When the time is right, we will cooperate with leading companies in the field of Data Analysis and Visualization based on accumulated experience, or explore more teaching and practice of open-source Data Analysis and Visualization software.

# 6 Specific Contents of the Reform

## 6.1 New Liberal Arts Curriculum System

In early 2019, East China Normal University designed and implemented the overall framework of the New Liberal Arts computer teaching reform, according to the guidance of Shanghai Municipal Education Commission on "Building a multi-level, multi-module, self-construction course teaching system" based on the main line of "Improving Information Literacy, strengthening Computational Thinking, and deepening integrated application", as shown in Fig. 2.

**Fig. 2.** New Liberal Arts computer course system

In view of the liberal arts major and some management majors in the University, it is divided into four degrees of difficulties and promoted step by step. Among them, IT practice course is an elective course for zero foundation students; College Information Technology-Data Analysis and Visualization and Data Analysis and Big Data Practice should be completed in the first year of University; while Computer Comprehensive Practice-Integration with majors course should be completed in any semester of sophomore or junior year. The course of University Information-Technology Data Analysis

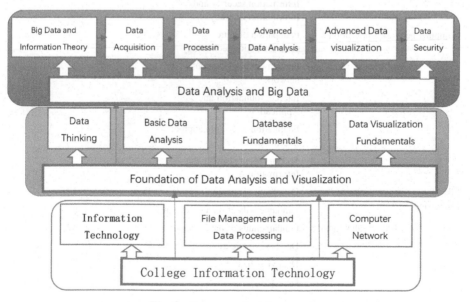

**Fig. 3.** Course content framework

and Visualization and the course of Data Analysis and Big Data Practice are given in two semesters, with 2 + 2 plus class hours per week per semester, including 2 online or offline theoretical teaching and 2 offline experimental teaching; at the same time, students are encouraged to conduct independent extracurricular learning with the help of mobile teaching platforms such as ELearning. The teaching content framework is shown in Fig. 3.

The first semester. Mainly includes four teaching modules: Information Technology foundation, Data and data file management, computer network foundation and application, and Data Analysis and Visualization basis. The teaching objectives are: to significantly improve students' Computational Thinking Ability and Information Literacy; to enable students to understand the operation principles of computer, communication and network systems; to enhance information awareness and cultivate good information ethics; to promote students to integrate Data Thinking with various disciplines, and initially have the ability to solve discipline problems and life problems by using Data Analysis methods and Data Visualization technology. The details are shown in Table 1.

**Table 1.** Curriculum system of the first semester

| Main knowledge modules | Knowledge points | Class Hour | |
|---|---|---|---|
| | | Theoretical | Experimental |
| Information Technology | Overview of Information Technology | 8 | 2 |
| | Computer system | | |
| | Development of Information Technology | | |
| | Information security and Information Literacy | | |
| Computer network Application | Foundation of data communication technology | 8 | 4 |
| | Fundamentals of computer network | | |
| | Internet foundation and Application | | |
| | Foundation and application of Internet of things | | |
| | Security technology in the Information Age | | |
| File Management & Data Processing | Data file management | 6 | 12 |
| | Fundamentals of data processing | | |
| Foundation of Data Analysis and Visualization | Fundamentals of data analysis | 8 | 12 |
| | Database application foundation | | |
| | Data visualization | | |
| | Data analysis practice | | |

The second semester. Mainly including Big Data and Information Theory overview, data acquisition, data processing, data analysis, data visualization, data security and data release. Teaching objectives: on the basis of the first semester course, continue to deepen and improve students' Data Literacy, systematically learn the complete process of solving Big Data problems, and strengthen the ability of using Data Analysis and Data Visualization technology to solve discipline problems and life problems. The details are shown in Table 2.

**Table 2.** Curriculum system of the second semester

| Main knowledge modules | Knowledge points | Class hour | |
|---|---|---|---|
| | | Theoretical | Experimental |
| Overview of Big Data and information theory | Basic concepts of Big Data | 2 | 0 |
| | Big Data support technology Introduction | | |
| | Basic knowledge of information theory | | |
| Data acquisition | Data sources | 4 | 2 |
| | Basic methods of data acquisition | | |
| | Information crawling | | |
| Data processing | Data cleaning | 6 | 8 |
| | Data conversion | | |
| | Data desensitization | | |
| | Data Integration | | |
| | Data reduction | | |
| Data analysis | Data type | 12 | 14 |
| | Function application | | |
| | Prediction and analysis of time series | | |
| | Regression analysis | | |
| | Cluster analysis | | |
| Data visualization | Foundation of Data Visualization | 4 | 4 |
| | Data Visualization tool | | |
| Data security and data release | Big data security concept | 2 | 2 |
| | Visual data sharing | | |

## 6.2 Content and Form Innovation

**Multiple Teaching Mode.** The course of the first semester is carried out in the Mixed Form of offline and online: teachers teach the key points of theory and practice offline, and students complete the experiment in the computer room; online students use the

ELearning platform to complete the preview before class, review after class, submit homework and answer questions after class online. Due to the epidemic situation, the course in the second semester has been changed to online completely. 11 teachers in the team prepare lessons collectively, work together, share lesson recording and question bank, and answer questions in cloud, providing a de facto MOOC environment for more than 1000 students in the school. Preview and pre-class test are arranged one week before class, and teachers can understand students' learning situation through background statistical function of ELearning. Instant messaging softwares are used in class to support the teacher-student interaction, and the intensive teaching is strengthened around the difficulties of the course. After class, homework is graded and comments are fed back through the ELearning platform to analyze the learning situation.

In the pure online teaching, teachers have paid more time and energy to answer questions, but on the whole, the efficiency of lesson preparation and teaching has been improved, and the effect of Flipped Classroom has been realized by using pure online method. What's more, comprehensive online produces comprehensive Data. Through the analysis of ELearning background data, we can get more granular user portraits of students, so as to design more accurate and personalized teaching content.

**Innovation of Teaching Content.** According to the principle of reform, the contents closely related to Big Data but rarely appeared in basic computer teaching in the past are introduced. For example, an introduction to Information Theory was arranged in the second semester. Information Theory is originally a professional course for senior undergraduates and postgraduates of majors such as Electronics, Communication, Computer and Automation. Systematic narration requires students to have sufficient Advanced Mathematics, Probability Theory and Mathematical Statistics foundation, which are not possessed by liberal arts students of zero basis of grade one. For this kind of "high-level" content, we try to avoid complex mathematical formulas, introduce a large number of practical cases, explain the key knowledge, help students inspire thinking and broaden their horizons.

## 7 Achievements of Reform Practice

In 2020, the global outbreak of the COVID-19 situation has unexpectedly promoted the Data Analysis and Big Data Practice course in the teaching reform from offline to online. All the teachers of the research group led the students to overcome the difficulties together with the people of the whole country and successfully completed the exploration and practice of the first round of the reform. According to the National Standard for Teaching Quality of Undergraduate Majors in Ordinary Colleges and Universities issued by the Ministry of Education in 2018, the reform performed very well in many aspects, such as goal achievement, social adaptability, condition guarantee, quality assurance effectiveness and result satisfaction, and initially achieved the goal of the reform.

### 7.1 Textbook Achievements

Teaching materials are important support of teaching. Up to now, four supporting textbooks have been completed and published in this reform, as shown in Fig. 4. They are University Information Technology, Data Analysis and Visualization Practice, Data Analysis And Big Data Practice and Experimental Guidance for Data Analysis and Big Data Practice. The training of Computational Thinking and Data Thinking of liberal arts students is sorted out, introduced and summarized from the outline and objectives. At the same time, a large number of practical cases closely related to the actual situation are provided.

**Fig. 4.** Published textbooks

## 7.2 Resource Outcomes

The course has established a set of teaching resources. The completed parts are shown in Table 3. The concept question bank is still expanding.

**Table 3.** Resource outcomes

| Types | Quantity |
| --- | --- |
| Presentation | College Information Technology, Data Analysis and Visualization Practice, Data Analysis and Big Data ( 1 set for each) |
| Video | Data Analysis and Visualization Practice, 80 segments, 270 min<br>Data Analysis and Big Data,101 segments, 1122.5 min |
| Question bank | College Information Technology, Data Analysis and Visualization Practice, 110 questions<br>Data Analysis and Big Data,150 questions |
| Resource platform | https://www.jsjjc.sh.edu.cn<br>https://elearning.ecnu.edu.cn/ |

## 7.3 Student Evaluation

National Standard for Teaching Quality of Undergraduate Majors in Ordinary Colleges and Universities requires that undergraduate teaching should be changed from "teaching well" to "learning well", and "learning well" is the most important thing for students' feelings and judgment. Every year after the end of the course, we will carry out a course questionnaire survey among students, and the questionnaire data in June this year gives the answer to the question: when answering the two key questions of "How much do you think this course will affect your future" and "How much have you learned from this course", compared with before the reform, students answered "significant" and "influential", and answered "great gains" and "rather great gains" proportion has increased significantly. The number of data with "great gain" has nearly doubled, while that of "small gain" has dropped to less than 1/3 of that before the reform, as shown in Fig. 5.

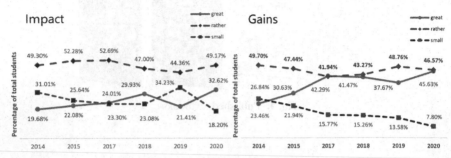

**Fig. 5.** The Course Impact and Gains

Considering that the course is the first time to carry out, and also comes across the COVID-19 situation, our teaching experience and teaching security system still have a lot of room for improvement. We will continue to organically combine normal monitoring with regular assessment to continuously improve.

### 7.4  Students' Works

In the final examination of Data Analysis and Big Data, we replaced the traditional closed book computer examination with the final course project. The final course project is completed by students' self-selected theme, team production of 3–5 people and collective defense.

In the project works of the final course, the liberal arts students have mastered the technical means and analytical methods, and "tell story by data". In the form of intuitive, vivid, multi-dimensional and interactive, the collected data with high credibility, time-liness and large amount of information are analyzed reasonably and logically. Some of the topics are related to professional research, others are related to interests and hobbies. They focus on current affairs and analyze the international situation. For example, Fig. 6 is a screenshot of the works analyzing the situation of China' new COVID-19 situation and the global epidemic situation changing over time; Fig. 7 is the work of analyzing the situation of movies in various countries; Fig. 8 is the screenshot of the work of analyzing the current situation of Chinese women' development; and Fig. 9 is the screenshot from aninteractive dynamic picture showing the bumpy career and broad-minded life of Su Shi, a great Chinese writer. The project was done by a student majoring in Chinese.

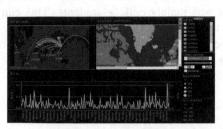

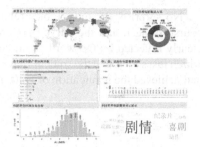

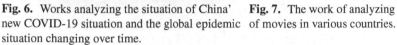

**Fig. 6.** Works analyzing the situation of China' new COVID-19 situation and the global epidemic situation changing over time.

**Fig. 7.** The work of analyzing the situation of movies in various countries.

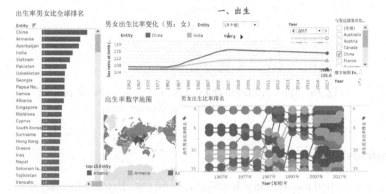

**Fig. 8.** The work of analyzing the current situation of Chinese women' development.

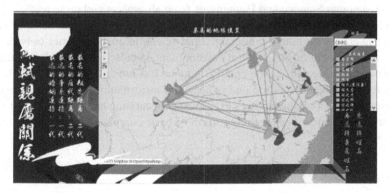

**Fig. 9.** An interactive dynamic picture showing the bumpy career and broad-minded life of Su Shi

No surprise, the most topics related to the epidemic situation were found in this course project, and the survey population was mainly college students. This can be clearly reflected by the word cloud chart of the theme of the works shown in Fig. 10, which just shows that the students have achieved the design goal of the curriculum reform and achieved the application of Data Thinking.

**Fig. 10.** The word cloud chart of the theme of students' project

## 8 Conclusion

In the future, we will use Artificial Intelligence and Data Science to analyze the teaching data obtained from offline, online and mixed modes, continuously accumulate teaching experience, gradually iterate the course content, design and implement more personalized and adaptive teaching plans. We will explore and practice the interdisciplinary education between Computer Science and Humanities from all aspects and more angles.

Apple Computer founder Steve Jobs once said "Computer science is a liberal art" [18]. In any era, "with both the humanity and scientific quality" is not only the actual needs of social, scientific and cultural development, but also the internal spiritual needs of the wise [19], especially in the era of Big Data. It is not only the call of the times for the new technological revolution and industrial transformation, but also the mission of the reform of basic computer courses to cultivate outstanding talents who can understand liberal arts and science. The teaching reform of basic computer science in liberal arts is always on the way.

## References

1. Huang, Q., Tian, X.: On the origin, characters, and paths of new liberal arts. J. Soochow Univ. (Educ. Sci. Ed.) **8**(02), 75–83 (2020)
2. Wu, Y.: Speech at the first plenary session of the Steering Committee on specialty setting and teaching in Colleges and Universities. https://jdx.cdtu.edu.cn/info/2042/3358.htm, Accessed 07 Mar 2020.

3. Tian, F.: What is new infrastructure construction. J. Orient. outlook **2020**(9), 8–13 (2020)
4. Xi, C.: Speed up the promotion of computer science education to be a pathfinder of data science education. J. Comput. Educ. **2020**(01), 2–8 (2020)
5. Li, X., Yang, X.: The connotation identification and training path of educational data thinking. J. Mod. Dist. Educ. Res. **31**(06), 61–67 (2019)
6. Ma, T., Sun, L.: Data literacy evaluation of college students from the perspective of information ecology. J. Inf. Sci. **37**(08), 120–126 (2019)
7. Hao, Y., Shen, T.: Research on the strategy and construction of data literacy and its cultivation mechanism. J. Inf. Stud. Theory Appl. **39**(01), 58–63 (2016)
8. Pink, D.H.: A Whole New Mind: Why Right-Brainers Will Rule. Riverhead Books, New York (2005)
9. Liu, S.: New liberal arts, thinking mode and academic innovation. J. Shanghai Jiaotong Univ. (Philos. Soc. Sci.) **28**(02), 18–22+34 (2020)
10. Jason wingard Forbes Jay-Z's Power of T-Shaped Leadership[J/OL]. https://www.forbes.com/sites/jasonwingard/2019/07/10/jay-zs-power-of-t-shaped-leadership/#25b8f72c22e5, 10 July 2019
11. Page, S.E.: The Difference: How the Power of Diversity Creates Better Groups, Firms, Schools, and Societies. Princeton University Press, Princeton (2007)
12. Zhou, A., Zhou, X.: Data professional training program and core curriculum system construction. J. China Univ. Teach. **2020**(06), 15–21 (2020)
13. May, C.: The Problem with "Learning Styles". Scientific American,https://www.scientificamerican.com/article/the-problem-with-learning-styles/, 29 May 2018
14. Yue, C.L., Bjork, E.L., Bjork, R.A.: Reducing verbal redundancy in multimedia learning: An undesired desirable difficulty? J. Educ. Psychol. (2013). https://xueshu.baidu.com/usercenter/paper/show?paperid=7c09d0ea52ea4c9d6b18daee2b400870&site=xueshu_se
15. Hermann, E.: Memory: A Contribution to Experimental Psychology. Martino Fine Books, Eastford (2011)
16. Ericsson, A., Pool, R.: PEAK: Secrets from the New Science of Expertise. Mariner Books, London (2017)
17. Wilson, R.C., Shenhav, A., Straccia, M., Cohen, J.D.: The eighty five percent rule for optimal learning. Nat. Commun. (2018). https://doi.org/10.1101/255182
18. Remembering Steve Jobs. https://www.npr.org/2011/10/06/141115121/steve-jobs-computer-science-is-a-liberal-art
19. Xu, B.: Liberal arts and science: master's way. J. Tsinghua Univ. (Phil. Soc. Sci.) **2001**(02), 10–12 (2001)

# Entity Coreference Resolution for Syllabus via Graph Neural Network

JinJiao Lin[1], Yanze Zhao[1], Chunfang Liu[1], Tianqi Gao[1], Jian Lian[2], and Haitao Pu[3($\boxtimes$)]

[1] Shandong University of Finance and Economics, Jinan, China
[2] Shandong Management University, Jinan, China
[3] Shandong University of Science and Technology, Jinan, China
pht@sdust.edu.cn

**Abstract.** Automatic identification of coreference and the establishment of corresponding model is an essential part in course syllabus construction especially for the comprehensive Universities. In this type of tasks, the primary objective is to reveal as much information as possible about the course entities according to their names. However, it remains a difficulty to most of the latest algorithms since the references to courses are commonly in line with the specifications of each University. Thus, it is important to link the course entities with similar identities to the same entity name due to the contextual information. To resolve this issue, we put forward a graph neural network (GNN)-based pipeline which was designed for the characteristics of syllabus. It could provide both the similarity between each pair of course names and the structure of an entire syllabus. In order to measure the performance of presented approach, the comparative experiments were conducted between the most advanced techniques and the presented algorithm. Experimental results demonstrate that the suggested approach can achieve superior performance over other techniques and could be a potentially useful tool for the exact identification of the entities in the educational scenarios.

**Keywords:** Knowledge point · Convolutional neural network · Coreference resolution

## 1 Introduction

Coreference resolution has become one of the most popular research fields for detecting the same entities in various practical scenes [1, 2]. And a large amount of algorithms have been provided to address the task of coreference resolution in natural language processing (NLP). However, none of them has been specifically designed for the entity discovery for the syllabus in University. For instance, in the work of [3] a learning method was presented to coreference resolution of noun phrases in unlimited text. One small and annotated corpus was leveraged as the dataset to produce a certain type of noun phrases like pronouns. Within this study, the entity types are not confined to specific categories. Kottur et al. [4] focus on the visual coreference resolution issue, which consists of

© Springer Nature Singapore Pte Ltd. 2021
W. Gao et al. (Eds.): FICC 2020, CCIS 1385, pp. 396–403, 2021.
https://doi.org/10.1007/978-981-16-1160-5_31

determining one noun phrase and pronouns whether refer to the same entity in a picture. A neural module network is employed for addressing this problem through using two elements: Refer and Exclude that could execute the coreference resolution in a detailed word level.

Since most of them focus on the association between each pair of entities rather than their contextual environment, it is also difficult for the previously proposed techniques to be adapted to the educational background. However, the rapid development of modern society proposes the higher requirements of students, the corresponding resources related to syllabus have grown greatly especially in the Universities. Therefore, the recognition of the same entities with similar or dissimilar names become much more complicated than ever before.

Meanwhile, the deep learning-based techniques have shown their performance in various NLP-oriented applications. For instance, Attardi et al. [5] propose an architecture of deep learning pipeline for NLP. A group of tools are built for creating distributional vector representations and addressing the NLP tasks in this work. In total, three techniques were introduced for embedding creation and two algorithms were exploited for the network training. And the convolutional network plays a vital role in this approach. In [6], a joint multiple-task model is introduced as well as a strategy for adapting its depth to the complexity of the tasks. Each layer contains the shortcut connection to both the word embedding and low-level predictions. One regularization with simple structure is used to implement the optimization of the objective function.

Recently, the graph neural network (GNN) could be a valuable deep learning model for implementing the coreference resolution tasks. Originally, GNN was used to deal with the non-Euclidean data. For instance, in the work of [7], a scalable approach based on a variant of convolutional neural network for semi-supervised classification within the graph structure data. Different from the traditional convolutional neural network, GNN is supposed to address the issues directly on graphs rather than the Euclidean data such as pixel images. And the original convolution operation is modified into the spectral graph convolution with the localized firstorder approximation. Both the local graph structure and the features of each node in the graph could be extracted with GNN. For clinical applications, a supervised GNN-based learning approach for predicting the products from organic reactions given the reactants, reagents, and solvents [8].

Base on the above analysis, we put forward a GNN-based pipeline trained by 1,312 pairs of entities for coreference resolution task. To note that all of the entities are extracted from the syllabus of the Universities in Shandong Province, China. The proposed GNN model adopts the spectral convolution operator as its primary computation unit. And each manually collected course entity is independently fed into the proposed GNN as one node of the whole graph of syllabus. Meanwhile, the similarity of each pair of entities and the corresponding adjacent relationship are taken as the characteristics of the nodes in the graph. In the trainings performed on the dataset, the parameters including the convolution and pooling layers' operators as well as the characteristics of the nodes could be optimized iteratively. With the labeled entities pairs, the trained GNN could be used to resolve the coreference in the given a new pair of course entities.

To measure the performance of the presented method, the comparison experiments were carried out on the samples of data collected manually between latest techniques and the presented approach. The corresponding results demonstrate that the presented GNN-based pipeline outperforms other techniques.

In general, this work has at least the following significance as:

- A GNN model is introduced to implement the coreference resolution task in course syllabus of Chinese University.
- The association between each pair of nodes in the presented graph structure could be used to represent the association between each pair of entities in the syllabus.
- Experiments on the real samples could demonstrate that the presented method is one potentially invaluable technique for coreference resolution.

The rest of this paper is as follows. In Sect. 2, we provide the concrete details of the presented approach. The results of the experiment are described in Sect. 3 and the conclusion is depicted in Sect. 4.

## 2  Methodology

### 2.1  Input of the Proposed GNN

According to Fig. 1, the proposed deep learning model adopts the similarity of each pair of entities and the corresponding adjacent relationship as its input. Each node in the input denotes one entity within the course syllabus and the link between each pair of nodes represents both the connection of the corresponding entities and their similarity. To note that the length of the link does not equal to the similarity of a pair of nodes.

### 2.2  Graph Convolutional Neural Network

1) Definition
It is assumed there are n entities (i.e. the course names) in general, $C_i = [C_1, C_2, ..., C_n]$. Each syllabus could be denoted by a matrix $C_i \in R^{m \times n}$, where $m$ is the number of course in each syllabus, n is the dimensionality for the feature vector extracted from the original course samples and $n_i \in \{0, 1\}$. The dataset from the syllabus could be represented with one weighted graph with a data structure of tuple $G = (V, E, W)$, where $V$ denotes the m nodes in the graph, E represents the whole group of edges in the same graph, and $W \in R^{m \times n_i}$ is the corresponding adjacency matrix. Meanwhile, is the weight assigned to the edge, which links $V_i \in V$ to $V_j \in V$. To note that the value of the association (the edge) denotes the similarity of the connected entities. Therefore, it has been set as one hyperparameter in the following experiments.

The convolutional operator illustrated in Fig. 1 approximately equals to the multiplication operation in the spectral domain and relates to the common convolution operator in the time domain. The whole process could be mathematically formulated as follows:

$$L = I_m - D^{-\frac{1}{2}} W D^{-\frac{1}{2}} \tag{1}$$

where D denotes the matrix degree and $I_m$ is the an identity matrix.

As mentioned by Defferrard et al. [9], the Laplacian matrix then could be represented by using Chebyshev polynomials:

$$T_k(L) = 2LT_{k-1}(L) - T_{(k-2)} \tag{2}$$

where $T_0(L) = 1$ and $T_1(L) = L$.

To note that a polynomial ordered by K can generate K filters without bias. And, the filtering of a signal with K filters could be implemented with:

$$o = g_\theta(L) * c = \sum\nolimits_k = O_k \theta_k T_k(\bar{L})c \tag{3}$$

where c denotes a course from the syllabus dataset for, $L = \frac{2}{\lambda_{max}} - I_d$ and $\lambda_{max}\lambda_{max}$ is the highest eigenvalue of the normalized L. Therefore, the output of the $l^{th}$ layer for each sample in a GNN can be formulated as:

$$O_c^l = \sum_{i=1}^{F_{in}} g_{\theta_i^l}(L)c_s^l, i \tag{4}$$

Where $F_{out}$ is the outcome filter and $F_{in}$ represents the inputfilter that would yield $F_{out} \times F_{out}$ vectors, $\theta_i^l \in R^k$ are the Chebyshev coefficients, and $\theta_{s,i}^l$ is the input feature map for sample c at layer $l$.

2) Network architecture

The structure of the proposed GNN is provided in Table. 1. Totally, it is composed of 5 convolutional layers. No pooling operation is used in the network architecture for conserving the completeness the extracted features. The dropout rates for 2nd, 3rd, 4th, and 5th convolutional layers are 0.4.

**Table 1.** Network architecture of the proposed GCNN.

| Table head | Layer | | | | | | |
|---|---|---|---|---|---|---|---|
| | Conv | Conv | Conv | Conv | Conv | Conv | Classifier |
| Channels | 32 | 32 | 64 | 64 | 128 | 128 | 2 |
| K-order | 9 | 9 | 9 | 9 | 9 | 90 | N/A |
| Stride | 1 | 1 | 1 | 1 | 1 | 1 | N/A |

The initial training rate of the proposed GNN is 0.001. The training is conducted with a fixed 600 steps. The learning rate would decrease by a factor of 0.5 once the validation accuracy drops in two consecutive rounds.

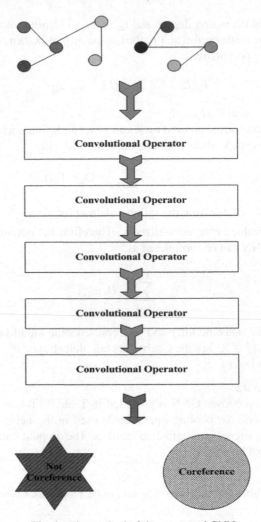

**Fig. 1.** The method of the presented GNN

## 3    Results and Discussion

The comparison experiments on the manually collated samples between latest techniques and the presented one to measure the performance of the presented GNN-based pipeline. And the results of the experiment as well as the analysis are provided in the following section.

### 3.1    Dataset

The proposed GNN is trained solely on samples data collected manually of course entities according to the syllabus in Universities of Shandong Province, China. In total, 1,312 pair of entities (600 of them are coreference) were manually collected from the raw materials.

Two educational experts were asked to perform the labeling tasks. Furthermore, none of the data augmentation techniques has been adopted to increase the diversity of the data samples due to the similarity of the course entities (only the course names). The adjacent matrix of the entities and the similarity of each pair of entities were both taken as the input the proposed GNN pipeline.

## 3.2 The Setting of the Hyperparamete

To determine the optimal setting of the hyperparameter as mentioned in Sect. 2.2, the classification experiments with different values of this hyperparameter from 0 to 1 with step of 0.1 were carried out and the corresponding accuracy is illustrated as Fig. 2.

Since the highest accuracy is achieved when we set $\lambda$ at 0.5, the value of 0.5 is adopted in the following experiments. Accordingly, the value 0.5 is used during process of training, testing, and evaluation.

In total, 70% of the samples are taken as the training set, 20% as the evaluation set, and the remaining are used as the testing set. The presented GNN has been fine-tuned by back propagation mechanism. Graphics Processing Unit (GPU), which has high performance, is employed in the presented GNN, and the learning rate of the Tensorflow deep learning platform is set as 0.01.

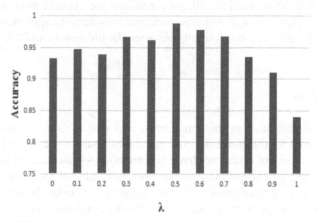

**Fig. 2.** Performance of the proposed GNN with different $\lambda$.

## 3.3 Experiments

To measure the performance of the presented GNN-based pipeline, the comparison experiments between latest [10–13] and the proposed techniques were carried out on the data samples collected manually.

As illustrated in Table 2, our technique outperforms other coreference resolution techniques in accuracy significantly.

**Table 2.** Performance comparison between latest and the presented GNN-based approach.

| Methods | Accuracy (%) |
|---|---|
| Lee et al. [10] | 83.25 |
| Meng et al. [11] | 87.03 |
| Pandian et al. [12] | 86.47 |
| Agarwal et al. [13] | 90.13 |
| Our method | 98.56 |

### 3.4  Analysis

According to performance comparison between the latest and ours on the data samples collected manually, we could observe that effectiveness of the proposed GNN-based approach. Through transferring the coreference resolution tasks into the neighboring relationship between each pair of nodes in the non-Euclidean graph, the introduced GNN could reveal both the association between each pair of entities and the corresponding similarities. Meanwhile, the accuracy obtained could satisfy the practical requirement for course syllabus.

The proposed GNN could significantly enhance the classification of coreference. Since we set different $\lambda$ to carry out the performance comparison experiments in Sect. 2.2, which is used to represent the similarity between the unknown similarity between one pair of entities.

## 4  Conclusion

The accurate identification of the coreference of a pair of course names is a potentially valuable tool for the automatic construction of course syllabus in Universities in China. A large amount of researches have paid attention to this area and have shown the effectiveness and efficiency of these works. However, most of the them did not aim at addressing the specific requirement of course syllabus. To bridge the gap, we propose a GNN-based network with transferring the coreference resolution issue into determining the node similarity in the graph. It offers an algorithm in an automatic manner.

This study offers at least the following contributions. First of all, a GNN designed for course syllabus scenarios is presented to implement the classification of coreference and non-coreference entities. Secondly, the original coreference resolution issue is transferred into a similarity measurement problem under the graph. Finally, the presented GNN outperforms other methods.

Next, we will go on study the extension of GNN and apply them in various fields, such as natural image processing [14], medical image processing [15] and [16].

**Acknowledgment.** Youth Innovative on Science and Technology Project of Shandong Province (2019RWF013), Postgraduate Education Reform Research Project of Shandong University of Finance and Economics (SCJY1911), Teaching Reform Research Project of Shandong University of Finance and Economics in 2020 (jy202011, Research on the Intelligent Teaching of Information Management and Information System -- Relying on the Intelligent Education Team, Study on the Reform of Curriculum Assessment Method in Shandong University of Finance and Economics).

# References

1. Adel, H., Schutze, H.: Impact of coreference resolution on slot filling. arXiv: Computation and Language (2017)
2. Uzuner, O., Bodnari, A., Shen, S., et al.: Evaluating the state of the art in coreference resolution for electronic medical records. J. Am. Med. Inform. Assoc. **19**(5), 786–791 (2012)
3. Soon, W.M., Ng, H.T., Lim, D.C., et al.: A machine learning approach to coreference resolution of noun phrases. Comput. Linguist. **27**(4), 521–544 (2001)
4. Kottur, S., Moura, J.M.F., Parikh, D., Batra, D., Rohrbach, M.: Visual coreference resolution in visual dialog using neural module networks. In: Ferrari, V., Hebert, M., Sminchisescu, C., Weiss, Y. (eds.) ECCV 2018. LNCS, vol. 11219, pp. 160–178. Springer, Cham (2018). https://doi.org/10.1007/978-3-030-01267-0_10
5. Attardi, G.: DeepNL: a deep learning NLP pipeline. In: North American Chapter of the Association for Computational Linguistics, pp. 109–115 (2015)
6. Hashimoto, K., Xiong, C., Tsuruoka, Y., et al.: A joint many-task model: growing a neural network for multiple NLP tasks. In: Empirical Methods in Natural Language Processing, pp. 1923–1933 (2017)
7. Kipf, T., Welling, M.: Semi-Supervised Classification with Graph Convolutional Networks. arXiv: Learning (2016)
8. Coley, C.W., Jin, W., Rogers, L., et al.: A graph-convolutional neural network model for the prediction of chemical reactivity. Chem. Sci. **10**(2), 370–377 (2019)
9. Defferrard, M., Bresson, X., Vandergheynst, P., et al.: Convolutional neural networks on graphs with fast localized spectral filtering. In: Neural Information Processing Systems, pp. 3844–3852 (2016)
10. Lee, K., He, L., Lewis, M., et al.: End-to-end neural coreference resolution. arXiv: Computation and Language (2017)
11. Meng, Y., Rumshisky, A.: Triad-based neural network for coreference resolution. arXiv: Information Retrieval (2018)
12. Pandian, A., Mulaffer, L., Oflazer, K., et al.: Event coreference resolution using neural network classifiers. arXiv: Computation and Language (2018)
13. Agarwal, O., Subramanian, S., Nenkova, A., et al.: Named person coreference in English news. arXiv: Computation and Language (2018)
14. Lian, J., et al.: Automated recognition and discrimination of human–animal interactions using Fisher vector and hidden Markov model. Signal Image Video Process. **13**(5), 993–1000 (2019)
15. Ren X, Zheng Y, Zhao Y, et al.: Drusen Segmentation from Retinal Images via Supervised Feature Learning. IEEE Access PP(99):1–1 (2017).
16. Lian, J., Zheng, Y., Jiao, W., Yan, F., Zhao, B.: Deblurring sequential ocular images from multi-spectral imaging (MSI) via mutual information. Med. Biol. Eng. Compu. **56**(6), 1107–1113 (2017)

# An Exploration of the Ecosystem of General Education in Programming

Yao Liu, Penglong Jiao, Wei Wang, and Qingting Zhu[✉]

East China Normal University, Shanghai 200000, China
qtzhu@cc.ecnu.edu.cn

**Abstract.** In the digital era, general education in programming faces the challenge of cultivating more graduates who have superior digital competence. To achieve this goal, the educational environment should carry out digital transformation. In this paper, we explore an ecosystem of general education in programming, in which teachers provide guidance and undergraduates are stimulated to learn. With the support of various information technologies, it integrates various teaching elements and provides an interactive learning environment with effective feedback. The deep teaching based on the ecosystem can spur the deep learning of students and cultivate students' digital thinking, programming thinking, data thinking and design thinking. In the teaching experiments, the ecosystem has achieved a satisfying effect, which is based on Shuishan online system.

**Keywords:** Ecosystem · General education in programming · Deep teaching · Smart learning techniques · Shuishan online system

## 1 Introduction

In the AI era, education information has entered the 2.0 version. With the support of various information technologies and abundant online learning resources, teaching and learning in the new era can improve information literacy of teachers and students, and transform the fusion development of education and information technology to the creative development [1, 2]. Smart education, which is guided by the growth of human wisdom, uses AI technology to promote the transformation of the learning environment, teaching methods, and education management [3].

In the digital era, the goal of computer general education for undergraduates is to train students who can explore, transform and innovate in the digital world. For this purpose, the educational environment should carry out digital transformation. Students could build a conceptual map of the digital world, cultivate digital thinking, programming thinking, data thinking and design thinking and promote digital competence in the digital environment. Programming is the most import part of computer general education for undergraduates.

In this paper, we build an ecosystem of general education in programming in which teachers provide guidance and undergraduates are stimulated to learn. With the support of various information technologies, it integrates various teaching elements, and provides an interactive learning environment with effective feedback. We have expanded the connotation of programming teaching by the ecosystem. After completing the introduction to programming basics, we continue to teach computer architecture, operating system, multimedia, data science.

At present, the theoretical research on university education reform is developing rapidly, but front-line teachers are still confused about how to implement the new teaching concept. The significance of this paper is to provide a practical experience on how to reform the general course of programming in terms of content, methods, environment, and evaluation.

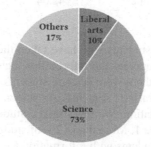

**Fig. 1.** Student category

The remainder of this paper is organized as follows. Section 2 analyzes the background and challenges of theoretical research on university education reform. Section 3 describes the five-level structure of the ecosystem of general education in programming. Section 4 illustrates three crucial factors for deep teaching: teaching content, deep teaching model and evaluation system. Section 5 summarizes the exploration and future work.

## 2  Background

### 2.1  Challenges

After higher education came into the stage of popularization, the scale of class is extended, which complicates the student group structure [4]. As for the general education curriculum of programming, the difference of computer skills between students is more prominent. The Chinese Ministry of Education has published *Criteria for Information Technology Courses in High School (2017 Edition)*, which has significantly increased the information literacy requirement for high school students. Information technology courses of high school with data and information as the core imparts high school students the latest developments in information technology. However, high schools implement the criteria differently. This means that freshmen probably have better information literacy and larger individual differences.

**Table 1.** A survey of freshmen information technology foundations

| Foundation (optional) | Percentage | Count |
|---|---|---|
| Software and hardware | 20.58 | 186 |
| Operating system | 13.05 | 118 |
| Office software | 51.99 | 470 |
| Internet application | 20.91 | 189 |
| Programming | 18.03 | 173 |
| Multimedia | 19.36 | 175 |
| No or little | 42.04 | 380 |

We took a survey of 904 students who have taken the general course in programming and investigated the categories and information technology foundations of them. As shown in Fig. 1, 73% of them were science students when they were high school students, and 10% were liberal arts students. As shown in Fig. 2, 18% of them had a certain programming foundation. We investigated students' information technology foundations, as shown in Table 1. More than 42% of students have never learned information technology. Among students with information technology foundations, most of them had office software foundation. Actually teachers have to face a class that has nearly a hundred students with different knowledge levels, and the polarization is expanding year by year.

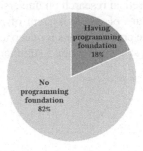

**Fig. 2.** Programming foundation of students

## 2.2  Related Work

In higher education, goals move beyond simple knowledge acquisition to promote students' engagement and high-level cognitive functions such as problem-solving and critical thinking, which are the characteristics of deep learning. Flipped Classed Model (FCM) based on online resources is considered one of the excellent ways for higher education, which can promote the development of high-level thinking. Based on the FCM, students can repeatedly watch online learning resources before class to preview the knowledge and discuss with teachers and classmates in the classroom to solve complex problems.

Ning et al. performed a meta-analysis on the 70 related research articles at home and abroad [6]. It was found that the FCM had positive effects on the three aspects of academic performance, cognitive skill, and emotional attitude, compared with the traditional teaching model.

However, there are some defects in the FCM. Zhu pointed out the ceiling of effect and the ceiling of cognition of the FCM, then proposed the concept of Smart Classroom and Smart Learning Space [7]. Through carrying out the teaching practice of the FCM in the college computer course, Ma et al. found there were still many challenges in the applicability of curriculum content and the adaptability of students during the implementation of the FCM [8].

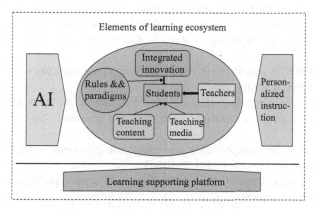

**Fig. 3.** A learning system based on education informatization 2.0

Based on the new requirements of the Education Informatization 2.0, Ma et al. designed a basic framework of the learning ecosystem as shown in Fig. 3, which combines rules, paradigms, teaching elements, and various information technologies [9]. To implement the framework above in real education, it is necessary to redesign the framework and refine the elements in the framework.

## 3    The Ecosystem of General Education in Programming

As a practical course, general education in programming is designed to cultivate students' cognitive ability, cooperative ability, and creative ability. Anderson et al. defined educational goals as knowledge, comprehension, application, analysis, evaluation, and creation [10]. Specially, the content of deep learning is focusing on analysis, evaluation, and creation. Deep learning and deep teaching are interdependent. Teachers implement deep teaching based on the learning ecosystem to stimulate the deep learning of students.

The purpose of this paper is to design a practical ecosystem of general education in programming. The ecosystem takes students as the main body, maximizes students' creativity and initiative, and encourages students to innovate and integrate knowledge. As the leader of teaching activities, teachers implement deep teaching and use information technology to provide students with personalized guidance. In the ecosystem, teachers and students use resources to learn, reconstruct, or create teaching content together. Figure 4 illustrates the structure of the ecosystem of general education in programming. The hierarchical ecosystem consists of five panes: learning support technologies, learning resources, instructional design, learning methods, and subjects.

## 3.1 Learning Support Technologies

Based on big data, AI, and other information technologies, learning support technologies can participate in all aspects of daily teaching and learning: acquisition of course resources, presentation of learning content, automatic evaluation of online question bank, interactive feedback of learning process, attendance management, etc.

The ecosystem is based on Shuishan online system developed by School of Data Science and Engineering of East China Normal University. The system is an integrated online education system including MOOC, KFCoding, OJ, Git, Vlab, Xboard, etc.

a) MOOC: MOOC is a platform in which teachers could organize online teaching and obtain students' learning data.
b) KFCoding: KFCoding is a training platform which provides a variety of programming practical environment, such as Linux terminal, Jupyter Notebook, VS code.
c) OJ: OJ is an online judge which can judge the programs submitted by students automatically and provide timely feedback.
d) Git: Git is a code collaboration platform that could support project collaboration and code hosting for students.
e) VLab: VLab is a virtual simulation that could provide a immersive storytelling architecture that helps students understand difficult concepts.
f) Xboard: Xboard is a visual tool that teachers could use the data generated by these platforms to track students' learning behavior and students could acquire their learning portraits.

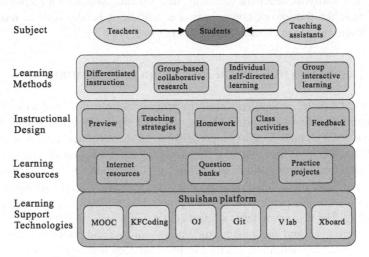

**Fig. 4.** Ecosystem of general education in programming

## 3.2   Learning Resources

Learning resources should not be limited to teaching notes provided by teachers. The Learning resources in the ecosystem consist of online resources, question bank, and practice projects.

a)   Internet Resources: Internet resources include the resources provided by teachers and the online discussion log. Internet resources can be provided in the form of MOOC (Massive Open Online Course), GitHub, wiki, Netease Cloud Class, etc. Electronic documents and short videos provided by teachers are the main learning resource for students. Besides, the questions and answers in discussion area, Wiki and social group also help students a lot.

b)   Question Bank: The quality of the question bank directly determines the effect of deep teaching. According to the learning stage, as shown in Fig. 5, the question bank can be divided into four independent sub question banks: pretest bank in the preview stage, exercise bank in the primary stage, program design bank in the improvement stage, and exam bank in the evaluation stage. Question types vary according to test objectives. At the preview stage, we focus on choice questions to check understanding of knowledge. At the primary stage, the main tasks are error correction and program completion. At the improvement stage, we focus on program design questions to evaluate students' algorithm design ability. At evaluation stage, we combine a variety of questions to evaluation students' comprehensive ability. The quality of the question bank directly determines the effect of students' autonomic learning.

c)   Practice Projects: After students have a certain programming ability, the teaching goal changes from primary cognitive knowledge to advanced cognitive ability. Practice projects are teaching activities to achieve this transformation. In a practice project, teachers put forward the basic requirements, and students determine the topic and implement it. During the project, students can acquire thinking and skills training, form the ability of independent learning and exploratory problem-solving, and eventually develop innovative thinking.

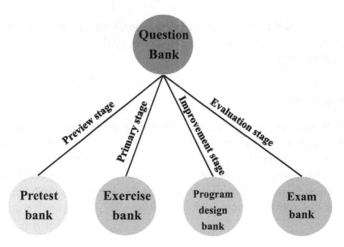

**Fig. 5.**  Contents of question bank

### 3.3  Instructional Design

With the support of resources and technologies, teachers need to build a deep teaching model based on specific teaching objects to promote students' deep learning. The instructional design includes preview, teaching strategies, homework, classroom activities, evaluation and feedback.

### 3.4  Learning Methods

Learning methods in the ecosystem include differentiated instruction, group-based collaborative research, individual self-directed learning, and group interactive learning. The goal of differentiated instruction is to allow students to master essential knowledge and core skills, which is the most fundamental teaching goal. Group-based collaborative research mainly cultivates students' comprehensive application ability. In individual self-directed learning, learners can choose learning resources independently according to their personal preferences and needs. Group interactive learning refers to the realization of the recycling of knowledge in the ecosystem through various forms of interaction in the ecosystem.

### 3.5  Subjects

Throughout the ecosystem, students are the mainstay and teachers are the leaders. The purpose of the ecosystem is to stimulate students to take an active part in the teaching process and make the ecosystem as a place for all teachers and students to develop. In the programming teaching, the most crucial difficulty is to provide effective feedback and timely help during the learning process. For this purpose, the teaching assistant is set up to help teachers tutor students. The teaching assistant can be a teacher, an excellent student in the class, or an AI robot.

## 4  Building the Ecosystem

In 2019, we started building the ecosystem of general education in programming from the preparation of teaching content, the implementation of deep teaching, and the comprehensive evaluation system, etc.

### 4.1  Teaching Content

We chose Python as the programming language for non-computer majors. As shown in Fig. 6, We designed five modules of the teaching content based on "data": simple data, composite data, multimedia data, class and object, network data. After completing the teaching of modules, we implement AI practices that students could learn machine learning and deep learning. Teaching were completed in two semesters, laying a solid foundation for developing students' AI capabilities.

## 4.2  Question Back Construction

The design of the question bank is the core content of the ecosystem. When designing the question bank, we must respect the cognitive rules of knowledge. In the preview stage, the exercises are mainly simple choice questions to check the students' general grasp of knowledge. In the primary stage, the exercises are mainly program completion questions and program correction questions to test students' grasp of the algorithm modes and the difficult points in classroom teaching. In the improvement stage, the exercises are mainly program design questions to assess students' programming skills.

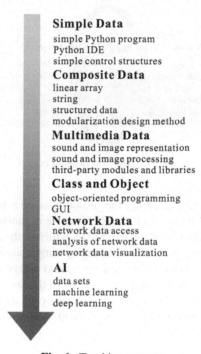

**Simple Data**
simple Python program
Python IDE
simple control structures

**Composite Data**
linear array
string
structured data
modularization design method

**Multimedia Data**
sound and image representation
sound and image processing
third-party modules and libraries

**Class and Object**
object-oriented programming
GUI

**Network Data**
network data access
analysis of network data
network data visualization

**AI**
data sets
machine learning
deep learning

**Fig. 6.** Teaching content

## 4.3  Deep Teaching

Eric et al. put forward the Deep Learning Cycle, dividing deep learning into five processes: preparation, knowledge construction, application of knowledge, evaluation and reflection [11]. Deep learning is a concentrating immersive learning which is interdependent with deep teaching. The deep teaching process we designed includes two parts: out-of-class learning and in-class learning. As shown in Fig. 7, the deep learning process spirals. The preparation and knowledge construction are completed from step 1 to step 4, and application of knowledge, evaluation and reflection are completed from step 5 to step 10.

Each student has a different level of cognition, especially in the face of college computer courses. The preview can help students with poor foundation to close the gap. Teachers can learn about students' cognitive level through pretest. The content of the pretest generally includes objective questions and simple programs. These questions can be automatically judged online, and the real-time feedback can be provided to help students evaluate the effect of the preview. Therefore, in the classroom, teachers only need to focus on the error-prone and the confusing content. This teaching method makes the core of teaching transfer from necessary knowledge to advance thinking ability.

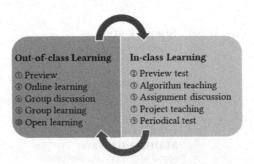

**Fig. 7.** Deep teaching process

As for the programming teaching, the difficulty is how to help students cultivate algorithmic thinking ability. In order of data type, we organize teaching content. Each data object has its own different algorithm mode, which is the core content of teaching. Beginners can start learning programs by applying simple algorithmic mode. After students have the certain practical ability, teachers could assign some classic AI projects in machine learning and deep learning to students. In the projects, students form group and collaborate to develop. In the classroom, teachers organize discussions about topic, technical difficulties, etc. to promote the implementation of the projects.

## 4.4   Evaluation System

Deep learning requires the continuity of evaluation process, the diversity of evaluation methods, and the diversity of evaluation subjects. Zhang et al. proposed that deep learning needs to monitor students' entire learning process. They addressed to apply a combination of diagnostic evaluation, formative evaluation and summative evaluation to check the completion of learning goals. So that teachers, students and partners could participate in the comprehensive evaluation [12]. The evaluation system of the ecosystem includes:

1)  Pretest results
2)  Assignment evaluation
3)  Stage test results
4)  Discussion Board Records
5)  Extra points for community Q & A

6)  Evaluation of group projects
7)  Practical ability test
8)  Final assessment

Timely evaluation and feedback are the keys to the success of practical teaching. The ecosystem can integrate information technology, teaching resources, and human attention to give each student timely help. Firstly, students can obtain the assessment results in real time from the online assessment system. Secondly, teachers could find students who do not actively participate in the ecosystem from evaluation, and provide them targeted counseling. Thirdly, evaluation could stimulate students' reflection and help them develop critical thinking.

## 5  Experiment

We have started the construction of the ecosystem of general education in programming since March 2019. We tried to use the online testing platform in some classes of March 2019, and fully launched the deep teaching based on the ecosystem in September 2019. In the second semester, we have to implement online teaching because of the 2019-nCoV. The ecosystem shows its strong support for online education. It has effectively improved the teaching quality of general education in programming and achieved the expected effect.

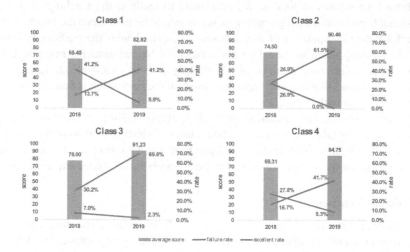

**Fig. 8.** Performance comparison of four classes in two years

### 5.1  Classroom Teaching

According to the deep teaching model, teachers do not need to spend classroom time on simple programming grammar teaching. The entire teaching progress is accelerated.

Students quickly complete the preliminary stage of programming learning, and can program some practical applications, such as image processing, text analysis, network data acquisition, etc. In this way, the basic knowledge of information technology is integrated into practical teaching. On the one hand, it greatly improves the learning interest of students, and on the other hand, it successfully realizes the expansion of teaching content.

### 5.2 Programming Practices

Intelligent evaluation platform supports real-time feedback for each practice session, which stimulates students' enthusiasm for programming. Before using intelligent evaluation platform, students spent less than one hour per week studying on average, and finished 1–2 programming questions per week on average. However, after using intelligent evaluation platform, students spent more than 3 h per week studying on the platform on average, and finished no less than 5 programming questions per week on average. In the online practice evaluation, the number of students with full marks increased, and the excellent rate improved significantly. With the help of learning resources in the ecosystem, students were able to successfully implement the practical projects. During the project display, a large number of outstanding works emerged. It showed that students had strong autonomous learning ability and innovative ability.

### 5.3 Teaching Results

We chose four classes (a total of 217 students) to analyze the teaching results. The assessment form is online programming assessment in the first semester. Figure 8 shows the performance comparison of four classes. Compared with the traditional teaching in 2018, the deep teaching based on the ecosystem helped students make significant progress. The failure rate has decreased strikingly and the excellent rate has increased notably. This shows that the deep teaching based on the ecosystem is helpful for solving the grade division within the class.

In the second semester, we carried out teaching reforms. The teaching content has new changes, including GUI, data crawling, data analysis, visualization, and machine learning. The assessment is to complete a GUI program for data analysis. The functionality and complexity of the works submitted by the students far exceeded expectations.

### 5.4 Lessons Learned

We conducted a questionnaire survey of the students in the above four classes. The survey result is shown in Table 2. More than 95% of students were satisfied with teaching content. The vast majority of students thought that the deep teaching in programming based on the ecosystem was helpful. However, compared with the students who enjoyed deep teaching, 13% of students preferred to follow teachers in the classroom, hoping that teachers would teach personally in more detail. Tracking these students, we found that their information technology foundation was almost zero in high school. Therefore, these students had some difficulties in accepting deep learning. In the future, we will adjust the teaching methods to focus more on the students with weak information technology foundation.

**Table 2.** Questionnaire survey

| Questions | Percentage |
| --- | --- |
| The course content is interesting | 95.6% |
| The ecosystem is useful for your future professional studies | 97.0% |
| Intelligent evaluation platform is helpful to improve your program practice ability | 100.0% |
| Using the learning resources provided by the ecosystem every week | 100.0% |
| Extracurricular learning resources are helpful | 90.0% |
| Deep learning is useful for your learning | 87.0% |

# 6  Conclusions

In the ecosystem of general education in programming, immersive learning could help students with different knowledge levels. And students could improve their cognitive, cooperative, critical, and creative abilities. Therefore, deep teaching based on the ecosystem is effective for practical teaching of programming education. It can also be popularized in more general education curriculums with large scale and diversified foundation of students.

In the future, with the accumulation of platform data, more AI applications can be integrated in the ecosystem to further improve the efficiency of teaching and learning, such as AI coach, adaptive learning and AI assistant.

**Acknowledgment.** This work was supported by the Shanghai General Computer Course Reform Project (No. 201940) and the Ministry of Education's University-Industry Collaborative Education Program(No. 201902146019).

# References

1. Lei, C.: Education Informatization: From the Version 1.0 to 2.0. J. East China Norm. Univ. (Educ. Sci.), **36**(01), 98–103 + 164 (2018)
2. Zhang, M., Lo, V.M.: Undergraduate computer science education in China. In: 41st ACM Technical Symposium On Computer Science Education (SIGCSE 2010), pp. 396–400. ACM Press, Milwaukee, Wisconsin (2010)
3. Cao, P.: Smart education: the educational reform at the age of artificial intelligence educational research. Educ. Res. **39**(08), 121–128 (2018)
4. Yan, G.: The undergraduate instruction: theory and experience, idea and evidence. J. East China Norm. Univ. (Educ. Sci.) **37**(06), 1–15 (2019)
5. Hornsby, D.J., Osman, R.: Massification in higher education: large classes and student learning. High. Educ. **67**(6), 711–719 (2014)
6. Ning, K., Xiaoqing, G., Wang, W.: The meta-analysis of the teaching effect of the flipped classroom——based on the 70 related research articles adopting random experiment or quasi-experiment. Mod. Educ. Technol. **28**(03), 39–45 (2018)
7. Zhu, Z.: New developments of smarter education: from flipped classroom to smart classroom and smart learning space. Open Educ. Res. **22**(01), 18–26 + 49 (2016)

8. Ma, X., Zhao, G., Wu, T.: An empirical study on the development of students' autonomous learning ability in FCM. China Educ. Technol. (07), 99–106 + 136 (2016)
9. Ma, X., Li, C., Liang, J.: Exploration of learning ecosphere from the perspective of educational informatization 2.0. Chin. J. ICT Educ. (22), 1–5 (2019)
10. Anderson, L.W., Krathwohl, D.R., Bloom, B.S.: A taxonomy for learning, teaching, and assessing: A revision of Bloom's taxonomy of educational objectives, Longman, (2001)
11. Jensen, E., Nickelsen, L.: Deeper Learning: 7 Powerful Strategies for In-Depth and Longer-Lasting Learning. Corwin Press, California (2008)
12. Zhang, X., Lv, L.: Constructing the deep teaching model for deep learning under the SPOC platform. China Educ. Technol. (04), 96–101 + 130 (2018)

# The New Theory of Learning in the Era of Educational Information 2.0—Connected Constructivism

Yazi Wang[1]([⊠]), Chunfang Liu[1], Yanze Zhao[1], Weiyuan Huang[2], Bizhen You[3], and Jinjiao Lin[1]

[1] School of Management Science and Engineering, Shandong University of Finance and Economics, Jinan, China
155350904@qq.com
[2] School of Marxism, Shandong University of Finance and Economics, Jinan, China
[3] School of International Education, Shandong University of Finance and Economics, Jinan, China

**Abstract.** Constructivism and Connectivism are two important schools of learning theory. The development of these two theories reflects the characteristics of the educational information era. This paper reviews the origin, development background and main viewpoints of constructivism and connectivism, and summarizes the characteristics of the two theories. In response to the changes in learning in the education information 2.0 era, on the basis of the original two learning theories, a new theory that is more suitable for learning in the Internet era is proposed: connected constructivism, and discusses the main points of the new theory.

**Keywords:** Constructivism · Connectivism · Connected constructivism

## 1 Introduction

The development of artificial intelligence affects human life and the world. The current internet, big data and other new theories and new technologies, as well as the overall advancement of the development of new generation artificial intelligence technological innovation, are triggering chain breakthroughs. Promoting the accelerated leap from digitization and networking to intelligence in all areas of the economy and society.

These also deeply affect the demand for talents and educational forms, including not only the methods of teaching and learning, but also the concept, culture and ecology of education, which in turn drives education informatization from the 1.0 era to the 2.0 era. The specific changes are as follows: knowledge fragmentation, uneven content; learning a specific technology or skill is difficult to survive in the future society; learning is networked and machined; artificial intelligence can provide data to master student learning behavior, etc. The knowledge that students need to master has changed from the original single subject knowledge to interdisciplinary knowledge, from solving a single problem to learning to solve complex problems. Students must learn to work with machines and the network to complete it. Teachers also have the original responsibility of imparting knowledge to training Students' lifelong learning ability, innovation ability, and use artificial intelligence to provide students with personalized teaching programs.

In this context, can existing learning theories guide education and learning practice? What changes should be made to adapt to the times? This is the question to be discussed next in this article.

Constructivism learning theory is a kind of learning theory that is widely used and has a mature system. The constant updating of artificial intelligence and information technology, a new view of learning theory appeared in 2005: Connectivism. Both theories have obvious characteristics of the times. Nowadays, in the era of education informatization 2.0, we need to reexamine and evaluate these two theories, and how to develop them to better guide educational practice. Especially in the outbreak of Corona Virus Disease 2019 at the beginning of 2020, in such a sudden situation of a global epidemic outbreak, schools are closed and cannot be taught offline, schools are closed without suspension, what changes should be made to our original learning theory to guide the current situation education and learning?

In view of the above thinking, this article innovatively puts forward the "connected constructivism" learning theory. The core of the theory is still constructivism, which integrates the concepts of networking, dynamics, connection, and circulation of connectivism in several aspects such as knowledge, learning, students, teachers, and teaching to reconstruct constructivism. That is to use the concept of connectivism to perfect constructivism. The model is as follows: Construct → Connect → Construct.

## 2 Constructivism

### 2.1 The Background of Constructivism

Constructivism learning theory is the core content of constructivism in education, and it is the further expand in learning theory from behaviorism to cognitivism. It is a theory about knowledge and learning and emphasizes learner initiative. And it believes that learning is completed in the social and cultural interaction, and learning is the process of learners generating meaning and constructing understanding based on original knowledge and experience. It believes that learning is a process of active construction.

Constructivism learning theory received widespread attention and universal acceptance in the 1990s. At that time, marked by the United States' plan to build an information superhighway, the United States and the world opened the era of modern information technology represented by the Internet. Under the background of this era, the society's demand for creative and individualized talents is increasingly strong. The view of knowledge, learning, teaching, student and teacher in constructivism conforms to the spirit of the times.

This theory occupies the dominant position of our teaching theory today. It reflects the modern humanistic view of people-oriented in the contemporary society, and also conforms to the requirements of the society for the cultivation of innovative talents.

## 2.2 The Main Points of Constructivism

**Constructivism View of Knowledge**

Constructivism emphasizes the dynamic nature of knowledge, and challenges the objectivity and certainty of knowledge to some extent. It is embodied in the following aspects:

(1) Knowledge only an interpretation, a hypothesis, not a completely certain representation of reality. It is not the final answer. And new explanations and hypotheses of it will appear with the progress of mankind.
(2) Knowledge cannot generalize the world rules. In specific problems, it cannot be used easily, but needs to be analyzed on a case-by-case basis.
(3) Knowledge cannot exist outside the specific individual as a entity. Although we use language symbols to give a certain external form for knowledge, learners will still understand and construct their own knowledge based on their own experience background.

**Constructivism View of Learning**

Constructivism learning view emphasizes the three aspects of active construction, social interaction.

(1) Active constructiveness of learning: learning is students construct their own knowledge, rather than a process of teachers transfer knowledge to students; learners are active information constructers, not passive information absorbers.
(2) Social interaction of learning: neither learner nor learning exists in isolation. Learning is the process of internalizing relevant knowledge and skills and mastering relevant tools through participation in a certain social culture.

**Constructivism View of Teaching**

Teaching is not to transfer things, but to build a certain surroundings and support to facilitate learners to spontaneously construct knowledge.

Scaffolding teaching: provides a conceptual framework. It can help learners to understand knowledge. Teachers can continuously raise students' intelligence from one level to another new higher level through the framework's supporting role. This teaching mode has five teaching links: building scaffolding (guiding exploration) - entering situation - independent exploration - collaborative learning - evaluation of learning effect.

(1) Anchored teaching: contextual teaching refers to the teaching based on contagious real events or problems. Knowledge and learning are associated with contextual activities. There are five teaching links in this teaching mode: creating situation - determining problems - autonomous learning - collaborative Learning - effect evaluation.

(2) Random access teaching: It means that learners can enter the same teaching content through different channels and different ways at will. This teaching mode has five teaching links: presenting the basic situation - entering learning randomly - thinking development training - group collaborative learning - learning effect evaluation.

(3) Cognitive apprenticeship: through this teaching mode, learners' higher-order thinking ability can be cultivated, that is, the ability of thinking, problem solving and complex tasks required for expert practice. Cognitive apprenticeship has six teaching links: modeling - contextual design - providing scaffolding - clear expression - reflection - removing scaffolding.

(4) Inquiry learning: refers to constructing knowledge based on problem-solving activities. In the teaching, it should be through meaningful problem situations that allow students to continuously discover and solve problems to learn knowledge related to the problem being explored, forming problem-solving skills and the ability to learn independently.

(5) Cooperative learning: it refers to that through discussion, exchange, point of view debate, they complement and modify each other, share the collective thinking results and complete the process of constructing the meaning of the learned knowledge.

## Constructivism View of Students

The main points are as follows:

(1) Constructivism emphasizes the richness of students' experience world, emphasizes the great potential of students, and points out that students do not enter the classroom with their heads empty.

(2) Emphasizes the differences in the world of student experience. Everyone has their own interests and cognitive styles.

## Constructivism View of Teachers

Constructivism regards teachers as helpers and collaborators of student learning.

(1) In teaching activities, through helping and supporting, teachers direct students to grow new knowledge and experience based on their original knowledge and experience, provide a ladder for students' understanding, and enable students to gradually deeper understanding of knowledge.

(2) To help students form the thinking and analysis of problems, inspire them to reflect on their own learning, gradually let students to their own learning to self-management, self-responsibility.

(3) Create a good, contextual, challenging, real, complex and diverse learning situation, encourage and assist students in learning through experiments, independent exploration, discussion, and cooperation.

(4) Organize students to conduct extensive exchanges with experts or practical workers in different fields, and provide strong social support for students' exploration.

## 2.3  Evaluation

**Constructivism Lacks the Concept of Connection**

Zhu Ronghua [10] and others analyzed the background of the generation of constructivism in "Analysis of the Background of the Generation and Rise of Constructivism Teaching Theory". Since the 90s of the 20th century, the world's science and technology have developed rapidly, the knowledge economy has emerged, and the information society has come to light.

Wang Zhuli [6] thought individual initiative play a key role in constructing cognitive structure, and they made a serious exploration on how to exert the initiative of the individual in cognitive process. He Kekang [12] emphasized in the "Constructivism Teaching Mode Teaching Method and Teaching Design" that in the constructivist teaching model, students are the active builders of the meaning of knowledge.

In summary, the background of constructivism view of knowledge and learning is the Internet era, and computer and network communication techniques are cognitive tools of this theory.

With the increasing popularity of cloud computing and the increasing maturity of artificial intelligence technology, it has promoted the transformation of information technology to the Internet of Things era. Especially under the integration of IoT+AI, everything has the ability to perceive. The way of learning has changed dramatically. How an individual explores the essence of things in massive amounts of information, distinguishes important information from non-important information, and how to perform distributed learning on the Internet is an area that the previous constructivism learning theory did not involve.

**Constructivism Does Not Mention that Learning Can Also Take Place Outside the Learners**

Chen Ping [9] believes that: different tendencies of constructivism pay attention to different aspects of knowledge construction. They either care about the knowledge construction realized by individuals in the process of dealing with their physical environment, or care about the interaction between individuals and social environment, but they all regard learning as a construction process, and explain the mechanism of knowledge construction by the interaction of old and new knowledge and experience.

Xue Guofeng [11] wrote in "Research on the Teaching Theory of Contemporary Western Constructivism" that social constructivists believe that knowledge is not only constructed by individuals in the physical and psychological environment, but also in the social environment. Therefore, learning is not only a process of individual construction and understanding, but also a process of social construction.

Jean [7] pointed out that the central point of social constructivism is that the individual constructed, unique, subjective meaning and theory can be developed only when they adapt to the social and physical world, because the main medium of development is social negotiation of meaning through interaction.In the age of Internet of things, the social structure has changed and everything is interconnected. Learning has become the formation of a network. Learning is no longer a process of learners' Internalization, but a process of external association at the same time.

## 3    Connectivism

### 3.1    The Background of Connectivism Theory

The current development of network technology has already broken the traditional ideological fetters that learning can only happen on campus. The form of learning is no longer single, and gradually presents an increasingly diverse development trend, and changes toward the direction of maximizing personal development benefits. Formal learning is being continuously supplemented by informal learning. People's pursuit of lifelong learning will inevitably further weaken the monopoly of formal learning in personal development, the nature of learning has changed.

In 2005, Canadian scholar George Simmons [1] put forward connectivism theory based on the results of his research on e-learning for many years. He made a comprehensive analysis of the learning behavior and its related mechanism when people are exposed to the network environment. He thinks that although the learning theories of behaviorism, cognitivism and constructivism still have their own unique and desirable points in explaining people's learning under specific circumstances. But even the most modern constructivism learning theory cannot effectively reveal the learning mechanism in the network era. So he expressed his view of knowledge and learning ecology with eight principles.

### 3.2    The Main Points of Connectivism

Connectivism learning theory has eight basic principles: (1) Learning and knowledge are based on the integration of various viewpoints; (2) Learning is the process of connecting nodes or information sources of different content; (3) Learning can be done by tools and equipment complete; (4) Compared with the mastery of current knowledge, maintaining the ability to learn is more important; (5) The maintenance of learning is based on cultivating and maintaining various connections; (6) In different fields, the ability to connect concepts or ideas is important; (7) The purpose of relevance learning is to make knowledge flow; (8) Decision making itself can be regarded as a learning process. The core idea includes several aspects:

**View of Knowledge**
Simmons believes that knowledge goes through a transformation process from classification and hierarchy to network and ecology, and knowledge can be described but not defined. Knowledge is an organization, not a structure. Today, knowledge organization mainly adopts dynamic network and ecology. In the Internet age, knowledge is created, spread, challenged, modified, perfected, updated and discarded by many people. Therefore, Simmons introduced the concept of knowledge flow, he thinks that knowledge circulates through the network like oil in a pipeline.

**View of Learning**
Connectivism holds that learning is the process of knowledge network formation, and it is consecutive. Nodes are external entities that can be used to build a network. The node may be a person, organization, library, website, book, magazine, database or any other

information source [2]. Creating external network nodes is one of the acts of learning, we connect and build information and knowledge sources in it. In the learning network, not all nodes will keep the connection continuously [3]. As an intelligent network, our mind will continually reshape and adjust to reflect the new environment and new information, and the nodes that are no longer valuable will be weakened gradually.

### 3.3 Evaluation

**The Theoretical Model of Connectivism Is Not Perfect Enough**
The theoretical basis of connectivism comes from the field of philosophy. It has not fully expanded from the field of philosophy to the field of education application research, and there is no perfect theoretical model. Therefore, it cannot illustrate and answer well to the current learning practice development.

**Connectivism Emphasizes the "Pipeline" Is More Important**
Wang Zhuli [4] questioned the proposition that "the pipeline is far more important than the knowledge in the pipeline". He thinks that content is also very significant. "Learning is essentially a personal matter. In the final analysis, the production, dissemination, application and innovation of knowledge still depend on people, not on machines and network technology alone" [6].

Therefore, critics consider that instead of giving up the active meaning construction of knowledge, learners should pay more attention to the construction work. As for connection, this can be done by the network and the machine. The author believes that paying attention to the pipeline and ignoring the "knowledge in the pipeline" does have the meaning of abandoning everything. After all, the knowledge construction is not only an approach for humans to understand the world, but also a way for humans to understand themselves.

In contrast, as the network becomes intelligent, the network is fully capable of taking on the job of building the connection better. For example, isn't the friend recommendation or song recommendation we encounter in social networks a manifestation of the intelligent network helping us establish connections with other nodes? Therefore, there is nothing wrong with the importance of pipeline advocated by connectivism. However, it takes the establishment and connection of pipelines as the main task of learners, which is worthy of further study.

## 4 New View-Connected Constructivism

Through the analysis of constructivism and connectivism above, it is found that some views of constructivism and connectivism are not suitable for the development of the current education informationization 2.0 era [13]. Constructivism has a perfect theoretical system, but lacks the view of all things connected. Therefore, the author thinks it is necessary to use the concept of connectivism to improve constructivism, and put forward the new theory of "connected constructiv-ism", the core is still constructivism, which integrates the network, dynamic, connection, circulation and other concepts of connectivism theory in the view of knowledge, learning, students, teachers and teaching to reconstruct constructivism. The model is as follows: Construct → Connect → Construct, the main points will be discussed:

## 4.1 Connected Constructivism View of Knowledge

In the age of internet of things, people, machines and things can be interconnected, and knowledge can be spread through this big network [14]. This is the network relation of knowledge, based on this characteristic, the "Construct → Connect → Construct" model is adopted to reconstruct the constructivism knowledge view.

(1) Knowledge is dynamic and networked. First of all, knowledge has gradually developed from static and independent disciplines to dynamic multi-disciplinary integration, and knowledge itself forms a network; Secondly, everything is interconnected, people and people, people and things, things and things will form a network. All the concepts and wisdom generated in this process can be quickly and dynamically transferred in this hyper linked network, constantly changing various forms, and finally becoming knowledge.

(2) The half-life of knowledge becomes shorter. Because of the interconnection of everything, the hyper-linking of the network, and the rapid spread of knowledge, the production and dissemination of knowledge may be in the same process, and subsequent revisions, improvements, updates, and abandonment are also fast.

(3) Fragmentation of knowledge. Since the knowledge stated in the first point may be generated at any nodes of the network, people may acquire knowledge from any node in the network, such as people, organizations, libraries, websites, books, magazines, databases or any other information sources. The knowledge is scattered, disordered and even repetitive, which makes it more difficult for learners to acquire complete and systematic knowledge.

(4) Overload of knowledge. In the background of artificial intelligence era, the whole world is in a huge network, any node in the network will produce information or knowledge, knowledge is in the state of explosion, it causes problems such as mass information, poor quality and low value of information, and also leads to information overload phenomenon.

## 4.2 Connected Constructivism View of Learning

The view of knowledge has changed, and the view of learning must change accordingly. Knowledge becomes dynamic, and knowledge is quickly spread and shared in the network, and connection becomes an important learning method.

(1) Learning is a process of active and systematic construction. The original constructivism view of knowledge only proposes that learning is actively constructive. The author puts forward a new view of knowledge: modern knowledge and information are overloaded and fragmented, learning tools are diversified, and learning is fragmented. If learners want to acquire systematic knowledge in a certain field, they must actively participate in the construction and actively carry out systematic learning.

(2) Learning is an connected process. It is the process of forming knowledge system or knowledge network. In the last point of view of knowledge, the author puts forward: in the era of artificial intelligence, knowledge is dynamic and networked; knowledge is fragmented, scattered in the form of fragments in each node of knowledge network. Then learning is to find the network connection point and the relevance between these things, connect, build new information and knowledge sources, and form a knowledge network.

(3) Learning is a lifelong undertaking. At present, the speed of data and knowledge generation and iteration is getting faster and faster, and the effect of knowledge creating value and wealth is becoming more and more obvious, so from now on, learning is life-long.

(4) Learning is a process of social interaction. In the Internet age, learning is not only the interaction between students and students, students and teachers, but also the interaction of information, which is accomplished through the interaction and cooperation between people and people, between people and machines, and between people and the network.

## 4.3   Connected Constructivism View of Teaching

Connected constructivism teaching view is a theory that is integrated on the basis of constructivism and relevance theory. Teaching is a series of activities composed of teachers, students, information technology, and learning resources [8]. First of all, teaching is based on the integration of various viewpoints and the construction of infectious real events or real problems. It is necessary to create a certain environment and support through information technology, connect nodes or information sources of different contents, establish online and offline interactive space, and promote learners to actively construct knowledge and maintain the significance of learning ability.

(1) With the continuous penetration and influence of emerging technologies such as artificial intelligence in education, big data for education, and "block chain +" education, teaching is not only limited to schools and classrooms, but also online and offline. Formal learning and informal learning, the rise of informal status, construct-connect-construct.

(2) "Internet +" classroom. Let smart phones and Internet enter the classroom instead of keeping them out of the classroom. Teaching practice shows that mobile phones have a very broad application prospects in schools and classrooms.

(3) Innovative education enters the classroom. In order to cultivate innovative talents, we must help learners to establish a unique and personalized knowledge system.

## 4.4   Connected Constructivism View of Students

At present, the daily data generation rate is 40 times faster than the growth rate of human beings, and it is still accelerating [5]. When a person establishes an organic connection with valuable information sources and constantly shares information, it is an important learning. Knowledge is as important as the Internet.

(1) According to the new view of learning, learning is mainly a process of forming knowledge network. As the main body of learning, students have subjective initiative and can actively construct their own learning objectives, directions, interests and hobbies, career planning. Through the Internet, learners can connect these, find the resources they need, and then integrate and construct their own knowledge network to serve learning goals, directions, hobbies.

(2) Learners are engaged in lifelong learning. Before the knowledge involved is relatively small, everyone only master a certain field of knowledge, the knowledge involved is relatively narrow, so we should constantly contact with multiple aspects of knowledge.

(3) The relationship between the learner and the technology becomes an equal relationship between the subjects. In the Internet era, learning is not just about students interacting with students and students interacting with teachers, but more about the interaction of information, which is accomplished through the interaction and cooperation between people, between people and machines, and between people and the network. Learners and intelligent machines form learning partners to learn from each other and complete knowledge production tasks together.

(4) The proportion of informal learning is increasing. As the development of Internet technology, many community learning, informal learning and nonacademic education are springing up.

## 4.5 Connected Constructivism View of Teachers

In the 21st century, the progress of science and technology has triggered educational reform, and the change of teachers' role is inevitable. This transformation is the embodiment of the deep integration and comprehensive correlation between artificial intelligence and education, thus forming new education ecology.

(1) The knowledge-based teaching role of teachers will change to educating people, from the education of students' knowledge to the comprehensive education of students. Artificial intelligence can collect various data of students and make statistics and analysis on students' learning problems, emotional problems and psychological changes, saving a lot of time for teachers. Teachers can put a lot of energy on education, care of the soul, spirit and welfare of students, spend more time interacting with students as equals, stimulate students' nature of seeking knowledge, enrich their hearts, and make students more creative and innovative in the AI era, and can develop in an all-round way in terms of morality, intelligence, physique and aesthetics.

(2) The role of a teacher is a designer and a server. Nowadays, artificial intelligence connects all these teachers' work, comprehensively analyzes, and reconstructs the evaluation of students. Based on this, teachers will help students design a learning and education plan that is most suitable for them. Every child is different. Teachers need to understand students and teach students according to their aptitude. The role of a teacher is a designer, a guide, a helper, and even a common partner in learning with students. Education is expected to get rid of the unreasonable reality of the same teaching materials, the same teaching steps, the same learning contents, the same

examinations and the students walking together, and truly promote personalized education.

(3) The future of education will usher in an age of collaborative co-existence between teachers and artificial intelligence. Teachers and AI will play to their respective strengths and collaborate to achieve personalized education, inclusive education, lifelong education and fair education, and promote the overall development of people.

## 5 Conclusion

Looking forward to the future, we will further learn and carry out the spirit of the 19th National Congress of the Communist Party of China, accelerate the modernization of education and the building of an educational power, and cultivate a new engine for educational innovation in the new era. Combining major national strategies such as "Internet +", big data, and new-generation artificial intelligence, based on emerging technologies, and relying on various smart devices and networks to actively carry out smart education innovation study and demonstrations, Focus on the role of the new generation of information technology in promoting "teaching intelligently, learning intelligently, management intelligently, and evaluation intelligently" on campus, and promote the innovative development of the educational information ecosystem.

**Acknowledgment.** Youth Innovative on Science and Technology Project of Shandong Province (2019RWF013), Postgraduate Education Reform Research Project of Shandong University of Finance and Economics (SCJY1911), Teaching Reform Research Project of Shandong University of Finance and Economics in 2020 (jy202011, Research on the Intelligent Teaching of Information Management and Information System -- Relying on the Intelligent Education Team, Study on the Reform of Curriculum Assessment Method in Shandong University of Finance and Economics).

## References

1. Simmons, G., Li, P.: Connectivism: a theory of learning in the digital age. Global Educ. Outlook (08), 11–15 (2005)
2. Qiu, C., Gao, A.: New progress of learning theory in the network era: connectivism theory. J. Guangdong Radio Telev. Univ. **19**(03), 2–7 (2010)
3. Liu, J.: Research on teaching and learning organization from the perspective of connectivism learning theory. Northeast Normal University, pp. 51–64 (2011)
4. Wang, Z.: Connectivism and new constructivism: from connectivity to innovation. J. Dist. Educ. **029**(005), 34–40 (2011)
5. Zhou, W.: A study of connectivism learning theory in the digital age. East China Normal University (2014)
6. Wang, Z.: A new view of knowledge and learning in the age of intelligence. J. Dist. Educ. **35**(03), 3–10 (2017)
7. Jean, P.: Epistemology of humanities. Central Compilation & Translation Press, Beijing (2002)
8. Yu, S.: The future role of AI teachers. Open Educ. Res. **24**(01), 16–28 (2018)

428     Y. Wang et al.

9. Chen, P.: "Learning by creating": the way of learning in the era of artificial intelligence. Jiangsu Educ. Res. (Z2), 4–8 (2019)
10. Zhu, R., Wang, W.: An analysis of the background of the emergence and rise of constructivism teaching theory. Hebei Vocat. Educ. 7(11), 80–81 (2011)
11. Xue, G., Wang, Y.: Comment on the teaching theory of contemporary western constructivism. High. Educ. Res. (01), 95–99 (2003)
12. He, K.: Constructivism teaching model, teaching method and teaching design. Journal of Beijing Normal University (Social Science edition) (05), 74–81 (1995)
13. Ministry of Education.: Education informatization 2.0 action plan (2018)
14. The State Council: Development plan of new generation artificial intelligence (2018)

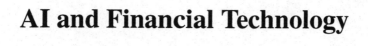

# AI and Financial Technology

# A Stock Index Prediction Method and Trading Strategy Based on the Combination of Lasso-Grid Search-Random Forest

Shaozhen Chen[1], Hui Zhu[1], Wenxuan Liang[2], Liang Yuan[1],
and Xianhua Wei[3(✉)]

[1] State Grid Xiong'an Fintech Corporation Co., Ltd., Beijing 100053, China
[2] College of Economics and Trade, University of International Business
and Economics, Beijing 100029, China
[3] College of Economics and Management, University of Chinese Academy
of Sciences, Beijing 100190, China
weixh@ucas.ac.cn

**Abstract.** This paper establishes stock index trading strategy by building stock index predicting model. First of all, this paper mainly reviews the application of machine learning in stock prediction, and constructs the L-GSRF stock index prediction model based on Lasso regression, grid search and random forest algorithm. After the input of textual and numerical data, the L-GSRF stock index prediction model greatly reduces the prediction error and improves the prediction accuracy in the photovoltaic (PV) stock index prediction, comparing with the traditional random forest and support vector machine (SVM) algorithm. In this paper, the trading strategy based on the prediction model has achieved a high annualized return. Finally, this study further clarifies the shortcomings of machine learning methods and future research directions.

**Keywords:** Machine learning · Lasso regression · Grid search ·
Random forest · Trading strategy

## 1 Introduction

The prediction of financial time series and constructing trading strategies have long been identified as an important but challenging topic in the research of financial market. Yang and Wu (2006) listed the analysis of time series data as one of the ten most challenging problems in data mining because of the uniqueness and complexity of time series data. Theoretically, the price of the stock is determined by a number of external factors, such as political events, the global economy, and traders' expectations and so on, so that financial time

Project 71932008 supported by NSFC: Research on the construction of a new generation of business intelligence system based on big data fusion.

series are non-linear, nonstationary, and irregular (Längkvist et al. 2014). The non-linearity is due to the fact that there are much noise and high-dimensional features in the time series; the irregularity is due to the fact that we can not fully grasp the causes of stock price changes; the nonstationarity is reflected in the changes in the mean value and variance of the stock price over time. The above characteristics make it difficult to study the changes of stock market returns.

For non-linear data prediction, machine learning enriches the research methods in the field of traditional economics, and it is changing the research methods of Economics (Athey 2017). Machine learning algorithm is "Data-Driven", which is continuously debugging the model by training a large amount of data to select the optimal model and parameters. Machine learning can extract abstract features from data, recognize hidden nonlinear relationships without relying on econometric assumptions and professional knowledge. In the era of big data, the advantage of machine learning is that it can improve the accuracy of the model, integrate interdisciplinary information and methods to help researchers analyze economic problems and make effective decisions. Table 1 lists the relevant research.

For non-linear data prediction, machine learning enriches the research methods in the field of traditional economics, and it is changing the research methods of Economics (Athey 2017). Machine learning algorithm is "Data-Driven", which is continuously debugging the model by training a large amount of data to select the optimal model and parameters. Machine learning can extract abstract features from data, recognize hidden nonlinear relationships without relying on econometric assumptions and professional knowledge. In the era of big data, the advantage of machine learning is that it can improve the accuracy of the model, integrate interdisciplinary information and methods to help researchers analyze economic problems and make effective decisions. Table 1 lists the relevant research.

Many of the above studies use text or numerical information to predict the stock price respectively. Studies combining time-series numerical variables and text-type variables are rare in China. Combining the two can incorporate the duration effect of past events into the model. (Such as the word "financial crisis" will have a lasting effect); and most models are trained based on macro market indexes or micro stock data, and less consider the specific domain knowledge of the meso-level industry influences. Therefore, machine learning methods still have room for improvement in establishing a safe and accurate stock prediction system.

This paper selects the photovoltaic industry as the research object. Photovoltaic industry is a immerging high tech industry in the 21st century. Due to its clean, safe and sustainable characteristics, photovoltaic industry is increasingly becoming a focused planning industry in various countries. It is expected to become another emerging industry with explosive production increase after information industry and microelectronics industry. Since 2009, photovoltaic building application demonstration project, the Golden Sun demonstration project and photovoltaic power station franchise bidding have been launched in China, which

**Table 1.** Related research on machine learning to predict stock price

| Author (Year) | Predicted objects (variable type, number) | Sample size (training set: validation set: test set) | Sample selection time period (frequency) | Machine learning methods | Error calculation method |
|---|---|---|---|---|---|
| Ding (2015) | S& P500 Index (News) | 664399 (4:1:1) | 2006.1–2013.11 (Day) | Event Embedded CNN | Matthews Correlation Coefficient (MCC) |
| Singh (2017) | Nasdaq Exchange Stocks (Value, 36) | 2843 (8:0:5) | 2004.8–2015.12 (Day) | PCA Dimension Reduction +DNN | RMSE |
| Akita (2016) | 10 stocks of the Tokyo Stock Exchange (News) | Unexplained (6:1:1) | 2001–2008 (Day) | LSTM | Simulated portfolio income |
| Chong (2017) | 38 KOSPI stocks in Korea (value) | 73041 (3:1:1) | 2010.01–2014.12 (Five minutes) | DNN | NMSE RMSE |
| Fischer (2018) | S& P 500 Index | Unexplained (3:0:1) | 1992.12–2015. 9 (Day) | LSTM | Diebold-Mariano (DM) test, simulating portfolio income |
| Zhong (2017) | S& P 500 ETF (Value, 60) | 2518 (14:3:3) | 2003.01–2013.05 (Day) | Unexplained (6:1:1) | Simulated portfolio benefits |
| Jang (2018) | Bitcoin (Value, 27) | Unspecified | 2011.12–2017.08 (Day) | Unexplained (6:1:1) | RMSE MAPE |
| Ran (2018) | 20 shares of A shares (News) | 1,447,500 (articles | 2008.01–2016.12 (Day) | Unexplained (6:1:1) | MSE |

Note: CNN: Convolutional Neural Network; PCA: Principal Component Analysis; DNN: Deep Neural Network; RMSE: Root Mean Square Error; LSTM: Long Short-Term Memory; NMSE: Normalized Mean Square Error; ANN: Artificial Neural Network; BNN: Bayesian Neural Network; MAPE: Mean Absolute Percentage Error; BPNN: Back-Propagation Neural Network; SVR: Support Vector Regression; MSE: Mean Square Error.

means that domestic photovoltaic development has officially started. With the implementation of opinions on promoting the healthy development of photovoltaic industry and a series of photovoltaic policy subsidies such as new photovoltaic regulations, China's photovoltaic industry is developing continuously. The installed capacity of photovoltaic power generation ranks first in the world for five consecutive years between 2013 and 2017. In the process of development, the photovoltaic industry is also facing problems such as financing difficulties. The traditional financing channels of photovoltaic power station mainly include state electricity price subsidy, self-raised funds, bank loan, and so on. As a heavy asset project, photovoltaic power station has high investment cost, large capital demand and long payback period. It is still difficult to obtain a large amount of capital through traditional financing channels. At present, it can only meet part of the capital demand. Therefore, the key to promote the further development of photovoltaic industry is to solve the financing bottleneck problem. The research on the photovoltaic industry in China's stock market is conducive to investors to grasp the characteristics of the fluctuation of the photovoltaic industry's stock price and make more effective allocation of photovoltaic assets.

This paper mainly reviews the application of machine learning in stock prediction and builds the L-GSRF stock index prediction system based on lasso regression, grid search and random forest. Compared with the traditional random dom forest and SVM algorithm, this algorithm has a better performance in the prediction of photovoltaic index. Then, based on this model, this paper constructs an index trading strategy of photovoltaic index. In the end, this study further clarifies the shortcomings of machine learning methods and future research directions.

## 2    Construction of L-GSRF Model

### 2.1    Theoretical Model of L-GSRF

L-GSRF Model is made up of Lasso regression, grid search and random forest algorithm.

Lasso regression is to find the regression coefficient that minimizes the loss function by modifying the loss function on the basis of linear regression. The method of modifying the loss function is to add the L1 regularization expression to the loss function. The specific method is as follows:

The expression of linear regression is $f(x) = w^T x + b$. The expression of the loss function is $J = \frac{1}{n} \sum_{i=1}^{n} (f(x_i) - y_u)^2$. The goal of linear regression is to find a set $(w^T, b)$ to minimize the loss function J. The method of Lasso regression is to add the L1 regularization term to the loss function. The modified loss function is:

$$J = \frac{1}{n} \sum_{i=1}^{n} (f(x_i) - y_i)^2 + \lambda \quad \omega|_1 \tag{1}$$

To minimize J, we can also rewrite formula (1) as:

$$\min_{\omega, b} \frac{1}{n} \sum_{i=1}^{n} (f(x_i) - y_i)^2$$
$$\text{s.t.} \quad \omega|_1 \le t \tag{2}$$

The solution of Formula (2) tends to produce a solution with a dimension of 0 under restricted conditions. If the eigenvalue with a weight of 0 does not contribute to the regression, then the feature with a weight of 0 can be directly removed so that the output value of the model remains unchanged. Therefore, Lasso regression can be used to reduce the dimension.

Decision trees are widely used in machine learning, but they are prone to overfitting, small noise is likely to change the tree's growth mode. Random forests can reduce the probability of over fitting by training decision trees in different subspaces, but increase some training deviations at the same time. In a random forest, each decision tree can only train part of the data, and the data is divided based on criteria such as Shannon Entropy and Gini Impurity. Gini Impurity can be expressed as:

$$g(N) = \sum_{i \neq j} P(w_i) P(w_j) \tag{3}$$

$P(w_i)$ here is the proportion of the i class. Another indicator for judging the effect of data division is Shannon entropy. Shannon entropy measures the degree of unpredictability of information contained in a branch of a decision tree. The Shannon entropy of the Nth branch can be expressed as:

$$H(N) = -\sum_{i=1}^{i=d} P(w_i) \log_2 (P(w_i)) \tag{4}$$

Where the d is the number of categories in the model, and $P(w_i)$ is the proportion of the i category. The optimal division result is to minimize the impurity and obtain the maximum information gain. The information gain can be expressed as:

$$\Delta/(N) = I(N) - P_L \times I(N_L) - P_R \times I(N_L) \tag{5}$$

I (N) represents the purity of the Nth branch (which can be expressed as Gini purity or Shannon entropy), $P_L$ is the proportion of branches to the left after division, and $P_R$ is the proportion to the right. $N_L$ and $N_R$ are the left and right branches of N.

In a random forest, parameters such as the number of trees and the number of randomly selected features of each decision tree are uncertain. Grid search is a method to find the optimal parameters. It optimizes by using exhaustive possible parameter values and cross-validation, and searches for the optimal parameters among the listed possible parameters. Cross-validation refers to the method of dividing the data set into n points, taking each one in turn as the test set, and every n-1 parts as the training set, and training the model multiple times to observe the stability of the model. The grid search algorithm will arrange and combine the possible values of each parameter, list all the combined results to generate a "grid", and then apply each group of parameters to different "grid" algorithms, using a cross-validation method and a certain scoring method. The training results are scored until the results of all parameter combinations are calculated in a loop, and the program is adjusted to the parameter combination of the best score.

## 2.2    Implementation of L-GSRF Model

In the random forest model, the parameters such as the number of trees and the pattern of random number generation all have the uncertainty of random selection. In the process of feature selection, if many features are not dimensionally reduced, then factors such as multicollinearity and data redundancy will greatly affect the accuracy and operation time of the model. Therefore, this paper builds the L-GSRF algorithm model.

The structure of the L-GSRF model is showed in Fig. 1. First, we standardized indicating data, then, input the predictive indicators to be selected into the Lasso regression model for dimensionality reduction, and retain the predictive indicators with non-zero regression coefficients. Next, based on the Grid Search algorithm, we select the value range of the parameters of the random forest, and determine the random forest model with the optimal score parameter combination after the loop iteration of the random forest algorithm. In the loop iteration, we use the cross-validation scoring method to substitute each parameter combination in the grid into the model. After obtaining the score, we compare it with the scores of other parameter combinations, and retain the parameter combination with the highest score as the parameter of the random forest model. Finally, we use the selected predictive factors to make a 40-day continuous multi-step prediction of the stock index through the L-GSRF model. We compare the prediction results with the original RF model and commonly used SVM and LSTM models, and verify the trading strategy performance of the model.

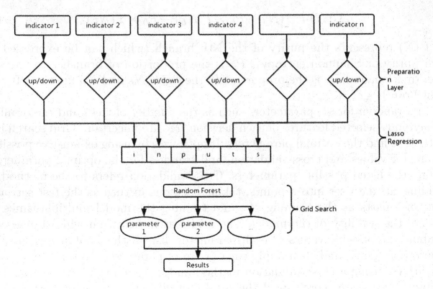

**Fig. 1.** Structure of L-GSRF model

The experimental steps of the L-GSRF prediction and trade algorithm are as follows:

Step 1: Obtain historical data of stock index and predictor, clean and standardize the data, and remove missing values;

Step 2: Divide training set and test set;

Step 3: Perform Lasso regression in the training set, and select the effective factor of the regression;

Step 4: Construct Random Forest algorithm using training set, and add Grid Search algorithm to build GSRF model.

Step 5: Test the model prediction performance with test set;

Step 6: Compare the L-GSRF prediction results with RF model and SVM model.

Step 7: Construct the L-GSRF trading strategy.

### 2.2.1   The Input of L-GSRF Model

At present, fundamental data, market data and technical indicators are mainly used as the characteristic input of the model in domestic and foreign research on the prediction of stock index. Fundamental data focuses on the political and economic data modeling and analysis of the impact on the long-term trend of the stock index. As the technical indicators of the index are calculated by the market data, the two have a significant correlation. More importantly, the technical indicators contain a lot of effective information other than the market data, such as identify patterns and trends that may indicate how stocks will behave in the future. We also used text-based review data collected in stock forum. The stock forum focuses on the relevant news of individual stocks, institutional research reports, and individual investor comments, which can comprehensively grasp market sentiment and indicate the trend of the stock. Taking them as explanatory variables of prediction models can help improve the performance of classification models.

In addition, previous literature results show that too many inputs will introduce too much noise in the learning process, and the training of the model may be affected by over-optimization. In this case, the model can achieve high performance on the in-sample data set, but get poor results on the out-of-sample data set. Therefore, we use Lasso Regression to reduce the input variables and avoid overfitting the model.

### 2.2.2   Performance Measurement

Performance measurements are used to measure the generalization ability of machine learning models. When comparing the performance of multiple models, different performance measures lead to different conclusions. In order to more clearly and effectively compare the predictive ability of the model, this paper selects two performance measurement, Root Mean Square Error (RMSE) & Mean Absolute Error (MAE) and accuracy rate.

Firstly, the RMSE and MAE calculation formula are as follows:

$$MAE = \frac{1}{m} \sum_{i=1}^{m} |y_i - \widehat{y_l}| \tag{6}$$

$$RMSE = \sqrt{\frac{1}{m} \sum_{i=1}^{m} (y_i - \widehat{y_l})^2} \tag{7}$$

Where $y_i$ is the true value and $\hat{y}_i$ is the predicted value.

Secondly, the accuracy rate is the most commonly used evaluation index in classification problems. For the prediction of stock index rise and fall, the following symbol definition is given to build the performance measurement of the model, and the calculation of accuracy rate is showed in formula 10:

(1) TP (True Positive): The number of samples that correctly predicted the rise.

(2) FP (False Positive): The number of samples that predicted a decline to a rise.

(3) TN (True Negative): The number of samples that correctly predicted the fall.

(4) FN (False Negative): The number of samples that forecast the rise to fall.

$$\text{Accuracy} = \frac{TP + TN}{TP + FP + TN + FN} \tag{8}$$

### 2.2.3   T-Test of Strategy Profit

In order to test the significance of profits, this paper adopts the common T-test method. Holding position is from buy to sell, sell to buy is short position. The corresponding time length and average daily yield are $n_h$, $n_s$, and $\mu_h$, $\mu_s$. $\Delta\mu$ is the difference between the average daily yields. When testing whether there is a significant difference between the average daily rate of return ($\mu_h$ or $\mu_s$) under the trading rules and the average daily rate of return $\mu$ over the entire sample period, the original hypothesis of the T test is $\mu_h = \mu$ or $\mu_s = \mu$, and the T-test statistics are:

$$t_h = \frac{\mu_h - \mu}{\sqrt{\frac{\sigma^2}{n_h} + \frac{\sigma^2}{n}}} \text{ or } t_s = \frac{\mu_s - \mu}{\sqrt{\frac{\sigma^2}{n_s} + \frac{\sigma^2}{n}}} \tag{9}$$

Where, n is the total number of observed values, $\sigma^2$ is the variance of the average daily return of the whole sample. If the average daily rate of return in the holding (short position) position is positive (negative) and significantly different from the average daily rate of return, then the trading strategy has good profitability.

If there is a significant difference between the average daily return rate $\mu_h$ in the holding position and the average daily return rate $\mu_s$ in the short position phase, the null hypothesis of the T test is $\mu_h = \mu_s$, and the T-test statistics are:

$$t_{hs} = \frac{\mu_h - \mu_s}{\sqrt{\frac{\sigma^2}{n_h} + \frac{\sigma^2}{n_s}}} \tag{10}$$

If $\mu_h$ is significantly greater than $\mu_s$, the trading strategy is valid.

## 3   Data Collection and Processing

### 3.1   Data Collection

The prediction object of this paper is the photovoltaic index (884045.WI), which is composed of 59 stock holdings. This paper selects the top 10 companies in the market value of the photovoltaic index component as the research object. The data source is divided into two parts, the first part is text-based data, from the 10-stock Oriental Fortune stock bar reviews (see Table 2). East Money Stock Forum was established in 2005 and is China's main financial and securities portal. This paper uses a crawler algorithm to collect all posts from November 2015 to May 2018, including the title, reading volume, number of comments, and posting time, a total of 392811 pieces of data. After deleting missing and duplicate values and garbled data, 358387 pieces of experimental data were obtained.

**Table 2.** Grab source and quantity of StockForum review data

| No. | Code | Name | Total market value (billions) | Industry | Total data volume | Experimental data volume |
|---|---|---|---|---|---|---|
| 1 | 601727.SH | Shanghai Electric Group Company Limited | 748.04 | Industrial | 37142 | 24656 |
| 2 | 601012.SH | LONGi Green Energy Technology Co., Ltd. | 475.55 | Information technology | 23766 | 20698 |
| 3 | 601877.SH | Zhejiang Chint Electrics Co., Ltd. | 471.59 | Industrial | 18058 | 12252 |
| 4 | 600438.SH | Tongwei Co.,Ltd. | 302.05 | Daily consumption | 23595 | 20934 |
| 5 | 600089.SH | TBEA Co., Ltd. | 249.24 | Industrial | 42233 | 39733 |
| 6 | 002506.SZ | GCL Energy Technology Co., Ltd. | 243.5 | Information technology | 125899 | 120962 |
| 7 | 600522.SH | Jiangsu Zhongtian Technology Co., Ltd. | 238.23 | Information technology | 57718 | 57656 |
| 8 | 002129.SZ | Tianjin Zhonghuan Semiconductor Co., Ltd. | 167.11 | Information technology | 41760 | 41641 |
| 9 | 300316.SZ | Mechanical & Electrical Co., Ltd. | 129.66 | Information technology | 10667 | 10660 |
| 10 | 603806.SH | Hangzhou First Applied Material CO., LTD. | 122.6 | Information technology | 11973 | 9195 |
| Total | | | | | 392811 | (358387) |

The second part of this paper uses Wind database as the data source. From November 2015 to May 2018, the data of component shares, photovoltaic index technical indicators and macro-level data are selected. The specific indicators are listed in Table 3.

**Table 3.** Types and abbreviations of predictors

| Type | Predictors | Abbreviations |
|---|---|---|
| Technical indicators | Closing price, volume, real amplitude trend indicator, momentum indicator, energy tide, trend index, relative | close, volume, ATR, MOM, OBV, CCI, RSI, MA |
| Macro-level index | Shanghai Composite Index, RMB Index, Brent Crude Oil Spot Price | SH1, CNY, OIL |

## 3.2  Data Processing

### 3.2.1  Text Data Processing

This paper uses text-based sentiment analysis technology to analyze the stock bar comment data. Sentiment analysis is also known as Opinion Mining. This technique can be used to determine the emotional state expressed by a piece of text. Text sentiment analysis mainly has two types of methods, namely sentiment dictionary and rule-based analysis methods (Ran et al. 2018), and the other is based on supervised machine learning methods. This type of method uses support vector machines, recurrent nerves models such as networks and convolutional neural networks, they can effectively improve the effectiveness of sentiment analysis. The current class libraries for sentiment analysis are NLTK, snownlp and so on. This paper selects PaddlePaddle for sentiment analysis of stock bar comment texts. PaddlePaddle is Baidu's open source machine learning platform, which includes Baidu's self-built dictionary, bidirectional single-layer LSTM analysis model for emotional orientation, and other open source programs. We use the PaddlePaddle to convert text data into three types: positive, neutral, and negative. The processed data statistics are as follows:

After obtaining the sentiment classification of stock bar reviews, we used the bullish index construction method proposed by Antwelier and Frank to compile

**Table 4.** Text data processing results

| Serial number | Code | Positive comment | Neutral comments | Negative comments |
|---|---|---|---|---|
| 1 | 601727.SH | 6039 | 13336 | 5282 |
| 2 | 601012.SH | 6157 | 9688 | 4854 |
| 3 | 601877.SH | 3709 | 5590 | 2954 |
| 4 | 600438.SH | 6140 | 10214 | 4581 |
| 5 | 600089.SH | 11025 | 19756 | 8953 |
| 6 | 002506.SZ | 31953 | 24202 | 64808 |
| 7 | 600522.SH | 16693 | 28396 | 12568 |
| 8 | 002129.SZ | 10681 | 8904 | 22057 |
| 9 | 300316.SZ | 3193 | 2351 | 5117 |
| 10 | 603806.SH | 2688 | 2063 | 4445 |
| Total | | 98278 | 124500 | 135619 |

the daily stock bar review sentiment index. The bullish index has been verified to be the most stable method for calculating affective tendencies, and has been adopted by a large number of studies (Längkvist et al. 2014). The calculation method is:

$$\text{sentiment}_{it} = \ln\left(\frac{1 + M_{it,pos}}{1 + M_{it,neg}}\right) \tag{11}$$

t is the observation time window, $M_{it,pos}$ is the number of positive comments of the i-th stock during the period of t, and $M_{it,neg}$ is the number of negative comments of the i-th stock during the period of t. Based on formula (11), we input the number of positive reviews and the number of negative reviews in a day into the model, and calculates the daily sentiment index of ten stocks. Then the sentiment index of each stock is weighted by market value according to the formula (12) to obtain a comprehensive emotional index $sentiment_{it}$ (Table 4):

$$\text{sentiment}_t = \sum_{i=1}^{10} \text{weight}_{it} \times \text{sentiment}_{it} \tag{12}$$

$weight_{it}$ is the ith stock's market value as a proportion of the total market value of the ten stocks in the *t-period*.

### 3.2.2 Numerical Data Processing

This paper selects the daily data of the 11 indicators listed in Table 3 from November 2015 to May 2018. After removing missing values, weighted integration of individual stock data, time matching and calculation of technical indicators, we obtained a total of 627 data. Descriptive statistics are shown in Table 5. Among them, ten stock data indicators are weighted by market value. In the technical indicators, the period of RSI, MOM, ATR, CCI and MA is set as 6 days, 12 days, 14 days and 6 days respectively.

**Table 5.** Descriptive statistics for data

| | Mean | Median | Standard deviation | Min | Maximum | Count |
|---|---|---|---|---|---|---|
| Close | 2466.7 | 2443.68 | 236.00 | 1952.32 | 2930.65 | 627 |
| Sentiment | −0.46 | −0.45 | 0.23 | −1.41 | 0.15 | 627 |
| Volume | 582360389 | 538892971 | 221618023 | 196317201 | 1620010434 | 627 |
| RSI | 51.73 | 51.44 | 19.34 | 4.48 | 98.82 | 627 |
| MOM | −1.17 | 8.83 | 152.42 | −745.67 | 428.41 | 627 |
| OBV | 2.2964E−10 | 2.478E-10 | 7709606825 | 906825270 | 3.4706E-10 | 627 |
| ATR | 46.75 | 37.11 | 23.23 | 22.36 | 118.00 | 627 |
| CCI | 10.54 | 23.98 | 97.81 | −200.00 | 200.00 | 627 |
| MA | 2468.07 | 2457.58 | 231.63 | 1987.66 | 2884.21 | 627 |
| SH | 3177.35 | 3168.90 | 197.12 | 2655.66 | 3651.77 | 627 |
| CNY | 117.89 | 116.97 | 2.31 | 113.86 | 123.89 | 627 |
| OIL | 51.67 | 50.09 | 10.79 | 25.99 | 80.30 | 627 |

## 4  L-GSRF Model Experimental Results and Analysis

### 4.1  Lasso Regression Result

This paper takes the first 587 data as the training set and the last 40 data as the test set. Lasso regression model was used in the training set data. The interpreted variable is the closing price, and the explanatory variables are sentiment index and trading volume. Technical indexes RSI, MOM, OBV, CCI, MA, Shanghai Composite Index, RMB Index and Brent Crude Oil Price. The penalty coefficient $\alpha$ is an important parameter in the Lasso regression model, and the regression results will be affected by different values of $\alpha$. This paper uses the BIC information criteria (Bayesian Information Code) to determine the optimal penalty coefficient $\alpha$ as 0.1661. The regression results are as follows:

**Table 6.** Lasso regression results

| Variable name | Regression coefficient |
| --- | --- |
| Sentiment | 7.4618031 |
| Trading volume | 0 |
| RSI | 0.80972198 |
| MOM | 0.05215402 |
| OBV | 0 |
| ATR | 0 |
| CCI | 0.21815196 |
| MA | 0.94396249 |
| SH | 0 |
| CNY | 0 |
| OIL | 0 |

From the regression results in Table 6, the regression coefficients of Sentiment, RSI, MOM, CCI and MA are not 0, and the regression coefficients for the other variables are 0, especially the macro indicators are all rejected. Macro indicators focus on national policies, model political and economic data, and analyze the impact on the long-term trend of stock indexes, which are mainly applicable to the relatively mature index trend prediction with a relatively long cycle. In this paper, short-term stock indexes are predicted, so fundamental analysis is not applicable to the model in this paper. Therefore, the above five indexes with non-zero coefficients are input into the L-GSRF model for stock price index prediction.

## 4.2   L-GSRF Model Prediction Results

### 4.2.1   MAE and RMSE Analysis

The results of the experiment are compared with the unadjusted RF model, adjusted GSRF model and SVM model. The MAE and the RMSE of each model are listed in Table 7.

Table 7. Error analysis of experimental results

| Serial number | Model | MAE | RMSE |
|---|---|---|---|
| 1 | RF | 6.01 | 8.07 |
| 2 | SVM | 4.06 | 6.60 |
| 3 | GSRF | 2.40 | 3.52 |
| 4 | L-GSRF | 1.43 | 2.32 |

From the experimental results, the unadjusted RF model and the SVM model fit the test set data poorly. The MAE and RMSE of RF model are 6.01 and 8.07 respectively, and the MAE and RMSE of the SVM model are 4.06 and 6.60, reflecting their low accuracy in data fitting. In the grid-adjusted GSRF model, MAE and RMSE are reduced to 2.40 and 3.52, and the errors are reduced by 60.06% and 56.38% respectively compared with the unadjusted RF model, indicating that grid-adjusted parameters have a significant improvement in the prediction of the model. In the L-GSRF model with Lasso regression, MAE and RMSE are reduced to 1.43 and 2.32, respectively. Compared with the unadjusted RF model, MAE and RMSE are reduced by 76.21% and 71.25%, respectively. Compared with the GSRF model, the two indexes are reduced by 40.42% and 34.09%, which shows that the L-GSRF model has greatly improved the performance of stock index prediction.

### 4.2.2   Accuracy Rate Analysis

In the accuracy analysis, we labeled the rise or fall of the index and forecast result as 1 and 0. If both the index and forecast result are rising, TP plus 1. If both the index and the forecast result are falling, TN plus 1. The results are shown in Table 8 (Fig. 2).

As can be seen from Table 8, the L-GSRF model has the highest accuracy in the test set of 40 samples. Among them, the prediction accuracy of L-GSRF model for rising is higher than that for falling. Accuracy decreases in sequence in GSRF model, RF model and common SVM model. This can also reflect the validity of L-GSRF model in predicting the rise and fall of stock index.

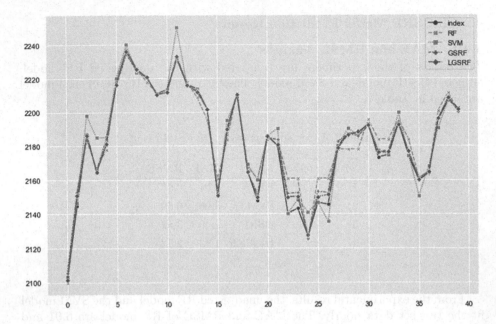

**Fig. 2.** Four models prediction result

**Table 8.** Error analysis of experimental results

| Serial number | Model | TP | TN | Accuracy rate |
|---|---|---|---|---|
| 1 | RF | 18 | 19 | 0.925 |
| 2 | SVM | 19 | 16 | 0.875 |
| 3 | GSRF | 20 | 18 | 0.95 |
| 4 | L-GSRF | 20 | 19 | 0.975 |

## 5   Trading Strategy Based on L-GSRF Model

Based on the L-GSRF prediction model, this paper constructs a simple photo-voltaic (PV) index daily frequency trading strategy: if the next trading day is expected to rise, it would buy at the opening price on next trading day; if it predicted a decline in the next trading day, it would sell at the opening price on next trading day. The 30-day data of the test set is used to do a backtest of the strategy. The annualized yield and yield curves are shown in Fig. 3, where the blue dotted line is the yield curve of the strategy and the red solid line is the yield curve of the PV index. From the comparison of annualized returns in Table 7, it can be seen that the return of the trading strategy based on the four models is better than that of the PV index. Among them, the yields of the RF model, the SVM model and the GDRF model are 73.52%, 72.9% and 74.11%, respectively, which are much higher than the annualized yield of 18.63% of the PV index. The trading strategy based on the L-GSRF model has achieved an annualized return of 78.94%, which is 323.73% higher than the annual return of the benchmark PV index. In the comparison chart of model yield, we can see that in Fig. 3, the yield curve of the L-GSRF model is smoother than the other three models, with minimal maximum drawdown, which reflects that it can better predict the trend in the declining market, with less time and higher accuracy (Table 9).

**Table 9.** Strategic annualized rate of return comparison

| No. | Model | Strategy annualized return | Annualized yield on photovoltaic index |
|-----|-------|---------------------------|----------------------------------------|
| 1 | RF | 73. 52% | 18.63% |
| 2 | SVM | 72. 90% | |
| 3 | GSRF | 74.11% | |
| 4 | L-GSRF | 78.94% | |

We conducted a T-test for the significance of strategy returns. As can be seen from Table 10, L-GSRF has the highest average daily return on holding position, reaching 0.1816%. The average daily yield of holding position and short position has a significant difference. It indicates that the timing of L-GSRF strategy is effective and it is more profitable than the other three strategies.

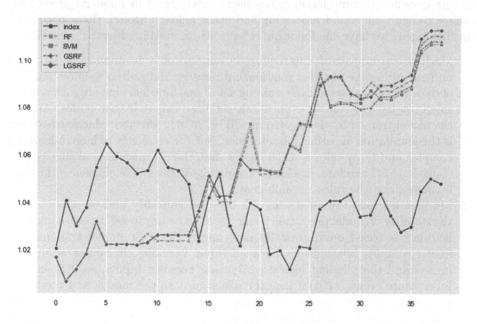

**Fig. 3.** Four strategies and benchmark annualized return (Color figure online)

**Table 10.** T-test for strategy profit

| | $n_h$ | $n_s$ | $\mu_h$ | $\mu_s$ | $\Delta_\mu$ | $\sigma_h$ | $\sigma_s$ |
|---|---|---|---|---|---|---|---|
| RF | 16 | 24 | 0.1646% | −0.2523% | 0.42%** | 0.49 | 0.10 |
| SVM | 20 | 20 | 0.0340% | −0.9370% | 0.97%** | 0.54 | 0.09 |
| GSRF | 19 | 21 | 0.1097% | −0.9486% | 1.06%*** | 0.52 | 0.12 |
| L-GSRF | 22 | 18 | 0.1816% | −0.5488% | 0.73%*** | 0.46 | 0.17 |

# 6    Research Conclusions and Reflections

In this paper, L-GSRF algorithm is constructed to predict the PV industry index by using 358, 387 text data from comprehensive stock comments and data from 12 economic indicators over 627 trading days, through data cleaning and feature extraction. It turns out that the L-GSRF algorithm can effectively improve the prediction accuracy of financial data compared to traditional RF, SVM and GSRF models. Based on the predicted results, the trading strategy of L-GSRF index is constructed. From the T-test of trading results and earnings, L-GSRF strategy shows better performance. The return of the L-GSRF strategy is 323.73% higher than the benchmark. The T-test shows that the holding position yield of L-GSRF is significantly higher than that of short position, achieving the highest average daily return.

At present, machine learning has been widely used in financial prediction, image recognition, sound processing and other fields. Through the experiments in this paper, we have the following reflections on machine learning algorithm:

1. Machine learning algorithms require the support of large amounts of data, and understanding the abstract relationship between data often requires millions of data as training sets. If the training set of machine learning algorithm is too small, we have to rely on explicit rule settings, otherwise the performance of the model trained by the algorithm will be greatly affected. Machine learning is the foundation of artificial intelligence, but the efficiency of human learning complex and fuzzy rules is much higher than that of machine learning system (Lake 2015). Therefore, machine learning algorithm is less efficient if the data is limited and the rules are ambiguous.

2. Machine learning cannot distinguish between causality and correlation, and the stability of model prediction results need to be improved. Because machine learning is based on historically determined data for model training and fitting complex relationships between input variables, it cannot accurately understand the inherent logical correlation between inputs, so it is easy to incorporate events without logical correlation into the model, and the learning presentation of unexpected events is not strong, which can easily lead to low model stability when predicting out-of-sample data.

3. Machine learning algorithm is currently not transparent, resulting in a lack of trust in it. Samek (2017) and Ribeiro (2016) have been discussing the opacity of neural networks. In large-scale neural network system, there are a large

number of hundreds of millions of neurons. As individual nodes of complex network, neurons are difficult to adjust parameters. In certain areas, such as financial investment or medical field, users of machine learning algorithm are more likely to understand the process of system decision-making, while due to the complexity of neurons and the weak tunability of individual node parameters, machine learning is characterized as a "black box" with low user trust.

In future research, we should focus on how to carry out effective machine learning in the case of insufficient data, how to use machine learning to effectively distinguish causality from correlations, and how to enhance machine learning transparency and enhance the credibility of machine learning. The discussion of these problems will help to improve the efficiency of machine learning algorithm and expand its application range.

# References

Ding, X., Zhang, Y., Liu, T., et al.: Deep learning for event-driven stock prediction. In: IJCAI, pp. 2327–2333 (2015)

Yang, Q., Wu, X.: 10 challenging problems in data mining research. Int. J. Inf. Technol. Decis. Making **5**(04), 597–604 (2006)

Tsai, C.F., Hsiao, Y.C.: Combining multiple feature selection methods for stock prediction: union, intersection, and multi-intersection approaches. Decis. Support Syst. **50**(1), 258–269 (2010)

Malkiel, B.G.: The efficient market hypothesis and its critics. J. Econ. Perspect. **17**(1), 59–82 (2003)

Phan, D.H.B., Sharma, S.S., Narayan, P.K.: Stock return forecasting: some new evidence. Int. Rev. Financ. Anal. **40**, 38–51 (2015)

Upadhyay, A., Bandyopadhyay, G., Dutta, A.: Forecasting stock performance in Indian market using multinomial logistic regression. J. Bus. Stud. Q. **3**(3), 16 (2012)

Gu, S., Kelly, B.T., Xiu, D.: Empirical Asset Pricing via Machine Learning. Chicago Booth Research Paper No. 18–04, 11 June 2018. SSRN: https://ssrn.com/abstract=3159577

Athey, S.: The impact of machine learning on economics. In: Economics of Artificial Intelligence. University of Chicago Press (2017)

Akita, R., Yoshihara, A., Matsubara, T., et al.: Deep learning for stock prediction using numerical and textual information. In: 2016 IEEE/ACIS 15th International Conference on Computer and Information Science (ICIS), pp. 1–6. IEEE (2016)

Fischer, T., Krauss, C.: Deep learning with long short-term memory networks for financial market predictions. Eur. J. Oper. Res. **270**(2), 654–669 (2018)

Chong, E., Han, C., Park, F.C.: Deep learning networks for stock market analysis and prediction: methodology, data representations, and case studies. Expert Syst. Appl. **83**, 187–205 (2017)

Zhong, X., Enke, D.: Forecasting daily stock market return using dimensionality reduction. Expert Syst. Appl. **67**, 126–139 (2017)

Ran, Y.F., Jiang, H.: Study on stock price prediction based on BPNN and SVR. J. Shanxi Univ. (Nat. Sci. Ed.) **41**(01), 1–14 (2018)

Jang, H., Lee, J.: An empirical study on modeling and prediction of bitcoin prices with Bayesian neural networks based on blockchain information. IEEE Access **6**, 5427–5437 (2018)

Singh, R., Srivastava, S.: Stock prediction using deep learning. Multimedia Tools Appl. **76**(18), 18569–18584 (2017)

Lee, H., Surdeanu, M., MacCartney, B., et al.: On the importance of text analysis for stock price prediction, LREC, pp. 1170–1175 (2014)

Uysal, A.K., Gunal, S.: The impact of preprocessing on text classification. Inf. Process. Manag. **50**, 104–112 (2014)

Kohonen, T.: An introduction to neural computing. Neural Netw. **1**(1), 3–16 (1988)

Roux, N., Bengio, Y.: Representational power of restricted Boltzmann machines and deep belief networks. Neural Comput. **20**, 1631–1649 (2008)

Antweiler, W., Frank, M.Z.: Is all that talk just noise? The information content of internet stock message boards. J. Finance **59**(3), 1259–1294 (2004)

Mengyuan, Wang, H., Wang, W.: The impact of online reputation on product sales: a fine-grained emotional analysis method. Manag. Rev. **29**(01), 144–154 (2017)

Längkvist, M., Karlsson, L., Loutfi, A.: A review of unsupervised feature learning and deep learning for time-series modeling. Pattern Recogn. Lett. **42**, 11–24 (2014)

Lake, B.M., Salakhutdinov, R., Tenenbaum, J.B.: Human-level concept learning through probabilistic program induction. Science **350**(6266), 1332–1338 (2015)

# Dynamic Copula Analysis of the Effect of COVID-19 Pandemic on Global Banking Systemic Risk

Jie Li[1,2] and Ping Li[1,2(✉)]

[1] School of Economics and Management, Beihang University, Beijing 100191, China
liping124@buaa.edu.cn
[2] Key Laboratory of Complex System Analysis, Management and Decision (Beihang University), Ministry of Education, Beijing 100191, China

**Abstract.** The ongoing COVID-19 pandemic has led to not only the loss of enormous lives, but the dramatic impact on global financial markets. By considering 29 global systemically important banks from four regions (North America, Europe, China, Japan), we employ the proposed truncated D-vine dynamic mixed copulas model to investigate the evolution of the systemic risk of global banking system during the COVID-19 pandemic period. From empirical results, as a worldwide shock, the COVID-19 pandemic does have increased the systemic risk of the global banking system. Specifically, the systemic risk level of the global banking sector was moderate during the period when the COVID-19 pandemic burst only in China, and increased rapidly when the virus spread over the world, then cooling down when emergency actions were taken by countries. In addition, the systemic risk contribution of banks in most regions (like North America, Europe, and Japan) under the similar epidemic situation during the COVID-19 period, seem to be not impacted by the evolution of the panic (as well as the systemic risk level), while the systemic risk contribution of Chinese banks kept falling due to its opposite situation to others in this period.

**Keywords:** COVID-19 pandemic · Truncated D-vine · Dynamic mixed copula · Systemic risk · Banking risk

## 1 Introduction

The COVID-19 pandemic have brought about the ongoing enormous costs in lives since its outbreak in December, 2019. Besides the loss in human lives, world economies also keep being dramatically affected by the coronavirus panic, such as, decrease in consumption, interruptions to the production, disruptions of global supply chains, unemployment, shutting down operations of companies, etc. Moreover, global financial markets have registered dramatic drop and the volatility is proved to increase (Zhang et al. [1], Albulescu [2]) and reach at level similar to the global financial crisis in 2008 (Fernandes [3]).

This work is supported by the National Natural Science Foundation of China (No. 71571008).

Therefore, one of emerging questions is that could COVID-19 pandemic be the source of a new financial crisis? or, how serious the COVID-19 pandemic affects the financial market of the world? On the other hand, since the banking system has been proved to be dominant role in the financial market, particularly during the crisis (Brownlees, and Engle [4]; Acharya et al. [5]), this paper investigates the impact of this worldwide epidemic on the global banking sector.

The fact that financial markets move more closely together during times of crisis is well documented, which is accompanied by the increase of the systemic risk. The systemic risk can be defined as the risk that the failure of a systemically important institution distress could cause the distress of the system. In other words, the systemic risk level could be a signal of the health situation of the financial system. Consequently, we analyze the evolution of the systemic risk of the global banking sector during the COVID-19 pandemic period.

The key point to account for the systemic risk of the global banking system is to character the (conditional) dependence among each bank by using the high-dimensional distribution. Hence, we employ the mixed vine copula model proposed by Weiβ and Scheffer [6] combing with the truncated structure (Brechmann et al. [7]), however, with time-variant parameters updated by the score-driven model of Creal et al. [8] and Harvey [9], which is called truncated D-vine dynamic mixed copulas.

The remainder of this article is structured as follows. Section 2 describes the model we proposed, and Sect. 3 presents the data description and marginal distributions for all series. In Sect. 4, we show our empirical results, and Sect. 5 concludes.

## 2   Methodology

In this section, we briefly describe our approach, truncated D-vine dynamic mixed copulas, to model the interdependence of d-dimensional observable variables, including the truncated D-vine copula specification and the dynamic mixture of time-varying copulas definition.

### 2.1   Truncated D-vine Copula

From Sklar's theorem [10], copulas can be used to separate a multivariate distribution into its marginals and the dependence structure which is fully captured by the copula function. Specifically, a $d$-dimensional cumulative distribution function F can be split in two parts, the marginal distribution function $F_i$ and a copula function $C$, as follows.

$$F(x_1, \ldots, x_d) = C(F_1(x_1), \ldots, F_d(x_d)). \qquad (1)$$

The probability density function $f$ can be represented by

$$f(x_1, \cdots, x_d) = c(F_1(x_1), \cdots, F_d(x_d)) \prod_{i=1}^{d} f_i(x_i) \qquad (2)$$

where $c$ is the copula density and $f_i$ are marginal densities. By the definition from Kurowicka and Cooke [11], the copula density function of the D-vine structure can be factorized as the product of bivariate copulas (called pair-copulas)

$$c(F_1(x_1), \cdots, F_d(x_d)) = \prod_{j=1}^{d-1} \prod_{i=1}^{d-j} c_{i,i+j|i+1,\cdots,i+j-1}(F(x_i|x_{i+1}, \cdots, x_{i+j-1}), F(x_{i+j}|x_{i+1}, \cdots, x_{i+j-1})) \quad (3)$$

where conditional marginal distributions can be calculated as follows

$$F(x|\mathbf{v}) = \frac{\partial C_{x v_j | \mathbf{v}_{-j}}(F(x|\mathbf{v}_{-j}), F(v_j|\mathbf{v}_{-j}))}{\partial F(v_j|\mathbf{v}_{-j})} \quad (4)$$

for a $k$-dimensional vector $\mathbf{v}$. Here $v_j$ is one arbitrarily chosen component of $\mathbf{v}$ and $\mathbf{v}_{-j}$ denotes the $\mathbf{v}$-vector, excluding this component. The specification can be given in form of a nested set of trees (Aas et al. [12]) shown in Fig. 1 which gives examples of a five-dimensional D-vine copula with five random variables, four trees and ten edges. Edges in each tree correspond to pair-copulas of two neighbor nodes corresponding to (conditional) marginal distributions.

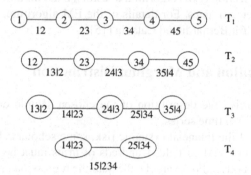

**Fig. 1.** D-vine copula with five variables

According to the idea that the most important dependencies in the vine structure are captured in first trees, Brechmann et al. [7] propose the truncated vine structure to allow for the best possible specification of first $K$ trees leaving all pair-variables in higher order trees independent. The D-vine copula density function in (3) truncated at level $K$ ( $< d$) can be reduced to.

$$c(F_1(x_1), \cdots, F_d(x_d)) = \prod_{j=1}^{K-1} \prod_{i=1}^{K-j} c_{i,i+j|i+1,\cdots,i+j-1}(F(x_i|x_{i+1}, \cdots, x_{i+j-1}), F(x_{i+j}|x_{i+1}, \cdots, x_{i+j-1})) \quad (5)$$

### 2.2 Dynamic Mixture of Time-Varying Copulas

Referring to Weiß and Scheffer [6], we employ mixed copulas (convex combinations of bivariate parametric copulas) as pair-copulas in the vine structure to solve the pair-copulas selection problem. However, instead of considering the static model, we employ the Creal et al.'s GAS model [8] to update the dynamic parameters. With $c_1(u_{1t}, u_{2t};$

$\theta_1),\ldots, c_g(u_{1t}, u_{2t}; \theta_{gt})$ being $g$ parametric copula densities, the corresponding mixed copula density is given by

$$c(u_{1t}, u_{2t}; \Theta_t) = \sum_{i=1}^{g} \pi_{it} c_i(u_{1t}, u_{2t}; \theta_{it}), \ \Theta_t = (\pi_{it}, \theta_{it}, i = 1, \ldots, g)$$

$$\Theta_{t+1} = \kappa + \mathbf{A} \Xi_t \nabla(\Theta_t | u_{1t}, u_{2t}) + \mathbf{B} \Theta_t$$

(6)

where $\pi_{it}$, $i = 1,\ldots, g$ are the non-negative quantities summing to one. $\nabla(\Theta_t | u_{1t}, u_{2t})$ is the score-driven variables corresponding to $\Theta_t$, and $\Xi_t$ is the positive definite scaling matrix (such as the identity matrix, see Oh and Patton [13]). $\kappa$, $\mathbf{A}$, $\mathbf{B}$ are coefficient matrices to be estimated. For the updating model of weights $\pi_{it}$, see Catania [14], and for the updating model of copula parameters $\theta_{it}$, see Bernardi and Catania [15]. Since the parameter vector $\Theta$ of the finite mixture model cannot be estimated via classical maximum likelihood estimation due to the incomplete structure of the data (see, e.g. Dempster et al. [16]; McLachlan and Peel [17]; Weiβ and Scheffer [6]), we employ the Expectation-Maximization (EM) algorithm to estimate the mixture parameters to avoid the biased parameter estimates. For details of the EM algorithm, see Kim et al. [18], Weiβ and Scheffer [6], Bernardi and Catania [15].

## 3 Data Description and Marginal Distribution

In this section, we give the description of the chosen sample data and estimate the marginal model on each time series.

In the research of the financial systemic risk, some scholars choose as many institutions as possible in a certain field. Hundreds of firms must be dealt with binary or hyper-dimensional models. For example, the quantile regression model is a widely used model to calculate CoVaR measuring the systemic risk contribution of one single institution to the system (e.g. Laeven et al. [19]). However, only considering the relevance between some institution and the system index by aggregating the rest, useful information about the system risk may be missed. Oh and Patton [13] propose the dynamic factor copula model to analyze the interdependence between 100 firms and facilitate the estimation of several systemic risk measures. Given the quite high-dimensional data, some restrictions on the homogeneity among variables must be imposed and it is not quite straightforward to balance the parsimony and effectiveness of the model. Therefore, we refer to the list of global systemically important banks (G-SIBs) identified by the Financial Stability Board (FSB, [20]), including 29 public large banks belonging to 11 countries all over the world.

Moreover, to investigate the systemic risk of the global banking sector, we consider the average weighted portfolio composed of 29 G-SIBs as the global banking system index. There exist two different commonly used weighting variables to define the system portfolio, the equal weight (Bernardi and Catania [15]) and the firm size (Adrian and Brunnermeier [21]). Nevertheless, the equally weighted portfolio neglects the heterogeneity among different institutions and the portfolio weighted by sizes may overestimate the scale of institutions. For this reason, we consider required levels of the additional capital buffers consisting of 5 categories of bank activities (size, interconnectedness, substitutability/financial institution infrastructure, complexity, cross-jurisdictional activity), for more details, see BCBS [22]. Table 1 shows the chosen banks, as well as the relevant information.

**Table 1.** List of banks

|    | Banks | Symbols | Buffers | Regions |
|----|-------|---------|---------|---------|
| 1  | JP Morgan Chase | JPM | 2.5% | US |
| 2  | Citigroup | CIT | 2.0% | US |
| 3  | HSBC | HSBC | | UK |
| 4  | Bank of America | BOA | 1.5% | US |
| 5  | Bank of China | BOC | | China |
| 6  | Barclays | BARC | | UK |
| 7  | BNP Paribas | BNP | | France |
| 8  | Deutsche Bank | DBK | | Germany |
| 9  | Goldman Sachs | GS | | US |
| 10 | Industrial and Commercial Bank of China | ICBC | | China |
| 11 | Mitsubishi UFJ FG | MUFJ | | Japan |
| 12 | Wells Fargo | WFC | | US |
| 13 | Agricultural Bank of China | ABC | 1.0% | China |
| 14 | Bank of New York Mellon | BNYM | | US |
| 15 | China Construction Bank | CCB | | China |
| 16 | Credit Suisse | CS | | Switzerland |
| 17 | Groupe Crédit Agricole | GCA | | France |
| 18 | ING Bank | ING | | Netherlands |
| 19 | Mizuho FG | MFG | | Japan |
| 20 | Morgan Stanley | MS | | US |
| 21 | Royal Bank of Canada | RBC | | Canada |
| 22 | Banco Santander | BS | | Spain |
| 23 | Société Générale | SG | | France |
| 24 | Standard Chartered | SC | | UK |
| 25 | State Street | SS | | US |
| 26 | Sumitomo Mitsui FG | SMFG | | Japan |
| 27 | Toronto Dominion | TD | | Canada |
| 28 | UBS | UBS | | Switzerland |
| 29 | UniCredit | UC | | Italy |

Notes: The fourth column "Buffers" denotes the required levels of additional capital buffers for each bank, and these percentages are used to calculate weights of each bank, and then, the banking system index

Considering the availability of the bank data and our focus on the COVID-19 pandemic, we use 29 banks' daily equity prices during June 3, 2019 to June 9, 2020 collecting from Bloomberg and all price series used in this paper are denominated in United States Dollars (USD).

We model time series of all banks' prices in log differences. From the summary statistics of the log-returns for all banks, of particular note is the nonzero skewness (positive for 27 series, and negative for 2 series) and excess kurtosis (greater than 3 for all series). The Jarque-Bera test rejects the null normality distribution for all the series. Both the ADF test and KPSS test show all time series are stationary. Next, we consider the most commonly used conditional mean and variance model for the financial time series, that is, AR(1)-GJR-GARCH(1,1) with the skewed students $t$ distribution for standardized residuals, to filter return series, as given by

$$x_{i,t} = \mu_i + \alpha_i x_{i,t-1} + e_{i,t},$$
$$e_{i,t} = \sigma_{i,t}\varepsilon_{i,t},$$
$$\sigma_{i,t}^2 = \omega_i + \beta_i \sigma_{i,t-1}^2 + \gamma_i e_{i,t-1}^2 + \lambda_i e_{i,t-1}^2 1\{e_{i,t-1} < 0\},$$
$$\varepsilon_{i,t} \sim \text{iid Skew } t(\nu_i, \psi_i). \tag{7}$$

## 4 Empirical Analysis

In what follows, we apply the truncated D-vine dynamic mixed copulas model described in Sect. 2 to exam the evolution of the systemic risk in global banking sector during the COVID-19 pandemic period.

### 4.1 Systemic Risk Measures

To quantify the systemic risk, we consider two popular measures, that is, the conditional value at risk (CoVaR) and the conditional expected shortfall (CoES) introduced by Adrian and Brunnermeier [21]. The CoVaR can be used to measure the spillover effects from some institution to the system conditional on this institution under its distress and the CoES is the expectation version of the CoVaR, as given by

$$P(r_{s,t} \leq \text{CoVaR}_{s|i,t}^{\beta|\alpha} | r_{i,t} \leq \text{VaR}_{i,t}^{\alpha}) = \beta \tag{8}$$

$$\text{CoES}_{s|i,t}^{\beta|\alpha} = E(r_{s,t} | r_{s,t} \leq \text{CoVaR}_{s|i,t}^{\beta|\alpha}) \tag{9}$$

where $r_{i,t}, r_{s,t}$ are equity returns of the institution $i$ and the system index, respectively. $\alpha$, $\beta$ are the predetermined confidence levels. As a direct consequence of the sub-additivity property (Mainik and Schaanning [23]; Bernardi et al. [24]), the average weighted version of CoESs of all institutions can be applied to measure the total systemic risk of the overall financial system, as shown below

$$\text{CoES}_{s,t}^{\beta|\alpha} = -\sum_{i=1}^{d} w_i \text{CoES}_{s|i,t}^{\beta|\alpha} \tag{10}$$

where $w_i$ denote weights associated with each institution in the system and negative CoESs ensure positive systemic risk levels. Besides, following Girardi and Ergün [25], Mainik and Schaanning [23], and Bernardi and Catania [15], we also consider the difference between the CoVaR (CoES) and their median value, as measures of the systemic risk contribution, as follows

$$\Delta\text{CoVaR}_{s|i,t}^{\beta|\alpha} = \frac{\text{CoVaR}_{s|i,t}^{\beta|\alpha} - \text{CoVaR}_{s|i,t}^{\beta|0.5}}{\text{CoVaR}_{s|i,t}^{\beta|0.5}} \cdot 100 \tag{11}$$

$$\Delta\text{CoES}_{s|i,t}^{\beta|\alpha} = \frac{\text{CoES}_{s|i,t}^{\beta|\alpha} - \text{CoES}_{s|i,t}^{\beta|0.5}}{\text{CoES}_{s|i,t}^{\beta|0.5}} \cdot 100 \tag{12}$$

where the median status is defined as $P(r_{i,t} \leq \text{VaR}_{i,t}^{0.5}) = 0.5$. The $\Delta$CoVaR and $\Delta$CoES quantify the percentage increase of the systemic risk conditional on a given distress event, to measure how the CoVaR and CoES change once a particular institution falls into financial distress.

## 4.2 Systemic Risk Level Analysis

We use 29 banks samples described in previous section to estimate the truncated D-vine dynamic mixed copulas model. The system index is not included to construct the vine structure due to that the system index is a key variable governing the interaction in the data set. This case may require to fit the C-vine copula (see, Kurowicka and Cooke [11]) consisting of bivariate copulas of the system index and each bank, resulting that CoVaR (CoES) of the bank $i$ may neglect the useful information between the bank $i$ and other banks, like binary models do.

The first tree of D-vine copula can be specified by the maximum sum of the absolute value of empirical Kendall's taus between two adjacent nodes, similar to the traveling salesman problem (TSP). The truncation level can be decided by the Akaike information criterion (AIC), for more detail, see Brechmann et al. [7]. As for mixed pair-copulas, we consider three copulas of different tail dependencies, Gaussian copula (tail independent), Gumbel copula (upper tail dependent), and Survival Gumbel copula (lower tail dependent), as mixture components to model the dependence structure of the sample in the empirical analysis. Figure 2 shows the estimated first tree in the vine structure of 29 banks, which identifies six clusters of economically and regionally similar variables, indicating the rationality of the estimated D-vine structure. Furthermore, Table 2 represents the estimated summary results of the truncated D-vine dynamic mixed copulas. Truncation level equal to 2 means that only pair-copulas in the first two trees are dynamic mixed copulas and those in higher order trees are independent.

In what follows, by the Monte Carlo simulation procedure, we use the estimated truncated D-vine dynamic mixed copulas model to simulate 10000 observations for each bank at each time point, and average weighted values (weights are computed using the Buffers in Table 1) of all banks' draws are considered as the simulated values of the banking system index. Employing the approach described before, the systemic risk measures CoVaR and CoES can be calculated by the simulation values of each bank and

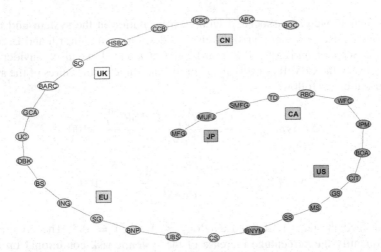

**Fig. 2.** The first tree of D-vine copula for equity time series of 29 banks

**Table 2.** Summary results of the truncated D-vine dynamic mixed copulas estimation

| Truncation level | Log-likelihood | No. of parameters | AIC | BIC | HQ |
|---|---|---|---|---|---|
| 1 | 3264.2 | 252 | −6024.3 | −5180.9 | −6105.9 |
| **2** | **3600.6** | **495** | **−6211.2** | **−4554.4** | **−6371.3** |
| 3 | 3746.8 | 729 | −6035.5 | −3595.5 | −6271.3 |

Notes: AIC, BIC, HQ are the Akaike Information criterion, Bayesian Information criterion, and Hannan- Quinn criterion, respectively. The best truncation level is chosen by the AIC, corresponding to 2.

the system index. Figure 3 shows the estimated CoES values in Eq. (10) over the sample period and we can find similar trends under different confidence levels, indicating the robust estimated results of the systemic risk level.

From Fig. 3, we can find that the systemic risk of the global banking sector significantly increases during the COVID-19 pandemic period, starting from February 25, 2020. It is worth noting that systemic risk levels maintain moderate during the preliminary outbreak of the coronavirus pandemic in China in January, 2020. Even though cases reported from China had peaked and began to fall in late February (WHO, 2020 [26]), the pandemic started to rapidly spread to other countries, accompanied by which financial markets across the world experienced the turbulent phase. The U.S. stock market plummeted over coronavirus fears on February 24, after the Dow Jones Industrial Average experienced the worst day in two years. Until March 18, the U.S. stock market triggered level 1 market wide circuit breakers during the opening hour four times in ten days. The STOXX Europe 600 index also dropped near 35% from February 21 to March 18. The magnitude of systemic risk levels experiences a giant increase, peaking on March 17, 2020 and then progressively cools down when most governments seal their borders to curb the virus's spread and spend large amounts of funds to tackle the

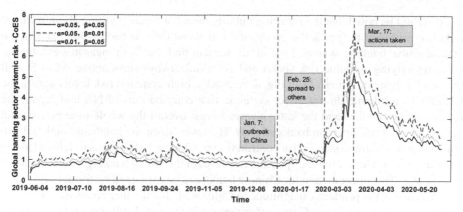

**Fig. 3.** Systemic risk level (CoES) of the global banking sector

epidemic and central banks announced the quantitative easing policy to stimulate their economies. Under the COVID-19 pandemic, the systemic risk of the global banking system depends on the performance of large percentages of banks across the world, not only that of some bank or banks in some region, like China.

### 4.3 Systemic Risk Contribution Analysis

In addition, using the $\Delta$CoVaR and $\Delta$CoES, we turn to analyze the ranking of different regions in terms of their systemic risk contributions. All 29 systemically important banks are classified to four groups geographically and economically, consisting of Europe (UK and EU), NA (US and Canada), China, and Japan. Figure 4 reports the average level of the $\Delta$CoVaR and $\Delta$CoES by group, and vertical dash lines represent the time points Jan. 7, Feb. 25, and Mar. 17, respectively.

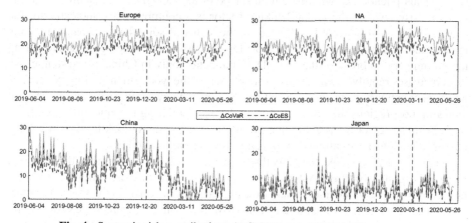

**Fig. 4.** Systemic risk contributions ($\Delta$CoVaR and $\Delta$CoES) of four groups

It should be noted that the resulting ranking does not mean anything about intrinsic riskiness profile of the particular group, and it is about the role each group plays in the overall global banking system. First of all, we can find that both measures exhibit the same underlying systemic risk signal and $\Delta$CoVaR always stays above $\Delta$CoES with the similar dynamic pattern from Fig. 4. Secondly, high systemic risk levels during the COVID-19 pandemic do not affect systemic risk contributions of NA and Japan, the former always highest and the latter always lowest overall the whole time period, and only affect that of European banks slightly. However, since the pandemic outbreaks in China in January, the systemic risk contribution of Chinese banks starts to fall and keeps the almost lowest level after the systemic risk level of the global banking system peaks in mid-March. This phenomenon for Chinese banks can be explained by the asynchronous performance of the pandemic magnitude in China with that in other countries. When the pandemic keeps spreading in China, other regions of the world still stay safe for the time being. More and more countries get to be affected by the rapid spread of the epidemic, while the situation in China turns to improve. Under the environment of the COVID-19 pandemic, the systemic risk contributions seem to be not impacted by the evolution of the systemic risk level for banks in most regions suffering the epidemic simultaneously expect for those in some region, like China, under the opposite condition.

## 5  Conclusion

This paper focuses on the systemic risk of the global banking system during the ongoing COVID-19 pandemic period. We propose a truncated D-vine structure with dynamic mixture of time-varying copulas to investigate the interdependence between 29 systemically important banks across the world. The estimated model is employed to calculate to two popularly used systemic risk measures CoVaR and CoES, which are considered to quantify the systemic risk level of the global banking sector and the systemic risk contribution of banks in different regions. We can conclude our empirical results into two relevant phenomena with the circumstance of the COVID-19 pandemic. First, the systemic risk level of the global banking sector is a comprehensive indicator for the overall system and will not be affected by the status of some bank or banks in some region, but by the status of banks in most regions. Although keeping moderate during the preliminary outbreak of the COVID-19 pandemic only in China, the systemic risk level increases rapidly when the virus spreads across the world, then cooling down once emergency actions are taken by countries. Second, the systemic risk contributions of banks in most regions, like NA, Europe, and Japan, suffering the epidemic simultaneously seem to be not impacted by the evolution of the systemic risk level, while those of banks in China keep falling due to its opposite situation to others.

As a worldwide shock, the COVID-19 pandemic do have increased the systemic risk of the global banking system. However, there seem to exist different influence mechanisms between the internal and external cause of the crisis in the global banking, even financial, system. For example, it seems to be ineffective to regulate the systemically important banks/financial institutions to avoid the financial crisis under the external pressure, like the coronavirus epidemic. For future researches, it could be interesting to investigate different operational mechanisms of the 08–09 global financial crisis and the

COVID-19 pandemic to benefit the understanding and the regulatory about the systemic financial risk.

# References

1. Zhang, D., Hu, M., Ji, Q.: Financial markets under the global pandemic of COVID-19. Financ. Res. Lett. **36**, 101528 (2020)
2. Albulescu, C.: Coronavirus and financial volatility: 40 days of fasting and fear. arXiv preprint arXiv:2003.04005 (2020)
3. Fernandes, N.: Economic effects of coronavirus outbreak (COVID-19) on the world economy. Available at SSRN (2020). https://ssrn.com/abstract=3557504
4. Brownlees, C., Engle, R.F.: SRISK: a conditional capital shortfall measure of systemic risk. Rev. Financ. Stud. **30**, 48–79 (2016)
5. Acharya, V.V., Pedersen, L.H., Philippon, T., Richardson, M.: Measuring systemic risk. Rev. Financ. Stud. **30**(1), 2–47 (2017)
6. Weiß, G.N.F., Scheffer, M.: Mixture pair-copula-construction. J. Banking Finan. **54**, 175–191 (2015)
7. Brechmann, E.C., Czado, C., Aas, K.: Truncated regular vines in high dimensions with application to financial data. Can. J. Stat. **40**(1), 68–85 (2012)
8. Creal, D., Koopman, S., Lucas, A.: Generalized autoregressive score models with application. J. Appl. Economet. **28**(5), 777–795 (2013)
9. Harvey, A.C.: Dynamic Models for Volatility and Heavy Tails: With applications to Financial and Economic Time Series. Cambridge University Press, Cambridge (2013)
10. Sklar, A.: Fonctions de Riépartition á n Dimensions et Leurs Marges. Publications de l'Institut Statistique de l'Université de Paris **8**, 229–231 (1959)
11. Kurowicka, D., Cooke, R.M.: Uncertainty Analysis with High Dimensional Dependence Modelling. Wiley, New York (2006)
12. Aas, K., Czado, C., Frigessi, A., Bakken, H.: Pair-copula constructions of multiple dependence. Insurance: Math. Econ. **44**, 182–198 (2009)
13. Oh, D.H., Patton, A.J.: Time-varying systemic risk: evidence from a dynamic copula model of CDS spreads. J. Bus. Econ. Stat. **36**(2), 181–195 (2018)
14. Catania, L.: Dynamic adaptive mixture models with an application to volatility and risk. J. Financ. Econ., 1–34 (2019)
15. Bernardi, M., Catania, L.: Switching generalized autoregressive score copula models with application to systemic risk. J. Appl. Economet. **34**, 43–65 (2019)
16. Dempster, A., Laird, N., Rubin, D.: Maximum likelihood from incomplete via EM algorithm. J. Royal Stat. Soc. Ser. B **39**(1), 1–38 (1977)
17. McLachlan, G.J., Peel, D.: Finite Mixture Models. Wiley, New York (2000)
18. Kim, D., Kim, J.M., Liao, S.M., Jung, Y.S.: Mixture of D-vine copulas for modeling dependence. Comput. Stat. Data Anal. **64**, 1–9 (2013)
19. Laeven, L., Ratnovski, L., Tong, H.: Bank size, capital, and systemic risk: some international evidence. J. Bank. Finance **69**, S25–S34 (2016)
20. FSB: 2019 list of global systemically important banks (G-SIBs) (2019). https://www.fsb.org/2019/11/2019-list-of-global-systemically-important-banks-g-sibs/
21. Adrian, T., Brunnermeier, M.K.: CoVaR. Am. Econ. Rev. **106**(7), 1705–1741 (2016)
22. BCBS. The G-SIB assessment methodology - score calculation (2014). https://www.bis.org/bcbs/publ/d296.htm
23. Mainik, G., Schaanning, E.: On dependence consistency of CoVaR and some other systemic risk measures. Stat. Risk Model. **31**(1), 47–77 (2014)

24. Bernardi, M., Maruotti, A., Petrella, L.: Multiple risk measures for multivariate dynamic heavy-tailed models. J. Empirical Financ. **43**, 1–32 (2017)
25. Girardi, G., Ergün, A.: Systemic risk measurement: multivariate GARCH estimation of CoVaR. J. Bank. Finance **37**(8), 3169–3180 (2013)
26. WHO: Coronavirus disease 2019 situation report-67. World Health Organization (2020). https://www.who.int/emergencies/diseases/novel-coronavirus-2019/situation-reports

# Real-Time Order Scheduling in Credit Factories: A Multi-agent Reinforcement Learning Approach

Chaoqi Huang[1], Runbang Cui[2], Jiang Deng[2], and Ning Jia[1(✉)]

[1] Tianjin University, Tianjin, China
jia_ning@tju.edu.cn
[2] QingDao Fantaike Technology Co., Ltd., Qingdao, China

**Abstract.** In recent years, consumer credit has flourished in China. A credit factory is an important mode to speed up the loan application process. Order scheduling in credit factories belongs to the np-hard problem and it has great significance for credit factory efficiency. In this work, we formulate order scheduling in credit factories as a multi-agent reinforcement learning (MARL) task. In the proposed MARL algorithm, we explore a new reward mechanism, including reward calculation and reward assignment, which is suitable for this task. Moreover, we use a convolutional auto-encoder to generate multi-agent state. To avoid physical costs during MARL training, we establish a simulator, named Virtual Credit Factory, to pre-train the MARL algorithm. Through experiments in Virtual Credit Factory and an A/B test in a real application, we compare the performance of the proposed MARL approach and some classic heuristic approaches. In both cases, the results demonstrate that the MARL approach has better performance and strong robustness.

**Keywords:** Multi-agent reinforcement learning · Order scheduling · Credit factory · Simulation system

## 1 Introduction

Recently, the consumer credit has achieved great success in China. A consumer finance company needs to approve the loan application submitted by the customer, and then determine whether to grant the loan. A credit factory is an important means to speed up the loan application process. Consumer credit is characterized by small quotas and high frequency. Considering these characteristics of consumer credit, many companies have introduced the credit factory mode.

The credit factory is a novel loan approval mode invented by Singapore's Temasek Group [1]. Credit factories have been adopted by many financial institutions, including Bank of China, China Construction Bank, and China Merchants Bank. The credit factory divides credit approval into several processes. Similar to a factory assembly line, the credit factory formulates standardized operations for each process. Credit factories

© Springer Nature Singapore Pte Ltd. 2021
W. Gao et al. (Eds.): FICC 2020, CCIS 1385, pp. 461–475, 2021.
https://doi.org/10.1007/978-981-16-1160-5_36

improve the efficiency of the credit approval process by introducing process standardization. More details about credit factories are available through the following URL: https://cf.bfconsulting.com/.

The purpose of credit factory is to improve the efficiency of credit application. One of the key components is order scheduling. Reasonable order scheduling can greatly reduce the application time. Time is extremely valuable for customers, especially in financial companies. And time is also a cost. Reduced loan application time means lower costs. Therefore, it enhances the company's cost advantage and facilitates customer loans.

The order scheduling problem has proven to be an np-hard problem [2]. In other words, the exact solution algorithms of order scheduling problems suffer from a non-polynomial increase in computation time. Although there are some similarities between them, the order scheduling problem in credit factories that we studied is different from classic job-shop problems. Firstly, the arrival time and processing time of the orders are unknown in advance. Secondly, order scheduling has the problem of delayed feedback. These differences are detailed in Sect. 3, and they make the order scheduling problem difficult to solve with traditional algorithms, such as mathematical optimization methods (e.g., mixed-integer linear programming) and meta-heuristics (e.g., genetic algorithms).

Recently, reinforcement learning (RL) has made significant progress in a variety of areas such as games [3], robotics [4], natural language processing [5]. Reinforcement learning has proved to be a powerful tool for online sequential optimized decision tasks. Considering the nature of the order scheduling problem in credit factories, we argue that reinforcement learning could be an effective method that has great potential to solve this problem.

In this paper, we propose a real-time order scheduling algorithm based on the MARL framework.

The main contribution of this research is two-fold:

- **MARL Framework.** We formulate the order scheduling problem as a MARL task. The loan approval process in credit factories is decomposed into several successive tasks. Each task can be modeled as a queue scheduling problem and is associated with one reinforcement learning agent. Agents cooperate by a reward assignment strategy and shared state generation, which will be introduced in Sect. 4.
- **Reward Mechanism and State Generation.** We explore a new reward mechanism, including reward calculation and reward assignment. To better calculate rewards, we make several improvements to the standard method. Furthermore, we design a reward assignment strategy to represent the correlations among agents. Besides, we use a convolutional auto-encoder and K-Means cluster analysis to generate the state, which can effectively represent the state of multi-agent.

We tested our MARL algorithm with different service capabilities and different system loads. Through experiments in Virtual Credit Factory and an A/B test in real application, we proved that our proposed MARL approach has better performance and strong robustness.

The rest of this paper is organized as follows: In the second section, we review related work. The third section introduces the problem formulation. The fourth section describes the MARL algorithm framework. The fifth section presents the experiments and results. The sixth section summarizes the paper.

## 2  Related Work

### 2.1  Single-Agent Reinforcement Learning

Xie et al. [6] studied the application potential of reinforcement learning for job shop scheduling problems, they quantified the state and action in many ways and defined many reward functions to investigate their influence on the algorithm. This work provided support for further investigations into applying reinforcement learning to job shop environments. Shahrabi et al. [7] combined reinforcement learning and the neighborhood search algorithm for application to job shop scheduling problems. But this work just used reinforcement learning to estimate the parameters of neighborhood search. In essence, this work used neighborhood search to solve the job shop scheduling problem. The superposition of reinforcement learning and neighborhood search caused too much computation and poor real-time performance.

Li et al. [8] applied Q-learning to an order scheduling system and showed that Q-learning outperformed the two benchmark policies, the First-Come-First-Serve (FCFS) policy and the threshold-heuristic policy. Waschneck et al. [9] studied the problem of semi-conductor production scheduling. They applied deep Q-learning to the problem and proved the effectiveness of the method. Shiue et al. [10] applied RL-based scheduling rule selection mechanism to job-shop scheduling problems and their results showed that the RL-based method outperformed the scheduling rules method and had good adaptability to environmental changes. All of these three works applied the single-agent reinforcement learning to scheduling problems. But for some decomposable problems, a single "super-agent" leads to a sharp increase in the dimension of agent inputs and outputs. Taking our study as an example, there were 4 processes in total. And each process had 4 actions, so the action space of a single "super-agent" was $4^4$.

The order scheduling problem in credit factory involves many processes. Instead of training a single "super-agent", MARL is more natural to model each process as a separate agent. MARL ameliorates the curse of dimensionality. Through the communication and cooperation among agents, MARL simulates the interaction among different processes in the credit factory. So MARL is a suitable method to solve order scheduling problem in credit factory. However, no one has ever studied the order scheduling problem in credit factory with MARL. Therefore, our proposed model is extended from single-agent to multi-agent reinforcement learning (MARL).

### 2.2  Multi-agent Reinforcement Learning

Gabel et al. [11] used MARL to solve job shop scheduling problems. This algorithm implemented decentralized scheduling and could be used for unknown situations without retraining. The arrival time and processing time of each job in this work were known. In contrast, order scheduling with unknown arrival time and processing time was more difficult. Qu et al. [12] Studied a manufacturing system that was able to handle multiple product types through multi-stages and multi-machines with dynamic orders, stochastic processing time and setup time. Their results demonstrated that the proposed MARL-based method provided better performance than most common scheduling rules. But this work did not specify how multiple agents communicate with each other to reflect the cooperative relationship. The state and action setting made this method only suitable for small-scale scheduling, and the numerical experiment in this work was also small-scale order scheduling. Liu et al. [13] used MARL for feature selection. This research used a graph convolutional network (GCN) to extract the state of reinforcement learning. The state extraction and cooperation between agents in this work provided support for the application of MARL to order scheduling.

Although the basic MARL algorithm had a wide range of applications, there were few studies on large-scale real-time order scheduling problems with multiple machines, multiple processes, random arrival times and processing times. Therefore, we aimed to design an effective MARL algorithm to solve the large-scale real-time order scheduling problem.

## 3   Problem Formulation

In this section, we introduce the typical process for a loan application order in credit factories. As shown in Fig. 1, an order needs to go through four processes (Processes 1–4) to complete the whole loan approval process. Process 1 (Classify) divides the documents of the order into different categories, such as ID cards, contracts, guarantee information, and so on. Process 2 (Slice) takes a screenshot of the key document information and stores it. For example, a name slice is obtained from the ID card. Process 3 (Input) enters the slice information generated in Process 2 into the database. Process 3 transforms unstructured data into structured data. Process 4 (Review) checks and confirms the information from different information sources. For example, it needs to be confirmed whether the name from the ID card matches the name from the loan contract.

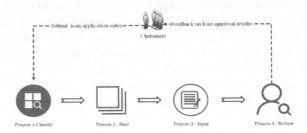

**Fig. 1.** The typical loan application process in credit factories

The approval process in credit factories can be modeled by four successive order queues. Figure 2 shows the essence of how a credit factory works. A customer arrives and sends a loan application order to the credit factory. In credit factories, each process consists of a queuing area and a processing area. There are several sorted orders in the queuing area. In the queuing area, these orders are sorted according to the scheduling rule, which is selected by RL. The processing area takes orders in turn from the queuing area. There are several workers in the processing area of each process. We assume all workers are homogeneous in credit factories. After leaving the processing area, the order is sent to the queuing area of the next process. The approval process continues until the order goes through all four processes. Because of the heterogeneity of orders, it is usually not efficient-maximum to use the naive first-in-first-out strategy, and this is where our problem begins.

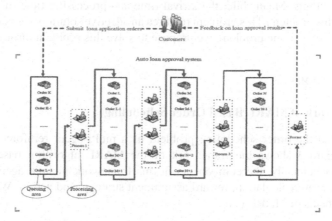

**Fig. 2.** Credit factory modeling and analysis

The problem we studied addresses two factors: 1) the timeout rate, which is the proportion of orders in a batch (like every 100 orders) that fail to finish in a limited time period (e.g. 60-min); and 2) the average waiting time. These two issues are designed to meet customer requirements from different perspectives. Consumer finance companies often promise customers one hour to make loans and regard low timeout rate as a competitive advantage. The lower the timeout rate, the less the company loses. However, an excessive pursuit of the low timeout rate may cause a decrease in total capacity. To avoid this situation, we add the second goal: the average waiting time.

The order scheduling problem is similar to classic job-shop scheduling problems, both of which are essentially sorting problems [14]. However, some critical differences between them hinder the application of traditional optimization techniques. The differences are shown in Table 1. Firstly, many traditional optimization methods are only suitable for static scheduling problems, in which job information, such as processing time, is known. But our order scheduling problem is a real-time dynamic scheduling problem. The arrival time and processing time of the orders are unknown in advance. In credit factories, the timeout rate is not a value that can be calculated immediately. After orders are scheduled, not all orders can complete the approval process in time. Therefore, order scheduling has the problem of delayed feedback.

Order scheduling is an np-hard problem and the large scale leads to a non-polynomial increase of computation time. Furthermore, real-time scheduling needs to complete the calculation in a short amount of time. The uncertainty of processing time makes this problem difficult to optimize. These reasons make it difficult to implement tradition

**Table 1.** The differences

| Name | Arrival and processing times are unknown | Feedback delay |
|---|---|---|
| Classic job-shop problems | No | No |
| Problem in this study | Yes | Yes |

accurate algorithms. Meanwhile, the arrival time and processing time of the order is unknown in this problem. This problem requires an adaptive solution. Considering these two characteristics of the problem, we propose to solve this problem through MARL.

## 4 Methodology

### 4.1 Framework of MARL Based Order Scheduling

We formulated the order scheduling problem as a multi-agent reinforcement learning task. Figure 3 shows the framework of the MARL algorithm based on order scheduling. Specifically, the components in this framework include agents, environment, state, reward calculation, reward assignment strategy, and action. We will detail these components of MARL later.

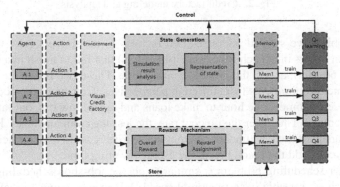

**Fig. 3.** Framework of MARL based order scheduling

The proposed MARL framework consists of four agents, each of which makes decisions and is trained according to the classic Q-learning algorithm. Every agent is responsible for the order scheduling of one of the processes mentioned in Sect. 3. At every time step, each agent takes actions based on their Q value tables. The Q value table takes the current state as the input and output action. Next the actions of all the agents are transferred to the environment. The environment schedules orders according to these actions. Then the next state and the overall reward are calculated. As mentioned in Sect. 3, the credit factory includes four successive order queues. Order processing in each queue is affected by its upstream and downstream order queues. If agents work independently, the mutual influence among order queues is difficult to interpret. Therefore, we need coordination between agents, which is demonstrated by reward assignment and state generation. To better calculate rewards, we propose several improvements to the standard method. Our reward assignment strategy assigns the overall reward to each agent according to the contribution. Moreover, an auto-encoder generates a multi-agent state. The reward assignment strategy and shared state demonstrate the cooperative relationship among multiple agents. Each agent updates the Q value table according to the state and its respective reward. Then the next round of iterations starts. The training of agents continues until there is convergence or several predefined criteria are met.

**Multi-Agent.** This order scheduling problem consists of four processes. Each process is modeled as a separate agent. Each agent controls the order scheduling of each process. The reward assignment strategy and shared state demonstrate the cooperative relationship among multiple agents.

**Environment.** The environment is the credit factory. To avoid physical costs during MARL training, we establish a simulator, named Virtual Credit Factory, to pre-train the MARL algorithm. Agents select actions and transfer them to the environment. The environment schedules orders according to these actions. Then, the environment converts the scheduling results into the reward and state.

**Action.** The action is to select a scheduling rule from the scheduling rules subset (SRS). This problem is a large-scale real-time order scheduling problem. Furthermore, it has the following two characteristics: 1) it is so complex that the resolution to optimality is impracticable; and 2) this problem must deal with a dynamic reality, but it takes too long to obtain the optimal solution. The scheduling rule approach can solve large-scale real-time order scheduling problems with low complexity and low latency. Thus, we choose the scheduling rule as the action. Four scheduling rules are selected to form the scheduling rules subset. All the scheduling rules are shown in Table 2. The selection of scheduling rules is based on a previous study [15].

**Table 2.** Scheduling rules in SRS.

| Abbreviation | Rule | Description |
| --- | --- | --- |
| FIFO | First in first out | |
| SST | Shortest slack time | slack time $=$ due date $-$ arrive time |
| EDD | Earliest due date | |
| LIFO | Last in first out | |

**State.** To extract the representation of state s, we explore a new strategy: auto-encoder based state representation. We map the latest order processing information matrix L into the low-dimensional representation matrix E using an encoder. Then, matrix E is applied to K-Means cluster analysis to obtain a final state class label. This state representation strategy is detailed in Sect. 4.2.

**Reward Calculation.** The reward function is essentially used to guide the learning agent toward its goal. Thus, we have to clarify the goal of this problem. As described in Sect. 3.1, our goal is to optimize the timeout rate and the average waiting time. We propose combining the timeout rate and the average waiting time as reward R. To better combine the average waiting time and the timeout rate, we make several improvements on the standard reward calculation method. In Sect. 4.3, the reward calculation method is discussed in detail.

**Reward Assignment Strategy.** We obtain the overall reward from the scheduling result. Then, we assign the overall reward to each agent as their respective reward. The assignment of the overall reward is based on the contribution of each agent to the timeout rate and average waiting time.

Firstly, we assign the overall relaxation time of each order to each process according to the processing time. Then, the relaxation time of the order in the k-th process (k = 1, 2, 3, 4) is obtained.

Secondly, from the scheduling result, information, such as the entry time, end time, and waiting time of the order in the k-th process, is obtained.

Thirdly, if the difference between the entry time and end time is greater than the relaxation time, then the order times out in the process; otherwise, it does not time out. In this way, the average waiting time and timeout rate of the order in the k-th process are obtained.

Finally, based on the average waiting time and the timeout rate of the k-th process, the overall reward is assigned to the k-th agent.

## 4.2 Reward Calculation

After the t step scheduling period is completed, the scheduling results of step t from the environment are obtained. Based on the scheduling results, we can calculate the average waiting time $\text{AWT}_t^{(s,a)}$ and timeout rate $\text{TR}_t^{(s,a)}$ of the orders in step t. Then, the performance measure $\text{PM}_t^{(s,a)}$ after taking the action a in the state s is calculated.

$$\text{PM}_t^{(s,a)} = \text{AWT}_t^{(s,a)} * \text{TR}_t^{(s,a)} \tag{1}$$

Next, $\text{PM}_t^{(s,a)}$ is compared with the mean performance in the state class label s. $\text{UCL}_{1\sigma}^s$ is defined as the one sigma upper confidence limit (i.e., 68.27%) for the mean performance in state class label s. If $\text{PM}_t^{(s,a)}$ is greater than $\text{UCL}_{1\sigma}^s$, then the learning agent receives a reward of +1. Moreover, $\text{LCL}_{1\sigma}^{(s)}$ is defined as the one sigma lower confidence limit (i.e., 31.73%) for the mean performance in state class label s. If $\text{PM}_t^{(s,a)}$ is less than $\text{LCL}_{1\sigma}^{(s)}$, then the learning agent receives a reward of −1; otherwise, the learning agent receives a reward of 0. This reward calculation method is described in Algorithm 1.

---

**Algorithm 1** Reward Calculation Method

---

**Input:** $PM_t^{(s,a)}$
**Output:** reward
1:   Initialize $UCL_{1\sigma}^s = PM_t^{(s,a)}$ and $LCL_{1\sigma}^s = PM_t^{(s,a)}$
2:   **for** $t \leftarrow 1$ to T **do**
3:       **if** $PM_t^{(s,a)} > UCL_{1\sigma}^s$ **then**
4:           reward = +1
5:       **else if** $PM_t^{(s,a)} < LCL_{1\sigma}^s$ **then**
6:           reward = -1
7:       **else**
8:           reward = 0
9:       **end if**
10:      **update** $UCL_{1\sigma}^s$ , $LCL_{1\sigma}^s$
11: **end for**

---

## 4.3 State Generation

In our experiments, one step of reinforcement learning determined the scheduling of a batch of orders. One batch contained 30 orders. The state representation from the processing information of the latest two batches of processed orders is extracted so that the state to reflect the current scheduling situation. The processing information of the order includes the arrival, waiting, processing, and delay time of orders in each process.

K-Means clustering is often used for state extraction in reinforcement learning but this algorithm does not work well for high dimensional data. High dimensional data are often transformed into low dimensional data where coherent patterns can be detected more clearly.

The auto-encoder can effectively reduce the dimensionality [16]. An auto-encoder contains an encoder and a decoder. The encoder can map high-dimensional input into low-dimensional representation. The decoder can reconstruct the output from the low-dimensional representation. The autoencoder minimizes the reconstruction loss between the original input and the reconstructed output. In this way, the auto-encoder can guarantee the quality of the information compression. So we use auto-encoder to reduce the dimensions of high dimensional data (60 * 5 = 300-dim), and then use K-Means to obtain the final state label. To derive accurate state representation from scheduling result, an auto-encoder based state representation method was developed.

Figure 4 shows how the state representation method works.

Step1: When extracting the state of step t +1, the order information of t and t−1 is selected as the latest order processing information. The Latest Order Processing Information Matrix L included processing information for 60 orders. The information of each order included the average waiting time for each process (I1–I4 in Fig. 4) and whether it had timed out (I5 in Fig. 4).

Step2: The dimension of the matrix L was 60 * 5. We used the convolutional auto-encoder to convert L into an Encoded Matrix E with a dimension of m* n.

Step3: We linked each column in matrix E together into the state vector $S_{t+1}$ with a length of m*n. Add $S_{i+1}$ to the state vector set SVS.

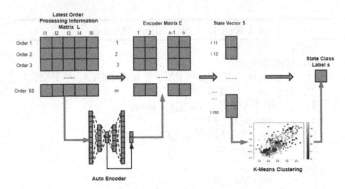

**Fig. 4.** Auto encoder-based state representation method

Step4: Using k-means clustering to classify the SVS when the number of state vectors in the SVS equals to g:

Requirements: State vector set SVS, the number of cluster centers $k (k < g)$;

a) Select k state vectors $S_1, S_2, \cdots, S_k$ as the initial cluster centers.
b) Calculate and find the shortest distance between any data object $S_i$ and all center objects according to $d(S_1, S_2) = \sqrt{(s_{111} - s_{211})^2 + (s_{112} - s_{212})^2 + \ldots (s_{1mn} - s_{2mn})^2}$, form a new data object set $D_h (1 \leq h \leq k)$ with $S_i$ and then remove $S_i$ from SVS.
c) Calculate and update the new center data object $S_h$ of set $D_h$ according to $S_h = \left( \sum_{i=1}^{|D_k|} a_{i11}/|D_h|, \ldots, \sum_{i=1}^{|D_k|} a_{imn}/|D_h| \right)$.
d) Repeat (b) until all data objects in SVS have divided into the corresponding data object set.
e) Form k data object set and k cluster centers.

Step5: Finally, when a new state vector $S_{t+1}$ appears, calculate the similarity between it and each state set $D_h$ according to $d(S_{t+1}, D_h) = d(S_{t+1}, S_h)$, then classify it to the state set with highest similarity and obtain final state class label s.

## 5    Numerical Experiments

### 5.1    Virtual Credit Factory

Current RL algorithms usually require many interactions with the environment. These interactions require high physical costs, such as real money, time, and poor user experience. To avoid physical costs, we built a simulator, named Virtual Credit Factory, which is based on a queuing simulation of Anylogic software (https://www.anylogic.com/). Then real data, including the arrival rate, processing time, system service capabilities, and so on, can be used to calibrate the Virtual Credit Factory, which aims to minimize the performance difference between real and virtual factories. The data was collected from the credit factory of Qingdao Fantaike Technology Co. Ltd. (https://www.fantaike.ai). Fantaike is a fast-growing fintech company. It is now providing loan approval services to several consumer finance companies, including ChangAn XinSheng, AVIC Trust, and BMW Automotive Finance. During 2019, the average daily order quantity was 334.76 in Fantaike's credit factory, which brings great pressure to the current scheduling system.

The RL strategy first conducted offline training in Virtual Credit Factory. Then, the offline RL strategy was obtained and applied to online scheduling in the real environment. Experimental results show that the offline strategy performed well in the real environment. The offline strategy could provide a good initialization and greatly speed up training in the real environment.

### 5.2 Performance Measures and Baseline Algorithms

To address our goals, the performance measures are related to the factors of 1) the timeout rate and 2) the average waiting time.

We compared the performance of our proposed MARL scheduling algorithm with those of four basic scheduling rules (FIFO, SST, EDD, and LIFO).

### 5.3 Experimental Settings

The experimental settings included order settings and allocation of system workers. In the order settings, the parameter we could control was the average time interval ($ARR_{mean}$) of the order. By changing $ARR_{mean}$, we can simulate the situation of different system loads. In addition, we could control the system service capability by changing the Allocation Of Workers (AOW) in the four processes. As mentioned early, the simulator RL strategy can provide a good initialization for RL in the actual environment. The number of workers in the actual environment is 10, so the number of workers we use in the simulator is also 10. In another of our studies, we concluded that when the total number of system workers was 10, the optimal AOW is [1, 5, 2, 2]. This means that there was one worker in process 1. Similarly, there were 5, 2 and 2 workers in processes 2, 3 and 4. In this configuration, the system processes the largest number of orders per unit time. Besides, we tried several different AOWs: [1, 6, 2, 1], [2, 4, 2, 2] and [2, 3, 3, 2]. By changing the AOW, we could prove that the MARL method is effective under different system service capabilities. Table 3 lists all the experimental settings.

**Table 3.** Experimental settings

|  | $ARR_{mean}$ | AOW | Remark |
|---|---|---|---|
| Setting 1 | 4.5 | [1, 5, 2, 2 ] | Changed $ARR_{mean}$ |
| Setting 2 | 4.0 | [1, 5, 2, 2] | |
| Setting 3 | 5.0 | [1, 5, 2, 2] | |
| Setting 4 | 4.5 | [1, 6, 2, 1] | Changed AOW |
| Setting 5 | 4.5 | [2, 4, 2, 2] | |
| Setting 6 | 4.5 | [2, 3, 3, 2] | |

## 5.4  Performance Measures and Baseline Algorithms

In Table 3, setting 1 lists the parameters of the benchmark setting. Figure 5 shows the performance of each algorithm under the benchmark setting. The dotted line describes the performance of the four scheduling rules. The solid line represents the average performance of MARL, while the corresponding shadow areas show standard deviations. The MARL algorithm introduces random seeds. To ensure the reliability of the experimental results, we obtained MARL results from ten independent simulation runs. As shown in Fig. 5, MARL demonstrates better performance than the scheduling rules.

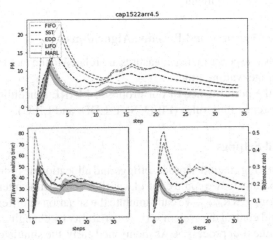

**Fig. 5.** Performance of each algorithm under the benchmark setting

## 5.5  Robustness Check

To observe the impact of system load on MARL, we made a variation on the average time interval (ARR$_{mean}$). A smaller ARR$_{mean}$ means there is a heavier system load. In the system load variation scenario, the ARR$_{mean}$ of experimental Setting 2 and Setting 3 were changed to 8/9 and 10/9 of the benchmark setting, respectively. The remaining parameters were the same as those in the benchmark setting. The setting 1, 2, 3 in Table 4 reflect the results of the system load adaptability tests.

To test the adaptability of the MARL algorithm, we made changes to AOW. For experimental Setting 1, Setting 4, Setting 5, and Setting 6, the AOW was different while the remaining parameters were the same. The setting 1, 4, 5, 6 in Table 4 reflect the results of the AOW adaptability tests. Table 4 shows the results of the two performances under six experimental settings. Moreover, the best performance scores in each setting are marked in bold.

In Setting 1, 3, 5, MARL is better than LIFO in both timeout rate (TR) and average waiting time (AWT). In Setting 2, 4, 6, MARL is worse than LIFO in TR, but better than LIFO in AWT. The reason for this result is that the performance measure in reward is AWT * TR, which makes the MARL algorithm sometimes sacrifice the TR to pursue a smaller AWT. Therefore, Table 4 shows that MARL has good adaptability to changes in system loads and service capabilities.

**Table 4.** Average waiting Time and timeout rate under different experimental settings

| AWT | FIFO | EDD | SST | LIFO | MARL | |
| --- | --- | --- | --- | --- | --- | --- |
| | | | | | Mean | Std |
| Setting 1 | 28.9 | 29.0 | 29.4 | 29.0 | **25.4** | 2.1 |
| Setting 2 | 184.0 | 184.8 | 186.5 | 207.2 | **180.2** | 31.1 |
| Setting 3 | 24.1 | 24.9 | 25.0 | 23.5 | **21.3** | 2.0 |
| Setting 4 | 64.0 | 61.0 | 65.0 | 72.0 | **58.1** | 5.3 |
| Setting 5 | 10.5 | 10.7 | 10.9 | 10.6 | **9.8** | 0.7 |
| Setting 6 | 169.3 | 170.1 | 170.5 | 170.2 | **151.6** | 4.1 |
| TR | FIFO | EDD | SST | LIFO | MARL | |
| | | | | | Mean | Std |
| Setting 1 | 0.199 | 0.192 | 0.167 | 0.127 | **0.112** | 0.011 |
| Setting 2 | 0.881 | 0.870 | 0.385 | **0.217** | 0.229 | 0.034 |
| Setting 3 | 0.157 | 0.132 | 0.154 | 0.123 | **0.105** | 0.011 |
| Setting 4 | 0.648 | 0.637 | 0.360 | **0.215** | 0.236 | 0.010 |
| Setting 5 | 0.060 | 0.050 | 0.048 | 0.059 | **0.039** | 0.004 |
| Setting 6 | 0.957 | 0.962 | 0.387 | **0.188** | 0.196 | 0.004 |

## 5.6  The Results of Online A/B Tests

We further carried out online A/B tests at the credit factory of Fantaike. The online tests were divided into three situations: busy, normal, and idle. Two groups of experiments were performed for each situation, the control group, and the treatment group. The control group used the best rule of the single scheduling rule, which can be regarded as the benchmark. The treatment group used the proposed MARL algorithm. For those new loan orders, we divide them into the treatment and control group by random sampling. In practice, one has equal probability (50%) to be assigned to the treatment and control. Then we test whether the AWT & TR of the treatment is different from those of the control.

Table 5 illustrates the results of the A/B tests. The MARL made marked improvements on Average Waiting Time (AWT) and Timeout Rate (TR) in all three situations. Furthermore, the average improvements were 7.57% and 5.87%, respectively.

**Table 5.** Results of Online A/B Tests. Delta (Δ) represents the improvement of the treatment group over the control group.

| Situation | Δ *AWT* | Δ *TR* |
|---|---|---|
| Busy | 4.21% | −3.24% |
| Normal | 11.30% | 12.10% |
| Idle | 7.20% | 8.75% |
| **Average** | **7.57%** | **5.87%** |

# 6 Conclusion

In this paper, we studied the optimization of the order scheduling in credit factories. After optimization, the maximum improvements of the average waiting time and timeout rate were 11.3% and 12.1%, respectively. We formulated the order scheduling problem in credit factories as a MARL task. The scheduling of each process was associated with one agent. A convolutional auto-encoder was applied to represent the multi-agent state. To better represent the reward, we made several improvements on the standard reward calculation method. Furthermore, a reward assignment strategy was designed to demonstrate the cooperative relationship among agents. To test the performance of MARL, we performed experiments in Virtual Credit Factory and an A/B test in a real application. The results showed that the proposed MARL method performed better than any other scheduling rule in most experimental setting.

# References

1. Liu, K., Ma, B.: China's Small Micro Enterprise Financing Problems and Countermeasures, dtem, no. icem, November 2016. https://doi.org/10.12783/dtem/icem2016/4033.
2. Wang, G., Cheng, T.C.E.: Customer order scheduling to minimize total weighted completion time. Omega **35**(5), 623–626 (2007). https://doi.org/10.1016/j.omega.2005.09.007
3. Silver, D., et al.: Mastering the game of Go without human knowledge. Nature **550**(7676), 354–359 (2017). https://doi.org/10.1038/nature24270
4. Hwangbo, J., et al.: Learning agile and dynamic motor skills for legged robots. Sci. Robot.4(26), eaau5872 (2019) https://doi.org/10.1126/scirobotics.aau5872.
5. Su, P.-H., et al.: On-line Active Reward Learning for Policy Optimisation in Spoken Dialogue Systems, arXiv:1605.07669 [cs], June 2016, Accessed 08 Jan 2020. https://arxiv.org/abs/1605.07669.
6. Xie, S., Zhang, T., Rose, O.: Online Single Machine Scheduling Based on Simulation and Reinforcement Learning. Simulation in Produktion und Logistik **2019**, 10 (2019)
7. Shahrabi, J., Adibi, M.A., Mahootchi, M.: A reinforcement learning approach to parameter estimation in dynamic job shop scheduling. Comput. Ind. Eng. **110**, 75–82 (2017)
8. Li, X., Wang, J., Sawhney, R.: Reinforcement learning for joint pricing, lead-time and scheduling decisions in make-to-order systems. Eur. J. Oper. Res. **221**(1), 99–109 (2012). https://doi.org/10.1016/j.ejor.2012.03.020
9. Waschneck, B., et al.: Deep reinforcement learning for semiconductor production scheduling, pp. 301–306 (2018)

10. Shiue, Y.-R., Lee, K.-C., Su, C.-T.: Real-time scheduling for a smart factory using a reinforcement learning approach. Comput. Ind. Eng. **125**, 604–614 (2018). https://doi.org/10.1016/j.cie.2018.03.039

11. Gabel, T., Riedmiller, M.: Adaptive reactive job-shop scheduling with reinforcement learning agents. Int. J. Inf. Technol. Intell. Comput. **24**(4) (2008)

12. Qu, S., Wang, J., Shivani, G.: Learning adaptive dispatching rules for a manufacturing process system by using reinforcement learning approach, pp. 1–8 (2016)

13. Liu, K., Fu, Y., Wang, P., Wu, L., Bo, R., Li, X.: Automating feature subspace exploration via multi-agent reinforcement learning. In: Proceedings of the 25th ACM SIGKDD International Conference on Knowledge Discovery & Data Mining - KDD '19, Anchorage, AK, USA, 2019, pp. 207–215 (2019). https://doi.org/10.1145/3292500.3330868.

14. Shl, G.: A genetic algorithm applied to a classic job-shop scheduling problem. Int. J. Syst. Sci. **28**(1), 25–32 (1997). https://doi.org/10.1080/00207729708929359.

15. Panwalkar, S.S., Iskander, W.: A Survey of Scheduling Rules. Oper. Res. **25**(1), 45–61 (1977). https://doi.org/10.1287/opre.25.1.45

16. Hinton, G.E.: Reducing the dimensionality of data with neural networks. Science **313**(5786), 504–507 ( 2006). https://doi.org/10.1126/science.1127647

# Predicting Digital Currency Price Using Broad Learning System and Genetic Algorithm

Nan Jing[1], Zhengqian Zhou[1], Yi Hu[1], and Hefei Wang[2(✉)]

[1] Shanghai University, Jiading, Shanghai 201899, China
[2] Renmin University of China, Beijing 100872, China
wanghefei@ruc.edu.cn

**Abstract.** With the development of the digital economy, the price of Bitcoin, which is the most representative digital currency, has fluctuated dramatically. Recent works have explored the volatility of the Bitcoin price and made predictions using financial time series models such as ARIMA. However, for high-frequency Bitcoin data, the financial time series models have poor prediction performance as they often do not accommodate the inherent characteristics of digital currency, such as blockchain information. Some other works in this topic use deep learning models, e.g., artificial neural networks. However, the complex structure and time-consuming training process of these deep learning models often incur low prediction efficiency for rapidly changing Bitcoin price. In this regard, Broad Learning System (BLS) is a new neural network that avoids the complex structure of hidden nodes in deep learning models by adding enhancement nodes in the input layer to improve training efficiency while delivering relatively high accuracy. Therefore, this work applies the broad learning system to predict the Bitcoin price and optimizes the prediction model with a genetic algorithm. Due to the lack of fundamental factors for digital currency, the proposed prediction model considers the macroeconomic variables and the information of Bitcoin blockchain, which is the underlying technology of bitcoin, as inputs. According to the experimental results, the BLS-based prediction model optimized with the genetic algorithm achieved a better performance than other machine learning models.

**Keywords:** Blockchain · Price prediction · Broad learning system · Genetic algorithm

## 1 Introduction

With the development of information technology, the digital economy has boomed rapidly worldwide. According to the United Nations Conference on Trade and Development [1], products and services in the field of Information Communications Technology (ICT) accounted for 6.5% of the Global Gross Domestic Product (GDP) by the end of 2017. The e-commerce, e-payment, Internet of things (IoT), Fifth generation (5G) wireless technology has changed public life. Electronic payment and mobile payment gradually replace the traditional paper-based payment, amongst which the birth of digital currency and blockchain technology has become a vital part of the digital economy.

© Springer Nature Singapore Pte Ltd. 2021
W. Gao et al. (Eds.): FICC 2020, CCIS 1385, pp. 476–488, 2021.
https://doi.org/10.1007/978-981-16-1160-5_37

As of March 2018, there were over 1,500 kinds of digital currencies with a total market value of $389.1 billion [2]. Amongst the digital currencies, Bitcoin is the supreme digital currency that occupies nearly half of the market share of the digital currency market.

The concept of Bitcoin originates from the "Bitcoin: A Peer-to-peer Electronic Cash System," published by Satoshi Nakamoto in 2009. Bitcoin is an electronic payment system based on peer-to-peer (P2P), also known as the digital currency. In Bitcoin's peer-to-peer network, users can transfer digital currencies using virtual digital addresses and earn rewards by block mining. In the Bitcoin transaction process, users do not need a third party to complete the transaction, so the decentralized transaction system dramatically reduces the information asymmetry [3]. Therefore, by its anonymity, decentralization, and non-tampering, Bitcoin has been supported by a large number of users.

Compared with traditional paper-based currency, digital currency costs less to issue and circulate and is more convenient and transparent in the transaction process. At present, in addition to Bitcoin, there are also other digital currencies such as Litecoin and Ethereum. As a critical financial product in the digital economy, the price of digital currency has been continuously rising in recent years, amongst which the price of Bitcoin has been increased beyond imagination. In 2010, the appearance of digital currency exchanges, including Mt.Gox [4] enabled Bitcoin to be traded 24 h a day. As shown in Fig. 1, the price of Bitcoin has a trend of drastic increase and decrease with intense volatility. Since Bitcoin has no fundamental factors and is different from other securities such as stocks, futures, and funds, the traditional pricing theory of financial products cannot explain the price changes of digital currencies. As the price of digital currencies is not controlled by authority or other third parties, the price of digital currency such as Bitcoin has intense volatility, which has attracted the attention of many scholars.

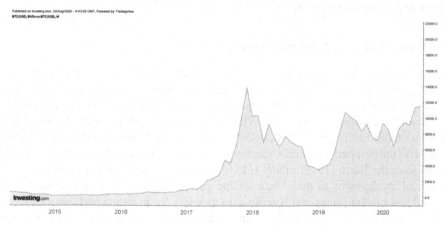

**Fig. 1.** Bitcoin price trend

These scholars have carried out researches on the volatility of the digital currency price, often using financial time series models such as GARCH(Generalized Autoregressive Conditional Heteroskedasticity) [5]. Different from traditional financial products, the price of digital currency has strong volatility. Besides, the policy of government also has a significant impact on the price of digital currency. Therefore, to improve the performance of the prediction model, some scholars adopt machine learning models for predicting the digital currency price. Deep learning models such as Bayesian neural networks (BNNs) [6] have shown its effect on predicting digital currency price. However, in the trend of high-frequency trading, the time-consuming model training process reduces the efficiency of the prediction model. Meanwhile, as a new neural network, the broad learning system (BLS) broadens the input layer and excludes the hidden layer to reduce the training time while maintaining good performance. Broad learning system has been applied to time series prediction [7], image recognition [8], and disease diagnosis [9].

Therefore, this work uses the broad learning system to predict the trend of the Bitcoin price. Due to the lack of the fundamental factors of the Bitcoin, we use the macroeconomic factors and blockchain information as the input of the prediction model. To improve the accuracy of the prediction results, this work also employs the genetic algorithms to optimize the prediction model, which tunes the hyper-parameters by evolutionary iterations. The remaining part of the paper proceeds as follows: Sect. 2 summarizes the recent work of digital currency price prediction and broad learning system. Section 3 introduces the proposed model and related methods. Section 4 demonstrates the experiment process and results. Section 5 summarizes this paper and gives the prospect of future work.

## 2   Literature Review

This work predicts the price of digital currency Bitcoin based on the Broad Learning System (BLS) and genetic algorithm. In this chapter, the authors will summarize the recent works of price prediction. Then the authors will introduce the broad learning system and its recent progress.

### 2.1   Digital Currency Price Prediction

Previous studies often use the existing financial time series models to predict the price volatility of digital currencies. For example, Katsiampa [5] investigated the ability of several GARCH (Generalized AutoRegressive Conditional Heteroskedasticity) models to explain the Bitcoin price volatility. They used the daily closing prices for the Bitcoin Coindesk Index from July 18, 2010, to October 1, 2016, and found that the AR-CGARCH model had the best performance. Ciaian et al. [10] derived an econometrically estimable model from the Barro (1979) model [11] for the gold standard and applied time-series analytical mechanisms to daily data for the period 2009–2015. Experimental results showed that market forces and Bitcoin attractiveness for investors and users had a significant impact on Bitcoin price but with variation over time. However, these estimates did not support previous findings that macro-financial developments were driving Bitcoin price in the long run. Besides, Bouri et al. [12] examined the relation between price returns and volatility changes in the Bitcoin market based on a daily database denominated in the US dollar. The aforementioned works reveal that more scholars tend to add relevant economic and financial variables into the traditional econometric model. By optimizing the model, better experimental results can be achieved. Nevertheless, digital currencies are different from traditional financial products. Thus, the validity of using a financial time series model to predict and analyze the price of digital currencies still needs further proof.

Recent works have also applied machine learning models for predicting the financial time series. Based on blockchain information, Jang and Lee [6] compared the Bayesian neural networks (BNNs) with other benchmark models on modeling and predicting the Bitcoin process. The experiment results showed that BNN performed well in predicting the Bitcoin price time series and explaining the high volatility of the recent Bitcoin price. Adcock and Gradojevic [13] used an artificial neural network (ANN) to forecast the Bitcoin returns based on past returns and straightforward technical trading rules. Furthermore, Mcnally et al. [14] predicted the Bitcoin price using Bayesian optimized recurrent neural network (RNN) and long short-term memory (LSTM). They also concluded that the nonlinear deep learning models performed better than ARIMA(Autoregressive Integrated Moving Average). Mallqui and Fernandes [15] employed ANN and support vector machine (SVM) algorithms to predict the maximum, minimum, and closing prices of the Bitcoin.

Regarding the regression experiments, the SVM algorithm obtained the best results for all predictions and both intervals. In a recent study, Chen et al. [16] used high-dimensional features with different machine learning algorithms (e.g., SVM, LSTM, random forest) to predict the Bitcoin daily price. Compared with the results for daily price prediction, the highest accuracy of 65.3% was achieved by SVM. However, the prediction of bitcoin price in this paper is for daily data instead of minute-level data. It is still arguable for using machine learning model in high-frequency price prediction of digital currency. Besides, Mudassir et al. [17] demonstrated high-performance machine learning-based classification and regression models for predicting Bitcoin price movements and prices in short and medium terms. Results indicated that the proposed models outperformed the existing models in the literature.

## 2.2 Broad Learning System

To predict the price volatility of digital currencies, machine learning models such as deep learning networks often have a time-consuming training process. As an alternative to deep learning networks, Broad Learning System (BLS) [18] removed the hidden layers in a deep structure and added enhancement nodes in the input layer to improve the fitting ability, as shown in Fig. 2. Previous works show that the BLS can obtain accurate results in short training time [19]. Liu and Chen [20] used the BLS in the classification of the public dataset MNIST. The experiment results showed that the BLS achieved an excellent performance in both precision and efficiency. Besides, recent studies have applied the BLS for time-series analysis and prediction. Xu et al. [21] proposed a novel recurrent broad learning system in time series prediction. Their study validated that recurrent broad learning system had better accuracy within a significantly short time.

Besides, Han [7] presented a unified framework for non-uniform embedding, dynamical system revealing, and time series prediction, termed as Structured Manifold Broad Learning System (SM-BLS). Simulation analysis and results show that SM-BLS has advantages in dynamic discovery and feature extraction of large-scale chaotic time series prediction. Chen et al. [22] provided a mathematical proof of the universal approximation property of BLS and compared the performances of the broad learning model in function

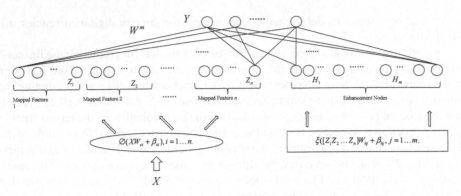

**Fig. 2.** Broad learning system

approximation, time series prediction, and face recognition. Experiments showed that the BLS had a universal approximation capability in different academic fields. Besides, the improved Broad Learning System is also applied in uncertain data modeling [23], disease diagnosis [9], and image recognition [8]. Compared to the deep learning models, BLS can achieve high prediction performance within a short training time. Therefore, it is applicable to predict the price volatility of Bitcoin amongst the various machine learning models.

### 2.3 Summary and Importance of the Proposed Work

The prediction of digital currency prices mainly adopts the traditional time series model. Yet, based on our findings as noted in this chapter, for the high-frequency digital currency price, machine learning methods are arguable. However, machine learning methods, such as deep learning models, suffer from a time-consuming training process and therefore often incur poor overall performance. As such, this work builds a digital currency price prediction model based on the broad learning system to improve the efficiency of high-frequency price predicting while maintaining relatively high accuracy. Meanwhile, this work will optimize the proposed model by employing a genetic algorithm to tune the parameters and further balance the prediction performance and generalization capacity of the prediction model.

## 3    A Prediction Model for Digital Currency Price

The prediction model in this work contains two parts, the broad learning system, and its optimization by genetic algorithm. As shown in Fig. 3, this work initializes the structure of the broad learning system and uses the genetic algorithm to tune the hyper-parameters. The optimization process includes encoding, selection, crossover, mutation, and fitness evaluation. After completing the iterations, we use the best hyper-parameters to build the prediction model. Then we trained and tested the model using K-fold cross-validation and evaluated the performance by the confusion matrix and its related metrics. For comparison, this work also builds the prediction model with other machine learning algorithms, and the experiment results will be presented in the next chapter.

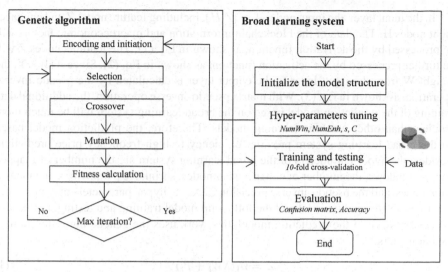

**Fig. 3.** Model structure

## 3.1  Price Prediction Based on the Broad Learning System

The inputs of the proposed prediction model are blockchain information and macroeconomic factors, and the output is the trend category of Bitcoin price. As shown in Fig. 4, the prediction model based on the broad learning system contains two layers, the input layer, and the output layer.

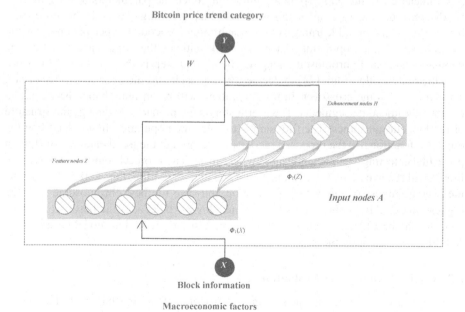

**Fig. 4.**  A prediction model based on the broad learning system

In the input layer, the input nodes $A = [Z|H]$, including feature nodes Z and enhancement nodes H. The data of the blockchain information and macroeconomic factors will be processed by the activation function, as shown in Eq. (1). The features nodes Z will be further processed by the activation function, as shown in Eq. (2). Since $A*W=Y$, the weight W from the input layer to the output layer is calculated by the pseudo-inverse operation, as shown in Eq. (3). With matrix pseudo-inverse operation, the additional data training of the prediction model based on the broad learning system will be much more efficient than other machine learning models. Therefore, the prediction model based on the broad learning system provides efficiency to high-frequency price prediction. Besides, the hyper-parameters of the broad learning system are the number of mapped feature windows, the number of enhancement nodes, shrinkage parameter s of enhancement nodes, and the pseudo-inverse precision C. As the hyper-parameters play important role in establishing the model and the follow-up model training and testing, to improve the performance of the prediction model, this work uses the genetic algorithm to tune the parameters.

$$Z = \emptyset_1(XW_i + \beta_i) \tag{1}$$

$$H = \emptyset_2(ZW_j + \beta_j) \tag{2}$$

$$W = [Z|H]^+ YA \tag{3}$$

### 3.2  Model Optimization Based on the Genetic Algorithm

The model optimization based on the genetic algorithm contains five steps. The first step is parameter encoding and population initiation. Since the parameters to be optimized are discrete numerical variables, this work uses binary coding to encode the four hyper-parameters of the broad learning system. Each jointed encoded hyper-parameter is the chromosome in the population. Then this work initiates the population based on the population size and chromosome length. The second step is the selection. This work uses the roulette algorithm [24] to decide whether each chromosome to survive or not. The third step is the crossover. In this step, there will be an interchange between two random chromosome segments. The fourth step is the mutation to change the structure of random chromosomes. The fifth step is to calculate population fitness based on the objective function. As the proposed model aims to predict the trend category of Bitcoin price, higher testing accuracy indicates higher fitness. Therefore, chromosomes with high fitness will survive in the population. At the end of the iteration, this work will calculate the average fitness of the current population and save the best chromosome with the highest fitness. After completing all iterations, this work decodes the best chromosome and uses the best hyper-parameters to train and test the prediction model based on the $K$-fold cross-validation method.

### 3.3  Model Training and Evaluation

This work uses the hold-out method in hyper-parameter tuning based on the genetic algorithm and $K$-fold cross-validation method [25] in training and testing the optimized model. The hold-out method divides the total dataset $D$ into two mutually exclusive sample sets. The training dataset $S$ and the testing dataset $T$ are divided at a certain percentage. Meanwhile, the $K$-fold cross-validation method divides the total dataset $D$ into $K$ sub-datasets. During the experiment, $K-1$ sub-datasets are used for model training, and the remaining one sub-dataset is used for model testing. After completing $K$ experiments, the final results are the average of $K$ experiments. As 10-fold cross-validation has been widely used in previous works, this work also set $K$ at 10.

**Table 1.** Confusion matrix

| | | True value | |
|---|---|---|---|
| | | Positive (Rise $Y = 1$) | Negative (decline $Y = -1$) |
| Predicted value | True (rise $Y = 1$) | TP | FP |
| | False (decline $Y = -1$) | FN | TN |

For the evaluation of the prediction model, this work uses the confusion matrix to evaluate the prediction results. The predicted value and true value of the output $Y$ is equal to 1(rise) or $-1$(decline). As shown in Table 1, this work calculates the *accuracy* based on the confusion matrix. Accuracy is equal to the correctly predicted sample number ($TP + TN$) over the total sample number in Eq. (4).

$$Accuracy = \frac{TP + TN}{TP + TN + FP + FN} \tag{4}$$

## 4 Experiment Result and Analysis

The experiment in this work includes four steps. As shown in Fig. 5, we collected data by python crawler and data interface. The collected data contains blockchain information, macroeconomic factors, and Bitcoin price. The detailed information will be presented in section B. Then, we processed the collected data, includes format adjusting, check for duplicate and missing values, data matching, and irrelevant variable deletion. After data processing, we completed the statistical analysis, multicollinearity analysis, and correlation analysis to have a better understanding of the data. With the processed data, we constructed the prediction model based on the broad learning system and genetic algorithm. For comparison, we also used Multilayer Perceptron (MLP), Back Propagation Neural Network (BPNN), Long Short Term Memory Recurrent Neural Network (LSTM-RNN), Multinomial Naive Bayes (MNB), and Random Forrest (RF) as a comparison. The final experiment results and evaluation will be shown in section C.

### 4.1 Experimental Environment

The experiment is conducted on a laptop computer with the Intel (R) Core (TM) i7 process that has 8 CPUs. The Random-Access Memory of the laptop is 8.00 GB. We used Python 3.6.5 to build, train, and test and prediction models, and the Integrated Development Environment (IDE) was the Jupyter notebook.

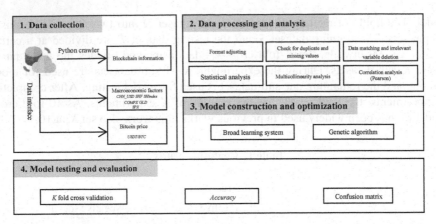

**Fig. 5.** Experiment process

### 4.2 Data Description and Analysis

We collected three types of data, blockchain information, minute-level Bitcoin price, and minute-level macroeconomics factors from December 7, 2017, to January 10, 2018, with a total of 5,621 data records. This work firstly collected the blockchain information from www.blockchain.com by python data crawler. There were 5,253 blocks, including 18 variables such as hash value, difficulty, bits, output, etc. Then we collected the minute-level Bitcoin close price USDT/BTC on Poloniex exchange by MultiCharts. For macroeconomics factors, we collected minute-level price and volume of SHF gold, Shanghai Composite Index, gold ETF, S&P500 index, Nasdaq Composite Index, the exchange rate of the yuan against the dollar by Wind, MultiCharts, IQFeed, OANDA.

We processed the collected data, including format adjusting, duplicate value check, missing value check, data matching, and irrelevant variable deletion. After the data processing, there were 5,621 data records with 15 input features and one Bitcoin price trend output. The statistical indexes of the 15 input features are shown in Table 2. After processing the data, we analyzed the processed data by using the Pearson correlation coefficient. As the prediction model aims to estimate the trend of Bitcoin price, the multicollinearity amongst these input features will not influence the prediction results.

### 4.3 Digital Currency Prediction Results and Analysis

We used the genetic algorithm to optimize the prediction model based on the broad learning system. As shown in Fig. 6, as the iterative number increasing, the testing accuracy of the prediction model increased. After 100 iterations, the testing accuracy stayed around 0.53.

Then we used the tuned hyper-parameters to train and test the prediction models based on different machine learning algorithms and 10-fold cross-validation. As shown in Table 3, the Accuracy of BLSGA (Broad Learning System and Genetic Algorithm) is superior to other machine learning models.

**Table 2.** Statistical analysis of input features

| Name | Mean | Std | Min | Median | Max |
|---|---|---|---|---|---|
| Difficulty | 1.79E + 12 | 1.46E + 11 | 1.59E + 12 | 1.87E + 12 | 1.93E + 12 |
| Transactions Number | 2288.87 | 566.58 | 1.00 | 2444.00 | 3506.00 |
| Block Output | 16892.34 | 10870.58 | 12.50 | 15245.27 | 108775.28 |
| Estimated Transaction Volume | 1744.77 | 1293.37 | 0.00 | 1391.97 | 13697.41 |
| Block Nonce | 2.12E + 09 | 1.17E + 09 | 1.88E + 07 | 2.19E + 09 | 4.27E + 09 |
| Transaction Fees | 4.81 | 2.36 | 0.00 | 4.31 | 13.72 |
| COMPX Close | 6962.27 | 94.68 | 6798.81 | 6946.64 | 7180.53 |
| GLD Close | 121.30 | 2.60 | 117.41 | 120.28 | 125.84 |
| GLD Volume | 12806.51 | 23051.23 | 100.00 | 6440.00 | 683151.00 |
| SPX Close | 2687.00 | 29.43 | 2633.21 | 2682.75 | 2759.08 |
| CHN/USD Close | 6.56 | 0.05 | 6.48 | 6.57 | 6.63 |
| SHF Close | 275.95 | 2.86 | 270.90 | 275.15 | 280.75 |
| SHF Volume | 141.42 | 251.69 | 0.00 | 64.00 | 6784.00 |
| SHindex Close | 3312.80 | 40.82 | 3267.92 | 3296.54 | 3413.90 |
| SHindex Volume | 6.72E + 06 | 3.65E + 06 | 8.41E + 05 | 6.92E + 06 | 1.33E + 07 |

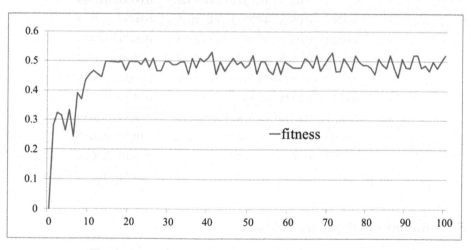

**Fig. 6.** Optimization process based on the genetic algorithm

For training efficiency, as shown in Table 4, except for MNB (Multinomial Naive Bayes), the BLSGA has the shortest training time and relatively highest accuracy. Therefore, taken together, the prediction model based on the broad learning system and genetic algorithm delivers high prediction accuracy within a relatively short time.

**Table 3.** Prediction accuracy of different machine learning models

| No. | BLSGA | MLP | BPNN | LSTM-RNN | MNB | RF |
|-----|-------|-----|------|----------|-----|-----|
| 1 | 0.5382 | 0.4902 | 0.4618 | 0.4618 | 0.4938 | 0.5133 |
| 2 | 0.5231 | 0.5427 | 0.4769 | 0.4769 | 0.4591 | 0.5249 |
| 3 | 0.5569 | 0.5214 | 0.4431 | 0.4431 | 0.5214 | 0.5142 |
| 4 | 0.5249 | 0.4626 | 0.4751 | 0.4751 | 0.4822 | 0.5552 |
| 5 | 0.5302 | 0.4982 | 0.4698 | 0.4698 | 0.4395 | 0.5338 |
| 6 | 0.5391 | 0.5196 | 0.4609 | 0.4609 | 0.4448 | 0.5178 |
| 7 | 0.5409 | 0.5267 | 0.4591 | 0.4591 | 0.5018 | 0.4929 |
| 8 | 0.5356 | 0.4786 | 0.4644 | 0.4644 | 0.4626 | 0.5267 |
| 9 | 0.5142 | 0.4822 | 0.4858 | 0.4858 | 0.5214 | 0.5409 |
| 10 | 0.5018 | 0.4858 | 0.4982 | 0.4982 | 0.5142 | 0.5480 |
| Mean | 0.5305 | 0.5008 | 0.4695 | 0.4695 | 0.4841 | 0.5268 |

**Table 4.** Training time(s) of different machine learning models

| No. | BLSGA | MLP | BPNN | LSTM-RNN | MNB | RF |
|-----|-------|-----|------|----------|-----|-----|
| 1 | 0.0359 | 0.1805 | 49.2859 | 752.0834 | 0.0030 | 0.0449 |
| 2 | 0.0259 | 0.0628 | 46.7774 | 744.3043 | 0.0030 | 0.0389 |
| 3 | 0.0249 | 0.0678 | 47.2023 | 748.1885 | 0.0030 | 0.0379 |
| 4 | 0.0329 | 0.0658 | 46.8821 | 764.2484 | 0.0020 | 0.0419 |
| 5 | 0.0309 | 0.0668 | 46.9629 | 678.5715 | 0.0020 | 0.0389 |
| 6 | 0.0459 | 0.0808 | 54.2180 | 773.9959 | 0.0030 | 0.0449 |
| 7 | 0.0329 | 0.0718 | 48.6888 | 764.1378 | 0.0030 | 0.0409 |
| 8 | 0.0359 | 0.0618 | 48.6788 | 763.6213 | 0.0020 | 0.0389 |
| 9 | 0.0269 | 0.0718 | 48.1742 | 761.3498 | 0.0020 | 0.0419 |
| 10 | 0.0369 | 0.0638 | 48.1098 | 743.2657 | 0.0010 | 0.0260 |
| Mean | 0.0329 | 0.0794 | 48.4980 | 749.3767 | 0.0024 | 0.0395 |

## 5  Summary and Future Work

This work predicted the price trend of the digital currency Bitcoin based on the broad learning system and genetic algorithm. The inputs of the model are blockchain information and macroeconomic factors, and the output is the price trend of Bitcoin price. The experiment results showed that the prediction model of the digital currency price using the broad learning system is superior to other machine learning models in both accuracy and efficiency.

In future work, the authors plan to add more macroeconomic factors and adopt feature mapping methods to improve the quality of the prediction model inputs. Besides, to improve the performance of the prediction model, the authors will employ feature extraction methods to process the input and output data. To validate the applicability of the proposed model, the authors also plan to apply the prediction model based on the broad learning system on the prediction of other digital currencies.

**Acknowledgement.** This work is supported by the National Natural Science Foundation of China (71531012). We would like to thank the anonymous reviewers and the editor for their helpful comments and valuable suggestions.

# References

1. NCTAD, Information Economy Report 2019 Value Creation and Capture: Implications for Developing Countries (2019)
2. Global Banking Research Group, Institute of International Finance, Bank of China, "Bank of China Global Banking Outlook Report (Q2 2018) (2018)
3. Yu, J.H., Kang, J., Park, S.: Information availability and return volatility in the Bitcoin Market: analyzing differences of user opinion and interest. Inf. Process. Manage. **56**(3), 721–732 (2019). https://doi.org/10.1016/j.ipm.2018.12.002
4. Brandvold, M., Molnár, P., Vagstad, K., Valstad, O.C.A.: Price discovery on Bitcoin exchanges. J. Int. Financ. Markets Inst. Money **36**, 18–35 (2015). https://doi.org/10.1016/j.intfin.2015.02.010
5. Katsiampa, P.: Volatility estimation for Bitcoin: a comparison of GARCH models. Econ. Lett. **158**, 3–6 (2017) https://doi.org/10.1016/j.econlet.2017.06.023.
6. Jang, H., Lee, J.: An empirical study on modeling and prediction of bitcoin prices with Bayesian neural networks based on blockchain information. IEEE Access **6**(99), 5427–5437 (2018). https://doi.org/10.1109/ACCESS.2017.2779181
7. Han, M., Feng, S., Chen, C.L.P., Xu, M., Qiu, T.: Structured manifold broad learning system: a manifold perspective for large-scale chaotic time series analysis and prediction. IEEE Trans. Knowl. Data Eng. **31**(9), 1809–1821 (2019). https://doi.org/10.1109/TKDE.2018.2866149
8. Jin, J., Liu, Z., Chen, C.L.P.: Discriminative graph regularized broad learning system for image recognition. Sci. China Inf. Sci. **61**(11), 1–4 (2018). https://doi.org/10.1007/s11432-017-9421-3
9. Shen, L., Shi, J., Gong, B., Zhang, Y., Dong, Y., Zhang, Q., An, H.: Multiple empirical kernel mapping based broad learning system for classification of Parkinson's Disease with transcranial sonography. In: Proceedings of the 2018 40th Annual International Conference of the IEEE Engineering in Medicine and Biology Society (EMBC), (2018)
10. Ciaian, P., Rajcaniova, M., Kancs, D.: The economics of Bitcoin price formation. Appl. Econ. **48**(19), 1799–1815 (2016). https://doi.org/10.1080/00036846.2015.1109038
11. Barro, R.J.: Money and the price level under the gold standard. Econ. J. **89**(353), 13–33 (1979). https://doi.org/10.2307/2231404
12. Bouri, E., Azzi, G., Dyhrberg, A.H.: On the return-volatility relationship in the Bitcoin market around the price crash of 2013. Social Science Electronic Publishing, p. 11 (2016). https://doi.org/10.5018/economics-ejournal.ja.2017-2.
13. Adcock, R., Gradojevic, N.: Non-fundamental, non-parametric Bitcoin forecasting. Phys. A Stat. Mech. Appl. **531**, 121727 (2019). https://doi.org/10.1016/j.physa.2019.121727

14. Mcnally, S., Roche, J., Caton, S.: Predicting the price of Bitcoin using machine learning. In: Proceedings-26th Euromicro International Conference on Parallel, Distributed and Network-Based Processing (PDP), Cambridge, United Kingdom, pp. 339–343 (2018). https://doi.org/10.1109/PDP2018.2018.00060.

15. Mallqui, D.C., Fernandes, R.A.: Predicting the direction, maximum, minimum and closing prices of daily Bitcoin exchange rate using machine learning techniques. Appl. Soft Comput. **75**, 596–606 (2019). https://doi.org/10.1016/j.asoc.2018.11.038

16. Chen, Z., Li, C., Sun, W.: Bitcoin price prediction using machine learning: an approach to sample dimension engineering. J. Comput. Appl. Math. **365**, 112395 (2019). https://doi.org/10.1016/j.cam.2019.112395

17. Mudassir, M., Bennbaia, S., Unal, D., Hammoudeh, M.: Time-series forecasting of Bitcoin prices using high-dimensional features: a machine learning approach. Neural Computing and Applications (2020). https://doi.org/10.1007/s00521-020-05129-6

18. Chen, C.L.P., Liu, Z.: Broad learning system: an effective and efficient incremental learning system without the need for deep architecture. IEEE Trans. Neural Networks Learn. Syst. **99**, 1–5 (2017). https://doi.org/10.1109/TNNLS.2017.2716952

19. Liu, Z., Zhou, J., Chen, C.L.P.: Broad Learning System: Feature extraction based on K-means clustering algorithm. In: Proceedings of 4th International Conference on Information, Cybernetics and Computational Social Systems (ICCSS). IEEE (2017). https://doi.org/10.1109/ICCSS.2017.8091501.

20. Liu, Z., Chen, C.L.P.: Broad learning system: structural extensions on single-layer and multilayer neural networks. In: 2017 International Conference on Security, Pattern Analysis, and Cybernetics (SPAC). IEEE (2018). https://doi.org/10.1109/SPAC.2017.8304264.

21. Xu, M., Han, M., Chen, C.L.P., Qiu, T.: Recurrent broad learning systems for time series prediction. IEEE Trans. Cybern. **50**(4), 1405–1417 (2020). https://doi.org/10.1109/TCYB.2018.2863020

22. Chen, C.L.P., Liu, Z., Feng, S.: Universal approximation capability of broad learning system and its structural variations. IEEE Trans. Neural Networks Learn. Syst. **3**(4), 1191–1204 (2019). https://doi.org/10.1109/TNNLS.2018.2866622

23. Jin, J.W., Chen, C.L.P.: Regularized robust broad learning system for uncertain data modeling. Neurocomputing **322**(1), 58–69 (2018). https://doi.org/10.1016/j.neucom.2018.09.028

24. Zhang, N., Yang, X., Zhang, M., Sun, Y.: A genetic algorithm-based task scheduling for cloud resource crowd-funding model. Int. J. Commun. Syst. **31**(1), e3394.1–e3394.10 (2018). https://doi.org/10.1002/dac.3394.

25. Xiong, Z., Cui, Y., Liu, Z., Zhao, Y., Hu, M., Hua, J.J.: Evaluating explorative prediction power of machine learning algorithms for materials discovery using k-fold forward cross-validation. Comput. Mater. Sci. **171**, 109203 (2020). https://doi.org/10.1016/j.commatsci.2019.109203.

# Selective Multi-source Transfer Learning with Wasserstein Domain Distance for Financial Fraud Detection

Yifu Sun, Lijun Lan$^{(\boxtimes)}$, Xueyao Zhao, Mengdi Fan, Qingyu Guo, and Chao Li

Cloud & Smart Industries Group, Tencent, China
smurflan@tencent.com

**Abstract.** As financial enterprises have moved their services to the internet, financial fraud detection has become an ever-growing problem causing severe economic losses for the financial industry. Recently, machine learning has gained significant attention to handle the financial fraud detection problem as a binary classification problem. While significant progress has been made, fraud detection is still a notable challenge due to two major reasons. First, fraudsters today are adaptive, inventive, and intelligent, making their fraud characteristics are too deep stealth to be detected by simple detection models. Second, labeled samples for training the detection models are usually very few as collecting large-scale training data needs a certain performance-time and is costly. To address the two problems, we propose a novel multi-source transfer learning approach with self-supervised domain distance learning for financial fraud detection problems. The core idea is to transfer relevant knowledge from multiple data-rich sources to the data-poor target task, e.g., learning fraud patterns from several other related mature loan products to improve the fraud detection in a cold-start loan product. Specifically, since the feature distribution discrepancy across domains may cause useless or even negative knowledge transfer, we propose self-supervised domain distance learning under the Wasserstein metric to measure the domain relevance/relationships between target and source tasks. The learned Wasserstein distance helps in selectively transferring most relevant knowledge from source domains to target domains. Thus it reduces the risk of negative transfer as well as maximizes the multi-source positive transfer. We conduct extensive experiments under multi-source few-shot learning settings on real financial fraud detection dataset. Experimental analysis shows that the inter-domain relationships learned by our domain distance learning model align well with the facts and the results demonstrate that our multi-source transfer learning approach achieves significant improvements over the state-of-the-art transfer learning approaches.

## 1 Introduction

Financial fraud can be defined as the intentional use of illegal methods of acquiring financial gain, e.g., case-out fraud in credit-card, money laundering, loan

© Springer Nature Singapore Pte Ltd. 2021
W. Gao et al. (Eds.): FICC 2020, CCIS 1385, pp. 489–505, 2021.
https://doi.org/10.1007/978-981-16-1160-5_38

default, and insurance fraud [1]. As financial enterprises are moving their services to the internet, financial fraud detection has become an ever-growing problem for the financial industry. Financial fraud detection is primarily formulated as a binary classification problem that aims at predicting whether an entity would be involved in fraudulent activity, e.g., a credit card transaction would be fraudulent, or a loan application in a bank would default.

Nowadays, machine learning and data mining approaches have increasingly gained attention in financial fraud detection problems. With the use of machine learning approaches, fraudulent patterns can be automatically mined from large datasets without the help of specific expert knowledge. However, as machine learning approaches heavily rely on rich datasets with abundant annotations, fraud detection in most financial systems is still a notable challenge. There are two main reasons. Firstly, fraudsters are continually evolving fraudulent methods and fast-moving from one financial service to another to hidden their illegal behaviors from detection models. Consequently, it is unrealistic to collect a comprehensive fraud dataset that includes diverse fraudulent patterns from single or few financial services or products. Secondly, annotated fraud samples are very few because of class unbalance and delayed performance. Take bank loan fraud as an example, knowing the label of a new borrower needs a certain performance time, usually 30 days or even several months. Therefore, it is both economically expensive and time-costly to collect a large-scale fraud dataset. The two reasons together drive the financial enterprises to explore default detection systems on very few labeled data.

To address the above problems, we propose a selective multi-source transfer learning approach with Wasserstein domain distance (WM-trans) for financial fraud detection. Our core idea is to transfer helpful knowledge from multiple different but related data-rich source tasks to data-poor target tasks. For example, in most banks, traditional personal instalment loan products like Mortgages, Auto Loan, Credit Card have massive labeled data. When banks are exploring new businesses like Apple Card Monthly Instalments some prior knowledge in the traditional product could help in detecting potential fraud at the cold-start period.

To complete the above task, a key challenge is transferring only positive fraud detection knowledge from related known source tasks while inhibiting negative knowledge transfer. However, existing multi-source transfer learning approaches usually are simply extended from single-source-single-target approaches by 1) fine-tuning a pre-trained model from a mix of source domain samples [2], or 2) learning a feature extractor that maps samples from different sources into a common feature space and exploring a domain-invariant classification model using adversarial training methods [3]. As both lines of works fail to take account of domain relevance during transfer learning, they might cause serious negative transfer in the financial fraud detection scenario. Figure 1 illustrates this multi-source transfer learning problem. Compared with mortgage and auto loan scenarios, customers from installment and credit cards scenarios are more likely to share similar characteristics, e.g., similar ages or consumption behaviors.

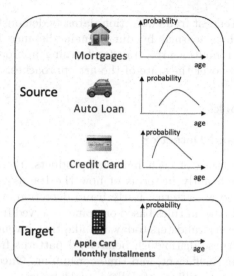

**Fig. 1.** Toy example

Therefore, increasing knowledge from credit card while reducing knowledge from mortgage and auto loan can maximally guarantee positive knowledge transfer to the model of target installment fraud.

Considering the above limitations, we propose the WM-trans system that addresses the multi-source transfer learning problem by introducing a self-supervised domain distance learning network under the Wasserstein metric. It can effectively measure the domain relevance/relationships between target and source tasks without any extra annotations and encourages fraud detection tasks from similar or relevant domains to present smaller Wasserstein distance. More importantly, the domain distance learner reduces the risk of negative transfer in our WM-trans system by selectively transferring the most relevant and useful knowledge from most related source tasks to target tasks with limited labeled data.

In summary, our contributions are concluded as follows:

- We extend the existing single-source-single-target transfer learning framework to a multi-source framework for financial fraud detection. The proposed framework can not only selectively transfer knowledge from large-scale source datasets to small-scale target dataset, but also has the capability of utilizing both labeled and unlabeled target samples to make effective sample utilization.
- We explore a self-supervised domain distance learning network to measure inter-domain relevance between target and source tasks. Unlike previous works that use tailored aggregation rules, our domain distance learning model can be reused on any target tasks without retraining on new target tasks.

- Experiments on Tencent fraud detection datasets demonstrate that the inter-domain relationships learned by our domain distance learning model are insightful and our multi-source transfer learning approach achieves significant improvements over the state-of-the-art approaches.

## 2  Related Work

### 2.1  Financial Fraud Detection

Given the varying nature of each financial products, the definition of financial frauds vary significantly in terms of how the fraudsters or anomalies act in distinguish behavioral patterns. Early fraud detection systems mainly focus on developing combinatorial rules based on some observed fraudulent behavioral patterns [4–6]. However, fraudsters today are adaptive, inventive, and intelligent. This makes mining those distinguish behavioral patterns from massive normal behaviors challenging, inefficient, and time-consuming. Once the rules are deciphered by the attackers, millions of dollars might be lost.

Due to the limitations of the rule-based methods, some classical machine learning models are introduced into the fraud risk area. The most famous and widely used one is logistic regression, known as the credit risk score method. Logistic regression automatically combines the features in linear to predict the probability of default. Although it is interpretable and flexible, logistic regression can not mine complex nonlinear features in training. Recently, more advanced machine learning approaches are introduced for financial fraud detection like SVM [7], non-parametric method [8], tree-based method [9], neural networks [10–12], and graph based method [13]. With the use of these machine learning approaches, intrinsic fraudulent pattern could be automatically mined from massive data. Commonly, these methods first extract statistical features using feature engineering methods from different aspects, like user-profiles and historical behaviors, and then train a classifier on the hypothesis feature space. The success of machine learning approaches mainly relies on large amount of labeled data. However, in most financial scenarios, annotated fraud samples are very few because of class unbalance and delayed performance. This problem is even more serious at the cold-start stage of a new financial product. In this context, solving the problem of training fraud detection model on limited labeled data becomes crucial for guaranteeing the benefits of financial enterprises.

### 2.2  Transfer Learning

Transfer learning aims at improving the performance of learners on target domains by leveraging knowledge from related source domains [14–16]. Classical transfer learning methods mainly focus on handling single source single target problem. One of the most famous transfer learning approaches is Domain Adversarial Neural Network (DANN) [17] and its extensions [18–22]. To reduce the distribution shift between source and target domains, domain adaption (DA) approaches explore domain-invariant structures and feature representations.

Recent efforts [2,3,23,24] extend early single-source-single-target transfer learning frameworks to multi-source transfer scenarios. Due to the domain shift among source domains, the critical challenge in multiple source transfer learning is the negative transfer, which is a phenomenon that the transferred knowledge may hurt the target learner if domains are irrelevant or have large domain discrepancy. One straightforward way is pre-training a starting model on a mix of all the source task samples and fine-tuning it on target task. However, as data volumes across source tasks can vary widely, simply combine multiple source data would result in the dominance of some larger-scale but less-related sources in the final transferred model. To tackle this problem, [25] proposed to rank the source tasks according to their relationships to the target. However, the computation of source-target relationships requires both source and target tasks have sufficient labeled data. Furthermore, the ranking system is tailored for individual target task, making it inflexible in real-work applications. In contrast, our proposed WM-Trans learns domain relevance in an self-supervised manner. The resulted domain distance learn can measure domain relevance between any two source and target tasks.

## 3   Methodology

In this section, we introduce our multi-source transfer learning approach, WM-Trans, for solving financial fraud detection problem.

### 3.1   Problem Formulation

The overall goal of our WM-Trans is selectively transferring good prior knowledge from multiple data-rich source tasks to target task that has very few labeled data, as well as overcoming negative knowledge transfer caused by domain shift between source and target tasks. Specifically, our WM-Trans achieves this objective by solving three sub-problems: 1) How to measure the domain relevance between source and target tasks. 2) how to transfer knowledge from each source task to the target task. 3) how to selectively combine transferred knowledge from multiple source tasks. Each of the following sections(B-D) will detail those three problems.

Let $\mathcal{D}_{tgt} = \{\mathcal{D}_{\mathcal{L}}, \mathcal{D}_{\mathcal{U}}\}$ be the target task containing both limited labeled data $\mathcal{D}_{\mathcal{L}}$ and a large volume of unlabeled data $\mathcal{D}_{\mathcal{U}}$. In the multi-source transfer learning setting, there are several source tasks with large-scale labeled datasets. Let $\mathcal{D}_{src} = \{\mathcal{D}_1, \mathcal{D}_2, ..., \mathcal{D}_S\}$ be the set of source tasks and $S$ is the total number of source tasks. All the source and target tasks can be treated as a binary classification problem and share the same label space, $b = \{0, 1\}$. $b = 0$ denotes normal entities, which could be a legitimate loan application, a faithworthy borrower, or a normal credit card transaction. Whereas, $b = 1$ indicates default or fraudulent entities. Note that $|D_{src,i}|$ is much larger than $|D_{tgt}|$, and source and target tasks may be drawn from different distributions with domain shift, such as instalment loan, credit card transaction, credit advance, auto loan, and so on.

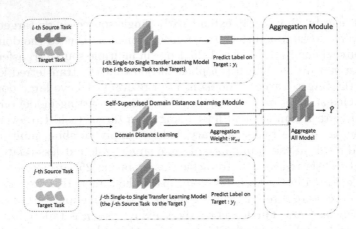

**Fig. 2.** Overall framework of our WM-Trans

Figure 2 illustrates the overall framework of our WM-Trans transfer learning approach, which contains three modules: self-supervised domain distance learning aiming at modeling domain relevance and relationships in a self-supervised manner, single-source-single-target (SSST) transfer learning with the goal of transferring target-specific knowledge from each source to the target task, and multi-source aggregation that is combining multiple SSST transfer models according to their domain relevance with the target task. Each of the three modules solves a problem that we just suggested before.

### 3.2 Self-supervised Domain Distance Learning Module

Motivated from the observation that irrelevant source tasks would bring about negative knowledge transfer, the key point in multi-source transfer learning is to find a metric that can effectively measure the domain relevance between source and target tasks. Usually, it can be formulated as a mathematical statistics problem that aim to measure how one probability distribution (e.g., source task) is different from a second (e.g., target task). One representative measurement is Kullback–Leibler (KL) divergence, which computes the directed divergence between two distributions, $\mathbb{P}$ and $\mathbb{Q}$, as:

$$D_{\mathrm{KL}}(\mathbb{P}\|\mathbb{Q}) = \sum_{x \in \mathcal{X}} P(x) \log \left( \frac{\mathbb{P}(x)}{\mathbb{Q}(x)} \right) \tag{1}$$

where, $\mathbb{P}(x)$ and $\mathbb{Q}(x)$ are the probability densities at $x$. It usually requires both compared distributions to be single-variate distributions on the same probability space. Therefore, directly applying KL-divergence to measure domain relevance would confront two problems: 1) the dataset of each fraud detection task is usually a multi-variate distribution on hundreds even thousands of dimensions of features; 2) the significance of each feature in measuring domain relevance also varies greatly.

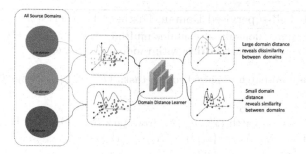

**Fig. 3.** Framework of self-supervised domain distance learning. It predicts the domain distance between any two fraud detection datasets.

To overcome the above problems, we proposed a self-supervised domain distance learning approach, which uses the Wasserstein metric to measure the domain discrepancy between any two fraud detection tasks. Figure 3 illustrates the framework of the domain distance learner. It aims at finding a metric that minimizes the domain distance prediction error among all sources on a hypothesis metric space $(H, \rho)$, where $\rho(x, y)$ is a distance function for two instances $x$ and $y$ in $H$. As a result, tasks from similar or related domains reveal short distance in $H$, whereas tasks from unrelated domains present long distance.

Let $\mathbb{P}$ and $\mathbb{Q}$ denote the datasets of two fraud detection tasks, the first Wasserstein distance between their data distributions is defined as a form of integral probability metric in Eq. 2

$$W_1(\mathbb{P}, \mathbb{Q}) = \sup_{\|f\|_L \leq 1} |\, \mathbb{E}_{x \sim \mathbb{P}}[f(x)] - \mathbb{E}_{x \sim \mathbb{Q}}[f(x)] \,| \tag{2}$$

where $\|f\|_L$ is the Lipschitz semi-norm, which is defined as $\|f\|_L = \sup | f(x) - f(y) | / \rho(x, y)$. In this paper, we use the first Wasserstein distance to measure the domain relevance of fraud detection tasks. In the first Wasserstein distance, $f(.)$ can be any Lipschitz-smooth nonlinear function, which is approximated by a multi-layer feedforward neural network in this work.

To enforce the Lipschitz constraint on the domain distance learner, we adopt the gradient penalty loss in [26] to constraint the gradients with norm at most 1.

$$\mathcal{L}_{grad}(\hat{X}) = \left(\left\|\nabla_{\hat{X}} f_d(\hat{X})\right\|_2 - 1\right)^2 \tag{3}$$

where $\hat{X} \leftarrow \{X_p, X_q\}$ is randomly sampled along straight lines between pairs of samples from the two compared tasks, $\mathbb{P}$ and $\mathbb{Q}$.

A detailed algorithm of self-supervised distance learning is given in Algorithm 1. The domain distance learner is a multi-layer neural network $f_d$ : $\mathbb{R}_d \rightarrow \mathbb{R}$ that maps each training sample into a real number with parameter $\theta_d$. In each training iteration, we randomly sample a mini-batch of source task pairs $\{\mathcal{D}_{i_1}, \mathcal{D}_{i_2}, d\}$ from the source task set $\mathcal{D}_{src}$. For each task, we further randomly sample $n$ data points to approximate its sample distribution. Note that

---

**Algorithm 1.** Self-supervised Domain Distance Learner

---

**Require:** The source domain set contains multiple source $\mathcal{D}_{src} = \{\mathcal{D}_1, \mathcal{D}_2, ..., \mathcal{D}_S\}$
1: Initialize domain distance learner with random weights $\theta_d$
2: **repeat**
3:     Sample a mini-batch of paired source tasks $\{\mathcal{D}_{i_1}^k, \mathcal{D}_{i_2}^k\}_{k=1}^m$ from $\mathcal{D}_{src}$, where $1 \leq i_1, i_2 \leq S$.
4:     **for** $k = 1, ..., m$ **do**
5:         sample a mini-batch $\{x_{i_1}^t, y_{i_1}^t\}_{t=1}^n$ from task $\mathcal{D}_{i_1}$ and a mini-batch $\{x_{i_2}^t, y_{i_2}^t\}_{t=1}^n$ from task $\mathcal{D}_{i_2}$, $X_{i_1} = \{x_{i_1}^t\}_{t=1}^n$, $X_{i_2} = \{x_{i_2}^t\}_{t=1}^n$
6:         $g_{i_1} \leftarrow f_d(x_{i_1})$, $g_{i_2} \leftarrow f_d(x_{i_2})$
7:         $\hat{X} \leftarrow \{X_{i_1}, X_{i_2}\}$
8:         $\theta_d \leftarrow \theta_d - \alpha_1 \nabla_{\theta_d} [\mathcal{L}_d(g_{i_1}, g_{i_2}; \theta_d) - \gamma \mathcal{L}_{grad}(\hat{X})]$
9:     **end for**
10: **until** $\theta_d$ converge

---

$\{\mathcal{D}_{i_1}, \mathcal{D}_{i_2}\}$ could be the same task, denoted as $d$. $d = 0$ indicates two sub-datasets are drawn from the same task, whereas $d = 1$ indicates the two are drawn from different tasks. In this work, we use $d$ as the ground truth label for training the domain distance learner in a self-supervised manner. For a category balance purpose, half of the task pairs are the same task, and half are different.

The final parameters $\theta_d$ of the domain distance learner is optimized with the domain critic loss $\mathcal{L}$:

$$\mathcal{L}_d(g_{i_1}, g_{i_2}; \theta_d) = -\Sigma_k^m (d_k \log(\frac{1}{n}\Sigma_{t=1}^n g_{i_1}^t - \frac{1}{n}\Sigma_{t=1}^n g_{i_2}^t) \\ + (1 - d_k) \log(1 - (\frac{1}{n}\Sigma_{t=1}^n g_{i_1}^t - \frac{1}{n}\Sigma_{t=1}^n g_{i_2}^t)) \tag{4}$$

where $n_p$ denotes the number of data in $X_p$, $n_q$ denotes the number of data in $X_q$. The ground truth distance $d^k$ implies whether or not the samples coming from the same task. $\frac{1}{n}\Sigma_{t=1}^n g_{i_1}^t - \frac{1}{n}\Sigma_{t=1}^n g_{i_2}^t$ is the estimated domain distance between two probability distribution $\mathbb{P}_{X_{i_1}}$ and $\mathbb{P}_{X_{i_2}}$, as shown in Eq. 2. The second part of Eq. 4 corresponds to the gradient penalty loss in Eq. 3.

Finally, once the domain distance learner is obtained, it can be applied to estimate the domain relevance between any two tasks with smaller domain distance indicating stronger domain relevance.

### 3.3   Single-Source-Single-Target Transfer Module

The objective of SSST transfer module is to transfer target-specific knowledge from each source task to the target task. To handle the domain discrepancy between source and target tasks, we adopt the adversarial training [17] to adapt knowledge from each source domain to the target domain. The core idea of domain adaption is finding domain-invariant feature representations. Therefore, classification models trained on the large-scale source samples can be directly used for the target task.

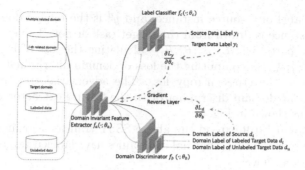

**Fig. 4.** Framework of single-source-single-target transfer learning.

As shown in Fig. 4, the domain adaption framework usually consists of three import components, i.e., feature extractor, label classifier, and domain discriminator:

- The feature extractor aims at learning a domain-invariant feature representation $f_a : \mathbb{R}^f \to \mathbb{R}^p$, that maps the instance to a $p$-dimensional representation with corresponding network parameter $\theta_a$.
- The label classifier predicts an instance is legitimate or fraudulent in the domain-invariant representation space, which is formulated as a binary classification problem $f_c : \mathbb{R}_f \to \mathbb{R}$ with network parameters $\theta_c$.
- The domain discriminator is to distinguish from which domain (source or target) an instance is drawn. It is also formulated as an binary classification problem in the domain-invariant representation space, $f_b : \mathbb{R}_f \to \mathbb{R}_2$ with network parameter $\theta_b$.

All the above functions $\{f_a, f_c, f_b\}$ can be approximated as a multi-layers neural network. To achieve the purpose of domain adaption, the feature extractor $f_a$ and classifier are trained to minimize the loss of label prediction as well as maximize the loss of domain discriminator $f_b$. Whereas, the domain discriminator $f_b$ learns to minimize the loss of domain discriminator $f_b$. The adversarial training objective is defined as:

$$
\begin{aligned}
&\mathcal{L}(\mathcal{X}_i, \mathcal{X}_u; \theta_a, \theta_b, \theta_c) \\
&= \mathcal{L}_c(\mathcal{X}_i; \theta_a, \theta_c) + \lambda \mathcal{L}_b(\{\mathcal{X}_i, \mathcal{X}_u\}; \theta_a, \theta_b)
\end{aligned}
\tag{5}
$$

$$
\mathcal{L}_c(\mathcal{X}_i; \theta_a, \theta_c) = -\frac{1}{K} \sum_{\mathbf{x}, \mathbf{y} \in \mathcal{X}_i} y \cdot \log f_c(f_a(x))
\tag{6}
$$

$$
\begin{aligned}
&\mathcal{L}_b(\{\mathcal{X}_i, \mathcal{X}_u\}; \theta_a, \theta_b) \\
&= -\frac{1}{2K} \sum_{\mathbf{x}, \mathbf{d} \in \mathcal{X}_i \cup \mathcal{X}_u} 1(y^d = l) \cdot log f_b(f_a(x^k))_l
\end{aligned}
\tag{7}
$$

where $\mathcal{X}_i$ and $\mathcal{X}_u$ respectively denote a batch of labeled data from the i-th source task and the unlabeled data from the target task, $K$ is the batch size, $y$ is the ground-truth label of a source instance, and $y^d$ is the binary domain indicator whether an instance is drawn from the target task or not. The overall objective consists of two parts, i.e., $\mathcal{L}_c(\mathcal{X}_i; \theta_a, \theta_c)$ calculating the loss of label prediction, and $\mathcal{L}_b(\{\mathcal{X}_i, \mathcal{X}_u\}; \theta_a, \theta_b)$ computing the loss of domain discriminator. Both parts are calculated as the cross-entropy loss. The adversarial training is achieved by connecting the domain discriminator to the feature extractor via a gradient reversal layer as shown in Fig. 4 .

Equation 8 and Eq. 9 extend the above SSST transfer learning process to a more common scenario that financial companies have labeled target samples but the number is very few.

$$
\begin{aligned}
&\mathcal{L}(\mathcal{X}_i, \mathcal{X}_l, \mathcal{X}_u; \theta_a, \theta_b, \theta_c) \\
&= \mathcal{L}_c(\{\mathcal{X}_i, \mathcal{X}_l\}; \theta_a, \theta_c) + \lambda \mathcal{L}_b(\{\mathcal{X}_i, \mathcal{X}_u\}; \theta_a, \theta_b)
\end{aligned}
\tag{8}
$$

$$
\mathcal{L}_c(\{\mathcal{X}_i, \mathcal{X}_l\}; \theta_a, \theta_c) = -\frac{1}{2K} \sum_{\mathbf{x}, \mathbf{y} \in \mathcal{X}_i \cup \mathcal{X}_l} y \cdot \log f_c(f_a(x))
\tag{9}
$$

where $\mathcal{X}_l$ denotes a mini batch of labeled target samples. The overall objective is to train the whole transfer learning network to not only minimize the label prediction loss on both the source task and the target task (shown in Eq. 9), but also maximize the domain discriminator loss (shown in Eq. 7).

The SSST transfer learning process is repeated for $S$ times, finally resulting in $S$ transfer models. Each SSST transfer model $m_i$ identifies knowledge shared between the i-th source task and the target task.

### 3.4   Aggregation Module

In this section, we introduce the aggregation module, which aims to selectively combine the $S$ SSST transfer models according to their domain relevance to the target task. Equation 10 computes the aggregation weights $W_{wd}$ upon the domain distance learned from the pre-trained domain distance learner:

$$
W_{wd}^i = \frac{1/(W_1^i)}{\Sigma_{i=1}^{S} 1/(W_1^i)}
\tag{10}
$$

where $W_1^i$ is the Wasserstein distance between the i-th source and the target task. $W_{wd}^i$ is the aggregation weight of the i-th SSST transfer model. A larger distance between the source and target tasks indicates weaker domain relevance. Therefore, $1/(W_1^i)$ would give those unrelated source tasks less weight during aggregation.

The aggregated prediction $\hat{y}$ is finally computed as the weighted sum of multiple SSST transfer model predictions. In this way, our WM-Trans selectively combines useful knowledge from multiple domains.

$$
\hat{y} = \Sigma_{i=1}^{S} W_{wd}^i \hat{y}_i
\tag{11}
$$

**Table 1.** Data size of financial fraud detection dataset

| Auto loan | | | Cash advance | | | Personal instalment loan | | |
|---|---|---|---|---|---|---|---|---|
| Source1 | Source2 | Target1 | Source3 | Source4 | Target2 | Source5 | Source6 | Target3 |
| 31219 | 43862 | 3000 | 5666 | 41717 | 3000 | 20460 | 25033 | 3000 |

## 4  Experiment

In this section, we extensively evaluate our proposed WD-Trans on a real-word fraud detection dataset from three aspects:

- In Sect. 4.2, we perform a qualitative evaluation to show that the domain distance learning module in our WD-Trans is able to identify the domain relevance/relationships between different fraud detection tasks.
- In Sect. 4.3, we compare our proposed multi-source transfer learning framework to baseline transfer learning approaches and show that our WD-Trans can bring significant performance improvements.
- In Sect. 4.4, we inspect how our WD-Trans ensure positive transfer while inhibiting negative transfer by a comparative analysis on domain distance and single-source-single-target transfer.

### 4.1  Experiment Settings

**Dataset.** We use a large fraud detection dataset provided by our financial partners. It consists of legal or fraudulent loan applications from three domains, i.e., personal instalment loan, cash advance, and auto loan. Personal instalment loans can be used for any personal expenses, and borrowers must repay with regularly scheduled payments or instalments. Cash advance is a short-term loan that borrowers receive cash to be used for any purpose. Auto loan is used with a designated purpose of helping borrowers afford a vehicle, but with the risk of losing the car if a borrower miss payments. Each domain has three datasets. We randomly select two from each domain to form the source tasks, and the remaining three form the target dataset. Table 1 provides sample sizes of both the source and target datasets, and each dataset has 5000 to 40000 labeled samples. For the privacy protection purpose, all the samples were anonymized, variable names were renamed, and values are normalized as a 457-dimensional feature vector with one extra binary label indicating whether a loan application is fraudulent or not. Note that the domain discrepancy between the nine fraud dataset can vary not only because of their loan types but also how a loan application is identified as fraud. For example, Days Past Due (DPD) is a common metric to determine a loan applicant is a fraudster, but different loan products may use totally different DPD values in their practices.

For a comprehensive evaluation purpose, each target dataset is further split into a training set, in-time (INT) test set, and out-of-time (OOT) test set. Notice that the in-time test samples are collected from the same period with the training samples. In contrast, the OOT samples are entirely separated from the training in-time data by consisting of data from a more recent period. For financial fraud detection, model performance on OOT set is an important metric to evaluate models' robustness on predicting out of time. To mimic the situation that the target task only has very few labeled samples, we randomly sample 2000 and 3000 fraud applications as two training sets to conduct twice experiments for each target dataset. The remaining instances are used as INT and OOT test examples for evaluation. For each target dataset, we repeat the evaluation process for 20 times and report the averaged performance on both INT and OOT sets.

**Compared Baseline Methods.** We compare our WM-Trans to one target-only baseline (i.e., XGBoost) and two multi-source transfer learning frameworks (i.e., Mix-Finetune, avg-Trans):

- **XGBoost**: It is an extended version on existing gradient boosted tree methods [27], and has been popularly used to solve the financial fraud detection problem.
- **Mix-Finetune**: Mix-Finetune is inspired by the widely used transfer learning method. It simply merges all available source samples to form a large dataset to train an initial DNN model. The purpose of transfer learning is achieved by fine-tuning the pre-trained DNN model on the target data, which only has very few labeled samples.
- **AVE-Trans**: In order to further inspect the effectiveness of W-distance, we simply average the predictions from multiple transfer models, which are trained in a single-source-single-target way. The most significant difference from our WM-Trans is avg-Trans treats each source with the same contribution to the target task, whereas our WM-Trans assign importance weight for each source according to their domain relevance to the target task.
- **WM-Trans**: WM-Trans is the abbreviation of our proposed multi-source transfer learning approach. It aggregates the predictions from multiple transfer models according to domain relevance captured by the Wasserstein distance in Algorithm 1.

**Setup.** For XGBoost, we did parameter tuning via an extensive grid search on parameters such as learning rate, sub-tree number, max tree depth, sub-sample columns, and so on. For Mix-Finetune and AVE-trans, we used the same neural network structure with our WM-trans model for a fair comparison. For WM-trans, we first trained the domain distance learning model only on the source datasets using SGD optimizer with a fixed learning rate of 0.001. The corresponding architecture is a three-layer feedforward neural network, DNN-457-128-64-1. Note that during inference, its input is a source-target dataset pair, and the output is a positive real number, indicating the corresponding domain distance. During the single-source-single-target transfer stage of our WM-trans, the network architectures for feature extractor, class predictor, and domain discriminator are respectively 456-256-128, 128-64-2, and 128-64-2. During multi-source transfer, each pair of feature extractor and class predictor compose a base model for the final multi-source aggregation. For the purpose of faster convergence, we

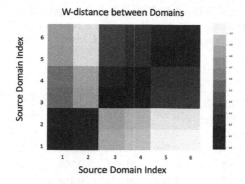

**Fig. 5.** Illustration of self-supervised domain distance learner on the Financial Fraud Detection dataset. The heat-map of domain relationships is evaluated by w-distances.

further initialize the neural network using a set of parameters coming from a Model-Agnostic Meta-learning model (MAML) [28] that is trained on the six sources.

### 4.2    Results on Domain Relationships

To inspect whether the Wasserstein distance can capture tasks' domain relevance/relationships, we compute the W-distance for any two of the six source datasets using the learned domain distance model. Figure 5 illustrates the covariance matrix of the source tasks, where darker color indicates smaller W-distance. Figure 5 shows that the source datasets collected from similar loan scenarios reveal smaller W-distance values. For example, it can be observed that Source1 and Source2, Source3 and Source4, as well as Source5 and Source6 are highly related to each other, respectively. This observation is consistent with the ground truth in Table 1. In addition, it is interesting to observe that the four source datasets from personal instalment and cash advance present stronger domain relevance when compared to the two datasets from auto loan. The reason might be (1) both personal instalment and cash advance loans are short-term loans while auto loan is a long-term loan, (2) when compared to auto loan, both personal instalment and cash advance loans involve a smaller amount of money usually for purposes of purchasing somethings. Therefore, loan applicants of personal instalment loan and cash advance share more common characteristics, e.g., age, income level, education background, repayment ability etc., thus revealing stronger domain relevance. In summary, the w-distance obtained from our domain distance learning model demonstrates insightful domain relationship information.

### 4.3    Comparison Results

Table 2 and 3 reports the comparison results with baseline approaches. The results are reported in terms of Kolmogorov-Smirnov (KS) values on both INT

**Table 2.** Comparison results with 2000 labeled target data. SSST denotes single-source-single-target transfer learning. Results are reported in KS values on both in-time (Test) and out-of-time (OOT) validation datasets. Best performance is in bold.

| | Model | Test KS | | | OOT KS | | |
|---|---|---|---|---|---|---|---|
| | | Target1 | Target2 | Target3 | Target1 | Target2 | Target3 |
| Target only | XGBoost | 0.296 | 0.414 | 0.181 | 0.152 | 0.292 | 0.130 |
| SSST | Source1 | 0.286 | 0.400 | 0.189 | **0.242** | 0.324 | 0.147 |
| | Source2 | 0.289 | 0.401 | 0.182 | 0.223 | 0.320 | 0.175 |
| | Source3 | 0.297 | 0.421 | 0.206 | 0.207 | 0.296 | 0.182 |
| | Source4 | 0.301 | 0.408 | 0.189 | 0.209 | 0.323 | 0.169 |
| | Source5 | 0.279 | 0.412 | 0.194 | 0.196 | **0.334** | 0.175 |
| | Source6 | 0.295 | **0.429** | 0.195 | 0.185 | 0.316 | 0.167 |
| Souce combine | Mix Fine-Tune | 0.294 | 0.399 | 0.188 | 0.196 | 0.307 | 0.155 |
| | avg-Trans | 0.287 | 0.408 | 0.185 | 0.185 | 0.315 | 0.162 |
| | WM-Trans | **0.307** | **0.429** | **0.204** | **0.224** | **0.329** | **0.183** |

and OOT validation datasets of the three target tasks. In financial fraud detection applications, KS test, which evaluates the difference between two distributions, i.e., positive-sample proportions and negative-sample proportions, is a significant measurement that estimates a model's ability to distinguishing badness and goodness samples. It can be observed that our WM-Trans achieves the best performance with a significant improvement margin over the baseline approaches. Table 2 and 3 firstly shows that the performance of Mix-Finetune is even worse than the XGBoost model only trained on target samples in most cases. The behind reason might be the data volume of source tasks is much larger than the target tasks, and simply mixing all the training samples may cause the final detection model overfits on some source tasks that have larger-scale training samples but their domain are quite different from the target task. In addition, it can also be observed that AVE-Trans surpasses both XGBoost and Mix-Finetune in most cases, especially on the OOT test set. This implies that instead of simply mixing all the source samples, averaging on single-source-single-target (SSST) transfer models indeed help in reducing the risk of negative influence caused by large-scale but unrelated source tasks. Last but most importantly, our WM-Trans outperforms ave-Trans on both the INT and OOT test sets. This suggests that weighting SSST models according to their domain relevance to the target task, indeed help in maximizing positive knowledge transfer during multi-source transfer learning.

### 4.4   W-Distance Vs. SSST Transfer

A key challenge of multi-source transfer learning is identifying source tasks that can bring performance gain to the target task while inhibiting source tasks with negative impact. To further inspect how our WM-trans work, we use the SSST

**Table 3.** Comparison results with 3000 labeled target data. SSST denotes single-source-single-target transfer learning. Results are reported in KS values on both in-time (Test) and out-of-time (OOT) validation datasets. Best performance is in bold.

| | Model | Test KS | | | OOT KS | | |
|---|---|---|---|---|---|---|---|
| | | Target1 | Target2 | Target3 | Target1 | Target2 | Target3 |
| Target only | XGBoost | 0.314 | 0.426 | 0.198 | 0.112 | 0.321 | 0.155 |
| SSST | Source1 | 0.321 | 0.400 | 0.210 | **0.250** | 0.330 | 0.166 |
| | Source2 | 0.312 | 0.420 | 0.206 | 0.233 | 0.330 | **0.182** |
| | Source3 | 0.320 | 0.425 | 0.210 | 0.190 | 0.299 | 0.172 |
| | Source4 | 0.320 | 0.423 | 0.216 | 0.198 | 0.334 | 0.168 |
| | Source5 | 0.313 | 0.426 | 0.209 | 0.169 | 0.342 | 0.171 |
| | Source6 | 0.314 | 0.437 | 0.210 | 0.177 | 0.329 | 0.164 |
| Souce combine | Mix Fine-Tune | 0.303 | 0.424 | 0.197 | 0.172 | 0.323 | 0.151 |
| | avg-Trans | 0.309 | 0.419 | 0.215 | 0.180 | 0.328 | 0.165 |
| | WM-Trans | **0.334** | **0.449** | **0.218** | 0.222 | **0.345** | 0.172 |

transfer learning performance shown in Table 3 as the baseline, and compare the correlation between the Wasserstein weight (W-weight) computed from our domain distance model and the SSST performance in Fig. 6. The heatmap shows a good alignment between W-weight and SSST performance, i.e., source tasks with larger W-weight show better single source single target transfer learning performance. This finding further suggests that our WM-trans has the ability to aggregate all the SSST models in a way of encouraging positive knowledge transfer while reducing the risk of negative transfer.

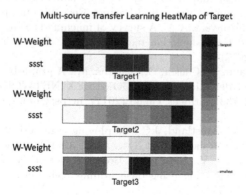

**Fig. 6.** Comparison between W-weight and SSST transfer learning performance. The darker color indicates stronger domain relevance computed from Wasserstein distance or better KS performance brought by SSST transfer learning. The W-weight shows good consistency with SSST transfer performance.

# 5    Conclusion

In this paper, we propose a multi-source transfer learning approach with self-supervised domain distance learning for financial fraud detection, named as WM-Trans. The core idea is to selectively transfer useful knowledge from multiple different but related data-rich source tasks to improve the model performance of target tasks that only have very few labeled data. Specifically, we first propose a self-supervised domain distance learning network, which can effectively measure the domain relevance between any two tasks based only on their training samples without any extra annotations. Next, our WM-Trans aggregates multiple single-source-single-target transfer models according to their domain relevance with the target tasks. Experimental results on a real financial fraud detection dataset showed that (1) our domain distance learning network is effective in modeling domain relevance; (2) Our WM-Trans achieves significant improvement over several representative transfer learning frameworks. The future work would focus on modeling domain relationships via simultaneously considering multiple tasks via graph learning approaches.

# References

1. Wang, D., et al.: A semi-supervised graph attentive network for financial fraud detection. In: 2019 IEEE International Conference on Data Mining (ICDM), pp. 598–607. IEEE (2019)
2. Cui, Y., Song, Y., Sun, C., Howard, A., Belongie, S.: Large scale fine-grained categorization and domain-specific transfer learning. In: Proceedings of the IEEE Conference on Computer Vision and Pattern Recognition, pp. 4109–4118 (2018)
3. Xu, R., Chen, Z., Zuo, W., Yan, J., Lin, L.: Deep cocktail network: Multi-source unsupervised domain adaptation with category shift. In: Proceedings of the IEEE Conference on Computer Vision and Pattern Recognition, pp. 3964–3973 (2018)
4. Adewumi, A.O., Akinyelu, A.A.: A survey of machine-learning and nature-inspired based credit card fraud detection techniques. Int. J. Syst. Assur. Eng. Manage. 8(2), 937–953 (2017)
5. Lebichot, B., Braun, F., Caelen, O., Saerens, M.: A graph-based, semi-supervised, credit card fraud detection system. COMPLEX NETWORKS 2016 2016. SCI, vol. 693, pp. 721–733. Springer, Cham (2017). https://doi.org/10.1007/978-3-319-50901-3_57
6. Jha, S., Guillen, M., Westland, J.C.: Employing transaction aggregation strategy to detect credit card fraud. Expert Syst. Appl. 39(16), 12650–12657 (2012)
7. Ravisankar, P., Ravi, V., Rao, G.R., Bose, I.: Detection of financial statement fraud and feature selection using data mining techniques. Decis. Support Syst. 50(2), 491–500 (2011)
8. Kumar, A., Bhatnagar, R., Srivastava, S.: Analysis of credit risk prediction using ARSkNN. In: Hassanien, A.E., Tolba, M.F., Elhoseny, M., Mostafa, M. (eds.) AMLTA 2018. AISC, vol. 723, pp. 644–652. Springer, Cham (2018). https://doi.org/10.1007/978-3-319-74690-6_63
9. Bhattacharyya, S., Jha, S., Tharakunnel, K., Westland, J.C.: Data mining for credit card fraud: a comparative study. Decis. Support. Syst. 50(3), 602–613 (2011)

10. Kirkos, E., Spathis, C., Manolopoulos, Y.: Data mining techniques for the detection of fraudulent financial statements. Expert Syst. Appl. **32**(4), 995–1003 (2007)
11. Bose, I., Wang, J.: Data mining for detection of financial statement fraud in Chinese companies. In: International joint Conference on e-Commerce, e-Administration, e-Society, and e-Education. International Business Academics Consortium (IBAC) and Knowledge Association (2007)
12. Zanin, M., Romance, M., Moral, S., Criado, R.: "Credit card fraud detection-through parenclitic network analysis. *Complexity* 2018, (2018)
13. Wang, B., Gong, N.Z., Fu, H.: Gang: detecting fraudulent users in online social networks via guilt-by-association on directed graphs. In: 2017 IEEE International Conference on Data Mining (ICDM), pp. 465–474. IEEE (2017)
14. Zhuang, F., et al: A comprehensive survey on transfer learning," arXiv preprint arXiv:1911.02685 (2019)
15. Zhang, L.: Transfer adaptation learning: a decade survey. arXiv preprint arXiv:1903.04687 (2019)
16. Luo, Y., Wen, Y., Duan, L.-Y., Tao, D.: Transfer metric learning: algorithms, applications and outlooks, arXiv preprint arXiv:1810.03944 (2018)
17. Ganin, Y., et al.: Domain-adversarial training of neural networks. J. Mach. Learn. Res. **17**(1), 2030–2096 (2016)
18. Tzeng, E., Hoffman, J., Saenko, K., Darrell, T.: Adversarial discriminative domain adaptation. In: Proceedings of the IEEE Conference on Computer Vision and Pattern Recognition, pp. 7167–7176 (2017)
19. Long, M., Cao, Z., Wang, J., Jordan, M.I.: Conditional adversarial domain adaptation. In: Advances in Neural Information Processing Systems, pp. 1640–1650 (2018)
20. Yu, J., et al.: Modelling domain relationships for transfer learning on retrieval-based question answering systems in e-commerce. In: Proceedings of the Eleventh ACM International Conference on Web Search and Data Mining, pp. 682–690. ACM (2018)
21. Shen, J., Qu, Y., Zhang, W., Yu, Y.: Wasserstein distance guided representation learning for domain adaptation. In: Thirty-Second AAAI Conference on Artificial Intelligence (2018)
22. Tzeng, E., Hoffman, J., Zhang, N., Saenko, K., Darrell, T.: Deep domain confusion: Maximizing for domain invariance, arXiv preprint arXiv:1412.3474 (2014)
23. Zhao, H., Zhang, S., Wu, G., Moura, J.M., Costeira, J. P., Gordon, G.J.: Adversarial multiple source domain adaptation. In: Advances in Neural Information Processing Systems, pp. 8559–8570 (2018)
24. Chen, Y.-C., Lin, Y.-Y., Yang, M.-H., Huang, J.-B.: CrDoCo: pixel-level domain transfer with cross-domain consistency. In: Proceedings of the IEEE Conference on Computer Vision and Pattern Recognition, pp. 1791–1800 (2019)
25. Afridi, M.J., Ross, A., Shapiro, E.M.: On automated source selection for transfer learning in convolutional neural networks. Pattern Recogn. **73**, 65–75 (2018)
26. Gulrajani, I., Ahmed, F., Arjovsky, M., Dumoulin, V., Courville, A.C.: Improved training of wasserstein GANs. In: Advances in Neural Information Processing Systems, pp. 5767–5777 (2017)
27. Chen, T., Guestrin, C.: XGBoost: a scalable tree boosting system. In: Proceedings of the 22nd Acm SIGKDD International Conference on Knowledge Discovery and Data Mining, pp. 785–794 (2016)
28. Finn, C., Abbeel, P., Levine, S.: Model-agnostic meta-learning for fast adaptation of deep networks. In: Proceedings of the 34th International Conference on Machine Learning, vol. 70, pp. 1126–1135 (2017) JMLR. org

# Author Index

Printed in the United States
by Baker & Taylor Publisher Services